BUSINESS, ECONOMICS

EDUCATION

ENERGY

ENVIRONMENT

(Continued on inside back cover)

INTRODUCTION
TO
STATISTICAL THINKING

INTRODUCTION TO STATISTICAL THINKING

E. A. MAXWELL
Trent University
Ontario, Canada

PRENTICE-HALL, INC., Englewood Cliffs, New Jersey 07632

Library of Congress Cataloging in Publication Data

Maxwell, E. A. (Edwin Arthur)
  Introduction to statistical thinking.

  Includes index.
    1.–Mathematical statistics. 2.–Probabilities.
  I. Title.
QA276.M374 1983   519.5   82-18533
ISBN 0-13-498105-7

Editorial/production supervisors:
    Ellen W. Caughey and Barbara Palumbo
Interior design: Ellen W. Caughey
Manufacturing buyer: John Hall

© 1983 by PRENTICE-HALL, INC., Englewood Cliffs, New Jersey 07632

Printed in the United States of America

10 9 8 7 6 5 4 3 2 1

ISBN 0-13-498105-7

PRENTICE-HALL INTERNATIONAL, INC., *London*
PRENTICE-HALL OF AUSTRALIA PTY. LIMITED, *Sydney*
EDITORA PRENTICE-HALL DO BRASIL, LTDA., *Rio de Janeiro*
PRENTICE-HALL CANADA, INC., *Toronto*
PRENTICE-HALL OF INDIA PRIVATE LIMITED, *New Delhi*
PRENTICE-HALL OF JAPAN, INC., *Tokyo*
PRENTICE-HALL OF SOUTHEAST ASIA PTE. LTD., *Singapore*
WHITEHALL BOOKS LIMITED, *Wellington, New Zealand*

FOR ROBIN AND JASON

CONTENTS

Appendices

PREFACE

This book has been written to provide an introduction to statistical methods for readers at an undergraduate level—studying natural and social sciences—who do not have a strong background in mathematics. The material is presented in an intuitive way, with as many topics as is reasonable developed through examples, so that readers should be able to grasp the general concepts of the statistical methods. The reader should thus be able to appreciate intuitively what is involved in the methodology and know what to expect from and how to interpret the results of applying the statistical tools introduced.

The presentation of the material is based on the assumption that the reader is capable of doing some elementary algebra; the material does not require calculus. In order to enhance the nonmathematical reader's ability to follow the development of the material, simple mathematical tools are included as they are required rather than in an appendix. There are many examples, practice problems, and exercises drawn from several areas of application, including simplified problems of academic research as well as general interest examples.

The material is presented in a traditional order, but there are options with regard to how the topics may be studied. Chapters 1 through 4 should precede all other chapters, although some instructors using this book may wish to deemphasize particular topics such as graphical procedures in Chapter 2, adjusting data in Chapter 3, and conditional probabilities in Chapter 4. After Chapter 4, various options exist—the most notable is the possibility of introducing ideas of regression, correlation, and analysis of variance very early.

The introductions to analysis of variance and regression and correlation are presented intuitively so that the first three sections of Chapter 10 and the first three sections of Chapter 11 require only the material in Chapters 2, 3, and 4. Although this unorthodox early introduction of two advanced topics may seem to be contrary to sound mathematical procedure, I have found it successful pedagogically. The fact that the students have not developed all of the formal foundations does not affect their grasp of the general ideas involved in these two topics. In fact, the earlier introduction helps to motivate the students to carry on through the foundations.

The book includes more than enough material for a full introductory course, thus providing instructors some choice with regard to the topics to be included in any particular course. I have resisted the temptation to list all of my suggestions for optional topics but leave that decision to individual instructors or readers.

At the end of each section, there are a few practice problems covering the basic topics included in the section. There are complete solutions to these problems immediately following the section exercises. The exercises at the end of each section provide immediate review of the material introduced in the section and practiced in the problems. There is also a set of review exercises at the end of each chapter to provide further review of the material covered in the chapter as a whole. In order to keep readers familiar with all of the material covered to any point in the book, there is also a set of general review exercises following each chapter. Again, I have avoided selecting exercises that I consider especially important or particularly difficult, but leave the selection of exercises to the discretion of individual readers and instructors.

Readers who wish to pursue extensions of statistical methodology beyond that presented in this book will be interested in the further readings listed in the appendices.

I would like to express my appreciation to the many students and reviewers whose comments on earlier drafts of the manuscript helped to improve the text from its early stages. Reviewers included Professors Mikel Aicken, Arizona State University; Robert Brown, UCLA; George Casella, Cornell University; James Friel, California State University; Marilyn Mays Gilchrist, North Lake College; T. D. Dwivedi, Concordia University, Montreal, Canada; William J. Anderson, McGill University, Montreal, Canada; and Gerald E. Rubin, Marshall University. The editorial staff at Prentice-Hall were particularly helpful. Funding assistance was provided by the Department of University Affairs (now the Ministry of Colleges and Universities) of the Province of Ontario, the National Sciences and Engineering Research Council of Canada, and Trent University. I am especially grateful for the many hours of difficult typing provided by Mavis Prior and Mary Snack. Most of all, I wish to thank my wife, Helen, who typed parts of the manuscript, provided examples and exercises, helped with the proofreading and editing and, even more importantly, kept me going throughout the project.

E. A. Maxwell
Lakefield, Ontario

INTRODUCTION
TO
STATISTICAL THINKING

INTRODUCTION

The purpose of this text is to help students to acquire the ability to analyze statistical information. As a comment on the importance of this ability, the writer H. G. Wells has been quoted as saying, "Statistical thinking will one day be as necessary for efficient citizenship as the ability to read and write." Indeed, we have now reached, if not long ago passed, that "one day" and our daily lives include a constant barrage of statistical information.

Much of the daily newspaper contains statistical information. Besides the usual performance measures found in the sports pages or the business section, numerical data are used as the basis for news articles or columns. For example, we can find an article in a metropolitan newspaper in which statistical data are used to support the claim that crime has increased in the metropolitan area. The reaction of many people to such an article is to become alarmed and to demand more "law and order." Instead of becoming immediately alarmed, however, we might study the data further to discover that the increase is in *reported crimes*. We might further discover that a higher proportion of crimes is now reported. This would account for some, but not necessarily all, of the increase. As a result of this further *statistical thinking*, we may not be as alarmed, but we might still exercise considerably more caution than previously.

Some information is valuable and affects everyday decisions, both major and minor:

> On the basis of statistical reports indicating potential reduced energy consumption with heat pumps, many homeowners have decided to heat and cool their homes with heat pumps.

Indications of increased demand for owned rather than rented apartments because of preferred tax deductions have lead to thousands of apartments across the United States being converted to condominiums.

A decision as to whether or not to carry an umbrella on a cloudy day is affected by a radio report on a meteorologist's statistical assessment of the chances of rain.

On the other hand, some information is quite frivolous and/or misleading. For example, the following heading appeared in a prominent newspaper: "Nude swimmers produce faster racing times." This heading, which is nothing more than an attention grabber, leads the reader to believe that swimming nude will produce faster times; however, on reading further, we discover that, of eight swimmers tested, only two improved while swimming nude.

Some statistics are presented to keep us informed on the state of the economy: current unemployment figures, average house prices, consumer price index, and so on. Some statistics keep us informed on people's attitudes on various issues: the results of a Gallup poll taken to determine how Canadians reacted to a 33% salary increase for federal members of parliament, or a poll by Ann Landers to determine how many of her North American readers would have children if they had the opportunity to make the decision over again.

Other statistics provide comparisons among segments of society. Data from an American Institute of Public Opinion survey of Americans and a similar Gallup poll survey of Canadians were combined to illustrate the fact that more Canadians than Americans would withhold strike privileges from workers in the public sector.

Care must be taken in reading statistical information. The Gallup polls were conducted scientifically by an experienced agency in such a way that there would, on the average, be an error of less than four percentage points in 19 of 20 such polls. On the other hand, the write-in poll taken by Ann Landers did not necessarily cover all possible biases, but might have tended to draw responses predominantly from people sufficiently annoyed with their own experience to want to warn others of the perils of raising a family.

Statistics are sometimes presented to explain actions that have taken place, to justify claims, or to influence decisions that will take place. An example of statistics used to explain or justify claims is an insurance company brochure produced to explain the need for increased automobile insurance premiums. The justification is based on a statistical demonstration of increased costs such as the average increase of $53 per claim on property damage and personal injury claims. Other examples include the use of traffic fatality data in a Connecticut study in which a reduction of fatalities is "explained" by a crackdown on speeding, or from the Netherlands in a study on the change over time in the percentages of different sentences imposed on persons convicted of offenses as evidence of an improved view of the courts on rehabilitative programs.

An example of statistics presented to influence future decisions is a cancer society brochure outlining the incidence of cancer, the progress in treatment, and the need for more funds to support further research. Other examples are the many advertisements referring to "statistical proof" of increased miles per gallon, greater moisture absorbing powers, fewer cavities, and so on.

In many cases, we are expected to draw certain conclusions from the data as presented. In other cases, the data would be too much for us to digest, perhaps even in summary form. We are offered advice or made subject to regulations on the basis of inferences drawn by others. We are advised not to smoke cigarettes because (after much controversy over interpretation of the data) scientists agree that statistical studies indicate that smoking may be harmful. We are also told that statistical studies indicate that wearing a seat belt reduces the chance of injury in an automobile accident; thus, we are advised to "buckle up." A study conducted in metropolitan schools indicated that such advice, even with backup data, would not necessarily lead to continued use of seat belts. As a result, some jurisdictions now have laws making the wearing of seat belts compulsory.

Statistics are not always presented directly to us and we are not always told that a statistical study was the basis for a particular action or recommendation; however, if we investigate, we soon find statistical studies in the background. Research on the Salk polio vaccine involved a very large experiment designed and analyzed according to current techniques in statistical methodology. The petroleum industry relies on statistical studies such as one by an oil and gas commission to determine relative frequencies of sizes of oil fields. Decisions on Great Lakes pollution control policy are based, in part, on statistical information such as the results of a survey of farmers to assess their reactions to various control policies and the anticipated effects of such policies on cornbelt agriculture. The reliability of our telephone has been determined on the basis of the durability of similar telephones selected at random and subjected to tests such as being exposed to extreme temperatures or dropped or shaken or having the cord stretched 250,000 times. Television programming available to Americans is determined by ratings obtained from surveys of sample viewers. Similarly, the programming available on the CBC television network is based (in part) on the results of continuous surveys of the viewing habits of a gradually changing group of 20,000 Canadians.

The study of statistical analysis of data leads to a better understanding of how data must be collected. Valid statistical inferences cannot be obtained from poorly designed experiments. Even subconscious reactions can affect the validity of results. For example, apparently promising results on the use of a neurohormone, beta-endorphin, in the treatment of psychiatric patients were not considered reliable because of the design of the study. Since placebos were not employed with the control group, the effect of the patients' attitudes to receiving treatments could not be separated from the treatment effect. Furthermore, the psychiatrist involved knew which patients should have responded because of treatment and could have subconsciously concentrated other therapeutic efforts on those patients.

A formal study of statistical methodology is, of course, not necessary to become aware of the prominence of statistics in a general sense such as found in many of the previous examples. It is necessary, however, for anyone who wishes to take part in the production of such statistics, or for anyone who might have to make intelligent decisions on the basis of statistical information.

The study of statistics is of particular importance for the student whose academic career will involve investigations requiring analysis of numerical data. The need for an appreciation of statistical methodology is not limited to graduate students embarking on thesis research, but extends to undergraduate students who must also read and write reports involving statistical analyses.

The subject of statistics consists of the collection, analysis, interpretation, and presentation of data. Data must be collected before they can be analyzed, but it is useful to first consider what data may be available and how they may be analyzed and interpreted. We will then be better prepared to determine how they should be collected. Aside from data collection, statistics may be considered to consist of two parts, descriptive statistics and statistical inference.

We will be concerned primarily with methods of statistical inference, that is, methods for studying partial information or sample data in order to draw conclusions of a general nature on the basis of particular cases.

The topics studied in the following chapters are sufficient to meet the basic needs. Although the techniques presented are relatively elementary and do not include the full mathematical development, they do cover a broad spectrum of statistical methodology. The material is sufficient to conduct many statistical investigations and provides a good foundation on which to build the required methodology for more involved studies and, of course, as a by-product, increases the ability to appreciate statistics as referred to in the previous examples.

The emphasis is on the intuitive aspects of the particular techniques and the logic of the methods of analysis. Mathematical theory is not developed in depth; however, statistical methodology and its foundation, probability theory, are deeply rooted in mathematics and the reader must be prepared to engage in some mathematics, such as elementary algebra, and to work with a certain amount of mathematical notation.

To assist readers who are not strong in mathematics, the techniques are usually introduced through examples and the more general mathematical approaches come as follow-up. Mathematical procedures—some quite elementary—are introduced as they are required to further assist such students. Readers who are stronger in mathematics may, accordingly, find some sections that they can skim over quite quickly.

In order to check their understanding of the material, readers should attempt the few practice problems at the end of each section and then compare their solutions with those that follow immediately after the review exercises. Readers should also do exercises from all the sets of exercises. The exercises at the end of sections provide immediate reinforcement and further practice on the techniques that have just been presented. The chapter review exercises provide practice on the chapter as a whole and help to provide practice in distinguishing between different technique requirements for similar procedures. Finally, the general review exercises should be attempted, as they help the student to retain familiarity with all the topics previously covered and provide practice in determining which procedures to apply in various situations.

SUMMARY DISTRIBUTIONS AND GRAPHICAL TECHNIQUES

2-1 INTRODUCTION

In this and the next chapter, we consider descriptive methods of statistics—the methods of summarizing and presenting data so that the essential information can be conveyed quickly and easily. In developing the techniques for reporting information in summary form, we must also take special care that the information conveyed is not misleading.

2-2 FREQUENCY DISTRIBUTIONS AND HISTOGRAMS

EXAMPLE 2-1   Suppose that, as the administrators of a social agency, we are interested in an "age profile" of unwed mothers in our region during a given time period. We are told that there were 88 births to unwed parents in the time period and that the ages of the mothers at the times of giving birth were as given in the following list:

18	23	31	27	18	42	17	23	17	15	17
17	16	16	26	39	20	19	18	21	21	16
18	20	13	17	21	18	30	20	23	35	36
22	16	23	17	18	19	24	21	15	27	36
20	21	25	16	18	20	19	24	19	21	20
21	16	18	34	19	22	24	25	20	18	26
16	22	20	19	20	22	22	22	20	38	19
23	19	15	14	20	25	23	24	20	19	20

We have available the complete information in this list of ages, but it is not very easy to obtain a clear picture from columns of figures. If we had 500 or 1000 figures, as is often the case, it would be almost impossible. We must consider methods for extracting and conveying the information available in the raw data. ▲

Frequency Distributions

The general impression from a set of data is much easier to obtain if we have the data summarized in a display indicating how many items in the data fall into each of a number of categories.

EXAMPLE 2-2   The following display represents a summary of the age data in Example 2-1:

Age of mother	Number of mothers in this age category
10–14	2
15–19	34
20–24	36
25–29	7
30–34	3
35–39	5
40–44	1

This presentation has been obtained by choosing a set of categories or *classes* and then counting the *frequencies* of occurrence of numbers in the data that fall into each class. For example, going through the columns of data, we find three ages in the interval from 30 to 34 (inclusive), namely 31, 34 and 30; thus, the *class* 30 to 34 has *frequency* 3. The display provides an indication of how the frequencies of occurrence are distributed across the class. ▲

A set of classes together with the frequencies of occurrence of values in each class in a given set of data is referred to as a *frequency distribution*.

The classes in this distribution were chosen in an attempt to satisfy the following conditions or "rules of thumb" required of a good frequency distribution:

1. It is necessary to avoid too many or too few classes. It is sometimes suggested that, if possible, the number of classes should be not less than 6 nor more than 15.

Too few classes leads to lost detail; too many produce too thin a spread in the data.

2. The classes in the distribution must include all of the data.

If we started the classes for the age data of Example 2-1 at 15 and finished at 34, we would lose some data and give the impression that no values occurred below or above these values.

3. Each item in the data should be included in only one class.

If we use classes 10 to 15, 15 to 20, 20 to 25, and so on, for the data of Example 2-1, then, whenever we find a 15, 20, 25, and so on, in the data, we do not know where to place it in the set of classes.

4. If possible, and practical, the classes should all be the same size. They should each cover the same range of values.

Each of the age classes in Example 2-2 includes five possible ages. If one of the classes included ten possible ages, its frequency would be disproportionately high and not compare easily with those of other classes. In considering the size of the classes, it is useful to keep in mind a set of classes that reads naturally, such as the set of classes based on the numbers 10, 15, 20, 25. We might have started with 13, the lowest value, and ended with 42, the highest; however, the numbers 13, 18, 23, 28 do not have the same "natural" appeal.

It is not always possible or practical to have classes of the same size. Sometimes it is necessary to use *open classes* such as "greater than 1000" or "other" at the end(s) of the frequency distribution.

EXAMPLE 2-3 If a granting agency provided a large number of organizations (say 115) with grants ranging from $100 to $1000, and provided only five organizations with larger grants, namely $1500, $2000, $2200, $3500, and $5000, then, in a frequency distribution, the five largest grants might be placed in a class "greater than $1000." ▲

As has been noted, these rules are not all strict rules; the first and fourth are guidelines. The rules as stated are based on quantitative classes and cannot all apply to classes based on qualitative characteristics such as color, brand, shape, or nationality. The second and third rules do apply to all cases, however, and, in all cases, the class definitions must be explicit. The following provides an illustration of a frequency distribution based on qualitative classes.

EXAMPLE 2-4 A survey of the specializations of specialist officers in the foreign service produced the following results:

Specialization	Number of officers with this specialization
Geographic area of service	39
Economics	22
Legal	21
Aid	7
International organization	5
Political	4
Other specializations	6

▲

A summarized presentation of the data has the advantage of being easy to read, but it also involves disadvantages, primarily the loss of the detailed knowledge of which individual values occurred and how often they occurred.

EXAMPLE 2-5 We know that three of the ages in Example 2-2 are in the class 30 to 34, but (from the distribution alone) we cannot determine the individual values for these ages. If we have the data in the summary form only, we cannot find specific values of interest such as the youngest age. From the distribution alone, the youngest age might be 10 (although, as we know from the raw data of Example 2-1, it is actually 13). ▲

Histograms

Although a frequency distribution improves upon the raw data in providing us with a general impression, we can go further and literally obtain a picture of the data through a graphical display of the frequency distribution. We construct a graphical presentation with the classes indicated on the horizontal scale and the frequencies on the vertical scale.

Upon each class of a frequency distribution as a base in a graph, we draw a rectangle with height equal to the frequency of the class. This graphical presentation of a frequency distribution is called a *histogram*.

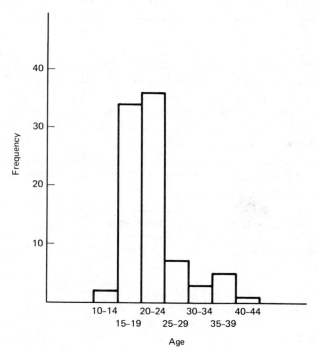

Figure 2-1 Histogram of age data.

EXAMPLE 2-6 The histogram for the frequency distribution of Example 2-2 is shown in Figure 2-1. ▲

Through the use of a frequency distribution or histogram, we can more easily obtain important information from the data.

EXAMPLE 2-7 The histogram in Figure 2-1 makes it apparent that the ages are, for the most part, in middle to late teens or early twenties, with very few ages younger than these and with gradually decreasing frequencies for late twenties or older. ▲

It is important to take care in interpreting summarized data since the impression that we obtain from the summarized data may miss some important detail. In some cases, different sets of classes may even given different impressions for the same data.

EXAMPLE 2-8 Using the values 15, 20, 25, and so on, as the highest rather than the lowest values in the classes for the data of Example 2-1 would produce the following frequency distribution, with histogram shown in Figure 2-2.

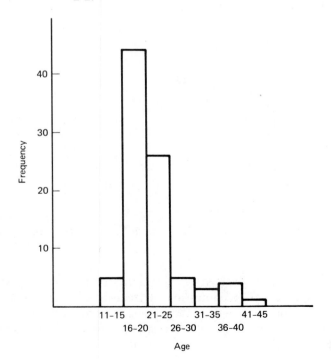

Figure 2-2 Histogram of age data.

Age	Frequency
11–15	5
16–20	44
21–25	26
26–30	5
31–35	3
36–40	4
41–45	1

The impression now is that of a generally younger group of mothers than appeared to be the case with the distribution in Example 2-2. This change is caused by shifting the thirteen 20's to a younger class, including them with the late teens. This is especially serious since 20 occurs more often than any other age. ▲

Neither of the distributions is necessarily better—in fact, as indicated below, they might both be improved upon. Both distributions indicate something we might have known anyway, that the majority of the ages are in middle to late teens or early twenties; however, neither provides much detail within this overall category.

EXAMPLE 2-9   In order to improve upon both distributions of Examples 2-2 and 2-8, let us consider violating one of our guidelines. We will reclassify the values in the second and third classes of the second distribution to produce the following:

Age	Frequency
11–15	5
16–17	13
18–19	18
20–21	20
22–23	12
24–25	7
26–30	5
31–35	3
36–40	4
41–45	1

We now have a much better picture of the detail of the ages from 16 to 25. ▲

Unfortunately, in the modification above we have introduced a new problem because of the different class sizes. The impression of frequencies obtained from histograms depends only partly on the heights of the rectangles; it depends primarily on the areas. It is necessary, therefore, to modify histograms representing frequency distributions with different class sizes to avoid distorted impressions.

EXAMPLE 2-10 Figure 2-3 represents an unmodified histogram for the frequency distribution of the ages as given in Example 2-9. The frequencies for the ages above 25 and below 16 appear very much out of proportion because the rectangles for these classes have bases that are 2.5 times larger than those for the five new classes and areas that are, therefore, not proportional to their class frequencies.

To modify the area impression so that the *areas are proportional to the frequencies*, we reduce the heights of the wider rectangles. To obtain new heights we divide the heights of the wider rectangles by 2.5 because each is 2.5 times the width of the smaller. We *cannot use the vertical scale now* with this height adjustment. Instead, we record the frequencies at the top of each rectangle. The final modified histogram is shown in Figure 2-4. ▲

> To provide for the correct interpretation of histograms,
> the areas must be proportional to the frequencies.

As we have seen with the age data, summarizing data can make information much easier to grasp; however, this is not without some sacrifice. Care must be taken in the way we summarize the data, and although we have rules to serve as guidelines,

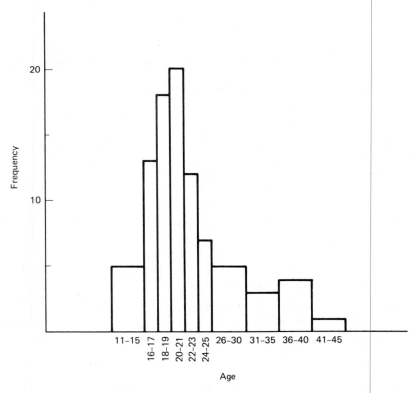

Figure 2-3 Incorrect histogram of age data.

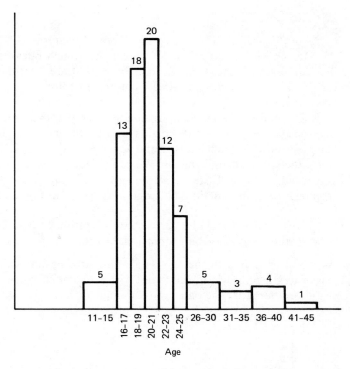

Figure 2-4 Correctly modified histogram of age data.

each case must be considered individually. In some cases, as illustrated above, we may consider it useful, or even necessary, to violate one or more of the guidelines.

We now define a few more technical aspects of frequency distributions for quantitative data.

Class Limits

The highest and lowest values that can occur within individual classes are referred to as the *upper class limits* and the *lower class limits*.

EXAMPLE 2-11   Consider the following distribution of the lengths of 115 logs measured to the nearest inch:

Length (inches)	Frequency	Length (inches)	Frequency
360–364	2	385–389	23
365–369	5	390–394	19
370–374	9	395–399	6
375–379	18	400–404	4
380–384	26	405–409	3

The lower limit of the first class is 360, the upper limit is 364; the lower limit of the second class is 365, the upper limit is 369. Lengths between 360 and 364 inches (inclusive) fall into the first class; lengths between 365 and 369 inches (inclusive) fall into the second class. Limits for the other classes are analogous. ▲

Class Boundaries

It appears in Example 2-11 that we have not accounted for lengths between 364 and 365 inches; however, as we are measuring to the nearest inch, actual lengths between 364 and 365 inches are rounded in the usual manner down to 364 or up to 365 and recorded as whole numbers.

> The value midway between the upper limit of one class and the lower limit of the next is called a *class boundary*. It is found by adding the upper limit of the first class to the lower limit of the next and then dividing by 2.

EXAMPLE 2-12   In the data of Example 2-11, the upper boundary of the first class is $(364 + 365)/2 = 364.5$. It is also the lower boundary of the second class. ▲

> Unlike limits, which belong to only one class, boundaries belong to two classes. The lower boundary of the first class and the upper boundary of the last class are exceptions. They are not midway between limits, but are defined analogously to boundaries between classes.

EXAMPLE 2-13   The boundaries for the data of Example 2-11 are 0.5 below or above limits. The lower boundary for the first class is thus 359.5 and the upper boundary for the last class is 409.5. ▲

> Also, unlike limits, boundaries do not occur as values for individual items in the data.

EXAMPLE 2-14   When we measure to the nearest inch, the value 364.5 does not occur. ▲

The method of calculating boundaries described above would not necessarily apply to data such as the ages of the unwed mothers in Example 2-1. The class 25 to 29 might include the ages of all the mothers who, at the time of giving birth, had passed their twenty-fifth birthdays but not yet reached their thirtieth. It would not necessarily stop at 29 years and 6 months. This case would require a new definition of boundary. If, however, age is taken to be age at the nearest birthday, the usual definition of a boundary would apply.

We had not previously noted it, but limits and boundaries were used in the construction of the histograms for the age data:

> The rectangles in histograms may be labeled with class limits, but the sides of the rectangles are actually drawn vertically above the class boundaries.

Class Marks

> Each class has a representative value or *class mark* which is the midpoint of the class. It may be calculated by adding the lower and upper class limits or boundaries and dividing by 2.

EXAMPLE 2-15 The class mark for the third class of Example 2-11 is $(370 + 374)/2 = 372$. ▲

> The class mark is not always a value that can occur in the data.

EXAMPLE 2-16 Considering the values in Example 2-1 to be age at the nearest birthday, and using the frequency distribution in Example 2-9, the class 16 to 17 has class mark 16.5, yet no age in the data can be 16.5 because the data are given as whole numbers only. ▲

Class Width

> The size of a class is referred to as the *class width* and is the difference between the class boundaries.

EXAMPLE 2-17 The frequency distribution of Example 2-9 has some classes with class widths of 2 and some with class widths of 5. For the log data of Example 2-11, the class widths are all 5. ▲

EXAMPLE 2-18 All of the foregoing definitions are illustrated further in the following distribution of measurements in centimeters of the inside diameters of a batch of bearings. The distribution has 12 classes and each class is referred to by number.

Class	Lower limit	Upper limit	Frequency	Class mark	Lower boundary	Upper boundary	Class width
1	1.270	1.279	3	1.2745	1.2695	1.2795	0.010
2	1.280	1.289	17	1.2845	1.2795	1.2895	0.010
3	1.290	1.299	28	1.2945	1.2895	1.2995	0.010
4	1.300	1.309	29	1.3045	1.2995	1.3095	0.010
5	1.310	1.319	11	1.3145	1.3095	1.3195	0.010
6	1.320	1.329	6	1.3245	1.3195	1.3295	0.010
7	1.330	1.339	7	1.3345	1.3295	1.3395	0.010
8	1.340	1.349	2	1.3445	1.3395	1.3495	0.010
9	1.350	1.359	6	1.3545	1.3495	1.3595	0.010
10	1.360	1.369	3	1.3645	1.3595	1.3695	0.010
11	1.370	1.379	1	1.3745	1.3695	1.3795	0.010
12	1.380	1.389	1	1.3845	1.3795	1.3895	0.010

▲

Using Relative Frequencies or Percentages

In considering the distribution of frequencies across classes, it is often of interest to know what *relative frequency* or *proportion* of the total falls into each class. Relative frequencies are usually converted to *percentages* for presentation.

EXAMPLE 2-19   The data in Example 2-5 included the reported specialization of 104 officers. Of these 104, we note that 39 specialized in the particular geographic area of their foreign service. The relative frequency is $39/104$, which produces the percentage $100 \times (39/104)\% = 37.5\%$. Similar calculations for the remaining officers produce the following percentage distribution:

Specialization	Percentage of 104 officers within this specialization
Geographic area of service	37.5
Economics	21.2
Legal	20.2
Aid	6.7
International organization	4.8
Political	3.8
Other specializations	5.8

▲

Using percentages instead of raw frequencies aids in the comparison of two distributions with different total frequencies. It is not necessarily useful to know that, in one distribution, 12 items fell into the first class, whereas in a second distribution, only three fell into that class. If, for example, there were four times as many items in the first distribution as in the second, then, proportionally, each distribution would have the same allocation to the first class. The relative frequency or percentage for the first class would be the same for each of the two distributions.

Comparisons using percentages can be obtained by parallel listing in a percentage format frequency distribution or by a "2-in-1 histogram" in which two rectangles are drawn over each class (sharing the base equally) and the rectangle heights are percentages instead of frequencies.

EXAMPLE 2-20 An employer pays hourly wages to male and female employees in both part-time and full-time positions. Hourly rates range from a minimum of $3.00 to a maximum of $11.00. A percentage breakdown of the hourly wages paid to 250 male employees and 80 female employees produced the following distribution, illustrated in Figure 2-5.

Hourly wage (dollars)	Percentage of 80 female employees	Percentage of 250 male employees
3.00–3.99	2.5	1.6
4.00–4.99	7.5	4.4
5.00–5.99	17.5	5.6
6.00–6.99	27.5	11.2
7.00–7.99	20.0	24.8
8.00–8.99	12.5	20.4
9.00–9.99	7.5	17.6
10.00–11.00	5.0	14.4

▲

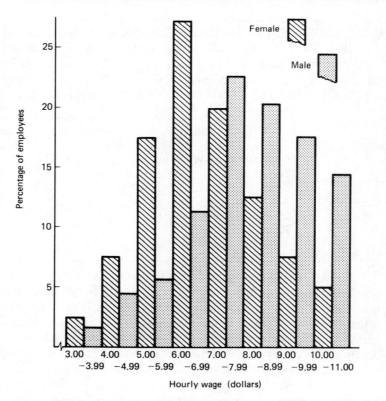

Figure 2-5 Employees earning various hourly wages.

P2-1 Recent studies have indicated that barrier islands off the Atlantic coast of the United States are migrating landward. Suppose that the following values in meters represent the shift landward over a 10-year period at 50 points on a barrier island:

12.2	22.3	19.9	8.0	13.7
11.7	17.0	10.8	3.7	9.6
16.8	8.6	4.6	11.7	9.5
25.6	1.1	2.0	10.6	17.8
21.2	8.0	9.8	14.0	15.8
23.3	12.9	13.1	14.1	18.5
22.7	9.9	14.2	22.7	16.9
27.3	17.1	26.5	19.8	22.5
33.8	29.8	11.2	7.8	27.5
30.2	23.2	10.4	13.1	31.3

(a) Form a frequency distribution for these data using classes 0.0 to 4.9, 5.0 to 9.9,..., 30.0 to 34.9. For each class, indicate the class limits, class boundaries, and class marks.
(b) What are the class widths?
(c) Construct a histogram to illustrate the frequency distribution in part (a).

P2-2 The instructor in an undergraduate course in elementary quantitative methods asked each student to indicate his or her major subject upon registering for the course. He obtained the following responses:

psychology	anthropology	psychology
psychology	biology	psychology
sociology	sociology	anthropology
geography	psychology	biology
political science	psychology	economics
geography	English	economics
economics	geography	geography
economics	biology	environmental studies
economics	anthropology	geography
political science	anthropology	environmental studies
sociology	geography	geography
psychology	biology	psychology
biology	biology	philosophy

(*cont.*)

psychology	sociology	psychology
psychology	psychology	biology
history	sociology	economics
economics	economics	economics
biology	political science	modern languages
geography	geography	geography
geography	geography	psychology

Construct a frequency distribution for the results.

P2-3 In an investigation of the possibility of a drug causing slight disorientation, the times in seconds required by 50 rats to complete a maze run after being injected with the drug were compared with the historical record of the times of 250 other "control" rats that had been tested prior to other experiments. The resulting times were as follows:

Time to complete maze run (seconds)	Number of control rats	Number of drugged rats
20–29	15	0
30–39	55	2
40–49	83	6
50–59	59	15
60–69	29	14
70–79	9	8
80–89	0	3
90–99	0	2

(a) Convert the frequencies in the distributions to percentages.
(b) Construct a 2-in-1 histogram of the percentages.

EXERCISES

2-1 The following values represent the ages at the nearest birthday of a number of pedestrian victims in nonfatal motor vehicle accidents:

34	53	13	62	57	4	84
35	9	57	8	83	17	78
64	5	67	20	25	8	82
20	67	19	24	90	18	14
65	51	2	38	12	30	9
6	78	3	54	16	4	72
1	38	24	5	45	45	58

(a) Form a frequency distribution with classes 1 to 5, 6 to 10, and so on, indicating class limits and class marks as well as class frequencies.

(b) Draw a histogram for the distribution.

2-2 (a) Modify the classes in Exercise 2-1 to 1 to 5, 6 to 15, 16 to 20, 21 to 25, 26 to 45, 46 to 65, and 66 to 90.

(b) Illustrate a frequency distribution of the data in this form with an appropriate histogram.

2-3 For a particular ailment, a number of prescription drugs are suitable and each is available at a number of different prices in different drugstores. A survey of prices charged for such a prescription drug produced the following results:

4.90	5.30	5.95	5.15	6.15	4.75	4.85	5.35	5.45
6.25	5.45	4.95	4.50	4.95	5.25	5.15	5.55	4.35
4.35	4.50	5.35	4.90	5.45	5.15	5.35	5.20	5.75
5.55	4.95	4.95	4.60	5.35	4.75	4.85	4.35	5.30
5.60	5.45	5.90	5.30	5.75	5.30	5.15	4.95	4.75
6.25	5.60	5.25	5.55	4.75	4.50	4.60	5.55	4.50
5.20	5.90	5.60	5.15	5.20	5.45	5.95	6.45	5.25

(a) Form a frequency distribution for these data.

(b) Illustrate the distribution with a histogram based on percentages rather than frequencies.

2-4 The following set of data represents the marks obtained by a class of students in a final examination:

54	50	15	65	70	42	77	33
55	92	61	50	35	54	75	71
64	10	51	86	70	66	60	60
71	64	63	77	75	70	81	48
34	73	65	62	66	75	60	85
64	57	74	60	63	85	27	67
79	37	66	45	58	67	77	53
23	61	66	88	58	8	73	76
81	83	62	75	69	68	66	71
66	67	50	76	60	45	68	74
70	68	69	66	73	67	73	

(a) Form a frequency distribution for these marks using classes 0 to 9, 10 to 19, and so on. For each class indicate class marks and boundaries as well as limits and frequencies.

(b) Illustrate the distribution with a histogram.

2-5 Suppose that the marks in Exercise 2-4 are converted to letter grades as follows:

80–100	A	excellent
75–79	B +	very good
70–74	B	good
60–69	C	acceptable
50–59	D	poor
0–49	F	failure

(a) Group the marks into numerical classes corresponding to the letter grades.

(b) Draw an appropriate histogram for this new distribution.

2-6 A survey of the destinations of 50 travelers embarking on winter holidays in Florida or the Caribbean area produced the following responses:

Florida	Florida	Florida	Florida
Bahamas	Florida	Jamaica	Bahamas
Bahamas	Carib. cruise	Barbados	Cayman Islands
Florida	Carib. cruise	Barbados	Cayman Islands
Florida	Carib. cruise	Barbados	Carib. cruise
Florida ﹅	Bahamas	Florida	Bahamas
Florida	Bahamas	Florida	Bahamas
Puerto Rico	Bahamas	Carib. cruise	Jamaica
Jamaica	Florida	Carib. cruise	Jamaica
Jamaica	Florida	Bahamas	Jamaica
Trinidad	Cuba	Bahamas	Florida
Cuba	Florida	Florida	Florida
Cuba	Florida		

Construct a frequency distribution for these responses expressing the frequencies as percentages.

2-7 In a study on riverbed material, grains of material were classified according to diameter. The percentage of a sample weight falling into each category was recorded before and a number of years after construction of a dam upstream from the measuring station. The following data represent the results of the study:

Grain size (millimeters)	Percentage of predam sample	Percentage of postdam sample
0.0–0.4	12	1
0.5–0.9	26	6
1.0–1.4	21	14
1.5–1.9	16	25

(*cont.*)	Grain size (millimeters)	Percentage of predam sample	Percentage of postdam sample
	2.0–2.4 ·	9	19
	2.5–2.9	6	13
	3.0–3.4	5	9
	3.5–3.9	4	6
	4.0–4.4	1	4
	4.5–4.9	0	3

Construct a 2-in-1 histogram to illustrate these results.

SOLUTIONS TO PRACTICE PROBLEMS

P2-1 (a) The frequency distribution is as follows:

Class					
Lower limit	Upper limit	Frequency	Lower boundary	Upper boundary	Class mark
0.0	4.9	4	−0.05	4.95	2.45
5.0	9.9	8	4.95	9.95	7.45
10.0	14.9	14	9.95	14.95	12.45
15.0	19.9	9	14.95	19.95	17.45
20.0	24.9	7	19.95	24.95	22.45
25.0	29.9	5	24.95	29.95	27.45
30.0	34.9	3	29.95	34.95	32.45

(b) In each case, the class width is 5.0.
(c) The histogram is illustrated in Figure P2-1.

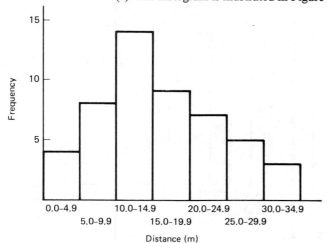

Figure P2-1 Histogram of landward shifts at 50 places on barrier islands.

P2-2 The following distribution represents one possible solution to Practice Problem 2-2. In view of the low frequencies (only 1 or 2), some subjects have been grouped as "other."

Subject	Number of students majoring in this subject
Anthropology	4
Biology	8
Economics	9
Geography	12
Political science	3
Psychology	13
Sociology	5
Other	6

P2-3 (a) The percentage distribution for the data is as follows:

Time to complete maze run (seconds)	Percentage of control rats	Percentage of drugged rats
20–29	6.0	0.0
30–39	22.0	4.0
40–49	33.2	12.0
50–59	23.6	30.0
60–69	11.6	28.0
70–79	3.6	16.0
80–89	0.0	6.0
90–99	0.0	4.0

(b) The histogram is shown in Figure P2-2.

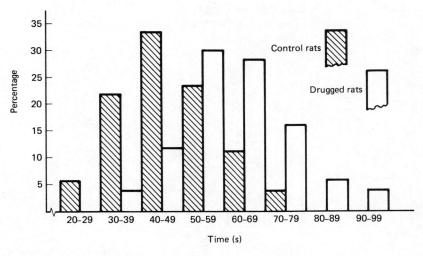

Figure P2-2 Maze times for control and drugged rats.

2-3 CUMULATIVE DISTRIBUTIONS

Cumulative Frequencies

Besides determining how many of the values are within individual classes in a frequency distribution, we may wish to determine how many items in the data are above or below a given value. To do so, we may either count values in the raw data or accumulate frequencies in a frequency distribution.

EXAMPLE 2-21 For the data in Example 2-1, we find, for example, that 12 of the mothers were 16 or younger, 36 were younger than 20, and 3 were over 37. With the data in Example 2-11, which is available in summary form only, we are limited as to the particular values that we can use. We cannot determine the number of logs that are less than 379 inches; however, we can determine the number that are less than 380 inches. We can use 380 since it is a lower limit and if we add up the frequencies in the preceding classes, we will have the number of logs with lengths less than 380, namely 34. We cannot use 379, an upper limit, since we do not know how many values in the fourth class are less than 379, nor how many are equal to 379. We can, however, use 379 for a slightly different accumulation. We can determine how many of the logs are 379 inches *or less* in length, namely 34. ▲

Cumulative Distribution

If we consider a full set of accumulated frequencies, we have a *cumulative distribution*. A distribution indicating how many values are less than each lower limit is a *less than* cumulative distribution. A distribution indicating how many values are equal to an upper limit or less is an *or less* cumulative distribution. Similarly, we can construct a *more than* cumulative distribution using upper limits or an *or more* cumulative distribution using lower limits.

EXAMPLE 2-22 A cumulative "less than" distribution for the frequency distribution in Example 2-9 is as follows:

Age	Frequency
Less than 11 years	0
Less than 16 years	5
Less than 18 years	18
Less than 20 years	36
Less than 22 years	56
Less than 24 years	68

Age	Frequency
Less than 26 years	75
Less than 31 years	80
Less than 36 years	83
Less than 41 years	87
Less than 46 years	88

▲

EXAMPLE 2-23   A cumulative "or more" distribution for the log data in Example 2-11 is as follows:

Length	Frequency
360 inches or more	115
365 inches or more	113
370 inches or more	108
375 inches or more	99
380 inches or more	81
385 inches or more	55
390 inches or more	32
395 inches or more	13
400 inches or more	7
405 inches or more	3
410 inches or more	0

▲

The distribution for the ages is formed by adding the frequencies of successive classes starting with the first class. The distribution for the log data is formed similarly but starting with the last class. Note that, in each case, a limit from an extra class is introduced with a corresponding frequency of 0 to indicate the appropriate end points of the distribution.

Cumulative distributions may also be constructed by accumulating percentages instead of frequencies.

In constructing cumulative distributions, we must ensure that we use the appropriate limits. As an alternative, we can base the distributions on boundaries. Since values equal to the boundaries do not occur in the data, the exclusion of the boundary value in a "less than" distribution will not alter the frequency from that found by including the boundary value in an "or less" distribution.

> Class boundaries may be used for both a "less than" or an "or less" distribution and also for both a "more than" and an "or more" distribution.

EXAMPLE 2-24   In the bearing data in Example 2-18, we determine that 11 measurements are *more than* 1.3495 cm and, since 1.3495 is a boundary value and cannot occur in the data, there are also 11 measurements that are 1.3495 cm *or more*.   ▲

Ogives

Graphical techniques can also be applied to illustrate cumulative distributions. Line rather than bar graph procedures are used.

> For graphs of cumulative distributions, cumulative frequencies or percentages are indicated on the vertical scale, and reference points (limits or boundaries) are indicated on the horizontal scale. The cumulative frequencies are plotted at points above the limits or boundaries and joined by lines to produce the graphical illustration of a cumulative distribution. This graph is referred to as an *ogive*.

EXAMPLE 2-25   Figure 2-6 shows the "less than" ogive for the bearing data in Example 2-18.   ▲

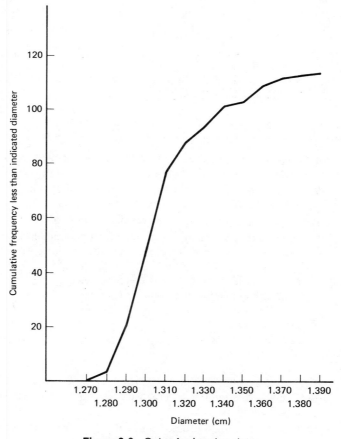

Figure 2-6   Ogive for bearing data.

P2-4 (a) Construct a cumulative "less than" distribution for the data in Practice Problem P2-1.

(b) Illustrate the distribution in part (a) with an ogive.

P2-5 In a study on whether forage from a region used for beef cattle that require a copper content of four parts per million would be suitable for dairy cattle requiring a copper content of ten parts per million, several samplings of alfalfa were analyzed for the copper content. The following values represent sample copper contents in parts per million.

10.5	4.7	8.4	9.0	9.6	11.0	9.4
5.9	10.2	5.2	8.7	7.9	9.6	8.4
5.6	7.2	10.6	10.2	7.5	7.2	8.8
8.0	6.2	6.5	6.4	12.3	10.3	9.9
7.7	9.8	7.1	9.2	9.9	6.4	7.8
8.0	7.8	11.8	8.2	7.5	8.6	7.9
10.9	9.4	9.4	8.4	7.6	7.8	11.1
9.2	9.9	8.5	8.9	10.3	7.2	7.4

Summarize these results with a cumulative "or more" percentage distribution based on values $4.0, 5.0, \ldots, 12.0$.

2-8 Convert the data in Exercise 2-1 to a cumulative "less than" distribution and draw the corresponding ogive.

2-9 Convert the data in Exercise 2-3 to a cumulative "more than" distribution and draw the corresponding ogive.

2-10 Draw an "or more" ogive for the distribution in Exercise 2-5.

2-11 Suppose that the air pressures in several automobile tires were checked after a number of trials over time and that the resulting values were grouped in the following classes: 22.00 to 22.19, 22.20 to 22.39, 22.40 to $22.59, \ldots, 25.80$ to 25.99. Which of the following could be determined from the frequency distribution? The number of tires with a pressure of

(a) more than 22.40

(b) less than 25.60

(c) 23.40 or more

(d) 24.79 or more

(e) less than 26.00

(f) 23.20 or less

2-12 Can the distribution in Practice Problem P2-2 be converted to a cumulative distribution? Why?

2-13 Can the distribution based on A, B + ,...,F in Exercise 2-5 be converted to a cumulative distribution? Why?

2-14 In some studies on human memory, subjects are tested for their ability to recall the particular placement and order of letters mixed in a sequence of digits. Each subject must keep stating an order until he or she finally has the order correct. The following frequency distribution represents the numbers of attempts made by 80 subjects to recall correctly a particular sequence.

Number of attempts to recall sequence correctly	Number of subjects
1	32
2	18
3	12
4	8
5	6
6	0
7	2
8	2

Illustrate these results with a cumulative "or fewer" ogive based on percentages rather than frequencies.

SOLUTIONS TO PRACTICE PROBLEMS

P2-4 (a) A "less than" distribution must be based on lower limits. Accumulating frequencies from the first class to the last produces the following cumulative distributions:

Landward shift (meters)	Cumulative frequency
Less than 5.0	4
Less than 10.0	12
Less than 15.0	26
Less than 20.0	35
Less than 25.0	42
Less than 30.0	47
Less than 35.0	50

(b) The ogive is shown in Figure P2-3.

P2-5 To prepare for the construction of the cumulative distribution, we first construct a frequency distribution with classes having $4.0, 5.0, \ldots, 12.0$ as lower limits and then accumulate the frequencies as follows:

Copper content (parts per million)	Frequency	Copper content (parts per million)	Cumulative frequency
4.0–4.9	1	4.0 or more	56
5.0–5.9	3	5.0 or more	55
6.0–6.9	4	6.0 or more	52
7.0–7.9	14	7.0 or more	48

(cont.)	Copper content (parts per million)	Frequency	Copper content (parts per million)	Cumulative frequency
	8.0– 8.9	11	8.0 or more	34
	9.0– 9.9	12	9.0 or more	23
	10.0–10.9	7	10.0 or more	11
	11.0–11.9	3	11.0 or more	4
	12.0–12.9	1	12.0 or more	1

The cumulative frequencies are then divided by 56 and multiplied by 100 to produce the following cumulative percentage distribution:

Copper content (parts per million)	Cumulative percentage
4.0 or more	100.0
5.0 or more	98.2
6.0 or more	92.9
7.0 or more	85.7
8.0 or more	60.7
9.0 or more	41.1
10.0 or more	19.6
11.0 or more	7.1
12.0 or more	1.8

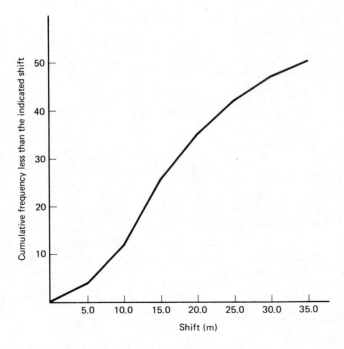

Figure P2-3 Cumulative distribution of landward shift of barrier islands.

2-4 OTHER GRAPHICAL TECHNIQUES

There are many other types of graphical displays of data. Illustrations of some of the different techniques for presenting data in graphical form are given in this section.

Frequency Polygon

Class frequencies may be illustrated in a graph similar to the ogive.

> Frequency heights are plotted above class marks and then joined by lines to produce a *frequency polygon*.

EXAMPLE 2-26 Figure 2-7 is a frequency polygon for the log data in Example 2-11. The graph may be "tied down," as is done in this case, by adding two extra classes with frequencies of 0. To save space and to give a better presentation, the horizontal scale does not start at 0 and run right through to 415, but is broken by a jagged line. This jagged part serves to point out the break but is not always necessary in a well-labeled graph. ▲

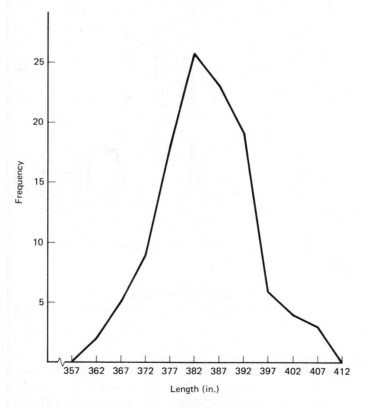

Figure 2-7 Frequency polygon for log data.

Bar Graph

A form of presentation in which the frequencies are represented by rectangles separated along the horizontal scale and drawn as bars is called a *bar graph*.

EXAMPLE 2-27   Figure 2-8 shows a bar graph for the frequency distribution in Example 2-2.   ▲

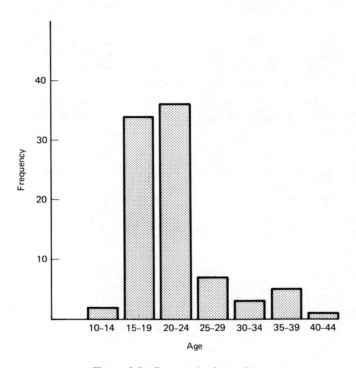

Figure 2-8   Bar graph of age data.

The bar graph has the advantage that, unlike a histogram or frequency polygon, it may be used for qualitative data.

EXAMPLE 2-28   Figure 2-9 shows a bar graph that illustrates the geographical distribution by water resource region of hydroelectric power generating plants in the United States (contiguous states only) in 1974.   ▲

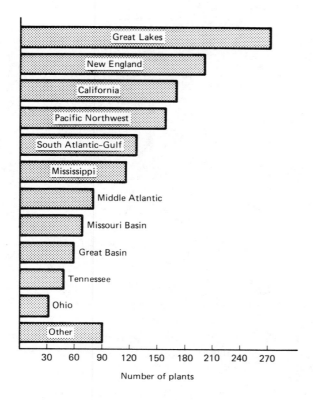

Figure 2-9 Geographical distribution of hydroelectric plants.

Other Illustrations

Other forms of illustration can be constructed to fit particular data.

> A *rose diagram* is a circular diagram suitable for illustrating *frequencies of directions*.

EXAMPLE 2-29 Figure 2-10(a) shows a rose diagram used by a geography student to illustrate frequencies of orientations of pebbles in a till fabric analysis of drumlins. In this diagram there is a problem with the area impression of the frequencies that requires some adjustment of the frequency scale, as shown in Figure 2-10(b). ▲

> Since area is proportional to the square of the radius, the frequencies, which are measured radially, are indicated in rose diagrams at distances proportional to their square roots.

Relative frequencies or percentages are often illustrated with a circle or "pie" that is cut into "slices" which are proportional in size to the percentages of the total

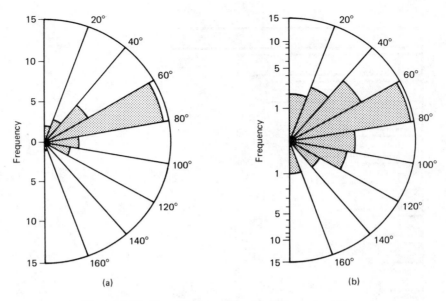

Figure 2-10 Rose diagrams for orientations of pebbles.

frequency falling into the various classes. To obtain the correct areas, the part of the circle allocated to each slice is equal to the appropriate percentage of the full circle (which is a full turn of 360°).

> Circular illustrations with slices proportional in area to frequencies are referred to as *pie charts*.

EXAMPLE 2-30 Figure 2-11 is a pie chart from the *Alberta Statistical Review* indicating the percentage distribution of the public assistance case load in Edmonton in 1975 by reason for assistance. Since "mental ill health" accounted for 7.64% of all cases, this category is represented by a slice formed from a partial turn of 7.64% of 360° or 27.5°. The other slices are formed through similar calculations. ▲

Frequencies or percentages are also often shown with pictorial displays.

> Pictorial displays of data are referred to as *pictographs*.

EXAMPLE 2-31 Pictographs are illustrated with the television chart shown in Figure 2-12 and the U.S. Bureau of Labor Statistics chart shown in Figure 2-13, in which the pictograph is also combined with an illustration of trend. The two pictographs illustrate the importance of considering area impression. In Figure 2-13 the black figures are approximately twice as high and twice as wide as the white figures, thus producing areas for the black figures that are approximately four times the area for the white figures. The reader of the chart may, at a glimpse, obtain the impression that the rate of unemployment for

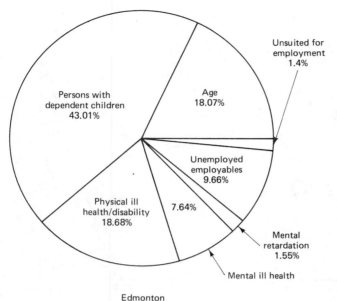

Persons with
dependent children
43.01%

Age
18.07%

Unsuited for
employment
1.4%

Unemployed
employables
9.66%

7.64%

Physical ill
health/disability
18.68%

Mental
retardation
1.55%

Mental ill health

Edmonton

Figure 2-11 Pie chart of public assistance caseload.

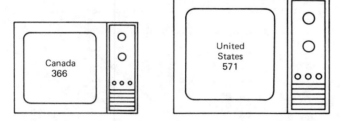

Canada
366

United
States
571

Number of televisions per 10,000 population 1974

Figure 2-12 Television ownership.

black workers is four times that for white workers—not twice as large, as is actually the case. ▲

> For correction interpretation, pictographs should be drawn so that the *areas* of the drawings are proportional to the frequencies they represent.

Although there is an area problem in the chart in Figure 2-13, the pattern of change in unemployment over time is quite clear from the well-labeled scales. The importance of labeling is more evident in the following example.

EXAMPLE 2-32   Figure 2-14 is similar to an unlabeled advertisement used by a brewery to indicate a sales increase. From the graph, however, it is not evident whether sales have increased by 1% or 50%, by 100 bottles of beer, or by 100,000 cases of beer. ▲

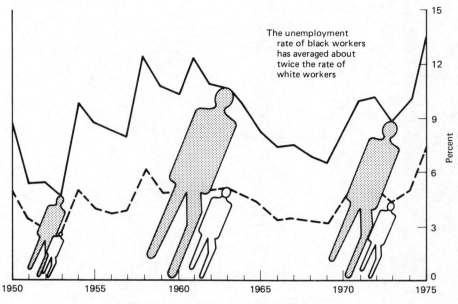

The unemployment rate of black workers has averaged about twice the rate of white workers

Unemployed as percent of civilian labor force by race

Figure 2-13 Unemployment trends for U.S. workers.

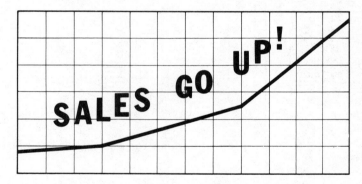

Figure 2-14 Sales chart.

Smooth Curves

Returning to graphs for frequency distributions, we may consider one other method of illustration—a smooth curve approximation to a histogram or frequency polygon. In such cases we may have an approximation only. With very large sets of data, or with infinitely large (hypothetical) sets of data (to be considered later), the smooth curve is a realistic illustration.

EXAMPLE 2-33 Figure 2-15 shows a histogram for the data of Example 2-11 with an approximate smooth curve superimposed on it. ▲

34 CHAPTER 2 *Summary Distributions and Graphical Techniques*

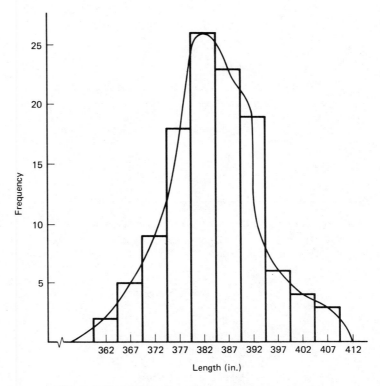

Figure 2-15 Histogram and smooth curve for log data.

It is helpful to use smooth curves as illustrations for discussions of general types of distributions or general properties of distributions.

Two basic forms or shapes of curves for distributions are *symmetric*, as shown in Figure 2-16, and *skewed*, as shown in Figure 2-17. Skewed distributions are said to be negatively or positively skewed, depending on whether the long "tail" of the distributions extends in the negative direction (to the left), or in a positive direction (to the right).

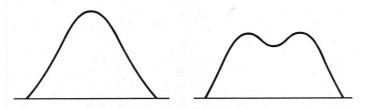

Figure 2-16 Symmetric distributions.

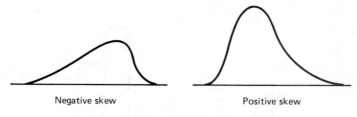

Negative skew   Positive skew

Figure 2-17   Skewed distributions.

P2-6   Construct a frequency polygon for the data in Practice Problem P2-5.

P2-7   A study on alleged grounds for marriage breakdown produced the following percentage analysis:

Alleged grounds	Percentage of grounds cited
Adultery	29.4
Physical cruelty	13.8
Mental cruelty	16.2
Other marital offenses	0.3
Addiction to alcohol	2.7
Separation for 3 years or more	33.8
Separation for 5 years or more	2.9
Other	0.9

Illustrate these data with a (horizontal) bar graph.

P2-8   The following distribution represents the results of a study on the degree of permanent rotational distortion of nine dozen shafts after being exposed to a temporary twisting:

Rotational distortion (degrees)	Frequency of occurrence
0–15	4
16–30	9
31–45	16
46–60	36
61–75	25
76–90	9
91–105	4
106–120	4
121–135	1

Illustrate these results with a rose diagram.

P2-9 The following percentages represent the responses to a survey question on which of five means allowed the respondent to achieve his or her most important goals:

Means to goals	Percentage
Work	57
Family	34
Friends	4
Church	4
Union	1

Illustrate these results with a pie chart.

P2-10 Draw a pictograph to illustrate the fact that the number of oil wells in a region had doubled in a 5 year period.

EXERCISES

2-15 Draw a frequency polygon for the frequency distribution in Exercise 2-1.

2-16 Suppose that a number of wind direction measurements produced the following results:

Direction	Frequency
N	12
NE	7
E	8
SE	6
S	4
SW	7
W	15
NW	10

Draw a rose diagram (full circle) to illustrate these results.

2-17 In a study of sources of air pollutant emissions, 61% of nitrogen oxides were attributed to transportation, 24% were attributed to stationary fuel combustion sources, 8% were attributed to industrial processes, and 7% were attributed to other miscellaneous sources. Illustrate the results of this study with a pie chart.

2-18 The following values represent the numbers of participants in a study on recreation:

Baseball or softball	190
Camping	259
Canoeing	173
Cross-country skiing	59
Curling	51
Downhill skiing	89

Golf	147
Horseback riding	83
Sailing	58
Skating	291
Snowmobiling	162
Squash	87
Swimming	638
Tennis	131

Draw a (horizontal) bar graph to illustrate the result of this study.

2-19 The following values represent the numbers of persons in prison per 100,000 population in selected countries:

Country (year)	Number of persons in prison per 100,000 population
United States (1970)	200.0
Australia (1972)	128.2
Canada (1972)	90.0
England and Wales (1971)	81.3
Sweden (1971)	61.4

Illustrate these data with a pictograph.

2-20 A study of the U.S. and Canadian destinations of Canadian oil indicated the following amounts in thousands of cubic meters per day.

Region	Cubic meters per day (thousands)
Ontario	86
Prairies and Northwest Territories	55
Eastern Canada	44
Central and eastern United States	42
British Columbia	26
Western United States	4

Construct a percentage-based bar graph to illustrate the above.

SOLUTIONS TO PRACTICE PROBLEMS

P2-6 The frequencies given in the solution to Practice Problem P2-5 are plotted above class marks to produce the frequency polygon shown in Figure P2-4.

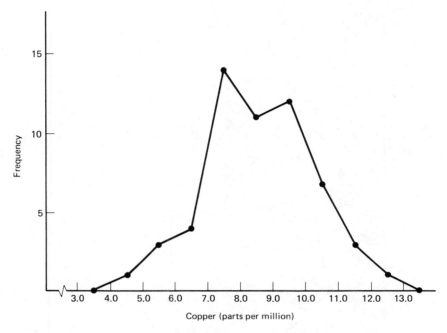

Figure P2-4 Frequency polygon for copper contents.

P2-7 The required bar graph is indicated in Figure P2-5. In this case, as is often done, the categories are rearranged in order of percentages. Furthermore, because of the low percentages, both "other" categories have been combined.

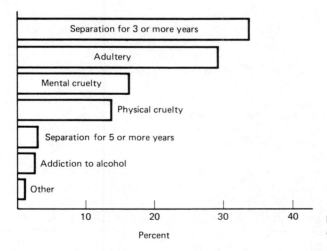

Figure P2-5 Percentage distribution of grounds for divorce.

P2-8 It was noted in the discussion on rose diagrams that the radii representing frequencies should be proportional to the square roots of the frequencies so that the areas would be proportional to the frequencies. The frequencies 4, 9, 16, 36,

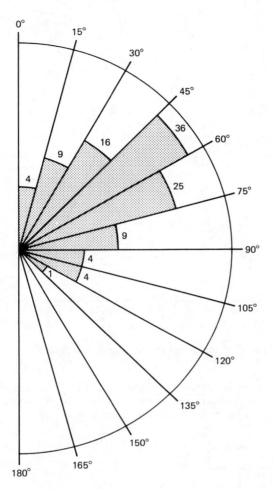

Figure P2-6 Rotational distortions of nine dozen shafts.

25, 9, 4, 4, and 1 are thus plotted at radii proportional to 2, 3, 4, 6, 5, 3, 2, 2, and 1, producing the rose diagram indicated in Figure P2-6.

P2-9 A full circle around the pie would be a rotation through 360°. Since "work" represented 57% of the responses, its share of the pie represents 57% of 360°, or 205.2° counterclockwise from an arbitrary starting point to a stopping point. Choosing to proceed counterclockwise is an arbitrary choice just as the starting point results from an arbitrary choice. A further 34% of 360° or 122.4° represents "family." We have a further 4% of 360° or 14.4° for each of "church" and "friends," and finally, 1% of 360° or 3.6° for "union," with the resulting pie chart shown in Figure P2-7.

P2-10 One possibility for the required pictograph is shown in Figure P2-8, in which a simple drawing of an oil rig is used to represent the number of oil wells. Since the number has doubled in area, the impression of the second drawing is double that of the first. This effect is achieved by making the height and width of the second drawing equal to $\sqrt{2}$ or 1.41 times those of the first.

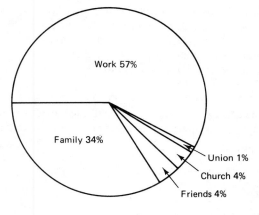

Work 57%

Family 34%

Union 1%

Church 4%

Friends 4%

Figure P2-7 Means to goals.

Figure P2-8 Increase in the number of oil wells.

2-5 CHAPTER REVIEW

2-21 The set of classes 0 to 9, 10 to 14, 15 to 19, 20 to 29, 30 to 34, 35 to 39, and 40 to 49 is not very suitable for a frequency distribution because _____.

2-22 A "less than" cumulative distribution must be based on boundaries or _____ limits.

2-23 Distributions of two different-sized sets of data from two different sources may be compared
(a) using the frequencies themselves
(b) using percentages instead of frequencies
(c) not at all

2-24 A histogram conveys the information of frequencies by use of heights and _____ _____.

2-25 To accommodate both a "less than" or an "or less" distribution, an ogive should be plotted at _____.

2-26 From a set of data listed in a frequency distribution with classes 0 to 24, 25 to 49,..., 150 to 174, 175 to 200, which of the following can be found?
(a) the number of items in the set of data that are greater than, or equal to, 50
(b) the number that are less than 174
(c) the number that are less than or equal to 75
(d) the number that are greater than 100
(e) the number that are greater than 24 and less than 100

2-27 A study on outstanding consumer credit indicated that the total was spread among the following principal sources according to the percentage share listed:

Banks	46%
Sales finance and consumer loan companies	19%
Credit unions	13%
Life insurance policy loans	6%
Department stores	6%
Other retailers	7%
Credit card issuers (other than banks) and public utilities companies	3%

Draw a pie chart to illustrate this allocation of the total outstanding credit.

2-28 Suppose that a pollution index can take values between 0 and 30 such that:

0–4	indicates very good
5–19	indicates acceptable
20–24	indicates poor
25–30	indicates dangerous

Suppose that a set of 100 readings of this index produced the following frequency distribution:

Index	Frequency
0–4	2
5–9	8
10–14	15
15–19	26
20–24	39
25–29	10

(a) Draw a histogram for these readings. What impression does it convey with regard to the general level of acceptability of the environment assessed?
(b) Suppose that the readings are grouped according to the ranges of degree of safety. For this grouping draw a histogram by plotting the class frequencies over the classes 0 to 4, 5 to 19, 20 to 24, and 25 to 29. What distortion of the data does this histogram convey? How could it be adjusted?

2-29 The following data represent a sample of readings of annual rainfall in inches for a given area:

29.0	25.2	25.3	31.9	32.9	28.7	24.0	30.2
25.2	28.0	25.2	28.1	25.8	29.1	34.8	36.5
26.6	26.0	33.3	25.7	28.9	26.8	23.5	34.4
28.5	24.9	26.0	29.1	28.6	30.6	30.9	

(a) Form a frequency distribution with classes 23.0 to 24.9, 25.0 to 26.9,..., 35.0 to 36.9.

(b) For this distribution give the class limits, class marks, and class boundaries.

(c) Draw a frequency polygon for this distribution.

(d) Form a cumulative "more than" distribution from this frequency distribution.

2-30 From 1967 to 1973, U.S. workers' hourly earnings increased by 73%. Figure 2-18 is intended to illustrate this change. If the 1967 illustration is correct, is the 1973 illustration also correct (in size)? If not, how should it be adjusted?

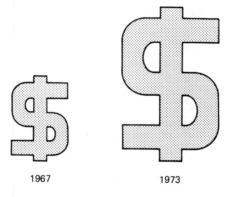

1967

1973

Figure 2-18 Change in hourly earnings.

2-31 Draw a "less than" ogive for the log data in Example 2-11 of Section 2-2.

2-32 Draw a frequency polygon for the bearing data in Example 2-18 of Section 2-2.

2-33 A study of the direction in which crabs go when released from under a bucket in crab races at a beach resort produced the following results. The directions represent the point of view of the observer.

Direction	Frequency
Straight ahead	20
Ahead—left	14
Left	9
Back—left	13
Straight back	8
Back—right	12
Right	9
Ahead—right	15

Illustrate these results with a rose diagram.

2-34 Use a pie chart to illustrate the following percentage breakdown of the responses to a survey question: What do you consider most important in a job?

Response	Percent
High income	12.4
Job security	7.2
Short hours	3.0
Opportunity for advancement	17.2
Sense of accomplishment	59.1
No response	1.1

2-35 Construct a bar graph to illustrate the following symptoms shown by psychiatric patients when assessed on a psychiatric rating scale.

Symptom	Number of patients
Emotional withdrawal	79
Conceptual disorganization	83
Mannerisms and posturing	59
Grandiosity	48
Hostility	48
Suspiciousness	82
Hallucinatory behavior	73
Motor retardation	38
Unusual thought content	86
Blunted affect	72

REPRESENTATIVE VALUES OR MEASURES

3-1 INTRODUCTION

In some cases, information may be presented in the form of raw data or partially summarized in frequency distributions. In other cases, the information may be reduced to one or more representative values or measures with statements such as:

A "typical" family in the subdivision has two children.

There are, on average, about 5000 deaths per year in France as a result of drunken driving.

American females aged 65 can expect to live another 17 to 18 years, but males aged 65 can expect to live only another 13 or 14.

98% of all coils of a particular cord will support 100 pounds before breaking.

Provincial average farm sizes range from a low of 60 acres in Newfoundland to a high of 850 acres in Saskatchewan.

60% of Illinois farmers sampled believed that crop rotation could reduce soil erosion.

Hydrangea plants will grow to heights of 3 to 10 feet.

Statistic

In this chapter we consider a few of the many possible representative values.

> Any numerical measure or characteristic of a set of data may be referred to as a *statistic*.

Later in this chapter we differentiate between two types of data, samples and populations. We will use the term "statistic" only when considering the data as a sample; we will introduce a new term when considering populations.

3-2 MEASURES OF LOCATION OR CENTRAL TENDENCY

One of the most familiar ideas of a representative value or statistic is the concept of an average. We very often think in terms of a vague measure of the "general level" of the values in a set of data, or of a "usual value" or a "typical value," and we call such a value an "average value."

Mean

> The most familiar average is the *arithmetic mean*, which we will refer to simply as the *mean*. It is obtained by summing all of the items in a set of data and then dividing the sum by the number of items.

EXAMPLE 3-1 If we have a sample of six bags of sugar with weights [in kilograms (kg)] of 2.272, 2.270, 2.274, 2.268, 2.270, and 2.272, we have a total of 13.626 kg of sugar, which, divided by 6, produces a mean of 2.271 kg of sugar per bag. ▲

As problems become more complex, we will find it useful to express calculations in general symbols or mathematical notation—a form of shorthand. We start introducing the notation now.

> In general, we consider a set of data to consist of n items, which we denote as $x_1, x_2, \ldots, x_n$, where the ith item in the set of data is denoted by x_i.

EXAMPLE 3-2 For the bags of sugar, $n = 6$, $x_1 = 2.272$, $x_2 = 2.270$, $x_3 = 2.274$, $x_4 = 2.268$, $x_5 = 2.270$, and $x_6 = 2.272$. ▲

> We use Σ (the Greek capital letter sigma) to denote summation. The expression $\sum_{i=1}^{n} x_i$ represents the sum of all the x_i's, with subscripts i taking all the values from 1 to n.

EXAMPLE 3-3 For the bags of sugar, $\sum_{i=1}^{n} x_i = \sum_{i=1}^{6} x_i = x_1 + x_2 + x_3 + x_4 + x_5 + x_6 = 2.272 + 2.270 + 2.274 + 2.268 + 2.270 + 2.2\,2 = 13.626.$ ▲

Modification of the expression allows for partial sums. For example, the sum of the weights of the second, third, and fourth bags of sugar may be expressed as $\sum_{i=2}^{4} x_i$.

Usually, we want the sum of all the values and then we simplify the expression to $\sum x$.

Next, we introduce the symbol $\bar{x}$ for the mean of the sample.

In general, for a sample of n values the mean is
$$\bar{x} = \sum x/n.$$

EXAMPLE 3-4 Let us consider the salaries of seven technicians working in a laboratory. There are two junior technicians, whose salaries are $10,200 and $9800; there are four technicians, whose salaries are $12,200, $12,200, $12,000, and $11,800; and one senior technician, the group supervisor, whose salary is $19,300. For this set of data, $n = 7$, $\sum x = \$87,500$, and $\bar{x} = \$87,500/7 = \$12,500$.
▲

Weighted Mean

EXAMPLE 3-5 As a different type of example, suppose that the prices of text books required for ten courses in a 2-year program are $10, $12, $14, $14, $10, $14, $12, $12, $14, and $14. We have $n = 10$ books with a total cost of $\sum x = \$126$ and a mean cost of $\bar{x} = \$126/10 = \$12.60$ per book. ▲

On occasion, a set of data with repeated values as in Example 3-5 can lead to the wrong calculation of the mean. We might consider that, for the ten books, there are only three possible prices, $10, $12, and $14, and that the mean is therefore $12. In such a calculation we have not considered the total cost—we have not given appropriate weights of importance to each of the three possible prices. For data with repeated values, we can find the correct mean from the raw data as above or by an alternative calculation assigning weights.

EXAMPLE 3-6 To each of the three possible prices in Example 3-5 we assign a weighting factor, which in this case is the number of books with the particular price. The weights are $w_1 = 2$, $w_2 = 3$, and $w_3 = 5$, and the total of the weights is $\sum w = 10$. To find the total cost, we multiply each possible price by the corresponding weight and then sum these products. Using a modification of the summation notation, we have
$$\sum wx = w_1 x_1 + w_2 x_2 + w_3 x_3$$
$$= (2 \times \$10) + (3 \times \$12) + (5 \times \$14) = \$126$$

We divide this weighted sum by the total weight 10 to obtain the mean price of $12.60 per book. ▲

A mean calculated from values x with corresponding weights w is called a *weighted mean*. The general expression for a weighted mean is

$$\bar{x} = \frac{\sum wx}{\sum w}$$

Note that the ordinary mean is a weighted mean with all weights equal to 1. The weights need not be whole numbers but may be fractions or percentages. In fact, the values being averaged might themselves be averages.

EXAMPLE 3-7 Consider the following data:

Region	Number of farms	Percent of all farms	Mean number of acres per farm
Atlantic	17,078	4.66	206
East	155,979	42.60	172
West	174,653	47.71	765
Pacific	18,400	5.03	316

If we take the mean of the four regional means as the overall mean farm size, we obtain $\bar{x} = 364.8$; however, complete information would indicate that the 366,110 farms have a total of 169,770,000 acres for a mean of 463.7 acres per farm. The latter value can also be found by calculating the weighted mean of the regional means, using the percent values as the weights. We calculate $\sum w = 100$ and $\sum wx = (4.66 \times 206) + (42.6 \times 172) + (47.71 \times 765) + (5.03 \times 316) = 46{,}374.79$. We thus find the weighted mean as $\bar{x} = \sum wx / \sum w = 46{,}374.79/100 = 463.7$. ▲

Median

For the bags of sugar in Example 3-1, the mean served quite well as the "average" weight—that is, it was a good representative value. When we use the mean as the average for the technicians' salaries in Example 3-4, however, we find that six of the seven technicians have salaries that are below the average. For the salaries, the mean is not as representative of the general level of the salaries.

This result indicates the effect that one or more extreme value(s) can have on the mean. For data with such extreme values, it is perhaps more appropriate to have an average that is more central in the data. In fact, we choose a very "central" value. We list the data in order of magnitude and choose the value in the center of the list as an average. This value is the *median*. We thus have as many values in the list that are above the median as there are below the median. If n is an even number, there is no central value in the data. In this case we define the median to be the mean of the two central values (after listing the data in order of magnitude).

> The *median* is denoted as $\tilde{x}$. After the data are listed in order of magnitude, the median is found as the value in the list in position $n/2 + 1/2$. When n is even, we define $\tilde{x}$, the value in position $n/2 + 1/2$, to be the mean of the values in positions $n/2$ and $(n/2) + 1$.

EXAMPLE 3-8　For the salaries of Example 3-4, $\tilde{x} = \$12{,}000$ is found in position 4.　　　　　　　　　　　　　　　　　　　　　▲

EXAMPLE 3-9　For the weights of Example 3-1, $n/2 + 1/2 = 3.5$. We define the median (the value in position 3.5) to be the mean of the values in positions 3 and 4; thus $\tilde{x} = (2.270 + 2.272)/2 = 2.271$. Alternatively, referring to values that actually occur in the data (2.271 does not), we might say that the median falls between 2.270 and 2.272.　　　　　　　　　　　　　　　　▲

Mode

EXAMPLE 3-10　For the $n = 10$ books of Example 3-5, the price $14 occurs more often than any other price. This price might be considered appropriate as another type of average value. We may consider $14 as an average or typical price because it occurs more frequently than the other values.　　　　　▲

> The value with the highest frequency in a set of data is called the *mode*.

The mode is not always a useful average, as many sets of data do not have a unique mode; for example, in the sugar data in Example 3-1 the highest frequency is 2, and both 2.268 and 2.272 occur with frequency 2—they are both modes.

Even if a set of data is unimodal (has a unique mode), the mode is not necessarily useful unless it has a relatively high frequency.

Comparing the Averages

The choice of average to use will depend on the type of data being considered and on what is desired of a representative value. When the data are unimodal and the mode stands out with a relatively high frequency, the mode is an appropriate average if we want the average to be a "typical" value. If we want an average that is representative of the general level, the mean or median is appropriate.

The mean is a reflection of the total quantity represented by the values and, as such, it is subject to the effects of extreme values. If the set of data includes extreme values, and we want an average that is more of a central value, the median is appropriate.

The mean is restricted to use with quantitative data. We cannot, for example, find the mean of five colors. The mode can be used for qualitative data—we can find the average color as the color that occurs most often. The median can be used for any data that can be ranked in some order. Ranking a number of cakes according to taste preference, we can find the average or median type of cake as that type in the middle position.

Using the mean or the median as the average, we may find no individual item in the data that is average. For the textbooks of Example 3-5, the mean price is $12.6 and the median price is $13; however, none of the books is either of these prices. Similarly, if we are told that the average dwelling is occupied by 3.5 people, we do not expect to find any individual dwelling that is so occupied; rather, this average is a measure of the ratio of the total population to the total number of dwellings.

In order to further understand these averages, it is helpful to look at their positions in graphs of frequency curves. In Figure 3-1, the mean, the median, and the mode (denoted by x^*) are indicated on the horizontal scale.

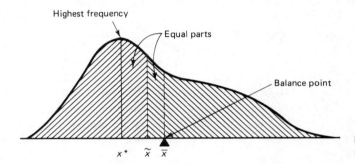

Figure 3-1 Positions of the mean, median, and mode.

The mode is the point at which the curve is the highest.
The median is the point that divides the area between the curve and the horizontal scale into two equal parts (since area represents frequency and the median divides the frequencies into two equal parts).
The mean is the point on the horizontal scale at which the shape enclosed by the curve balances.

The three averages considered tend to be central in the distribution of the data; hence, they are often referred to as *measures of central tendency*.

Figure 3-2 shows two frequency curves with similar shapes. They are different in that one represents data with higher values and hence higher averages. They are located at different positions on the horizontal scale as represented by the different positions of the indicated averages (means). This illustration demonstrates the use of averages to indicate the location of a distribution of data.

The term *measure of location* is often used for a value such as a mean, a median, or a mode.

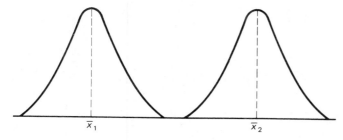

Figure 3-2 The mean as a measure of location.

It should be noted that "measure of location" and "measure of central tendency" are synonymous.

Figure 3-3 provides further illustrations of the positions of averages or measures of location for symmetric distributions.

For symmetric distributions, the mean and the median are equal.

The distribution on the right is not only symmetric, it is also unimodal.

For unimodal symmetric distributions, the mean, the median, and the mode are all equal.

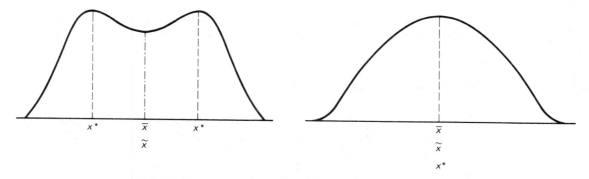

Figure 3-3 Mean, median, and mode in symmetric distributions.

Figure 3-4 represents positively (left distribution) and negatively (right distribution) skewed distributions. For these cases, the mean, the median, and the mode are different. To help remember their relative positions in these cases, we note that they occur in reverse alphabetical, or alphabetical, order, respectively. We note also that the mean is pulled farthest into the tail of a skewed distribution because of the influence of extreme values in the data.

Other Measures

Two other familiar measures of location are *quartiles*, dividing the data into four equal parts, and *percentiles*, dividing the data into 100 equal parts.

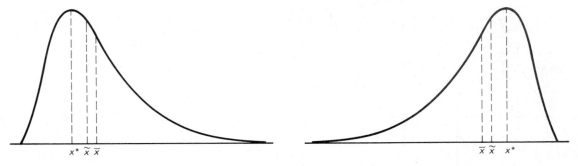

Figure 3-4 Mean, median, and mode in skewed distributions.

Statistics often quoted are the 10th and 90th percentiles, which cut off the lower 10% and the upper 10% of a set of data, or the first and the third quartiles (i.e., the 25th and 75th percentiles). The second quartile or 50th percentile is, of course, the median. Other quantiles may be found in a similar fashion.

In a set of n values, we may place the ith percentile at position $i \times n/100 + 1/2$ or the jth quartile at position $j \times n/4 + 1/2$.

EXAMPLE 3-11   The position of the 75th percentile (i.e., the third quartile) in a set of 255 values is $75 \times 255/100 + 1/2 = 191.75$ (i.e., $3 \times 255/4 + 1/2$); that is, the 75th percentile is three-fourths of the way from the 191st value to the 192nd value, or one-fourth of the way from 192nd to the 191st and is thus equal to $x_{191} + (3/4)(x_{192} - x_{191})$.   ▲

EXAMPLE 3-12   In the age data for unwed mothers in Example 2-1 of Section 2-2, there are 88 values. The 80th percentile is found in position $80 \times 88/100 + 1/2 = 70.9$ and is thus nine-tenths of the way from the 70th to the 71st value. Since, after listing the data from smallest to largest, we find that each of these two values is equal to 24, the 80th percentile is found to be 24.   ▲

PRACTICE PROBLEMS

P3-1 The following values represent the results of several measurements on the concentration of calcium in a stream [in milligrams per decimeter (mg/dm)]:

  14 8 9 8 13 13 9 8 12 11 10 7 10 12 12

(a) Calculate the mean concentration.
(b) Calculate the median.

P3-2 In a study on the physical capabilities of certain anthropoid primates, the amount of rotation through which a forearm joint could move was noted for 12 subjects of the hominoid genera. The following values represent the results (in degrees):

154 151 170 163 156 153 159 140 160 147 150 166

Find (a) the mean and (b) the median for these data.

P3-3 The following numbers represent the responses received from a number of students when asked how many times they had consulted material in the library since the start of term:

3 9 5 3 1 3 2 2 8 3 4 0 1 3 3 6 1 3 7 4

Determine the mode for these data.

P3-4 In order to establish a base time required to read and type symbols in a recognition study, a psychologist ran several trials on a number of occasions with each subject and recorded the times in milliseconds (msec). The following values represent the results from ten occasions for one of the subjects:

Occasion	Number of trials	Mean response time (msec)
1	25	950.08
2	20	894.90
3	50	978.26
4	50	831.52
5	30	982.10
6	30	868.30
7	20	901.80
8	50	796.72
9	10	1013.60
10	25	966.12

Calculate the mean response time for this subject.

P3-5 Suppose that the following values are the total times in minutes that 16 battery packs operated before requiring recharging:

814 793 775 746 758 729 749 760
736 804 764 756 778 728 745 780

Determine the third quartile for these values.

P3-6 Determine the 90th percentile for the data in Practice Problem P2-1.

3-1 Suppose that in a study on injuries in a given sport, the total number of league games missed by the players for each of ten teams in a league were as follows (e.g., if six players each miss two games, that produces a total of 12 missed games):

62 118 123 170 83 183 72 91 68 93

(a) Calculate the mean number of missed games per team.
(b) Find the median number of missed games per team.

3-2 Upon contacting several service garages for an estimate of the cost of repairing an engine problem, a newspaper reporter obtained the following estimates:

$35 $100 $50 $50 $75 $45 $50 $65 $40 $50 $40 $60 $50 $55 $75

Determine the mean, median, and mode for these estimates.

3-3 In a conditioning study, a psychologist observed the percentages of total time that experimental rats spent at a feeding tube when the tube contained a sucrose solution. The following values represent the percentages recorded:

22 16 21 22 26 13 19 19 22 18

20 17 27 26 22 19 24 30 16 20

29 24 20 26 20 22 21 20 21 18

Determine the mean and median for these values.

3-4 Determine the mean and the mode for the unwed mothers' ages given in Example 2-1 of Section 2-2.

3-5 As of May 1974, the number of hydroelectric power generating plants in eight selected U.S. Water Resources Council regions and their mean capacities [in megawatts (mW)] were as follows:

Region	Number of plants	Mean capacity (mW)
Arkansas–White–Red	22	88.1
California	172	41.2
Great Lakes	271	14.8
Middle Atlantic	79	16.9
Missouri Basin	69	48.8
Ohio	35	40.9
Pacific Northwest	160	129.9
Upper Colorado	21	63.1

Calculate the overall mean capacity for the 829 plants in the selected regions.

3-6 Determine the first and third quartiles for the set of marks in Exercise 2-4.

3-7 Consider the marks in Exercise 2-4 converted to grades as in Exercise 2-5. Determine the average of the grades if "average" is interpreted as most typical. What is the name of this average?

3-8 Determine the 10th and 90th percentiles for the prescription drug price data of Exercise 2-3.

3-9 In a survey on the taste of a new natural foods snack, the numbers of people indicating various possible responses were as follows:

Response	Number
Excellent	25
Good	27
Fair	28
Not good	10
Poor	7
Very poor	3

(a) What are the median response and the modal response?

(b) Why can the mean response not be calculated?

3-10 A monitoring process at a lead refinery is used to measure the daily emission levels from the plant into the atmosphere. The following values represent daily emission levels in micrograms for 10 weeks (the plant operates 5 days a week):

31.2	12.4	28.3	7.4	18.6	9.9	5.6	5.7	27.7	3.1
8.2	10.4	10.4	3.5	4.5	8.3	3.8	4.0	5.6	5.6
3.0	14.5	5.8	5.5	9.7	3.9	4.7	6.2	6.1	7.4
4.6	8.3	12.2	9.6	7.2	6.8	3.2	6.9	2.5	2.7
5.9	5.2	11.7	4.8	4.4	7.3	4.4	5.0	5.5	1.0

(a) Find the mean and the median for these data.

(b) Which of these two measures is more suitable as an indicator of the general level of the major central portion of the data?

(c) Which of these two measures is more suitable as an indicator of the total lead emission from the refinery?

SOLUTIONS TO PRACTICE PROBLEMS

P3-1 (a) We have $n = 15$ values producing a sum $\Sigma x = 156$, and the mean is $\bar{x} = 156/15 = 10.4$.

(b) Rearranging the $n = 15$ values from smallest to largest produces:

$$7 \quad 8 \quad 8 \quad 8 \quad 9 \quad 9 \quad 10 \quad 10 \quad 11 \quad 12 \quad 12 \quad 12 \quad 13 \quad 13 \quad 14$$

and the value in position $n/2 + 1/2 = 8$ is 10. We thus have the median $\tilde{x} = 10$.

P3-2 (a) The $n = 12$ values produce a sum $\Sigma x = 1869$ and mean $\bar{x} = 1869/12 = 155.75$.

(b) Rearranging the $n = 12$ values from smallest to largest produces:

$$140 \quad 147 \quad 150 \quad 151 \quad 153 \quad 154 \quad 156 \quad 159 \quad 160 \quad 163 \quad 166 \quad 170$$

and the value in position $n/2 + 1/2 = 6.5$ is halfway between the values in positions 6 and 7. The value is halfway between 154 and 156 and the median is thus $\tilde{x} = (154 + 156)/2 = 155$. Alternatively, we might just report that the median falls between 154 and 156.

P3-3 The value 3 occurs more frequently than any other value in the data and is the mode.

P3-4 Each of the individual mean times must be weighted according to the number of trials involved. These weights are

$$25 \quad 20 \quad 50 \quad 50 \quad 30 \quad 30 \quad 20 \quad 50 \quad 10 \quad 25$$

and the weighted mean is

$$\bar{x} = \frac{\Sigma wx}{\Sigma w} = \frac{(25 \times 950.08) + (20 \times 894.90) + \cdots + (25 \times 966.12)}{25 + 20 + \cdots + 25}$$

$$= \frac{279{,}812}{310} = 902.62$$

P3-5 For $n = 16$ values, the third quartile is in position $3 \times n/4 + 1/2 = 3 \times 16/4 + 1/2 = 12.5$. It is halfway between the values in positions 12 and 13. Rearranging the data from smallest to largest produces 728 729 736 745 746 749 756 758 760 764 775 778 780 793 804 814. The third quartile is then $(778 + 780)/2 = 779$. We might also just report that the median falls between 778 and 780.

P3-6 For $n = 50$ values, the 90th percentile is in position $90 \times n/100 + 1/2 = 90 \times 50/100 + 1/2 = 45.5$. It is halfway between the values in positions 45 and 46. From the frequency distribution in the solution to Practice Problem P2-1, we note that the values in the class 25.0 to 29.9 occupy positions 43, 44, 45, 46, and 47 because there are 5 values in this class and 42 values preceding this class. Selecting the values between 25.0 and 25.9 inclusive from the data and arranging them in order produces 25.6, 26.5, 27.3, 27.5, and 29.8. The 45th and 46th values are 27.3 and 27.5. We may thus report the 90th percentile to be between 27.3 and 27.5 or we might give the specific value for the 90th percentile as $(27.3 + 27.5)/2 = 27.4$.

3-3 MEASURES OF VARIATION OR DISPERSION

Summarizing data with a single representative value or average may involve too much of a sacrifice of information.

EXAMPLE 3-13   As an extreme example, consider a very poor heating/cooling system which operates as follows. If the temperature drops to 17°C, the heating cycle starts and increases temperature continuously at a fairly constant rate until the temperature soon reaches 25°C. When the temperature reaches 25°C, the cooling cycle starts reducing the temperature at a fairly constant rate until the temperature soon reaches 17°C.

As a result of this system, we have a very uncomfortable constant change in temperature with short equal time periods of too hot, too cold, and comfortable. If we summarize with the average (mean or median) temperature, we obtain 21°C. This single statistic gives the impression of a comfortable temperature. With no other information, the amount of variation in the temperature is not disclosed.   ▲

Because of problems such as that in Example 3-13, it is necessary to include in our information another statistic to indicate the degree of variability in the data.

Statistics that measure the variability in data are called
measures of variation or dispersion.

Figure 3-5 illustrates two possible, quite different, frequency curves which have the same means, the same medians, and the same modes. The two curves, however, display considerable differences in the amount of variation or dispersion in the data.

An attempt is often made to indicate the degree of variability through the use of more location measures, such as the 10th and the 90th percentiles. Although helpful, this approach is not without some weaknesses. Figure 3-6 shows two

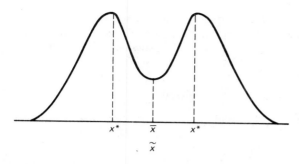

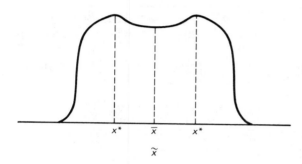

Figure 3-5 Different distributions with equal averages.

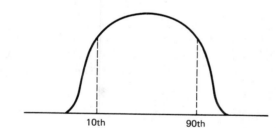

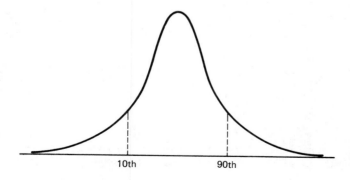

Figure 3-6 Different distributions with equal 10th and 90th percentiles.

frequency curves with equal averages and with equal 10th and equal 90th percentiles. These two curves are, however, very different. The data represented in the lower distribution cover a larger range of values from lowest to highest value.

Range

In this concept of the total range of values included, we have one of the simplest measures of variation.

> The *range* is defined as the difference between the highest and the lowest values in a set of data.

EXAMPLE 3-14   For the temperatures of Example 3-13, the range is $25°C - 17° = 8°C$. The range indicates the uncomfortable variation in temperature. ▲

Knowing the range alone is not very informative, especially if there is no indication of whether the data may be extremely skewed.

> A useful expansion on the notion of range is to report the smallest and largest values in the data in a phrase such as: "The values range from (lowest value) to (highest value)."

EXAMPLE 3-15   For the temperatures in Example 3-13, the values range from 17°C to 25°C.   ▲

Although the range does measure the total spread in the data, it does not indicate whether the values are uniformly spread or perhaps clustered around the mean. Figure 3-7 illustrates two different frequency curves that have equal means and equal ranges (including equal lowest and equal highest values), but quite

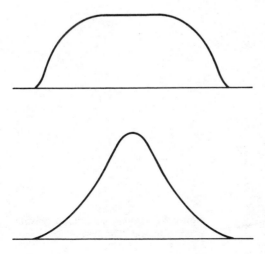

Figure 3-7   Different distributions with equal means and ranges.

different patterns of variation about the mean. For this reason, the range is not a fully informative measure of variation.

Mean Deviation

The next step is to find a measure of the variation about the mean. For each member of the data x_i, we consider its position relative to the mean by finding the difference $x_i - \bar{x}$. For a value that is 1.5 above (i.e., greater than) the mean, we have $x_i - \bar{x} = 1.5$; for a value that is 2.3 below the mean, we have $x_i - \bar{x} = -2.3$, a negative number. For data clustered near the mean, the differences will tend to be small; for data greatly dispersed about the mean, the differences will tend to be large. Accordingly, to obtain a measure of the total variation in the data, it is appropriate to find an average of these differences.

For the weights in Example 3-1, these differences are: 0.001, -0.001, 0.003, -0.003, -0.001, and 0.001, and their average (mean or median) is 0.000. In fact, if we use this mean of the differences, then for any data we will always obtain the same value, 0. This result can be justified intuitively from the definition of the mean position. It can also be proven algebraically with a further development of the summation notation as follows.

To find the mean of all the differences, we must first find their sum, which we write as $\sum_{i=1}^{n}(x_i - \bar{x})$ or simply $\sum(x_i - \bar{x})$. The summation sign may be carried through the sum of differences since $\sum(x_i - \bar{x}) = (x_1 - \bar{x}) + (x_2 - \bar{x}) + \cdots + (x_n - \bar{x}) = x_1 - \bar{x} + x_2 - \bar{x} + \cdots + x_n - \bar{x}$, which may be rearranged as $x_1 + x_2 + \cdots + x_n - \bar{x} - \bar{x} - \cdots - \bar{x} = (x_1 + x_2 + \cdots + x_n) - (\bar{x} + \bar{x} + \cdots + \bar{x})$. The expression may then be written as $\sum x - \sum \bar{x}$, which may be rewritten as $\sum x - n\bar{x}$, because the second part of the expression is the sum of n identical values $\bar{x}$ and equals $n\bar{x}$. Further, since $\bar{x} = \sum x/n$, we may write the expression as

$$\sum x - \left(n \times \frac{\sum x}{n}\right) = \sum x - \sum x = 0$$

Thus, we finally have $\sum(x - \bar{x}) = 0$, and the mean of the differences is, therefore, $\sum(x - \bar{x})/n = 0/n = 0$, as previously stated.

We may overcome this problem by considering only the magnitude of each difference, ignoring negative signs.

If we convert negative signs to positives, we have the *absolute value* of each of the differences or deviations from the mean. We denote absolute value with two vertical bars. The absolute value of $x - \bar{x}$ is written $|x - \bar{x}|$.

EXAMPLE 3-16   For numerical values, we have $|1.5| = 1.5, |-2.3| = 2.3$.   ▲

If we take the mean of the absolute values of the deviations, we will not have the problem of an automatic 0. We will have a positive measure of the general magnitude of the deviations of the values about the mean. This measure is called the *mean deviation*.

The *mean deviation* is equal to $\dfrac{\sum |x - \bar{x}|}{n}$.

EXAMPLE 3-17   For the weights of Example 3-1, the mean deviation is

$$\frac{\sum |x - \bar{x}|}{6} = \frac{|0.001| + |-0.001| + |0.003| + |-0.003| + |-0.001| + |0.001|}{6}$$

$$= \frac{0.001 + 0.001 + 0.003 + 0.003 + 0.001 + 0.001}{6}$$

$$= 0.0017 \qquad\qquad \blacktriangle$$

Standard Deviation and Variance

The mean deviation is easy to understand, but it is not often used because absolute values do not possess certain mathematical properties desired for more sophisticated statistical procedures. Instead, we consider a similar measure.

We first find an average squared deviation. Since the square of any real number is positive, squared deviations eliminate the problems of negative deviations, just as absolute values do. The average that we use is similar to a mean except that we divide the total by $n - 1$ instead of n. For each deviation, $x_i - \bar{x}$, we find the square, $(x_i - \bar{x})^2$. We sum these squared deviations and then divide by $n - 1$.

The division by $n - 1$ instead of n makes this average consistent with many similar measures used in statistical work. In many procedures which we will encounter in later chapters, sums of squares are converted to mean squares by dividing by a value which we refer to as the degrees of freedom. In this case we have n differences of the form $x_i - \bar{x}$. These values are not all independent because, as we have already noted, they always sum to 0. In fact, given any $n - 1$ of them, we can always determine the nth. We have $n - 1$ "free" choices and the nth difference is then determined on the basis of these $n - 1$. The n differences thus exhibit $n - 1$ degrees of freedom.

The measure found by dividing the sum of squared deviations by $n - 1$ is an average *squared* deviation. Its square root is an average deviation providing information similar to that provided by the mean deviation. This square-root measure is called the *standard deviation*.

The *standard deviation* is denoted as s and is defined as

$$s = \sqrt{\frac{\sum (x - \bar{x})^2}{n - 1}}$$

The value we have before we take the square root is referred to as the *variance* and is

$$s^2 = \frac{\sum (x - \bar{x})^2}{n - 1}$$

EXAMPLE 3-18 For the weights in Example 3-1, the variance is

$$s^2 = \frac{(0.001)^2 + (-0.001)^2 + (0.003)^2 + (-0.003)^2 + (-0.001)^2 + (0.001)^2}{5}$$

$$= \frac{0.000001 + 0.000001 + 0.000009 + 0.000009 + 0.000001 + 0.000001}{5}$$

$$= 0.0000044$$

The square root or standard deviation is $\sqrt{0.0000044} = 0.0021$. ▲

In order to describe data in very brief summary form, the statistics most often presented are the mean and the standard deviation or, equivalently, the mean and the variance.

EXAMPLE 3-19 For the weights in Example 3-1, we have found the summary statistics to be a mean of 2.271 kg and a standard deviation of 0.0021 kg or, using symbols, $\bar{x} = 2.271$ kg and $s = 0.0021$ kg. ▲

Calculating Formulas

If we return to the age data in Example 2-1 of Section 2-2, we have a larger set of data. For this set of data, $n = 88$, $\Sigma x = 1894$, and $\bar{x} = 21.522727\ldots$. In order to take the differences $x - \bar{x}$, we must round $\bar{x}$ at some point and thus introduce a rounding error in each $(x - \bar{x})^2$, a possible error in the sum $\Sigma(x - \bar{x})^2$, and hence a possible error in the final s^2 or s.

The amount of calculation can be reduced considerably with the following calculating formula:

$$\Sigma(x - \bar{x})^2 = \Sigma x^2 - \frac{(\Sigma x)^2}{n}$$

The sum of the squared deviation is equal to the sum of the squares of the original values minus the square of the sum of the original values divided by the number of values.

Using algebra, this calculating formula is developed as follows. We may rewrite each squared difference as $(x - \bar{x})^2 = x^2 - 2x\bar{x} + \bar{x}^2$ and, carrying the summation sign through, we have $\Sigma(x - \bar{x})^2 = \Sigma x^2 - 2\Sigma x\bar{x} + \Sigma\bar{x}^2$. The second term becomes $2\bar{x}\Sigma x$ because $\bar{x}$ is a (constant) common factor in the sum, and the third factor becomes $n(\bar{x})^2$ because it is the sum of n identical items, $\bar{x}^2$.

Finally, rewriting $\bar{x}$ as $\Sigma x/n$, we may rewrite the expression as

$$\Sigma x^2 - 2\bar{x}\Sigma x + n\bar{x} \times \bar{x} = \Sigma x^2 - \left(2 \times \frac{\Sigma x}{n} \times \Sigma x\right) + \left(n \times \frac{\Sigma x}{n} \times \frac{\Sigma x}{n}\right)$$

$$= \Sigma x^2 - \frac{2(\Sigma x)^2}{n} + \frac{(\Sigma x)^2}{n} = \Sigma x^2 - \frac{(\Sigma x)^2}{n}$$

which produces the calculating formula.

A *calculating formula* for the variance is

$$s^2 = \frac{\sum x^2 - (\sum x)^2/n}{n - 1}$$

or when top and bottom are multiplied by n,

$$s^2 = \frac{n\sum x^2 - (\sum x)^2}{n(n - 1)}$$

These are the expressions that we will use for calculating s^2. The original expression is important as a definition to help understand the concept of variance as an average squared deviation from the mean. For calculating purposes it is generally useful only for small sets of data, or for data with very simple means involving no rounding error. Taking square roots of the foregoing expressions, we obtain *calculating formulas* for the standard deviation.

Calculating formulas for the standard deviation are

$$s = \sqrt{\frac{\sum x^2 - (\sum x)^2/n}{n - 1}}$$

or

$$s = \sqrt{\frac{n\sum x^2 - (\sum x)^2}{n(n - 1)}}$$

EXAMPLE 3-20   For the age data in Example 2-1, we calculate

$$\sum x^2 = 18^2 + 17^2 + 20^2 + 13^2 + \cdots + 20^2$$
$$= 324 + 289 + 400 + 169 + \cdots + 400$$
$$= 43{,}648$$

and

$$(\sum x)^2 = (1894)^2 = 3{,}587{,}236$$

Using the second calculating formula for the standard deviation, we calculate

$$s = \sqrt{\frac{88 \times 43{,}648 - (1894)^2}{88 \times 87}} = 5.76$$

▲

We always use the positive square root for the standard deviation as we want a positive measure of dispersion in the data. The other measures of variation that we have discussed are automatically positive. The range must be positive because it is the highest minus the lowest. The variance is positive because it comes from a sum of squares. This fact is not immediately apparent from the calculating formula, but it is obvious when we consider the original defining formula.

Interpreting the Standard Deviation

Measures of location were interpreted graphically by indicating their positions on the horizontal scale in graphs of frequency curves. The range was also apparent as the total length of the horizontal scale covered by the frequency curve. The standard deviation, which is a measure of how the values are dispersed about or positioned relative to the mean, can be interpreted graphically by indicating the position of values that are this "average distance" or standard deviation away from the mean. The value that is at the position at a distance equal to the standard deviation above the mean is $\bar{x} + s$. Similarly, the value that is this distance below the mean is $\bar{x} - s$. More extreme values that are twice the standard deviation from the mean are $\bar{x} - 2s$ and $\bar{x} + 2s$.

Figure 3-8 illustrates a frequency curve with $\bar{x}$ and other points indicated on the horizontal scale. For this set of data, the variation about $\bar{x}$ is indicated in units of s.

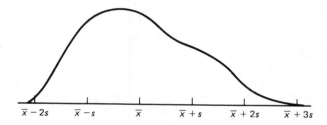

Figure 3-8 Indication of variation about the mean in units of the standard deviation.

This set of data has a range of about five standard deviations, $(x + 3s) - (x - 2s) = 5s$. The same relationship does not hold true for all sets of data, but for most (although not all) data, the range is roughly three to five times the standard deviation.

EXAMPLE 3-21 For the age data in Example 2-1, the range is $42 - 13 = 29$. This is just over five times the standard deviation, which was found to be 5.76 in Example 3-20. For the data in Example 3-1, the range is $2.274 - 2.268 = 0.006$ and is just under three times the standard deviation. ▲

Using this approximate relationship, the range, which is easy to find, can be used to provide a rough check on the calculation of the standard deviation.

We can note other aspects of the spread in terms of the standard deviation by referring to the particular data illustrated in Figure 3-8. Over half of the values are within 1 standard deviation of the mean, most of the values are within 2 standard deviations of the mean, and almost all are within 3 standard deviations of the mean.

Chebyshev's Theorem

Although other frequency distributions will have different patterns of variation, it is true of all sets of data that *at least 88.9%* of the values are within 3 standard deviations of the mean; *at least 75.0%* are within 2 standard deviations; and *at least*

55.6% are within 1.5 standard deviations of the mean. These results are all due to the following theorem.

> *Chebyshev's theorem* states that, for any set of data, *at least* 100 $(1 - 1/k^2)$% of the values are within k standard deviations of the mean, where k is any number greater than 1.

EXAMPLE 3-22 For the age data in Example 2-1, suppose that, without enumerating directly, we would like some idea of the number of ages within 2.5 standard deviations of the mean. According to Chebyshev's theorem, *at least* $100 \times (1 - 1/(2.5)^2)$% or 84% of the values are within 2.5 standard deviations of the mean. [In this case we have the raw data, so we may check the results of Chebyshev's rule. With $\bar{x} = 21.5$ and $s = 5.76$ the values within 2.5 standard deviations of the mean are the values between 7.1 and 36.9, which are found as $21.5 - (2.5 \times 5.76)$ and $21.5 + (2.5 \times 5.76)$. Direct enumeration indicates that there are 83 ages or 94.3% of the values in this range.] ▲

EXAMPLE 3-23 To obtain an idea of how many of the ages in Example 2-1 are within 10 years of the mean, we note that 10 years is equal to $10 \div 5.76 = 1.7$ standard deviations; thus, using $k = 1.7$ in Chebyshev's theorem, there are *at least* $100 \times [1 - 1/(1.7)^2]$% or 64.4% of the ages within 10 years of the mean. (Again, as a check, direct enumeration indicates that there are actually 81 ages or 92% of the values within 10 years of the mean.) ▲

Chebyshev's theorem is very important in helping us to interpret the standard deviation in an intuitive sense. From the knowledge of the mean and the standard deviation, we gain a general impression of where the major portion of a set of data is located and of how much variation there is in the data. A small standard deviation indicates that the values are clustered close to the mean. A large standard deviation indicates that the values are quite varied about the mean.

Coefficient of Variation

Whether or not the variation is to be considered to be large or small depends not only on the standard deviation, but also on the general magnitude of the values in the data.

EXAMPLE 3-24 If we have a number of steel reinforcing rods whose lengths have a mean of 5 m and a standard deviation of 0.5 cm (0.005 m), we might consider these values to have relatively little variation. If, however, we have a number of bearings whose diameters have a mean of 1.0 cm but still a standard deviation of 0.5 cm, we might consider the values to have a great deal of variation. ▲

To determine whether a set of values has much variation or whether a number of measurements are very precise, we require the mean as well as the standard deviation. These two values can be combined in a single value that measures the standard deviation proportional to the mean. We call this value the *coefficient of variation*.

The *coefficient of variation* is equal to

$$100\left(\frac{s}{\bar{x}}\right)\%$$

EXAMPLE 3-25   For the reinforcing rods in Example 3-24, the coefficient of variation is $100 \times (0.005/5)\% = 0.1\%$. The standard deviation is only one-tenth of 1% of the mean. This set of data has relatively very little variation. For the bearings, the coefficient of variation is $100 \times (0.5/1.0)\% = 50\%$. This set of data is, relatively, quite varied. ▲

P3-7 In a study on response to sound intensity detection under hypnotic suggestion of impaired hearing, a number of subjects indicated the following thresholds (in decibels):

$$61 \quad 46 \quad 59 \quad 46 \quad 72 \quad 37 \quad 68 \quad 56 \quad 50$$

(a) Calculate the range for these data.
(b) Calculate the mean deviation.
(c) Calculate the variance and standard deviation as defined and using a calculating formula.

P3-8 In a study on age and body fat, the percentages of fat in a sample of men aged 40 to 49 produced a mean of 26.00 and a standard deviation of 6.19.
(a) At least what proportion of the men had percentages of fat between 15.00 and 37.00?
(b) If 150 men were included in the sample, at most how many had fat percentages that differ from the mean by more than 15.00?

P3-9 In a study on the variability of traffic over a bridge being considered for upgrading, a sampling of daytime traffic rates produced a mean of 32.0 and a standard deviation of 9.6. Calculate the coefficient of variation.

3-11 In a study on the variability of oil viscosity, 12 quarts of oil of the same grade were sampled and their viscosities (in centipoise) were

$$218 \quad 231 \quad 228 \quad 219 \quad 227 \quad 235 \quad 226 \quad 227 \quad 221 \quad 226 \quad 233 \quad 221$$

(a) Calculate the mean deviation for the viscosities.

(b) Calculate the variance and the standard deviation first by using the defining expression and then by using the calculating formula.

(c) Calculate the coefficient of variation.

3-12 Calculate and compare the range and the standard deviation for the prescription drug prices in Exercise 2-3.

3-13 A pressure gauge is to be considered sufficiently precise for the range of low pressures that it will be used to measure if it exhibits a standard deviation of at most 0.0020 kPa in a sampling of 20 repeat measurements of the same pressure. If 20 such measurements produced the values

| 0.173 | 0.175 | 0.167 | 0.171 | 0.172 | 0.168 | 0.166 | 0.172 | 0.173 | 0.169 |
| 0.172 | 0.173 | 0.173 | 0.171 | 0.174 | 0.167 | 0.169 | 0.170 | 0.172 | 0.174 |

is the gauge sufficiently precise?

3-14 If a set of 1000 package weights has a mean of 5.00 lb and a standard deviation of 0.08 lb, at least how many of the packages must have weights between 4.80 and 5.20 lb?

3-15 Suppose that 60 subjects recorded the following times to respond to an audio stimulus as the volume increased:

10	8	12	12	13	17	11	11	16	14	18	17	19	33	26
12	26	11	30	17	15	12	11	12	16	12	17	18	19	7
10	31	10	29	23	16	13	11	13	15	6	17	15	21	13
32	12	26	17	14	22	16	11	23	19	25	16	28	27	18

(a) Using Chebyshev's theorem, determine how many (at least) of the times should be within 9 of the mean.

(b) By direct enumeration, determine how many of the times are within 9 of the mean.

3-16 Suppose that measurements on the depth of the sand layer in various areas on the bottom of a cove produced the following values:

| 4.9 | 3.2 | 3.7 | 3.6 | 2.7 | 4.9 | 3.8 | 5.3 | 4.5 | 3.5 | 3.3 | 4.1 |
| 4.6 | 4.0 | 2.8 | 0.5 | 0.7 | 3.9 | 1.4 | 1.2 | 3.7 | 3.9 | 3.3 | 2.9 |
| 3.4 |

Determine the coefficient of variation for these values.

3-17 A set of eight values has produced deviations from the mean such that $x_1 - \bar{x}$, $x_2 - \bar{x}, \ldots, x_7 - \bar{x}$ are, respectively, -3.2, 6.3, 4.1, -1.0, -4.8, 2.3, and 0.7.

(a) What does $x_8 - \bar{x}$ equal?

(b) Find the mean deviation for the original eight values.

(c) Calculate the variance and standard deviation for the original eight values.

(d) Determine the range for the original eight values.

P3-7 (a) The lowest value is 37 and the highest is 72. The range is $72 - 37 = 35$.

(b) Since $n = 9$ and $\Sigma x = 495$, the mean is $\bar{x} = 495/9 = 55$. The mean deviation is then

$$\frac{\Sigma |x - \bar{x}|}{n} = \frac{|61 - 55| + |46 - 55| + \cdots + |50 - 55|}{9}$$

$$= \frac{|6| + |-9| + \cdots + |-5|}{9}$$

$$= \frac{6 + 9 + \cdots + 5}{9} = \frac{82}{9} = 9.11$$

(c) From the definition, the variance is

$$s^2 = \frac{\Sigma (x - \bar{x})^2}{n - 1}$$

$$= \frac{(61 - 55)^2 + (46 - 55)^2 + \cdots + (50 - 55)^2}{8}$$

$$= \frac{6^2 + (-9)^2 + \cdots + (-5)^2}{8}$$

$$= \frac{36 + 81 + \cdots + 25}{8} = \frac{1022}{8} = 127.75$$

Since $\Sigma x^2 = 61^2 + 46^2 + \cdots + 50^2 = 28{,}247$, the calculating formula produces

$$s^2 = \frac{\Sigma x^2 - (\Sigma x)^2/n}{n - 1}$$

$$= \frac{28{,}247 - (495^2/9)}{8} = \frac{1022}{8} = 127.75$$

$$s = \sqrt{127.75} = 11.30$$

P3-8 (a) Since the mean is 26.00, values between 15.00 and 37.00 are within 11.00 of the mean. Since the standard deviation is $s = 6.19$, these values are within $k = 11.00/s = 11.00/6.19 = 1.777$ standard deviations of the mean. By Chebyshev's theorem, the proportion within this distance of the mean is at least $1 - 1/k^2 = 1 - 1/1.777^2 = 1 - 1/3.158 = 1 - 0.317 = 0.683$.

(b) Values that differ from the mean by *no more than* 15.00 are within $k = 15.00/6.19 = 2.423$ standard deviations of the mean. By Chebyshev's theorem, at least $100(1 - 1/k^2)\% = 100 \times [1 - 1/(2.423)^2]\% = 100 \times [1 - (1/5.871)]\% = 83.0\%$ of the values are no more than 15.00 from the mean; hence, at most 17.0% are more than 15.00 from the mean. As 17.0% of 150 is 25.5, we can say that at most 25.5 or, in whole numbers, at most 25 of the men have fat percentages which differ from the mean by more than 15.00.

P3-9 The coefficient of variation is $100(s/\bar{x})\% = 100 \times (9.6/32)\% = 30\%$; that is, the standard deviation is equal to 30% of the mean.

3-4 STATISTICS FOR FREQUENCY DISTRIBUTIONS

Mean

EXAMPLE 3-26   Suppose that we now wish to find the mean length of the logs in Example 2-11 of Section 2-2. We cannot find the total of the lengths because the values have been grouped into classes and we do not know the individual lengths. We can, however, find an approximate total. For the first class, 360 to 364, the class mark 362 can be used as a representative value. If we assume that the two logs in the class have a mean length equal to this value, their total length is approximately equal to $2 \times 362 = 724$. If we find the corresponding subtotals for all the classes and take their sum, we will have the approximate total for the complete set of data as

$$(2 \times 362) + (5 \times 367) + (9 \times 372) + \cdots + (3 \times 407) = 44{,}185$$

The number of lengths in the data is 115, the total of the class frequencies, and the approximate mean is therefore $44{,}185/115 = 384.2$. ▲

In general, for grouped data (data grouped in a frequency distribution), we let x_i and f_i represent the class mark and the frequency, respectively, for the ith class. The approximate total is then $\sum_{i=1}^{m} x_i f_i$ (where there are m classes) or simply $\sum xf$, and the number of values is $n = \sum_{i=1}^{m} f_i$ or simply $\sum f$.

The approximate mean for grouped data is

$$\bar{x} = \frac{\sum xf}{\sum f} = \frac{\sum xf}{n}$$

This mean is calculated in the same manner as a weighted mean. It is the weighted mean of the class marks using the class frequencies as the weights.

Median

EXAMPLE 3-27   Suppose that we also wish to find the median for the log data in Example 2-11. Again, we must use an approximation. We want a value that will divide the frequency into two equal parts. It is helpful to use a histogram in the development of the approximation. In the histogram in Figure 3-9, area represents frequency. We have a total frequency of 115 and therefore the area included in the histogram is 115 units. The median must be placed on the horizontal scale at the point that divides the area into two equal parts of $115/2$ or 57.5 units of area. (Note that this is $n/2$, not $n/2 + 1/2$ as used to find the position of the median in a discrete list of individual values.)

If we accumulate frequencies and hence units of area starting with the first class and adding on the frequencies corresponding to successive classes, then, after the first four classes, we have a total frequency (area) of 34; we have still not reached the midpoint. If we include the entire fifth class, we have a

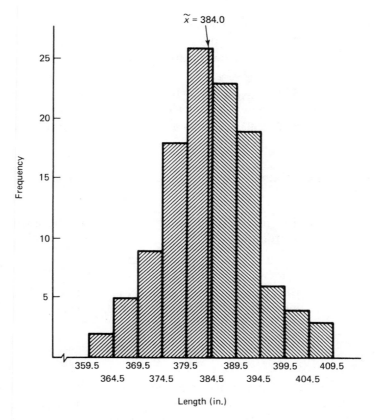

Figure 3-9 Median for log lengths.

total of 60; we have passed the midpoint. We must therefore include only a fraction of the fifth class.

After the fourth class, we have a total of 34 units of area and require 23.5 more to produce 57.5. The fifth class has a frequency of 26, that is, the area of this class is 26 units; thus, the fraction of the fifth class that we require is 23.5/26. To acquire an area of 23.5 units we must go this fraction of the way across the base of the rectangle for this class. The base of the rectangle is equal to the class width, 5; therefore, we must go a distance (23.5/26) × 5 across the fifth class from its starting point, which is the lower boundary, 379.5. This takes us to the point 379.5 + [(23.5/26) × 5] = 384.0. This point is the midpoint in terms of the area and since area represents frequency, we use this as the midpoint in the frequencies, that is, as the median. ▲

In general, for grouped data, the median is found from the following steps:

1. Determine from the frequency distribution which class includes the median.

2. Denote the lower boundary of this class as l.
3. Determine the area a that we require from this class as $n/2$ minus the frequency accumulated in the previous classes.
4. Denote the frequency of this class as f.
5. Denote the class width as w.
6. Calculate the approximate median of the data as $\tilde{x} = l + \left(\dfrac{a}{f} \times w \right)$.

EXAMPLE 3-28 For the median found in Example 3-27, $l = 379.5$, $a = 23.5$, $f = 26$, and $w = 5$. ▲

The median can also be found from the alternative expression

$$\tilde{x} = u - \left(\frac{a'}{f} \times w \right)$$

where u is the upper boundary of the class including the median and a' is the area required if we start accumulating with the last class.

Percentiles

The techniques described above can also be used to determine approximate percentiles, quartiles, and so on. For example, to find the ith percentile we accumulate frequencies to a point at which we will have accumulated an area in a histogram of $i \times (n/100)$. The point is determined as with the median.

EXAMPLE 3-29 Suppose that we wish to find the approximate 90th percentile for the bearing data in Example 2-18 of Section 2-2. Since there are 114 values, we must accumulate $(90/100) \times 114 = 102.6$. We require the 101 values from the first seven classes plus a further 1.6 from the eighth. Since the eighth class has a total frequency of 2, we must go a fraction 1.6/2 of the total distance from the lower boundary 1.3395 to the upper boundary 1.3495. This total distance is the class width 0.010, and we finally find the approximate 90th percentile as

$$1.3395 + [(1.6/2) \times 0.010] = 1.3475 \quad ▲$$

Modal Class

Since the actual mode can occur in a class with a relatively low frequency (all of the values in one class may be equal, for example), it is not useful to try to find an approximate mode. Instead, we introduce the concept of a *modal class*.

We refer to the class with the highest frequency as the *modal class*.

EXAMPLE 3-30 For the log data in Example 2-11 again, the modal class is 380 to 384, which has a frequency of 26. ▲

Standard Deviation

An approximate standard deviation can also be found by using approximate totals. With individual values, calculating expressions for the standard deviation were developed as

$$s = \sqrt{\frac{n\sum x^2 - (\sum x)^2}{n(n-1)}} \quad \text{and} \quad s = \sqrt{\frac{\sum x^2 - (\sum x)^2/n}{n-1}}$$

For grouped data, we replace $\sum x$ with the approximate total $\sum xf$ as used with the approximate mean. By similar arguments, the approximate sum of the squares for the entire set of data is $\sum x^2f$, or, more explicitly, $\sum_{i=1}^{m} x_i^2 f_i$. We still have $n = \sum f$.

A calculating expression for the approximate standard deviation for grouped data is

$$s = \sqrt{\frac{n\sum x^2f - (\sum xf)^2}{n(n-1)}} \quad \text{or} \quad s = \sqrt{\frac{\sum x^2f - (\sum xf)^2/n}{n-1}}$$

EXAMPLE 3-31 Again, for the log data in Example 2-11, we have $\sum xf = 44{,}185$ and $n = 115$ from Example 3-26, and we now calculate

$$\sum x^2f = \left[(362)^2 \times 2\right] + \left[(367)^2 \times 5\right] + \cdots + \left[(407)^2 \times 3\right]$$
$$= 16{,}986{,}655$$

and

$$s = \sqrt{\frac{115 \times 16{,}986{,}655 - (44{,}185)^2}{115 \times 114}} = 9.37$$

▲

PRACTICE PROBLEMS

P3-10 The following values represent a portion of the results of an investigation of various characteristics of low-income families:

Age of head (nearest birthday)	Number of families
15–24	6
25–34	17
35–44	21
45–54	16
55–64	14
65 or older	22

(a) Find the median age of the family head.

(b) Why can the mean age not be found?

P3-11 The following data represent the results of several readings on radioactive fallout content in whole milk. The radioactive content is the amount of strontium 90 measured in picocuries per liter (pCi/ℓ).

Strontium 90 content (pCi/ℓ)	Frequency
7.5–7.9	1
8.0–8.4	3
8.5–8.9	8
9.0–9.4	27
9.5–9.9	22
10.0–10.4	20
10.5–10.9	14
11.0–11.4	4
11.5–11.9	1

(a) Calculate the mean for these data.

(b) Calculate the variance and standard deviation for these data.

P3-12 The following data represent the weights in kilograms which produced breakage in 360 sample tests on a production of nylon fishing line.

Breaking point (kg)	Frequency	Breaking point (kg)	Frequency
Less than 4.00	3	4.50–4.59	67
4.00–4.09	7	4.60–4.69	60
4.10–4.19	18	4.70–4.79	41
4.20–4.29	27	4.80–4.89	24
4.30–4.39	40	4.90–4.99	15
4.40–4.49	53	5.00 or more	5

(a) Determine the first and third quartiles for these data.

(b) Determine the 10th and 90th percentiles for these data.

3-18 Suppose that a sample of 600 travel agencies produced the following distribution of receipts in dollars:

Total receipts (dollars)	Number of agencies
Under 10,000	101
10,000–24,999	142
25,000–49,999	165
50,000–99,999	114
100,000–249,999	53
250,000–499,999	15
500,000–999,999	7
1,000,000 or more	3

(a) Determine the median for this distribution.

(b) Determine the mean for those agencies with receipts of less than $1,000,000. (Round the class marks to whole dollars.)

(c) Determine the overall mean if the mean of those with receipts of $1,000,000 or more is $2,313,500.

3-19 Suppose that measurements on ice thickness over a given region produced the following distribution of values in centimeters.

Thickness (cm)	Frequency
0–4	3
5–9	8
10–14	18
15–19	16
20–24	26
25–29	20
30–34	7
35–39	2

(a) Determine the mean thickness.

(b) Calculate the standard deviation for this distribution.

3-20 Height measurements in centimeters for 200 plants a fixed length of time after germination produced the following distribution:

Height (cm)	Frequency
25–29	9
30–34	12
35–39	15
40–44	24
45–49	33
50–54	37
55–59	41
60–64	23
65–69	6

(a) Calculate the mean height.

(b) Calculate the first and third quartiles.

3-21 The following distribution represents the aggregate scores achieved by 300 subjects on a series of skill tests:

Score	Frequency	Score	Frequency
150–159	4	210–219	59
160–169	8	220–229	44
170–179	22	230–239	10
180–189	27	240–249	12
190–199	51	250–259	3
200–209	60		

For this distribution, calculate
(a) the mean
(b) the standard deviation
(c) the 10th and 90th percentiles

3-22 The following cumulative data represent the results of 1310 incomes determined in a sample survey:

Income	Cumulative Frequency
Less than $2000	210
Less than $4000	420
Less than $6000	670
Less than $8000	890
Less than $10,000	1090
Less than $15,000	1230
Less than $20,000	1270
Less than $50,000	1300
$50,000 or more	10

("Less than $2000" means up to and including $1999.99. Other values are analogous.)
(a) Find the median income.
(b) Rounding class marks to the nearest dollar, calculate the mean income for those with incomes less than $50,000.
(c) Assuming that those with incomes over $50,000 had a mean income of $150,000, find the overall mean.
(d) Interpreting "average" as the mean, and using in reverse the methods for obtaining the median, determine the approximate proportion of incomes that are below average.

SOLUTIONS TO PRACTICE PROBLEMS

P3-10 (a) The median must partition the total frequency (in the area concept) into two equal parts of $n/2 = 96/2 = 48$. Accumulating frequencies, we note that the median must be in the class 45 to 54. The accumulated frequency from previous classes is $6 + 17 + 21 = 44$ and the required additional frequency is $a = 48 - 44 = 4$. Since the class lower boundary, frequency, and width are, respectively, $l = 44.5$, $f = 16$, and $w = 10$, we find the median as $\tilde{x} = l + [(a/f) \times w] = 44.5 + [(4/16) \times 10] = 47$.
(b) Since the last class is an open class with no class mark or other indicated representative value, we cannot approximate the average age within that class. As a result, we cannot find the mean for the full set of data.

P3-11 (a) The class marks are 7.7, 8.2, 8.7,..., 11.7. With these class marks as x and the frequencies f as given, we find $\Sigma xf = (7.7 \times 1) + (8.2 \times 3) + \cdots + (11.7 \times 1) = 974.0$, $n = \Sigma f = 100$, and $\bar{x} = \Sigma xf/n = 974.0/100 = 9.74$.
(b) Since $\Sigma x^2 f = (7.7^2 \times 1) + (8.2^2 \times 3) + \cdots + (11.7^2 \times 1) = 9544.10$, we find the variance as $s^2 = [\Sigma x^2 f - (\Sigma xf)^2/n]/(n - 1) = (9544.10 - 974.0^2/100)/99 = 0.5792$ and the standard deviation as $s = \sqrt{0.5792} = 0.761$.

P3-12 (a) The first quartile must cut off the lower quarter of the total frequency and must (in the area concept) cut off the lower $n/4 = 360/4 = 90$. We accumulate 55 from the first four classes and require an additional $a = 90 - 55 = 35$ from the $f = 40$ in the fifth class 4.30 to 4.39 with lower boundary $l = 4.295$ and width $w = 0.10$. The first quartile is then $l + [(a/f) \times w] = 4.295 + [(35/40) \times 0.10] = 4.383$.

For the third quartile we must cut off the lower $(3 \times n)/4 = 270$. The third quartile is found to be in the class 4.60 to 4.69. As above, we determine $a = 270 - 215 = 55$, $l = 4.595$, $f = 60$, $w = 0.10$, and the third quartile is then $4.595 + [(55/60) \times 0.10] = 4.687$.

(b) Proceeding as in part (a), we determine that the 10th percentile which must cut off the lower $(10 \times 360)/100 = 36$ must be in the class 4.20 to 4.29 and is $4.195 + [(8/27) \times 0.10] = 4.225$.

Similarly, the 90th percentile which must cut off the upper 36 or lower 324 is in the class 4.80 to 4.89 and is $4.795 + [(8/24) \times 0.10] = 4.828$.

3-5 ADJUSTING OR CODING DATA — STANDARDIZED VALUES

On occasion, it may be necessary or convenient to adjust the data, perhaps changing inches to feet or converting crop yields to differences between the yields and a standard. We will consider the effect on summary statistics of two types of adjustments, relocating and rescaling.

Relocating Data

EXAMPLE 3-32 The following distribution of values represents the number of attempts that 60 subjects required to learn a specific task.

Number of attempts required	Number of subjects requiring this many attempts
6	2
7	3
8	7
9	12
10	15
11	5
12	8
13	4
14	4

Suppose that we wish to measure performance in relation to a previously established standard of 10 attempts. For example, 7 is 3 less than the standard and is converted to -3; 12 is 2 more than the standard and is converted to 2. We convert each value by subtracting 10, producing the distribution of adjusted data that follows:

Number of attempts minus 10	Number of subjects
− 4	2
− 3	3
− 2	7
− 1	12
0	15
1	5
2	8
3	4
4	4

Figure 3-10 illustrates histograms for the distribution for both the original and the adjusted data. The histograms are located at different positions on the horizontal scale. Subtracting 10 from each value has had the effect of moving the distribution a distance of 10 to the left. Adding 10 to each of the new values returns the distribution a distance of 10 to the right. Both such changes are changes of location.

▲

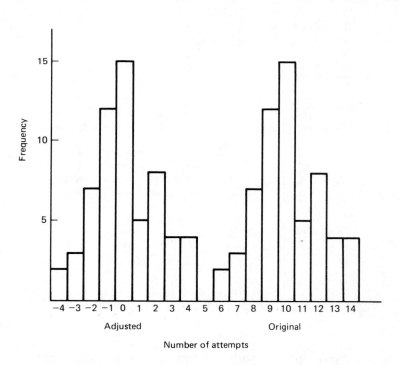

Figure 3-10 Histograms of numbers of attempts.

Adjusting the data by adding a constant value to, or subtracting a constant value from, each member of the data is referred to as *relocating* the data. Such a change is a *relocation*.

Since we have changed the location by subtracting 10, we expect measures of location to change accordingly.

In general, if we add (subtract) a constant value to (from) each member of a set of data, measures of location will be increased (decreased) by the addition (subtraction) of that same constant value.

EXAMPLE 3-33 The mean, median, and mode for the original data in Example 3-32 are 10.1, 10, and 10, respectively. For the adjusted data, they are 0.1, 0, and 0, respectively, as expected. ▲

Except for the fact that the histograms in Figure 3-10 are at different locations, they look exactly the same. The patterns of variation within the data have not changed, and we expect the measurement of variation to be unchanged.

In general, adding (subtracting) a constant value to (from) each member of a set of data does not affect measures of variation.

EXAMPLE 3-34 The range of the original data in Example 3-32 is $14 - 6 = 8$ and, for the new data, it is $4 - (-4) = 8$. For the original data, the standard deviation is

$$\sqrt{\frac{6356 - (606)^2/60}{59}} = 2.00$$

For the adjusted data, it is

$$\sqrt{\frac{236 - (6)^2/60}{59}} = 2.00$$

The measures of variation are unchanged, as expected. ▲

The adjustment resulted in a slightly easier calculation of the standard deviation. This result illustrates one of the most common reasons for adjusting data, to make calculations easier. By relocating data, we may work with easier numbers and still obtain the same values for measures of variation.

EXAMPLE 3-35 The following values provide a more extreme example. Consider the data 11,203, 11,215, 11,208, 11,217, 11,209, and 11,206. For these values $\Sigma x = 67{,}258$; $(\Sigma x)^2 = 4{,}523{,}638{,}564$; and $\Sigma x^2 =$

$125,507,209 + \cdots + 125,574,436 = 753,939,904$. The standard deviation is $s = \sqrt{(6 \times 753,939,904 - 4,523,638,564)/(6 \times 5)} = \sqrt{860/30} = 5.35$.

If we subtract 11,200 from each value, we obtain 3, 15, 8, 17, 9, and 6 with $\Sigma x = 58$; $(\Sigma x)^2 = 3364$; $\Sigma x^2 = 9 + 225 + \cdots + 36 = 704$; and $s = \sqrt{(6 \times 704 - 3364)/(6 \times 5)} = \sqrt{860/30} = 5.35$ as before.

If we subtract a more central value such as 11,210, we obtain -7, 5, -2, 7, -1, and -4 with $\Sigma x = -2$, $(\Sigma x)^2 = 4$, $\Sigma x^2 = 144$, and $s = \sqrt{(6 \times 144 - 4)/(6 \times 5)} = \sqrt{860/30} = 5.35$ again. ▲

Not only do such adjustments make work easier for hand calculations, but they may also make results more accurate, even when using a calculator or computer. Using fewer digits reduces the number of possible keypunching errors and smaller numbers may also eliminate rounding errors in a calculator or computer.

EXAMPLE 3-36 Rounding all figures in Example 3-35 to eight digits produces $s = 5.48$ instead of 5.35 with the original values. With the adjusted data, none of the figures exceeds eight digits; therefore, there would be no rounding error introduced. ▲

Rescaling data

EXAMPLE 3-37 The following data represent a set of distances measured in meters (m): 12,900, 13,100, 13,700, 12,500, 13,000, 12,800, 12,900, 13,600, 13,200, and 13,100. These values have a mean of 13,080 m. The range is 1200; the variance is 128,444; and the standard deviation is 358. Suppose that we wish to divide these distances by 1000 to convert them to kilometers (km): 12.9, 13.1, 13.7, 12.5, 13.0, 12.8, 12.9, 13.6, 13.2, and 13.1. The values in the data will change, but the physical distances that they measure will still be the same. We thus expect the average distance (mean) to still be 13,080 m, which will now be expressed as 13.08 km. ▲

> In general, if we divide (multiply) all of the values in a set of data by a constant value, all measures of location will also be divided (multiplied) by that same constant value.

EXAMPLE 3-38 Calculation of the mean of the adjusted distance data in Example 3-37 confirms the value 13.08. ▲

The difference between the shortest and longest distances is still 1200 m but is now expressed as 1.2 km. All other differences between distances must, similarly, be divided by 1000 when expressed in kilometers. and their squares must therefore be divided by the square of 1000. Accordingly, we might expect average distances to be divided by 1000 and average squared distances to be divided by the square of 1000.

If we divide (multiply) all of the values in a set of data by a constant value, all measures of variation such as the standard deviation will also be divided (multiplied) by this constant value and all squared measures of variation such as the variance will be divided (multiplied) by the square of this constant value.

EXAMPLE 3-39 Direct calculation of s^2, s, and the range for the new data in Example 3-37 produces $s^2 = 0.1284$ (which is $128{,}444/1000^2$), $s = 0.358$ (which is $358/1000$), and, as previously noted, the range is 1.2. ▲

In the particular adjustment for the distances, the division by 1000 has *changed the scale* of measurement from meters to kilometers.

Whenever we adjust data by multiplying or dividing by a constant, we refer to such an adjustment as *rescaling*.

In some cases we will require both a relocation and a rescaling to produce the adjusted data.

EXAMPLE 3-40 Consider 12 beakers in which crystals dissolving in liquid are generating heat. The temperatures in degrees Fahrenheit (°F) at a certain time after introducing the crystals are: 132.5, 134.6, 133.8, 130.7, 131.6, 132.4, 133.8, 132.9, 133.3, 133.4, 131.9, and 132.4. The mean temperature is 132.78°F and the standard deviation is 1.08. If we wish to convert the temperatures from Fahrenheit to Celsius, we must subtract 32 and then multiply by 5/9.

If we consider the adjustment in stages, the first stage is the relocation. From our general results for relocating data, if we subtract 32, the mean should become 100.78 (i.e., $132.78 - 32$) and the standard deviation should be unchanged at 1.08.

The second stage of the adjustment is a rescaling. If we adjust the interim data by multiplying by 5/9, the mean should become $100.78 \times (5/9) = 55.99$, and the standard deviation should become $1.08 \times (5/9) = 0.60$. The new values are 55.83, 57.00, 56.56, 54.83, 55.33, 55.78, 56.56, 56.06, 56.28, 56.33, 55.50, and 55.78, and the mean and the standard deviation are 55.99 and 0.60, as expected. In terms of the complete adjustment, the new statistics may be calculated as $(132.78 - 32.00) \times (5/9) = 55.99$ and $1.08 \times (5/9) = 0.60$ ▲

General relationships in mathematical notation are as follows: If we adjust data by the change $y = k(x - c)$, then $\bar{y} = k(\bar{x} - c)$ and $s_y = |k|s_x$; or if $y = (x + c)/k$, then $\bar{y} = (\bar{x} + c)/k$ and $s_y = s_x/|k|$. To reverse the adjustments we note that, for example, in the first case, $x = (y/k) + c$, leading to $\bar{x} = (\bar{y}/k) + c$ and $s_x = s_y/|k|$, and so on.

Coding a Frequency Distribution

As previously illustrated, data may be adjusted to simplify calculations. Such adjustments of data can also be applied to simplify calculations for data grouped in a frequency distribution.

Adjusting data in a frequency distribution is often referred to as *coding* data.

EXAMPLE 3-41 Suppose that we wish to calculate the mean and the standard deviation for the diameters in Example 2-18 of Section 2-2. To do so, we must first calculate $\sum xf = (3 \times 1.2745) + (17 \times 1.2845) + \cdots$ and $\sum x^2 = [3 \times (1.2745)^2] + [17 \times (1.2845)^2] + \cdots$. We can simplify the calculations by subtracting, from each class mark, a class mark such as 1.3045, which is fairly central in the bulk of the data, to produce new class marks: -0.0300, -0.0200, $-0.0100, \ldots, 0.0700, 0.0800$. These values can be adjusted further by dividing by the class width 0.010 to produce the coded values $-3, -2, \ldots,$ $7, 8$. In summary, we then have:

Class	Original class mark, x	Coded class mark, y	Class frequency, f
1	1.2745	-3	3
2	1.2845	-2	17
3	1.2945	-1	28
4	1.3045	0	29
5	1.3145	1	11
6	1.3245	2	6
7	1.3345	3	7
8	1.3445	4	2
9	1.3545	5	6
10	1.3645	6	3
11	1.3745	7	1
12	1.3845	8	1

From the coded values, $\sum yf = 44$, $\sum y^2 f = 624$, and $n = \sum f = 114$; thus, $\bar{y} = 44/114 = 0.386$ and $s_y = \sqrt{[(114 \times 624) - 44^2]/(114 \times 113)} = 2.318$. By the general rules for adjusted data, $\bar{y} = (\bar{x} - 1.3045)/0.010$ and $s_y = s_x/0.010$; thus, $\bar{x} = 0.010\bar{y} + 1.3045 = 1.3084$ and $s_x = 0.010 s_y = 0.0232$. From very simple calculations based on coded data, we have found the mean and standard deviation of the original bearing data to be 1.3084 cm and 0.0232 cm. ▲

Standardized Values

There is a special adjustment of data involving relocation and rescaling that converts the values in a set of data to standardized values. In general, with the adjustment $y = (x - c)/k$, if we choose $c = \bar{x}$ and $k = s_x$, we obtain $\bar{y} = 0$ and $s_y = 1$. Such an adjustment, subtracting the mean and dividing by the standard deviation, converts each value into the number of standard deviations that it is above or below the mean. A value that is converted to 1.8 is 1.8 standard deviations above the mean; a value that is 2.3 standard deviations below the mean is converted to -2.3; and so on. This adjustment puts all sets of data into a standard form.

> The converted values, $(x - \bar{x})/s$, are referred to as *standard units* or *standard scores*. Standard scores have a mean of 0 and a variance and standard deviation of 1.

By converting each value to its relative position in its own set of data (by converting to standard scores), we can make comparisons across sets of data. Individual values are compared in terms of their standard scores.

EXAMPLE 3-42 Two students in different classes with different instructors may be compared in terms of their standard scores. Suppose that one student obtains a mark of 78 in a class that has a mean of 60 and a standard deviation of 9, while a second student obtains a mark of 82 in a class with a mean of 63 and standard deviation of 10. The first mark 78 is converted to a standard score of $(78 - 60)/9 = 2$ (i.e., it is 2 standard deviations above the mean); whereas, the other mark, 82, is converted to $(82 - 63)/10 = 1.9$ (i.e., it is only 1.9 standard deviations above the mean). On the basis of standard scores, we may declare the first student to be more outstanding. ▲

There are other methods of comparison, but mathematical results are well developed for standard units and comparisons based on standard units are quite common in statistical inference.

PRACTICE PROBLEMS

P3-13 After analyzing the financial gains on a number of transactions conducted for a client, a broker has determined that the gains produced a mean of $1582 and a median of $1270 and were distributed with a standard deviation of $463. The client's net gain is subsequently determined by subtracting a broker's fee and other fixed costs which total $170 per transaction. What are the mean, median, and standard deviation for the net gains?

P3-14 An analysis of production in a meat packing plant has indicated that the dressed weights in pounds (lb) of hogs produced a mean of 170.8 and a variance of 10.24. For reporting purposes, values must be converted to kilograms (1 lb = 0.4536 kg). What are the mean and variance of the dressed weights in kilograms?

P3-15 In studies on the brain, comparisons between the left and right hemispheres are made in terms of size and function. In one study, the number of cells in each hemisphere was determined. The following values represent the number of cells in the left hemisphere from 12 subjects:

$$2350 \quad 2640 \quad 2370 \quad 2280 \quad 2690 \quad 2730$$
$$2220 \quad 2360 \quad 2290 \quad 2380 \quad 2540 \quad 2460$$

(a) Calculate the mean and standard deviation for these values.

(b) Subtract 2000 from each of the values and then divide the resulting values by 10. Calculate the mean and standard deviation for these final values. Adjust these statistics to produce the mean and standard deviation of the original values and compare these statistics to those obtained in part (a).

(c) Repeat part (b), subtracting 2400 instead of 2000.

P3-16 Code the class marks for the data in Practice Problem P3-11 to produce new values $-4, -3, -2, \ldots, 3, 4$. Calculate the mean, variance, and standard deviation for these coded values. Adjust these statistics to produce the corresponding values for the original data and compare these to the results obtained in Practice Problem P3-11.

P3-17 Piecework studies have shown that workers using equipment manufactured by company X can produce daily quantities with a mean of 105 and a standard deviation of 2.5. Workers using the more modern company Y designed equipment produce daily quantities with a mean of 110 and a standard deviation of 2. Jones produced 112 units on company X equipment and Smith produced 115 units on company Y equipment.

(a) Convert these two outputs to standard scores.

(b) Which worker's output was more outstanding?

3-23 If the median salary of an employee group is currently $18,500 and each member of the group is to receive an increment of $500 plus 3% of current salary, what will be the new median salary?

3-24 Samples of package weights have been considered to be sufficiently uniform if the standard deviation did not exceed 2.5 ounces (oz). If the package weights are to be expressed in metric units and the same limit on the standard deviation is to apply, what value in grams (g) must the standard deviation not exceed? ($1oz = 28.35$ g.)

3-25 Code the data in Exercise 3-20 to produce adjusted class marks of $-4, -3, -2, -1, 0, 1, 2, 3,$ and 4 and find the mean and standard deviation for the coded values. Apply the appropriate adjustment to these two values to produce the mean and standard deviation of the original heights.

3-26 (a) Calculate the standard deviation of the following set of data using one of the calculating formulas. In the squares, sums, and so on, round any values to eight digits.

21.39825	21.40016
21.40168	21.40124
21.40231	21.40083
21.39978	21.39912
21.39893	21.40171

(b) Subtract 21.40000 from each value, then multiply by 100,000. Find the standard deviation of these values by the same method as in part (a) and then convert this to the standard deviation for the original values.

3-27 The following distribution represents the weights in grams of a compound precipitated from a fixed quantity of a liquid in which the compound was dissolved:

Dissolved weight (g)	Frequency
1.70–1.74	2
1.75–1.79	3
1.80–1.84	8
1.85–1.89	10
1.90–1.94	14
1.95–1.99	17
2.00–2.04	15
2.05–2.09	11
2.10–2.14	7
2.15–2.19	2
2.20–2.24	1

(a) Code the data and find the mean and standard deviation for the coded values.

(b) Adjust the coded mean and standard deviation to produce the mean and standard deviation for the original weights.

(c) Determine the coefficient of variation for the original weights.

3-28 The following data represent the marks obtained by students in a mathematics class and by students in a geography class:

Mathematics		Geography	
Student	Mark	Student	Mark
Allen	84	Allen	85
Anderson	57	Boynton	58
Bacon	69	Collison	74
Collison	64	Davis	70
Dionne	78	Delucci	70
Duncan	67	Duncan	75
Flagler	69	Edwards	71
Gibson	68	Farrow	82
Greaves	63	Foster	68
Hagen	76	Gibson	69
Harper	70	Hughes	79
Hughes	83	Hyne	82
Hucolic	58	Johnson	76
Jardine	60	Kelly	62
Jones	64	Lee	88
Lee	65	Mason	66
Mason	60	McCall	70
McGrath	64	Meyer	68
Meisner	71	North	65
Morgan	63	Oswald	64
Nichols	63	Parker	58

(*cont.*)

Mathematics		Geography	
Student	Mark	Student	Mark
North	71	Reid	63
Oullet	52	Robinson	54
Plant	61	Scott	77
Robinson	53	Smith	55
Smith	66	Taylor	63
Spencer	73	Thompson	62
Taylor	76	Underwood	86
Vanderburg	68		
Williams	74		

(a) Find the mean and standard deviation for each class.

(b) Convert the marks in mathematics and in geography to standard scores for students who are in both classes.

(c) In which class is Allen's mark relatively higher? In which class is Robinson's mark relatively lower? In which class is Gibson's mark relatively higher?

(d) Which mark is relatively most outstanding, Allen's mark in mathematics or Lee's mark in geography?

3-29 If n values x_i are changed to n values y_i by the change $y_i = x_i + c$ for some constant value c, show algebraically that $\bar{y} = \bar{x} + c$ and $s_y = s_x$.

3-30 If n values x_i are changed to n values y_i by the change $y = kx$ for some constant value k, show algebraically that $\bar{y} = k\bar{x}$ and that $s_y^2 = k^2 s_x^2$.

SOLUTIONS TO PRACTICE PROBLEMS

P3-13 Subtracting $170 is a relocation affecting only the mean and median but not the standard deviation. $170 is subtracted from the original mean and median to produce new values $1412 and $1100, respectively. The standard deviation is still $463.

P3-14 The change to kilograms is a rescaling achieved by multiplying by 0.4536. The new mean is the old mean multiplied by 0.4536 or 77.47 kg, and the new variance is the old variance multiplied by 0.4536^2 or 2.107.

P3-15 (a) The data produce $n = 12$, $\Sigma x = 29{,}310$, and $\Sigma x^2 = 71{,}906{,}100$. The mean and standard deviation are $\bar{x} = 29{,}310/12 = 2442.5$ and $s = \sqrt{(71{,}906{,}100 - 29{,}310^2/12)/11} = 169.605$.

(b) The adjustment $y = (x - 2000)/10$ produces new values 35, 64, 37, 28, 69, 73, 22, 36, 29, 38, 54, and 46 such that $\Sigma y = 531$, $\Sigma y^2 = 26{,}661$, $\bar{y} = 531/12 = 44.25$, and $s_y = \sqrt{(26{,}661 - 531^2/12)/11} = 16.9605$. Adjusting these statistics to original values produces $\bar{x} = 10\bar{y} + 2000 = 2442.5$ and $s_x = 10s_y = 169.605$, as before.

(c) The adjustment $y = (s - 2400)/10$ produces $-5, 24, -3, -12, 29, 33, -18,$ $-4, -11, -2, 14,$ and 6 with $\Sigma y = 51$ and $\Sigma y^2 = 3381$. We then determine $\bar{y} = 4.25$ and $s_y = 16.9605$. The original mean and standard deviation are $\bar{x} = 10\bar{y} + 2400 = 2442.5$ and $s_x = 10s_y = 169.605$, as before.

P3-16 Subtracting the central class mark 9.7 from the original class marks and then dividing by the class width 0.5 produces new class marks $y = (x - 9.7)/0.5$, which are $-4, -3, -2, -1, 0, 1, 2, 3,$ and 4, as required. We then find $\Sigma yf = 8$ and $\Sigma y^2 f = 230$. Since $n = \Sigma f = 100$ as before, we find that $\bar{y} = 8/100 = 0.08$ and $s = \sqrt{(230 - 8^2/100)/99} = \sqrt{2.3168} = 1.522$. The corresponding statistics for the original values are $\bar{x} = 0.5\bar{y} + 9.7 = 9.74$ and $s_x = 0.5s_y = 0.761$, as before.

P3-17 (a) Jones's output becomes a standard score of $(112 - 105)/2.5 = 2.8$; Smith's becomes $(115 - 110)/2 = 2.5$.

(b) Jone's output was more outstanding because it produced the larger standard score.

3-6 POPULATIONS AND SAMPLES

There is one further aspect of describing data to consider—something that is quite important for inference. We differentiate between two types of sets of data.

Populations and Samples

If the set of data that we are considering includes all the values that are of interest to us in a particular study, we refer to the set of data as a *population*.

If the set of data is only part of the entire collection of values of interest to us, we refer to the set of data as a *sample*.

Depending on our interests, any particular set of data can be either a sample or a population.

EXAMPLE 3-43 If the 114 bearings referred to in Example 2-18 of Section 2-2 are all of the bearings in a given batch and are the only ones of current interest, the set of diameters is a population. If, however, we are interested in a batch of 2000 bearings and these 114 are taken from that batch, the set of diameters is a sample. ▲

In later chapters on inference, we describe the characteristics of samples and, from them, estimate the values of the characteristics of populations. Although it is appropriate to refer to characteristics of any set of data (i.e., sample or population) as statistics, we will use a different term when referring to population characteristics.

Statistics and Parameters

We define a *statistic* as any numerical measure or characteristic of a *sample*.

We define a *parameter* as any numerical measure or characteristic of a *population*.

All of the notation used previously applies to samples. We introduce new notation to distinguish population parameters from sample statistics. Corresponding sample and population values are denoted as follows:

Characteristic	Sample notation	Population notation
Number of members	n	N
Mean	$\bar{x}$	μ (the Greek lowercase letter mu)
Median	$\tilde{x}$	$\tilde{\mu}$
Standard deviation	s	σ (the Greek lowercase letter sigma)
Variance	s^2	σ^2

The median is still determined as the middle value in the full set of data. The mean, variance, and standard deviation are defined and calculated as follows:

$$\mu = \frac{\Sigma x}{N}$$

$$\sigma^2 = \frac{\Sigma (x - \mu)^2}{N}$$

$$= \frac{\Sigma x^2}{N} - \left(\frac{\Sigma x}{N}\right)^2$$

$$= \frac{\Sigma x^2}{N} - \mu^2$$

$$\sigma = \sqrt{\frac{\Sigma (x - \mu)^2}{N}} = \sqrt{\frac{\Sigma x^2}{N} - \mu^2}$$

The use of N as the divisor for σ^2 differs from the use of $n - 1$ for s^2. Division by N makes the population variance consistent with the variance of random variables considered in Chapter 5.

Some statisticians prefer a population variance based on dividing by $N - 1$. This calculation produces what is often called a finite population variance.

The alternative *finite population standard deviation* is denoted as S and the variance as S^2, where

$$S = \sqrt{\frac{\Sigma (x - \mu)^2}{N - 1}} = \sqrt{\frac{\Sigma x^2 - (\Sigma x)^2 / N}{N - 1}}$$

Other aspects of data will have appropriate expressions for populations.

The coefficient of variation for a population is $100(\sigma/\mu)\%$

Standard scores for a population are $(x - \mu)/\sigma$. As with samples, populations that have been standardized have mean 0 and standard deviation 1.

PRACTICE PROBLEMS

P3-18 The records at a nursing station include the blood pressures at time of admission of all of the patients currently in the wards supervised from this station.
(a) If a nursing supervisor is interested only in the blood pressures of patients currently in the wards and consults the records at the nursing station, do the data form a population or a sample?
(b) If the nursing supervisor has available only the records of current patients but is interested in the blood pressures of all the patients who have been in the wards in the previous year, do the data available form a population or a sample?
(c) What symbols should be used to denote the mean and standard deviation for the available data in each of parts (a) and (b)?

P3-19 Considering the data in Practice Problem P3-15 to be a small population, determine the mean and standard deviation. Convert the value 2220 to a standard score.

EXERCISES

3-31 Considering the marks for the classes in Exercise 3-28 to be populations, indicate, with appropriate notation, the population sizes, means, and variances.

3-32 A collector has 10 paintings whose values in thousands of dollars are 5, 10, 7, 15, 30, 9, 12, 35, 25, and 22. Considering this collection as a population, determine the mean and standard deviation and convert each value to a standard score.

3-33 Verify that the standard scores in Exercise 3-32 do produce a mean and standard deviation of 0 and 1 when considered to be a population.

3-34 Prove algebraically that if a population of values x is changed to a population of standard scores by the change $z = (x - \mu)/\sigma$, the population of z's will have mean 0 and standard deviation 1.

3-35 Prove algebraically that

$$\frac{\Sigma(x - \mu)^2}{N} = \frac{\Sigma x^2}{N} - \mu^2$$

P3-18 (a) Since the data available represent all the data of interest, they form a population.

(b) Since the available data represent only part of the data of interest, they represent a sample.

(c) In part (a) the mean and standard deviation are denoted as μ and σ, but in part (b) they are denoted as $\bar{x}$ and s.

P3-19 Considering the data in Practice Problem P3-15 to be a population, we have $N = 12$, $\Sigma x = 29,310$, and $\Sigma x^2 = 71,906,100$. The mean and standard deviation are $\mu = \Sigma x/N = 29,310/12 = 2442.5$ and $\sigma = \sqrt{\Sigma x^2/N - \mu^2}$ $= \sqrt{71,906,100/12 - 2442.5^2} = 162.385$. The value 2220 is converted to $(2220 - 2442.5)/162.385 = -1.37$. (As in Practice Problem P3-15, the results could have been obtained from simpler calculations by adjusting the data.)

3-7 CHAPTER REVIEW

3-36 The difference between the highest and lowest values in a sample is called the _____.

3-37 Using summation notation, the quantity $(x_1 + x_2 + x_3 + x_4 + x_5 + x_6)$ may be written as _____.

3-38 By our convention, the symbol for the variance of a population is _____.

3-39 If a set of data has mean 27 and each value in the set of data is increased by 5, the new mean will be _____.

3-40 A parameter is
(a) a description of a sample
(b) a quantity that varies from sample to sample
(c) a characteristic of a population
(d) any of these

3-41 Of the three basic measures of location, the one most affected by an extreme value is the
(a) mean
(b) median
(c) mode

3-42 To find the median from grouped data using the area of a histogram, we consider the beginnings and endings of the classes to be plotted at _____.

3-43 The sum of the deviations of all the values from the mean is always equal to _____.

3-44 To reduce the variance to half its original value, all of the numbers in a set of data must be divided by _____.

3-45 A numerical description or characteristic of a sample is called a _____.

3-46 For the rainfall data in Exercise 2-29, find the mean and median rainfall. Also find the variance and the range of the annual rainfall values.

3-47 The following data represent a sample of "rating" scores assigned to a new product by a number of samplers.

(a) Calculate three average scores.
(b) Is one of the averages more (less) appropriate than the others? If so, why?
(c) Calculate the standard deviation.

6	10	10	7	6	10	6	7	9	9
3	5	10	6	8	7	5	7	6	9
6	3	4	8	7	8	8	8	6	5
9	10	5	9	6	6	7	7	4	7
4	10	6	7	6	6	4	9	6	9
9	3	7	4	10	8	5	6	7	8
10	8	5	8	10	5	8	6	7	4
8	9	8	10	8	6	9	7	6	6

3-48 Suppose that, in a performance study conducted over a given period of time, a number of machines of each of six different models of machinery were used and the numbers of malfunctions noted. For the first model, eight machines produced a mean of 1.25 malfunctions per machine. For the second model, 16 machines produced a mean of 0.75 malfunction per machine. For the third, fourth, fifth, and sixth models, 14, 10, 10, and 12 machines, respectively, produced means of 0.5, 6.3, 0.1, and 3.25 malfunctions per machine. Considering all of the models, what was the mean number of malfunctions per machine?

3-49 The following data represent a sample of dosages in grains (gr) of a drug required to produce a given reaction.

Dose (gr)	Frequency
7.0–7.9	5
8.0–8.9	8
9.0–9.9	9
10.0–10.9	21
11.0–11.9	33
12.0–12.9	17
13.0–13.9	9
14.0–14.9	5
15.0–15.9	1

(a) Apply an appropriate coding to the data and find the mean and standard deviation.
(b) Adjust the results of part (a) to produce the mean and standard deviation for the original values and hence find the coefficient of variation.
(c) Determine the median dosage.
(d) Calculate the 10th and 90th percentiles.
(e) Convert dosages of 8.5 and 13.4 to standard scores.

3-50 If a member of a population has a value that becomes 2.3 when converted to a standard score, what does Chebyshev's theorem say about the proportion of the population that is closer than this value to the mean?

3-51 The following distribution represents a sample of salaries in a given job classification:

Salary (thousands of dollars)	Frequency	Salary (thousands of dollars)	Frequency
10.5–10.9	23	13.5–13.9	127
11.0–11.4	37	14.0–14.4	36
11.5–11.9	14	14.5–14.9	12
12.0–12.4	42	15.0–15.4	19
12.5–12.9	59	15.5–15.9	6
13.0–13.4	80		

(a) Adjust the data in the distribution and find the mean and standard deviation of the adjusted data; hence, find the mean and standard deviation of the original data.

(b) Using Chebyshev's theorem, find at most what proportion of values in the distribution should be more than $1500 from the mean.

(c) Determine the first and third quartiles.

3-52 An agency providing cultural grants produced the following annual summary statistics from data on a sample of grantees.

Type of organization	Number of organizations	Mean number of performances	Mean attendance per performance
Live theatre	25	320	362
Music	10	148	985
Dance	5	92	1120
Opera	2	36	1030

(a) Calculate the overall mean number of performances per organization.

(b) Calculate the overall mean attendance per performance.

3-8 GENERAL REVIEW

3-53 If a population is negatively skewed, the mode is
(a) less than the mean
(b) equal to the mean
(c) greater than the mean
(d) impossible to classify relative to the mean

3-54 Class marks are the reference points for
(a) an ogive
(b) a histogram
(c) a frequency polygon
(d) a pie chart

3-55 Because the variance is formed from a sum of squares, it must always be _____ .

3-56 Write the following instructions using mathematical summation notation: "Subtract the sample mean from each member of the sample, then square all of these differences and take a total of these squares." _____.

3-57 Using a frequency distribution is a means of
(a) making data more accurate
(b) summarizing data
(c) distorting data

3-58 The mode is the highest value in a set of data.
(a) True
(b) False

3-59 The heights in frequency polygons are plotted above the _____.

3-60 For a set of data with highest value 53.8 and lowest value 22.7, a variance of 60.4 would appear to be
(a) too low
(b) reasonable
(c) too high

3-61 If all of the values in a set of data are doubled, the coefficient of variation will be
(a) doubled
(b) unchanged
(c) halved

3-62 The classes 0 to 5, 5 to 10, 10 to 15, 15 to 20, 20 to 25, 25 to 30, and 30 to 35 are not suitable for a frequency distribution because _____.

3-63 The range of a sample is
(a) a statistic
(b) a parameter
(c) a measure of location
(d) possibly negative

3-64 Consider a competition in which the scores of one large group of competitors have a mean of 134 and a standard deviation of 32, and those of a second independent group have a mean of 120 and a standard deviation of 28. Suppose that one of the competitors in the first group has a score of 182 and one in the second group has a score of 176. Which is the more outstanding score? Why?

3-65 In the following table, referral sources are given by 196 clients for referral to a hospital psychological service. Draw a bar graph for these data. Is it possible to draw a cumulative distribution polygon for these data? Is it of any value to do so? What does this indicate about cumulative distributions?

Source of referral	Frequency
Psychiatrist	141
Family doctor	27
Public service agency	15
Industrial or school counselor	6
Clergy	5
Self	2

3-66 Suppose that an investor has a portfolio of stocks of a number of alcoholic beverage companies and that the stocks changed in value over a particular time period. For each company the initial stock value, the final value, and the number of shares held by the investor are as listed.

Company	Initial value	Final value	Number of shares
Distiller I	$31.40	$28.85	200
Distiller II	32.25	29.75	200
Brewery A	19.25	19.00	100
Brewery B	2.50	2.75	400
Vineyard X	10.25	9.00	300
Vineyard Y	9.75	8.75	300

Determine the mean change per share.

3-67 Suppose that the following is a sample of fuse lives for a particular type of fuse:

42	78	78	69	68	84	45	62	60
59	81	87	24	78	115	122	84	90
77	72	73	95	78	36	42	80	65
61	75	57	51	54	42	30	109	74
72	51	81	75	112	60	94	39	29

(a) Find the mean and median fuse lives.
(b) Find the standard deviation and the range of these values.
(c) Form a frequency distribution with classes such as 20 to 29.
(d) For this distribution give the class marks, class limits, and class boundaries.
(e) Form a cumulative "less than" distribution from this frequency distribution.
(f) Draw a "less than" ogive for this distribution.

3-68 A study on the distance traveled to work by 150 employees produced the following data.

Distance traveled (miles)	Number of employees
Less than 1	11
Less than 2	21
Less than 3	32
Less than 4	43
Less than 6	69
Less than 8	98
Less than 10	110
Less than 15	133
Less than 20	137
Less than 50	146
50 or more	4

(Distances are determined to the nearest tenth of a mile and less than 1 means up to and including 0.9; other values are analogous.)

(a) Give this information in the form of a frequency distribution.

(b) Indicate two difficulties with this form of frequency distribution.

(c) Give class limits, class marks, class boundaries, and class widths for each class, where possible.

(d) Find the approximate (grouped) median.

(e) Determine the 10th and 90th percentiles.

(f) What difficulty prevents the calculation of a mean and standard deviation?

(g) Find the approximate mean distance of those traveling less than 50 miles.

(h) Assuming that the mean distance of those traveling more than 50 miles is 70 miles, find the approximate mean distance for the full sample.

(i) Interpreting "average" as mean, find the approximate proportion of distances that exceed the average.

3-69 The following data represent the percentages of total 1973 milk production for North, Central, and South American nations except for Bolivia, Nicaragua, and Caribbean Islands other than Cuba.

Nation	Percent of total production
United States	61.1
Brazil	9.6
Canada	8.8
Argentina	6.9
Mexico	4.2
Colombia	2.7
Others	6.7

(a) Draw a pie chart for these data.

(b) Draw a suitable pictograph for these data.

INTRODUCTION TO PROBABILITY AND SAMPLING DISTRIBUTIONS

4-1 INTRODUCTION

To make inferences of a general nature about a population based on the characteristics of a particular sample, it is necessary to know how well or how poorly the sample might represent the population. For example, a sample mean may be greater or smaller than the population mean; it may be almost equal to, or quite different from, the population mean. Samples are subject to chance variation and we cannot predict exactly how they will turn out. We can, however, determine what is "likely" to happen or what is "unlikely" to happen.

We use probability to measure how likely or unlikely it is that certain sample values or sample statistics will occur. The specification of all possible values for a particular statistic, together with the corresponding probabilities or relative frequencies of occurrence, produces a sampling distribution for that statistic.

In this chapter the concepts of probability and sampling distributions are introduced. More detailed discussion of probability distributions and sampling distributions is deferred until Chapter 5.

4-2 ELEMENTARY PROBABILITY

Probability is a measure of uncertainty and represents an attempt to formalize such statements as "The odds are that it will rain today"; "In all probability, this artifact belongs to the second culture"; "There is no chance of a victory." Instead of using such imprecise statements, it is preferable to have a system that assesses the uncertainty involved in various statements in terms of a precise numerical scale ranging from "impossible" to "certain."

Probability

The scale that is used ranges from 0 to 1, with impossible assessed as 0 and certain as 1. All other degrees of uncertainty are assessed as numbers between 0 and 1. These numerical values are probabilities.

In a general sense, probabilities are interpreted as in the following statements: An event with probability close to 0 is very unlikely to occur; an event with probability close to 1 is almost certain to occur; an event with probability 0.5 is just as likely to occur as it is not to occur.

In a more precise numerical sense, the easiest type of interpretation of probability is based on the "physical opportunity" of the event in question relative to the opportunities of all possible events.

EXAMPLE 4-1   If we roll a balanced die, it may show any face from 1 to 6. Because the die is balanced, each face has the same opportunity of showing and, hence, each face has the same probability. The total probability is the probability of certainty (i.e., 1). Each of the six faces has an equal share of this total; thus, each has probability 1/6.   ▲

EXAMPLE 4-2   If a bag contains 100 beads that are identical except for color and, after shaking the bag, we draw out one bead blindly, each bead has equal opportunity and, hence, equal probability, 1/100. If 12 of the beads are red, the event "a red bead is drawn" has the combined probability of 12 of the 100 beads and, hence, has probability 12/100.   ▲

As with statistics, there are special terms and mathematical notation used in the study of probability.

> Any operation subject to chance, such as rolling a die or drawing a bead or taking a sample, is called a *trial*.
> Each of the most elementary possible results of the trial is an *outcome*.
> Any designated outcome or combination of outcomes is an *event* and events are denoted by capital letters.
> The probability of an event E is denoted as $P(E)$.

EXAMPLE 4-3   With the die, there are six possible outcomes corresponding to the six faces of the die, and the statement "a 2 shows on the die" represents both an outcome and an event. It may be denoted as T. Since T is obtained from one of six equally likely outcomes, $P(T) = 1/6$.   ▲

EXAMPLE 4-4   With the beads, there are 100 possible outcomes corresponding to the 100 beads and the statement "a red bead is drawn" represents an
event that includes 12 outcomes. This event might be denoted as R. Note that,
to obtain R, we require only 1 of the 12 outcomes that satisfy the condition of
the event and $P(R) = 12/100$.   ▲

General Rules

In the sense of the foregoing examples, the probability of an event is the ratio of the
number of possibilities that produce that event to the total number of possibilities,
assuming that all of the possibilities are equally likely.

In mathematical notation, we can express the definition of probability as a
ratio as follows:

> If a trial has a total of $a + b$ possible outcomes, all of
> which are equally likely, such that a of the outcomes
> produce the event E, the probability for the event E is
> defined to be

$$P(E) = \frac{a}{a + b}$$

EXAMPLE 4-5   Consider the ordinary deck of 52 playing cards consisting of
four sets or "suits," designated "spades," "diamonds," "hearts," and "clubs."
Each suit consists of 13 cards labeled 2, 3, 4, 5, 6, 7, 8, 9, 10, J (for jack), Q (for
queen), K (for king), and A (for ace). The deck is shuffled so that the cards are
in random order and then a player makes a random selection of one card. If
the card is labeled with a number less than 5, the player wins a bet. The
player's choice of a card may be considered as a trial with $a + b = 52$ possible
outcomes. The event E, "the card is labeled with a number less than 5," is
produced by any one of $a = 12$ outcomes. (There are three numbers less than 5
in each of the four suits.) The probability that the player wins the bet is then
the probability of the event E or

$$P(E) = \frac{a}{a + b} = \frac{12}{52} = \frac{3}{13}$$ ▲

A probability defined as above is, as required, a number between 0 and 1
inclusive. If $a = 0$, E is impossible and $P(E) = 0$; if $b = 0$, E is certain and
$P(E) = 1$. In mathematical notation, we have

$$0 \leqslant P(E) \leqslant 1$$

The sign " $\leqslant$ " is a variation of " $=$ " and indicates that the value on the left of the
sign is less than or equal to the value on the right. The rule expressed in this

mathematical notation makes two statements: that P(E) is greater than or equal to 0 and that P(E) is less than or equal to 1.

> An event E with $a = 0$ is a *null event* and is designated $\varnothing$. The null event $\varnothing$ represents an impossibility; thus, P($\varnothing$) = 0.

EXAMPLE 4-6 Obtaining a card labeled "23" from an ordinary deck as defined in Example 4-5 is impossible and is thus a null event with probability 0. ▲

> The collection of all possible outcomes of a trial is called the *sample space*. An event E with $b = 0$ represents the entire sample space and is designated S. This event S is certain to occur (if we perform the trial, we must obtain some outcome); thus,

$$P(S) = 1$$

EXAMPLE 4-7 The numbers 1 through 6, inclusive, form the complete sample space for the roll of a die and the probability of obtaining a number between 1 and 6, inclusive, has probability 1. ▲

> Each event E has a complementary event E' or "not E" which consists of all the outcomes not included in E. With E defined as in the general trial above, there are b outcomes of a total of $a + b$ that produce E', so that

$$P(E') = \frac{b}{a + b}$$

Combining the probabilities for E and E' produces the expected result, P(E) + P(E') = [$a/(a + b)$] + [$b/(a + b)$] = $(a + b)/(a + b) = 1$. Subtracting P(E') from both sides of the equation P(E) + P(E') = 1 produces the general result

$$P(E') = 1 - P(E)$$

EXAMPLE 4-8 In Example 4-2 we have $a = 12$ red beads and $b = 88$ beads of other colors, for a total of $a + b = 100$ beads. The general rule produces P(R) = $a/(a + b) = 12/100$, as before. We may find the probability of a nonred directly as P(R') = $88/100$ or as $1 - $ P(R) = $1 - 12/100 = 88/100$. ▲

Although we may not know whether the bead in Example 4-8 will definitely be red or not (it is neither certain nor impossible), we may numerically assess the degree of certainty or uncertainty that is attached to the drawing of a red bead.

Chance

The method of defining probabilities described above provides a precise meaning for the numerical value of a probability. If an event has probability 3/4 or 0.75, then 75% of the possible outcomes produce the event and only 25% produce the complementary event.

The use of a percentage is quite common in statements of probability and the term *chance* is often used when probabilities are expressed in percentage terms.

EXAMPLE 4-9 In Example 4-2 we say that the *probability* of our drawing a red bead is 12/100 or 0.12, but we say that there is a 12% *chance* of our drawing a red bead. ▲

We can obtain an intuitive general feeling for probabilities by considering the chance concept. Most people can interpret a 30% chance, a 95% chance, and so on. A 0% chance, corresponding to a probability of 0, is easily interpreted as the lowest possible value—an event cannot have a −15% chance of occurring. Similarly, a certain event with probability 1 has a 100% chance of occurring—an event cannot have a 175% chance of occurring.

Odds

Besides probability or chance, uncertainty can be expressed in another manner, through the use of odds.

EXAMPLE 4-10 Consider drawing a card at random from a well-shuffled ordinary deck of 52 playing cards as previously defined. Suppose that we are interested in the event "a diamond is drawn." The probability of this event is $13/52 = 1/4$, as 13 of the 52 cards are in the suit "diamonds." Since there are three times as many nondiamonds as there are diamonds, there are three outcomes against the event for each outcome in favor. This relationship is expressed as *odds* such that the *odds against* drawing a diamond are 3 to 1, or the *odds in favor* of drawing a diamond are 1 to 3. ▲

In a general trial with $a + b$ equally likely outcomes such that a produce the event E and b produce the event E', the *odds in favor* of E are a to b and the *odds against* E are b to a.

Odds are usually simplified to the smallest possible whole numbers.

EXAMPLE 4-11 The odds against the drawing of a diamond in Example 4-10 are expressed as 3 to 1, not 39 to 13. ▲

Given odds, we can use previous considerations to obtain probabilities.

EXAMPLE 4-12   If the odds are 3 to 2 against an event, there are three outcomes against the event for every two in favor; thus, of five outcomes, two produce the event, giving the probability as 2/5.   ▲

> In general, if the odds are c to d against E (or d to c in favor of E), $P(E) = d/(c + d)$.

We may also obtain odds from the probabilities for an event and its complement.

> In general, the odds *in favor* of E may be found by arithmetically simplifying the expression "$P(E)$ to $P(E')$." The odds *against* E may be found by simplifying $P(E')$ to $P(E)$.

EXAMPLE 4-13   For a balanced die, the odds in favor of the die showing a 2 are 1/6 to 5/6 or, after simplifying, 1 to 5.   ▲

Interpreting Probability

The interpretation of probability based on the physical structure of the trial provides a precise meaning for probability, chance, or odds. There are, however, many situations in which this interpretation is not applicable.

EXAMPLE 4-14   A gardener transplants a small tree and is interested in the probability that the tree will survive the transplanting. In the gardener's view, this is not a tree selected at random from a number of trees, some of which are destined to survive and some of which are not. The physical interpretation of probability does not apply. Instead, the gardener considers a statistical record of previous similar transplants and notes that 87% resulted in survival. Accordingly, this tree is assumed to have an 87% chance of survival; that is, the probability of survival is 0.87.   ▲

> A probability determined on the basis of a statistical record may be referred to as a *statistical probability*. Probabilities of this type follow the same rules as physical probabilities, with the record of previous occurrences being analogous to the list of possible outcomes.

In some cases there is neither a statistical record nor a physical opportunity model and subjective considerations are used instead to produce *subjective probabilities*.

EXAMPLE 4-15   An entrepreneur wishes to assess the probability of success of a new product before placing it on the market. No similar product has been sold before so that there is no statistical record of previous introductions. Neither of the previous (objective) methods of obtaining a probability can be applied. In this case, the entrepreneur might use a subjective probability based on personal opinion. If, for example, the entrepreneur has four times as much confidence in success as failure, the subjective odds in favor of success are 4 to 1 and the subjective probability of success is found by converting the odds to the probability 4/5.　　　　　　　　　　　　　　　　　　　　　　　　　　　▲

To this point, probability has been concerned only with uncertainty about future events. We may also measure the uncertainty of something that has already occurred.

EXAMPLE 4-16　Suppose that a card has been drawn at random from a well shuffled deck of cards and placed face down on the table and that we are interested in the probability that the card is a diamond. Since the card has already been selected, its suit is no longer subject to chance as in examples of the previous type. As the card has already been drawn, it is already a diamond or it is not. In such circumstances, we consider the probability that was appropriate before the drawing, namely 1/4, and we use this value again as the measure of uncertainty.　　　　　　　　　　　　　　　　　　　　　　▲

In this manner we obtain probabilities as measures of uncertainty for events that have already taken place but are still unknown.

For different circumstances we determine and interpret probabilities in different ways to help overcome uncertainty. It is important, however, to note that no matter how probabilities are obtained, they only measure, and do not eliminate, uncertainty.

EXAMPLE 4-17　We know that if a card is drawn at random from a well-shuffled ordinary deck of playing cards, the probability that it will be the ace of spades is very small—it is only 1/52. Nevertheless, we do not know with certainty that the card will not be the ace of spades. We may, however, be considerably more confident that it will not be the ace of spades than that it will.　　　　　　　　　　　　　　　　　　　　　　　　　　　　　▲

There is still another interpretation of probability to consider—the *long-run frequency interpretation*. According to this interpretation, if a trial is repeated a large number of times, the proportion of times that a given event will occur will be very close to the probability of the event. More specifically, if a trial is repeated a large number of times, there is a high probability that the proportion of repeats producing a specific event will be very close to the probability of that event. Accordingly, even though single trials may be subject to considerable uncertainty, overall proportions resulting from a large number of repeats can be predicted quite accurately.

EXAMPLE 4-18　We cannot predict the outcome of a single roll of a balanced die with certainty; however, if the die is rolled several hundred thousand times, we can be very confident that approximately one-sixth of the rolls will be 1's, approximately one-sixth will be 2's, and so on.　　　　　　　　　　　　▲

Subsequent statistical work will involve uncertainties which we will not be able to eliminate, but the probabilities involved will assist us in determining whether something is usual or rare, depending on whether its probability is close to 1 or 0. The probabilities will also help us to determine how confident we may be in the truth of various statements. As well, even though individual results may involve considerable uncertainty, with long-run interpretations, aggregate results can be considered quite accurate.

PRACTICE PROBLEMS

P4-1 A small carton contains 12 cans of soft drinks consisting of three regular colas, two diet colas, two regular ginger ales, two diet ginger ales, two regular orange, and one diet orange. The only part of each can visible is the unlabeled top. One can is to be selected at random. Consider the selection of a single can to be a trial.
(a) Provide a description of the collection of possible outcomes for this trial.
(b) Provide a listing of possible events for this trial.

P4-2 A can of soft drink is chosen in the trial in Practice Problem P4-1 and the type of soft drink noted.
(a) What is the probability that the soft drink chosen will be a cola?
(b) What is the probability that the soft drink chosen will be a diet drink?
(c) What is the chance that a can of orange will be chosen?
(d) What are the odds against the choice of a ginger ale?

P4-3 The dean of science has noticed that, according to past records, only 55% of the students who begin a program successfully graduate from the program 4 years later. If we choose a name at random from the list of beginning students, what might we consider to be the probability that the student will successfully graduate from the program in 4 years?

P4-4 If a political candidate has only a 35% chance of winning an election, what are the odds against election for that candidate?

P4-5 Cargos of dangerous materials sufficient to jeopardize large areas are transported regularly through populated areas. It is estimated that about 1000 such cargos pass through each individual large metropolitan area every week and that for each cargo there is only 1 chance in 1 million of an accident.
(a) Considering the transportation of one cargo through one area as an individual trial and considering ten metropolitan areas over a 10-year period, how many trials are there in total?
(b) With the long-run frequency interpretation of probability, about what proportion of these trials should result in an accident?
(c) How many accidents would this be?

EXERCISES

4-1 Suppose that one representative is to be selected at random from 25 sales agents of whom 21 are men and 4 are women. The selection of the agent is to be considered as a random trial.

(a) Describe the outcomes and two complementary events for the trial.

(b) What is the probability that a man will be selected?

(c) What are the odds against the selection of a woman?

4-2 If there are 2000 tickets in a lottery with only one prize and five of the tickets belong to lottery organizers, what are the odds against the event that an organizer wins the prize?

4-3 A single container of a compound is to be chosen randomly from a collection of 215 containers of which 10 are contaminated. What is the chance that a contaminated container will be chosen?

4-4 If we believe that the odds are 5 to 1 against our *failing* in a project, what would we believe to be the probability of *success*?

4-5 If 4% of the seats in a classroom are suitable for left-handed people and a left-handed student is assigned a seat randomly, what is the probability that the seat will be suitable?

4-6 An insurance company assigns a probability of 0.001 to the possibility of a claim being made against an individual policy.

(a) If a large number of policies are underwritten, about what percentage of them can be expected to result in a claim?

(b) About how many claims can be expected from 10,000 policies?

4-7 Over the past season of play in a league, the Spitfires and the Mustangs met in 20 games with the Spitfires winning 12 and the Mustangs 8. The two teams tied for first place in the league and must play a single game to decide the championship.

(a) On the basis of the statistical record, what is the probability that the Spitfires will be the champions?

(b) What are the odds against the Mustangs in the championship game?

4-8 An illustrator has offers of employment from three advertising agencies, Image Makers, Guaranteed Sales, and New Look. These three agencies are the only ones left bidding on a major contract and the pay will be highest with whichever agency wins the contract. The illustrator believes that the odds are 3 to 2 in favor of New Look's winning the contract. What is the greatest chance that the Image Makers agency would have of winning the contract?

4-9 A laboratory technician, in haste to leave for the weekend, mistakenly placed a male rabbit into a pen with four female rabbits. After leaving the laboratory, the technician realized the error but, being late, merely returned and removed the rabbit closest to the pen opening without checking that it was male. If the rabbits in the meantime had been moving about the pen, what is the probability that the technician left a male rabbit for the weekend with three female rabbits?

SOLUTIONS TO PRACTICE PROBLEMS

P4-1 (a) We may consider this trial to have 12 possible outcomes corresponding to the 12 individual cans that may be selected.

(b) Events may be considered to be the flavor selected or whether the soft drink is a diet or regular drink. Events may then be "cola," "ginger ale," "orange," or perhaps "diet," "regular." Events may also be combinations such as "diet cola," "regular orange," and so on.

P4-2 (a) Since 5 of the 12 soft drinks are cola, the probability is 5/12.

(b) Since 5 of the 12 soft drinks are diet drinks, the probability is 5/12.

(c) Since 3 of the 12 drinks (i.e., 25% of the drinks) are orange, there is a 25% chance that an orange will be selected.

(d) Since 4 of the 12 drinks are ginger ale, the odds are 8 to 4 or, more simply, 2 to 1 against the choice of ginger ale.

P4-3 Basing probabilities on the statistical record, the student has a 55% chance and, hence, a probability of 55/100 or, more simply, 11/20 of graduating successfully in 4 years.

P4-4 With a 35% chance of election, there is a 65% chance of nonelection. The odds against election are then 65 to 35 or simply 13 to 7.

P4-5 (a) There will be 52,000 trials each year or 520,000 trials in the 10 years for each city and thus 5,200,000 trials in total.

(b) We would expect about one one-millionth of these trials to result in an accident.

(c) This proportion of 5,200,000 is 5.2. We would thus expect about five accidents.

4-3 PROBABILITIES FOR COMBINED EVENTS

A further grasp of the general nature of probability can be obtained by considering the calculation of probabilities for combined events.

EXAMPLE 4-19 Suppose that a card is to be drawn at random from a well-shuffled deck of 52 ordinary playing cards. Let A be the event "a face card (jack, queen, or king) or an ace is drawn," and let B be the event "a number less than 5 is drawn." We then have $P(A) = 16/52 = 4/13$, and $P(B) = 12/53 = 3/13$. Consider the probability of obtaining a face card or an ace or a number less than 5. Since there are 16 face cards or aces and 12 cards less than 5 for a total of 28, the probability is $28/52 = 7/13$. ▲

> The event formed by obtaining any outcome from either of the two events A and B is denoted by A or B. We have $P(A$ or $B)$ as the probability of obtaining event A or event B.

EXAMPLE 4-20 With the cards above we found that $P(A$ or $B) = 7/13$. In this case the probability is the sum of the probabilities for the two original events; $P(A$ or $B) = P(A) + P(B)$. ▲

Addition Rule for Mutually Exclusive Events

Example 4-20 illustrates the procedure applied in the general case for which two original events have no outcomes in common. The events A and B in the example have no outcomes in common because no card is both a face card or an ace and less than 5.

Events with no common outcomes are said to be *mutually exclusive*. If A and B are mutually exclusive events,

$$P(A \text{ or } B) = P(A) + P(B)$$

This additional rule works for cases such as Example 4-20 but will not work for two events that do have common outcomes. For events with common outcomes, a more general rule is required.

General Addition Rule

EXAMPLE 4-21   Suppose that a balanced die is to be rolled. Let A be the event that an even number shows and let B be the event that a number greater than 4 shows. For this case, $P(A) = 1/2$, $P(B) = 1/3$ and, since there are four numbers that are even or greater than 4, $P(A \text{ or } B) = 2/3$. If we consider the sum as given in the rule above, $P(A) + P(B) = 5/6$, not $2/3$. The rule does not apply in this case because the events are not mutually exclusive. Both events A and B include the outcome "6"; thus, $P(A)$ and $P(B)$ both include probability $1/6$ for this number. For this reason, the sum $P(A) + P(B)$ exceeds $P(A \text{ or } B)$ by $1/6$. Accordingly, to obtain $P(A \text{ or } B)$ from $P(A) + P(B)$, we must subtract $1/6$, the probability of the common outcome.   ▲

When there are common outcomes such that the events A and B can occur simultaneously, A or B is interpreted as A or B or both.

With the die as above, A or B denotes the collection of numbers that are even or greater than 4 or both even and greater than 4.

The event that includes all of the outcomes common to both A and B is denoted as A and B; thus, $P(A \text{ and } B)$ denotes the probability of obtaining both events together.

EXAMPLE 4-22   In the die example the outcome "6" represents the event A as well as B. It is thus the event A and B. In this case $P(A \text{ and } B) = 1/6$.   ▲

The value of $P(A \text{ and } B)$ is the amount by which the sum $P(A) + P(B)$ exceeds $P(A \text{ or } B)$ and must be subtracted to produce the correct value for $P(A \text{ or } B)$. The subtraction of this probability $P(A \text{ and } B)$ for the common outcome(s) produces the general rule:

For any events A and B,

$$P(A \text{ or } B) = P(A) + P(B) - P(A \text{ and } B)$$

EXAMPLE 4-23   With the die above, $P(A \text{ or } B) = 1/2 + 1/3 - 1/6 = 2/3$, as found directly in Example 4-21.   ▲

This general rule applies in all cases. If A and B are mutually exclusive, they have no common outcomes; thus, A and $B = \emptyset$ and $P(A \text{ and } B) = P(\emptyset) = 0$.

EXAMPLE 4-24 In Example 4-19, in which A and B are collections of cards with no cards in common, $P(A \text{ and } B) = 0$ and $P(A \text{ or } B) = 4/13 + 3/13 - 0 = 7/13$, as before. ▲

Exhaustive Events

If A and B exhaust the sample space (i.e., between them, they include all of the outcomes in the sample space), they are said to be *exhaustive*.

Since exhaustive events include all possibilities, it is certain that at least one of them must occur.

For exhaustive events A and B, $P(A \text{ or } B) = 1$.

Combining the rules for mutually exclusive events with the rule above produces another general rule:

If A and B are mutually exclusive and exhaustive, $P(A) + P(B) = 1$.

We have already used this result when considering the event E and its complement E', which are mutually exclusive and exhaustive.

EXAMPLE 4-25 As a further illustration of the general rules, consider two sources of error in an experimental assessment, machine error and human error. In 5% of the assessments, a machine error is made; in 8%, a human error is made; and in 3%, both types of errors are made. Letting M denote a machine error and H a human error, we have $P(M) = 0.05$, $P(H) = 0.08$, and $P(M \text{ and } H) = 0.03$. We then determine the probability that any given assessment will involve an error as $P(M \text{ or } H) = P(M) + P(H) - P(M \text{ and } H) = 0.05 + 0.08 - 0.03 = 0.10$. If we let E denote an error of either type (i.e., E is M or H), we may rewrite the above result as $P(E) = 0.10$ and, by subtracting $P(E)$ from 1, we find the probability that any given assessment will be error free as $P(E') = 1 - P(E) = 1 - 0.10 = 0.90$. ▲

More Than Two Events

The addition rules can be extended to more than two events. For the case of three events, A, B, and C, notation is as follows: $P(A \text{ or } B \text{ or } C)$ is the probability of obtaining at least one of the three events. $P(A \text{ and } B \text{ and } C)$ is the probability of obtaining all three of the events. The events are said to be exhaustive if all of the outcomes in the sample space are included among them. They are said to be mutually exclusive (meaning mutually exclusive in pairs) if A and B have no outcomes in common, A and C have no outcomes in common, and B and C have no outcomes in common.

Addition rules for three events A, B, and C are as follows: In general,

$$P(A \text{ or } B \text{ or } C) = P(A) + P(B) + P(C) - P(A \text{ and } B)$$
$$- P(A \text{ and } C) - P(B \text{ and } C)$$
$$+ P(A \text{ and } B \text{ and } C)$$

If A, B, and C are *mutually exclusive*,

$$P(A \text{ or } B \text{ or } C) = P(A) + P(B) + P(C)$$

If A, B, and C are *exhaustive*,

$$P(A \text{ or } B \text{ or } C) = 1$$

If A, B, and C are *mutually exclusive and exhaustive*,

$$P(A) + P(B) + P(C) = 1$$

The results are extended in the same manner for four or more events. The last rule above is extended to one which we will discuss later when discussing random variables.

If $E_1, E_2, \ldots, E_n$ are n mutually exclusive and exhaustive events,

$$P(E_1) + P(E_2) + \cdots + P(E_n) = 1$$

That is,

$$\sum_{i=1}^{n} P(E_i) = 1$$

The rules for combined events are perhaps easier to follow through the use of Venn diagrams such as those shown in Figures 4-1 and 4-2, in which the rectangles represent the entire sample space S and the regions within S represent the events A and B, with the areas of the regions proportional to the probabilities of the events that they represent.

In Figure 4-1, in which A and B are mutually exclusive, the area corresponding to A or B is the area covered by the diagonal lines and is the sum of the areas of the two regions. For this case, we have the first result: $P(A \text{ or } B) = P(A) + P(B)$.

In Figure 4-2, A and B are not mutually exclusive. The area corresponding to A or B is again the area covered with diagonal lines and is equal to the area corresponding to A plus the area corresponding to B minus the area with double diagonals, which was included with both the area for A and the area for B and must, therefore, be subtracted once so that it will only be counted once. This diagram illustrates the general rule for the probability of A or B: $P(A \text{ or } B) = P(A) + P(B) - P(A \text{ and } B)$. Similar diagrams can be drawn to illustrate the other rules (see Exercise 4-15).

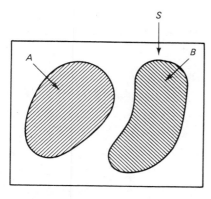 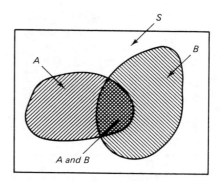

Figure 4-1 A and B mutually exclusive.

Figure 4-2 A and B not mutually exclusive.

P4-6 Refer to the selection of a single can of soft drink from the carton in Practice Problem P4-1. Let C, O, G, D, and R denote, respectively, cola, orange, ginger ale, diet drink, and regular drink. How many outcomes are there in each of the following events? *C or O, R and G, G or D or O, C and G, D or R*.

P4-7 Of 15 pups handled by veterinary students, 8 were vaccinated against rabies and 9 were vaccinated against distemper. Of those pups vaccinated, 5 were vaccinated against both rabies and distemper. One of the pups is to be selected at random. Let R and D denote vaccination against rabies and vaccination against distemper. For the pup selected at random, find

(a) $P(R)$

(b) $P(D)$

(c) $P(R \text{ and } D)$

(d) $P(R \text{ or } D)$

(e) the probability that the pup was not vaccinated

P4-8 A purchaser of a car radio has a catalog including 12 models, all of the appropriate specifications. The models are not identified as to origin, but four are a German make, five are Japanese, and three are American. Let G, J, and A denote German, Japanese, and American. If a radio is selected at random from the catalog, find

(a) $P(G \text{ or } J)$

(b) $P(J \text{ or } A)$

(c) $P(G \text{ or } A)$

P4-9 A psychologist has interviewed 24 possible subjects for an experiment involving first experience with hypnosis in the treatment of a nervous disorder. In the interviews it was discovered that six of the subjects were unsuitable because they had previous experience with hypnosis (denoted by H) and 12 were unsuitable because of the nature of their disorders (denoted by D). As well, four were unsuitable because they were acquainted with the therapist (denoted by A). Of those found unsuitable, three had experienced hypnosis and had an inappropriate disorder, two were acquainted with the therapist and had an inappropriate disorder, and two were acquainted with the therapist and had previous experience with hypnosis. Furthermore, of these, one was rejected on all three grounds. All of the subjects were still assembled and one was chosen at random by an assistant unaware of the interviews. What is the probability that an unsuitable candidate was chosen?

P4-10 The odds against winning for each of the three competitors in a round-robin tournament are given as 3 to 2, 3 to 1, and 5 to 1. Do the corresponding probabilities obey the appropriate rule?

EXERCISES

4-10 An indecisive person trying to arrange a vacation cruise has decided to make a random selection from ten cruise ships. Four of the ships are of Norwegian registry; two of these sail from Miami, Florida, and one each sails from Ft. Lauderdale, Florida, and San Juan, Puerto Rico. Two of the ships are Italian, with one sailing from Miami and the other from San Juan. Two are Greek, with one sailing from Miami and one from San Juan. Two are Liberian, with one sailing from Ft. Lauderdale and one from San Juan.
(a) What is the probability that the ship chosen will be Italian or Greek?
(b) What is the probability that the ship chosen will not be Norwegian?
(c) What is the probability that the ship chosen will be Norwegian or will sail from San Juan (or both)?
(d) What is the probability that the ship will sail from Florida?

4-11 What is the expression for calculating $P(A \ or \ B \ or \ C \ or \ D \ or \ E)$ if A, B, C, D, and E are mutually exclusive?

4-12 What is the value of $P(D)$ if A, B, C, and D are mutually exclusive and exhaustive events such that $P(A) = 1/5$, $P(B) = 1/4$, and $P(C) = 1/3$?

4-13 Suppose that of the plants studied in a particular unfavorable area, 15% had been damaged by disease, 20% had been damaged by insects, and 17% had been damaged by rodents. Suppose further that 6% had been damaged by insects as well as disease, 3% had been damaged by rodents as well as disease, and 7% had been damaged by both insects and rodents (all of these plants were included in the first 15%, 20%, and 17%). No plant suffered all three forms of damage. For any plant selected at random, find
(a) the probability that the plant was damaged
(b) the probability that the plant suffered two types of damage
(c) the probability that the plant suffered from disease but had no other damage
(d) the probability that the plant suffered only one type of damage

4-14 Students are required to concentrate their studies in a single subject or in each of two subjects but may take courses for interest in other subjects. In a given class of 60 students in a particular subject, 21 are concentrating in a single subject, 42 are concentrating in the particular subject of the class, and 12 of these are concentrating only in that subject. What is the probability that a student selected at random from the class will be
(a) concentrating in two subjects?
(b) concentrating in a single subject or the particular subject of the course?
(c) concentrating in two subjects or taking the particular subject for interest?

4-15 Draw a Venn diagram to illustrate the general rule for P(*A or B or C*).

4-16 A sports analyst offered the following odds against ultimately winning the playoffs for the Stanley Cup for each of the following eight teams entering the quarter finals:

Team	Odds against winning
Boston Bruins	4 to 1
Buffalo Sabres	5 to 2
Chicago Black Hawks	10 to 1
Minnesota North Stars	8 to 1
Montreal Canadiens	3 to 2
New York Islanders	5 to 1
New York Rangers	10 to 1
Philadelphia Flyers	3 to 1

Do the corresponding probabilities obey the appropriate rule?

4-17 Twelve manuals are available to trace circuit problems. We have four types of circuits, *A*, *B*, *C*, and *D*. Type *A* are included in four manuals, *B* in three manuals, *C* in four, and *D* in two. Of these manuals, one includes both types *A* and *C* but no others; one includes both *B* and *C* but no others; and one includes all four types. There are no combinations other than those indicated or implied in the listings described above.
(a) How many manuals include both of types *B* and *D*?
(b) How many manuals include all three of types *A*, *B*, and *C*?
(c) If a manual is selected at random, what is the probability that it will include none of the types?

4-18 Past records indicate that 10% of past springs have included an unusually rapid snowmelt, producing flooding. In 12% of past springs, extremely heavy rainfalls have produced flooding. Furthermore, 7% of past springs (included in the 10% and 12% above) have suffered both. What are the odds against flooding in an individual spring?

SOLUTIONS TO PRACTICE PROBLEMS

P4-6 Counting cans of soft drinks in each classification, we find that the numbers of outcomes in *C or O*, *R and G*, *G or D or O*, *C and G*, and *D or R* are, respectively, 8, 2, 9, 0, and 12. Note that since there are no outcomes in *C and G*, and because all of the outcomes are in *D or R*, we may write *C and G* = ∅ and *D or R* = *S*.

P4-7 (a) P(*R*) = 8/15, by direct count
(b) P(*D*) = 9/15, by direct count
(c) P(*R and D*) = 5/15, by direct count
(d) P(*R or D*) = P(*R*) + P(*D*) − P(*R and D*)
  = 8/15 + 9/15 − 5/15 = 12/15 = 4/5
(e) 1 − P(*R or D*) = 1/5

P4-8 By direct count we find P(G) = 4/12, P(J) = 5/12, and P(A) = 3/12. Since the events are mutually exclusive, we find the required probabilities as

(a) P(G or J) = P(G) + P(J) = 9/12 = 3/4

(b) P(J or A) = P(J) + P(A) = 8/12 = 2/3

(c) P(G or A) = P(G) + P(A) = 7/12

P4-9 Letting the choice of each subject be an outcome, we have 24 equally likely outcomes. From the information given, the events H, D, A, H and D, A and D, A and H, and H and D and A include, respectively, 6, 12, 4, 3, 2, 2, and 1 outcomes. The probability that the chosen subject is unsuitable is

$$P(H \ or \ D \ or \ A) = P(H) + P(D) + P(A) - P(H \ and \ D)$$

$$- P(A \ and \ D) - P(A \ and \ H) + P(H \ and \ D \ and \ A)$$

$$= \frac{6}{24} + \frac{12}{24} + \frac{4}{24} - \frac{3}{24} - \frac{2}{24} - \frac{2}{24} + \frac{1}{24} = \frac{16}{24} = \frac{2}{3}$$

P4-10 The corresponding probabilities of winning for each of the three competitors are 2/5, 1/4, and 1/6. Since one competitor must win, and only one can win, we have a set of mutually exclusive and exhaustive events and the corresponding probabilities must add to 1. Since 2/5 + 1/4 + 1/6 = 24/60 + 15/60 + 10/60 = 49/60, which is not 1, the probabilities do not obey the appropriate rule.

4-4 MORE PROBABILITIES FOR COMBINED EVENTS: INDEPENDENCE, CONDITIONAL PROBABILITY

In determining the general rule for finding P(A or B) as P(A) + P(B) − P(A and B), we did not consider methods of calculating P(A and B), the probability of obtaining both events together. In some cases, direct enumeration of all of the outcomes in A and B may be appropriate, but in many cases other procedures will be available and often required.

EXAMPLE 4-26 Suppose that we toss a coin and roll a die. Let A be the event "the coin shows heads" and let B be the event "the die shows a 4." Through direct enumeration, we could find the 12 equally likely possibilities "heads and 1," "heads and 2,"..., "tails and 6," of which only one produces A and B, thus determining that P(A and B) = 1/12.

We might also argue that with repeated rolls of the die, a proportion 1/6 of the rolls would be expected to result in "4" (i.e., in the event B). We might also expect that 1/2 of all tosses of a coin would result in "heads" (i.e., in the event A). Putting the two results together, and noting that the result of tossing the coin should not affect the result of rolling the die, and vice versa, we expect 1/6 of all combined tosses to produce B and 1/2 of these to also produce A. We expect 1/2 of 1/6 or 1/12 of all combined trials to produce A and B. These proportions are just the corresponding probabilities and we have combined them to produce P(A and B) = (1/2) × (1/6) = 1/12, as before. We have found the probability of obtaining A and B together as P(A and B) = P(A) × P(B). ▲

Independent Events

The latter procedure used in Example 4-26 is a procedure that can be used for any events that are independent.

> Two events are defined to be independent if the occurrence of either does not alter the probability of occurrence of the other.

EXAMPLE 4-27 When drawing a single card from an ordinary deck, the events "spade" and "king" are independent. There are four kings in a full deck of 52 cards and one king in the suit of 13 spades. In either case, we have a proportion of $1/13$. Similarly, one of four kings is a spade or 13 of the 52 cards (i.e., $1/4$ of the full deck) are spades. ▲

EXAMPLE 4-28 As another illustration, if two arbitrators are selected randomly, one from each of two mixed groups of men and women, the event "a woman is chosen from the first group" is independent of the event "a woman is chosen from the second." ▲

Multiplication Rule for Independent Events

For cases involving independent events, we have the following multiplication rule:

> For two independent events A and B,
> $$P(A \text{ and } B) = P(A) \times P(B)$$

EXAMPLE 4-29 The probability that the card in Example 4-27 will be both a spade and a king may be found as $(1/4) \times (1/13) = 1/52$. (This result could also be found by direct enumeration.) ▲

EXAMPLE 4-30 In the selection of two arbitrators in Example 4-28, if 40% of the first group are women and 55% of the second group are women, the probability of a woman from each group is $0.40 \times 0.55 = 0.22$. ▲

Compound Event Structures

In some cases an event may be a combination of events which are already combinations of other events. Combinations of probability rules are often required for such compound events.

EXAMPLE 4-31 In the arbitrator selection in Example 4-30, consider the event "a man and a woman are chosen." This event is a combination of two combined events, a man from the first group and a woman from the second, or a woman from the first and a man from the second. Let W_1, W_2, M_1, and M_2

denote "woman from first," "woman from second," "man from first," and "man from second." The required probability is then $P([M_1 \text{ and } W_2] \text{ or } [W_1 \text{ and } M_2])$, which is $P(M_1 \text{ and } W_2) + P(W_1 \text{ and } M_2)$ because $M_1 \text{ and } W_2$ and W_1 and M_2 are mutually exclusive. By independence, $P(M_1 \text{ and } W_2) = P(M_1) \times P(W_2)$ and $P(W_1 \text{ and } M_2) = P(W_1) \times P(M_2)$. Finally, the required probability is $[P(M_1) \times P(W_2)] + [P(W_1) \times P(M_2)] = (0.6 \times 0.55) + (0.4 \times 0.45) = 0.51$.

▲

More Than Two Independent Events

The multiplication rule for independent events extends very simply to three or more independent events. For three events we have the following:

> If A, B, and C are all independent of each other (i.e., the probability of any one is not affected by the occurrence of any combination of the others), then
>
> $$P(A \text{ and } B \text{ and } C) = P(A) \times P(B) \times P(C).$$

EXAMPLE 4-32   With the die and coin in Example 4-28, let us also consider drawing a card from a well-shuffled deck of ordinary playing cards. Let C denote the event a heart is drawn; then $P(C) = 1/4$. With A and B as previously defined, we obtain the probability that the coin shows heads, the die shows 4, and the card is a heart as $P(A \text{ and } B \text{ and } C) = P(A) \times P(B) \times P(C) = (\frac{1}{2}) \times (\frac{1}{6}) \times (\frac{1}{4}) = \frac{1}{48}$.

▲

The probability of obtaining several independent events simultaneously is the product of their individual probabilities.

> For n events $E_1, E_2, \ldots, E_n$ all of which are independent of each other,
>
> $$P(E_1 \text{ and } E_2 \text{ and } \cdots \text{ and } E_n) = P(E_1) \times P(E_2) \times \cdots \times P(E_n)$$

Nonindependent Events — Conditional Probability

When events are related and the occurrence of one affects the probabilities for others, different procedures must be applied.

EXAMPLE 4-33   A box contains 15 beads such that 2 are wooden and colored red, 4 are wooden and colored green, 5 are plastic and red, and 4 are plastic and green. We denote the events "wood," "plastic," "red," and "green" as W, P, R, and G, respectively. We then find that $P(W \text{ and } R) = 2/15$, but $P(W)P(R) = (6/15) \times (7/15) = 14/75$. Also, $P(P \text{ and } G) = 4/15$, but $P(P)P(G) = (9/15) \times (8/15) = 24/75$.

▲

The previous multiplication rule does not apply because the events combined are not independent. The wooden beads in Example 4-33 consist of more green than red; the plastic beads consist of more red than green. If a bead is known to be wooden, the probability of its being red is 1/3. Such a probability is a *conditional probability*.

> A conditional probability is written as $P(A|B)$ and is read *the probability that the event A will occur (or has occurred) given the condition that the event B will occur (or has occurred)* or simply *the probability of A given B*.

EXAMPLE 4-34   In Example 4-33, 1/3 of the wooden beads are red and $P(R|W) = 1/3$. Similarly, since four of the nine plastic beads are green, $P(G|P) = 4/9$.   ▲

As a condition of independence, conditional probabilities involving independent events are just ordinary probabilities.

> If A and B are independent,
>
> $$P(A|B) = P(A)$$
> $$P(B|A) = P(B)$$

Conditional probabilities are used in a more general multiplication rule for events that are not independent.

EXAMPLE 4-35   For the beads in Example 4-33, the probability that a bead is red given that it is known to be wooden is combined with the probability of the bead's being wooden to produce $P(W\ and\ R)$ as $(6/15) \times (1/3) = 2/15$. Intuitively, six-fifteenths of the beads are wooden and one-third of these are red; thus, the proportion of beads that are wooden and red is one-third of six-fifteenths or two-fifteenths; that is, $P(W\ and\ R) = (6/15) \times (1/3) = 2/15$, as found by direct enumeration. In probability notation, we have found that $P(W\ and\ R)$ is $P(W) \times P(R|W)$.   ▲

General Multiplication Rule

Example 4-35 illustrates the general rule for finding the probability of two events occurring together.

> In general, for any two events A and B,
>
> $$P(A\ and\ B) = P(B)P(A|B)$$
>
> or
>
> $$P(A\ and\ B) = P(A)P(B|A)$$

EXAMPLE 4-36 In Example 4-33, the probability that a bead is plastic and green is $P(P \text{ and } G) = P(P) \times P(G|P) = (9/15) \times (4/9) = 4/15$, as found by direct enumeration. ▲

The general rule also works for independent events because, if A and B are independent, $P(A|B) = P(A)$ and $P(B|A) = P(B)$.

Quotient Rule for Conditional Probability

The general multiplication rule may also be written in quotient forms suitable for calculating conditional probabilities.

Conditional probabilities are found as

$$P(A|B) = \frac{P(A \text{ and } B)}{P(B)}$$

and

$$P(B|A) = \frac{P(A \text{ and } B)}{P(A)}$$

EXAMPLE 4-37 Of several people complaining about headaches, 42% took a headache pill and received relief within an hour. In total, 56% of the people with headaches received relief within an hour. If one person chosen at random from all of those with headaches received relief within an hour, the probability that he or she had taken a headache pill may be found from the quotient rule. Let P denote the event that a pill was taken and let N denote the event that the headache is no longer present. The desired probability is then found as

$$P(P|N) = \frac{P(P \text{ and } N)}{P(N)} = \frac{0.42}{0.56} = 0.75.$$ ▲

In the latter form, we can illustrate the quotient rule with the Venn diagram in Figure 4-3 in which probability is proportional to area such that the total area of the sample space is 1 and $P(A) = $ area of A, and so on. Suppose that we are given the

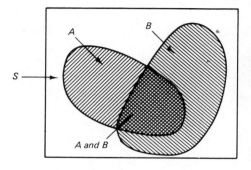

Figure 4-3

condition A, that is, we are restricted to outcomes in A. Then, to obtain B, we must obtain an outcome in the combined event A *and* B. The probability of obtaining such an outcome is the area of A *and* B divided by the total possible area, which in this case is the area of A (we are restricted to A). Since the areas are probabilities, we obtain $P(A$ *and* $B)/P(A)$, as before.

Compound Event Structures

As with independent events, rules may have to be combined to deal with compound event structures for dependent events.

EXAMPLE 4-38 Suppose that we have boxes of light bulbs of three brands, labeled B_1, B_2, and B_3. Boxes of the first brand contain 8 red bulbs and 12 white. Boxes of the second brand contain 5 red bulbs and 20 white. Boxes of the third brand contain 18 red and 12 white. Stored in a film-processing darkroom, we have an inventory of five boxes of brand B_1, three of brand B_2, and two of brand B_3. A box is to be chosen at random in the dark and a bulb is to be drawn at random from this box to replace the dead bulb in the overhead lamp. Only the red bulbs are acceptable (denoted A). Using a white bulb could damage the film. We wish to know the probability that the chosen bulb will be an acceptable red bulb so that the film will be safe; that is, we want to find $P(A)$.

An acceptable bulb can be obtained in three ways, in conjunction with choosing a box of brand B_1, a box of brand B_2, or a box of brand B_3; thus, $P(A) = P([B_1$ *and* $A]$ *or* $[B_2$ *and* $A]$ *or* $[B_3$ *and* $A])$. Since the events B_1, B_2, and B_3 are mutually exclusive, the events B_1 *and* A, B_2 *and* A, and B_3 *and* A are mutually exclusive and $P(A) = P(B_1$ *and* $A) + P(B_2$ *and* $A) + P(B_3$ *and* $A)$.

Using the general multiplication rule for $P(B_1$ *and* $A)$, $P(B_2$ *and* $A)$ and $P(B_3$ *and* $A)$, we obtain $P(A) = P(B_1)P(A|B_1) + P(B_2)P(A|B_2) + P(B_3)P(A|B_3)$.

We have ten boxes, of which five are brand B_1, producing $P(B_1) = 5/10$. If we draw from a box of this brand, we draw from 8 red and 12 white; thus, $P(A|B_1) = 8/20$. Accordingly, we obtain $P(B_1$ *and* $A) = P(B_1)P(A|B_1) = (5/10) \times (8/20) = 1/5$. Through similar calculations for drawing from B_2 and B_3 we finally obtain $P(A) = [(5/10) \times (8/20)] + [(3/10) \times (5/25)] + [(2/10) \times (18/30)] = 19/50$. ▲

Example 4-38 has illustrated two other general considerations.

Partition of the Sample Space

In Example 4-38 the brands B_1, B_2, and B_3 formed a *partition* of the sample space of all of the bulbs into three mutually exclusive and exhaustive events. Any set of n mutually exclusive and exhaustive events are said to form a partition.

If $B_1, B_2, \ldots, B_n$ are mutually exclusive and exhaustive events, they are said to form a *partition* of the sample space.

As illustrated in Example 4-38, when a partition exists, the probability of some other event A is found by combining the probabilities of that event in combination with each of the events forming the partition.

If the events $B_1, B_2, \ldots, B_n$ form a partition of the sample space, then for some other event A,

$$\begin{aligned}
P(A) &= P(B_1 \text{ and } A) + P(B_2 \text{ and } A) + \cdots + P(B_n \text{ and } A) \\
&= \left[P(B_1) \times P(A|B_1)\right] + \left[P(B_2) \times P(A|B_2)\right] \\
&\quad + \cdots + \left[P(B_n) \times P(A|B_n)\right] \\
&= \sum_{i=1}^{n} P(B_i) \times P(A|B_i)
\end{aligned}$$

Bayes' Rule

Using this new expression for $P(A)$ in the denominator of the quotient rule for a conditional probability produces a more general expression. This result is known as *Bayes' rule*.

If $B_1, B_2, \ldots, B_n$ are mutually exclusive, exhaustive events, some of which may occur in conjunction with event A, then, for any individual $j = 1, 2, 3, \ldots, n$,

$$P(B_j|A) = \frac{P(B_j) \times P(A|B_j)}{\displaystyle\sum_{i=1}^{n} P(B_i) \times P(A|B_i)}$$

EXAMPLE 4-39   Suppose now that we draw a bulb as indicated in Example 4-38, and we do obtain an acceptable red bulb and then wish to know the probability that the box was of brand B_1. We now want to find $P(B_1|A)$. By the general rule,

$$P(B_1|A) = \frac{P(B_1 \text{ and } A)}{P(A)} = \frac{P(B_1)P(A|B_1)}{\sum P(B_i)P(A|B_i)}.$$

$$= \frac{(5/10) \times (8/20)}{\left[(5/10) \times (8/20)\right] + \left[(3/10) \times (5/25)\right] + \left[(2/10) \times (18/30)\right]}$$

$$= \frac{1/5}{19/50} = \frac{10}{19}$$

The calculations above can also be obtained through the use of a diagram, as illustrated in Figure 4-4. The full rectangle represents the full sample space of all possible outcomes. The ten horizontal divisions represent

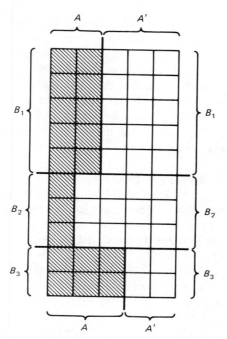

Figure 4-4 Bulb, brand diagram.

the ten possible boxes. The top five correspond to B_1, the next three to B_2, and the last two to B_3.

The least common denominator, 5, is then determined for each of the three conditional probabilities, $P(A|B_1)$, $P(A|B_2)$, and $P(A|B_3)$ to produce these as 2/5, 1/5, and 3/5, respectively. The five vertical divisions correspond to the acceptable red, nonacceptable white classifications, with acceptable on the left and nonacceptable on the right. Within B_1 there are two on the left and three on the right, since with a box of brand B_1 there is a probability 2/5 of obtaining an acceptable red. Similarly, for B_2 and B_3, there are separations with one and three divisions on the left, respectively.

The horizontal and vertical divisions produce a total of 50 squares, each of which has a probability representation equal to 1/50. There are 10 squares corresponding to a box of brand B_1 and an acceptable red; thus, $P(B_1 \text{ and } A) = 10/50 = 1/5$, as before. There are 19 squares (indicated with diagonal lines) corresponding to acceptable red bulbs; thus, $P(A) = 19/50$, as found before. Of the 19 squares corresponding to acceptable red, 10 also correspond to a box of brand B_1; thus, $P(B_1|A) = 10/19$, as before. ▲

Updating Probabilities

In some cases, information obtained from a random trial may be used to produce conditional probability statements, which reflect an updating of our measure of uncertainty.

EXAMPLE 4-40   Suppose that we are interested in a given type of gambling device in a particular establishment. We know that there are two indistinguishable variations of the device, A and B. With a device of type A, there is a 20% chance of winning. With a device of type B, there is only a 5% chance of winning. Furthermore, both types are such that individual trials on a given device are independent. Suppose also that we know that 90% of the devices are of type B, and 10% are of type A. Initially, we wish to know the probability of winning if we pick one device at random and try it. Subsequently, if we do win, we will wish to update our information to find the probability of a second win from the same device.

Let W denote a win. We can then summarize our initial information as probabilities: $P(A) = 0.10$, $P(B) = 0.90$, $P(W|A) = 0.20$, and $P(W|B) = 0.05$. Initially, we wish to know $P(W)$. Since a win can be obtained with either of the (mutually exclusive) types of devices that form a partition, we find that $P(W) = P(A \text{ and } W) + P(B \text{ and } W) = P(A)P(W|A) + P(B)P(W|B) = (0.10 \times 0.20) + (0.90 \times 0.05) = 0.065$.

Now, suppose that we have chosen a device and tried it and we have won. We now wish to know the probability that we will win a second time if we try the same machine again. The probability, as a measure of our uncertainty, is not the same as it was for the first trial since we now have additional information or evidence that should cause us to believe more strongly in a device of type A. The device that we have chosen has produced a win and that is more likely to occur with a type A than with a type B. To incorporate this evidence in the calculation of $P(W)$, we place the original values for $P(A)$ and $P(B)$ with "updated" values conditional on the information from the first trial, namely with $P(A|W)$ and $P(B|W)$. Since we have already used the general partition rule to find $P(W)$, we may express Bayes' rule in the form of the simpler quotient rule to find

$$P(A|W) = \frac{P(A \text{ and } W)}{P(W)} = \frac{0.10 \times 0.20}{0.065} = \frac{20}{65}$$

and

$$P(B|W) = \frac{P(B \text{ and } W)}{P(W)} = \frac{0.90 \times 0.05}{0.065} = \frac{45}{65}$$

(Note that our current probability on A is $20/65 = 0.3077$ or slightly more than three times the original value of 0.10.) Using these "updated" probabilities on A and B, we produce an updated probability of another win on the second trial given a win on the first. We substitute $P(A|W)$ for $P(A)$ and $P(B|W)$ for $P(B)$ in the calculation of $P(W)$ to produce the new updated probability.

$$P(\text{second win} | \text{first win}) = \left(\frac{20}{65} \times 0.20 \right) + \left(\frac{45}{65} \times 0.05 \right) = 0.096 \quad \blacktriangle$$

P4-11 A committee charged with reviewing a program of study in political economy is to include two students. One student is to be selected at random from a political science class consisting of 15 third-year students and 5 fourth-year students. The other is to be selected at random from an economics class consisting of 20 third-year students and 10 fourth-year students.

(a) What is the probability that both students will be third-year students?

(b) What is the probability that both will be fourth-year students?

(c) What is the probability of obtaining a third-year student and a fourth-year student?

P4-12 An auditor is to select one personal account at random from each of three bank branches as a quick check on agreement between the bank and its clients. In the first branch 0.1% of the clients believe the account balance to be different from that shown by the bank. The corresponding figures for the other two branches are 0.5% and 0.2%.

(a) What is the probability that there will be agreement for none of the three accounts sampled?

(b) What is the probability that there will be agreement for all three of the accounts sampled?

P4-13 The records of a zoo show that 40% of animals born in the zoo are a result of inbreeding and that only 55% of these survive the first 6 months. On the other hand, of the 60% not resulting from inbreeding, 70% survive the first 6 months. Suppose that a newborn animal is randomly selected.

(a) What is the probability that it resulted from inbreeding and will survive the first 6 months?

(b) What is the probability that it did not result from inbreeding and will survive the first 6 months?

(c) What is the probability that it will survive the first 6 months?

(d) If it survived the first 6 months, what is the probability that it resulted from inbreeding?

P4-14 In the submission of an annual tax return, a taxpayer has applied for the allowance of certain expenses for which the legitimacy is subject to interpretation. Applications must be considered by one of three review boards operating on a rotation schedule. The first board, consisting mainly of individuals who are finishing long terms and have become cynical, approves only 25% of such claims. The second board, consisting mainly of midterm individuals, approves 50% of such claims. The third board, consisting mainly of newer members dubious about refuting claims, approves 70% of such claims. There is a 20% chance that the application will be considered by the first board, a 50% chance that it will be considered by the second board, and a 30% chance that it will be considered by the third board.

(a) What is the probability that the application will be approved?

(b) If the application is approved, what is the probability that it was considered by

(1) the first board?

(2) the second board?

(3) the third board?

(c) The review board has just started considering applications and will continue for a fixed time period. The application submitted was a test case to determine the advisability of submitting a similar application involving more money. If the first application was approved, what is the probability that the second will also be approved if it is submitted in the same time period?

4-19 A red die, a green die, a penny, a nickel, and a dime are all tossed together. What is the probability that both dice show 6 and all of the coins show heads?

4-20 According to past records, 91% of captured dogs do not have a dog tag and 60% of these are destroyed. What is the probability that a captured dog will not have a tag and will be destroyed?

4-21 A planter is to be planted with three plants—one from each of three containers. Each container has a different variety of plant in it. The plants are randomly mixed and none are blooming. Of the 8 plants in the first container, 3 will produce white blooms and 5 will produce red blooms. Of the 12 plants in the second container, 7 will produce white blooms and 5 will produce red. Of the 18 plants in the third container, 10 will produce white blooms, 6 will produce red, and 2 will produce orange.
(a) What is the probability that the planter will have all white blooming plants?
(b) What is the probability that the planter will have two red blooming plants and one orange?

4-22 One of the containers in Exercise 4-21 is to be selected at random and one of the plants randomly selected for a single pot.
(a) What is the probability of choosing a red blooming plant if the first container is selected?
(b) What is the probability of choosing the first container and a red blooming plant?
(c) What is the probability of choosing a red blooming plant?
(d) If the selected plant blooms red, what is the probability that it came from the first container?

4-23 (a) If the plant selected in Exercise 4-22 blooms red, what is the probability that a second plant selected at random from the same container will also bloom red?
(b) If the plant selected in Exercise 4-22 blooms white, what is the probability that a second plant chosen from the same container will also bloom white?
(c) If the plant selected in Exercise 4-22 blooms orange, what is the probability that a second plant chosen from the same container will also bloom orange?

4-24 A review panel is to be formed by randomly selecting one representative from each of four groups. The first group consists of 12 men and 8 women; the second consists of 10 men and 15 women; the third consists of 15 men and 15 women; and the fourth consists of 15 men and 5 women.
(a) What is the probability that the panel will consist entirely of men?
(b) What is the probability of choosing
(1) a woman from the first group and men from each of the others?
(2) a woman from the second group and men from each of the others?

(3) a woman from the third group and men from each of the others?

(4) a woman from the fourth group and men from each of the others?

(c) Using the results of part (b), determine the probability of choosing a panel that consists of one woman and three men.

(d) Using the results of parts (a) and (c), determine the probability that the panel will include at least three men.

4-25 Suppose that a salvage company is to receive a contract for a salvage operation. The operation is selected at random from six sites. The six salvage sites are such that there is one in the coastal waters of each of British Columbia, California, Florida (Gulf of Mexico), Georgia, Newfoundland, and Texas. The salvage operations on the west coast can be assumed to have a 65% chance of producing success; the operations on the east coast have a 55% chance of success; and the operations in the Gulf of Mexico have a 75% chance of success.

(a) What is the probability that the contract will be assigned to the west coast and that the operation will be successful?

(b) What is the probability of an assignment to the east coast without success?

(c) What is the probability of an assignment to Canadian waters and a successful operation?

(d) What is the probability of a successful operation?

(e) Given the information that the operation was successful, what is the probability that the assignment was to an American site?

4-26 A restaurant has two suppliers of house wine. One supplier provides a European *vin ordinare*. The other provides a North American table wine which is supposed to be very similar to the European wine. Containers of house wine are selected at random from a stock of which 60% is European and 40% North American. A particular customer's capabilities are such that he can correctly identify the European wine 95% of the time that it is provided unidentified and he can correctly identify the North American wine 90% of the time that it is provided unidentified.

(a) If the customer arrives at the restaurant and has a glass of house wine, what is the probability that the wine is European?

(b) What is the probability that the customer will receive North American wine and correctly identify it?

(c) If the customer orders a glass of house wine and tries to identify it, what is the probability that he will be correct?

(d) If he orders a glass of wine, tastes it, and identifies it as North American, what is the probability that it actually is North American?

4-27 The probability that a previously conditioned experimental rat will complete a maze run in the required time is 0.85. The probability that a novice rat will complete the run in time is 0.10. Through mishap, ten conditioned rats and two novice rats have become mixed. In an attempt to classify the rats, the rats are to be tested in the maze. If one of the 12 rats is selected at random and fails to complete the run in the required time, what is the probability that it is a novice?

4-28 Suppose that 37% of drivers acquired their driving licenses within one year of becoming eligible and 60% of these were charged with a traffic offense in their first year. A further 45% acquired licenses within the next 10 years and 20% of these were charged with an offense in their first year. The remaining 18% acquired licenses after the first 11 years and 35% of these were charged within the first year.

(a) What is the probability that a driver who received a license after the first year, but before the twelfth, was not charged within the first year of driving?

(b) What is the probability that a driver selected at random acquired a license within a year of becoming eligible and was charged within the first year?

(c) What is the probability that a driver selected at random was not charged within the first year?

4-29 On the basis of the information in Exercise 4-28, determine the probability that a driver charged with an offense in the first year of driving received a license within a year of becoming eligible.

SOLUTIONS TO PRACTICE PROBLEMS

P4-11 (a) The probabilities of selecting a third-year student from each of the political science and economics classes are, respectively, $15/20 = 3/4$ and $20/30 = 2/3$. Selections from the two classes are independent; thus, the required probability is $(3/4) \times (2/3) = 1/2$.

(b) Proceeding as in part (a), the probability of two fourth-year students is found as $(1/4) \times (1/3) = 1/12$.

(c) The required probability is the probability of a third-year student from the political science and a fourth-year student from the economics class, or vice versa. Since these two events are mutually exclusive, we add their probabilities to obtain the required probability as $[(3/4) \times (1/3)] + [(1/4) \times (2/3)] = 5/12$.

P4-12 (a) Since the account selected in one bank is independent of those selected in the others, we use the independence rule to find the probability as $0.001 \times 0.005 \times 0.002 = 0.00000001$.

(b) Using the three complementary probabilities, we find the probability of agreement in all of these accounts as $0.999 \times 0.995 \times 0.998 = 0.99201699$.

P4-13 Let I denote inbred, N denote not inbred, S denote survival through the first 6 months, and F denote failure to survive. The required probabilities are then found as:

(a) $P(I \text{ and } S) = P(I)P(S|I) = 0.4 \times 0.55 = 0.22$

(b) $P(N \text{ and } S) = P(N)P(S|N) = 0.6 \times 0.70 = 0.42$

(c) $P(S) = P(I \text{ and } S) + P(N \text{ and } S) = 0.64$

(d) $P(I|S) = P(I \text{ and } S)/P(S) = 0.22/0.64 = 0.34375$

P4-14 Let B_1, B_2, and B_3 denote consideration by the first, second, and third boards, respectively. Let A denote approval of the application. The required probabilities are then found as follows:

(a) $P(A) = P(B_1 \text{ and } A) + P(B_2 \text{ and } A) + P(B_3 \text{ and } A)$
$= P(B_1)P(A|B_1) + P(B_2)P(A|B_2) + P(B_3)P(A|B_3)$
$= (0.20 \times 0.25) + (0.50 \times 0.50) + (0.30 \times 0.70)$
$= 0.05 + 0.25 + 0.21 = 0.51$

(b) (1) $P(B_1|A) = P(B_1 \text{ and } A)/P(A) = 5/51$
(2) $P(B_2|A) = P(B_2 \text{ and } A)/P(A) = 25/51$
(3) $P(B_3|A) = P(B_3 \text{ and } A)/P(A) = 21/51$

(c) We use the probabilities in part (b) as updated probabilities to replace $P(B_1)$, $P(B_2)$, and $P(B_3)$ in the expression for $P(A)$ in part (a). We then find the probability of the second acceptance as $[(5/51) \times 0.25] + [(25/51) \times 0.50] + [(21/51) \times 0.70] = 0.558$.

4-5 PROBABILITY GRAPHS AND OUTCOMES IN A CONTINUUM

For statistical analyses, we will be interested in events that are represented by numbers. The graphical techniques used to illustrate frequency distributions for quantitative data can also be used to illustrate probabilities for numerical events.

Area as Probability

EXAMPLE 4-41   Figure 4-5 shows the very simple histogram for the set of possible outcomes of rolling a balanced die. Each of the numbers 1 through 6 corresponds to one of the six faces of the die and has frequency 1. The total frequency is 6 and with the area representation for frequency we have six units of area.

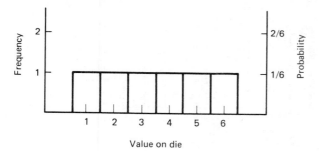

Figure 4-5 Histogram/probability graph for a die.

To this histogram, we have added another vertical scale. This scale, on the right side, measures relative frequency or probability instead of frequency. Each possible value 1 through 6 has one-sixth of the total frequency and thus has relative frequency or probability 1/6. In converting to relative frequencies or probabilities, we have divided the individual frequencies by the total frequency 6; hence, we have converted the total area to 1 for the probability interpretation. The total area now corresponds to the total probability. The rectangle over each value has area 1/6 corresponding to the probability for the value. ▲

EXAMPLE 4-42   For another example, consider a trial to be the rolling of two balanced dice. The value of interest is the sum of the numbers showing on the two dice. There are 36 possible outcomes for the trial. The 36 corresponding sums are indicated in the following table. For example, the value 5 in the second row and third column is the sum obtained from rolling a 3 on the first die and a 2 on the second.

	First die					
Second die	1	2	3	4	5	6
1	2	3	4	5	6	7
2	3	4	5	6	7	8
3	4	5	6	7	8	9
4	5	6	7	8	9	10
5	6	7	8	9	10	11
6	7	8	9	10	11	12

Figure 4-6 shows a histogram for the possible values of the sum with a relative frequency or probability scale on the right. ▲

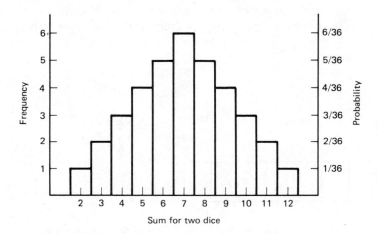

Figure 4-6 Histogram/probability graph for two dice.

Again, using the right-hand scale, total area is adjusted to 1 (the total probability) and the area of the rectangle over each value is equal to the probability for the value.

> In probability graphs, *areas are probabilities*. For a graph to be a probability graph, it must have a *total area of 1*. Since probabilities are not negative, the graph must not cross below the horizontal scale.

As for frequencies in a histogram, areas are important in interpreting graphs of probabilities.

EXAMPLE 4-43   In Figure 4-6 we can see from the relative magnitudes of the rectangles that the most probable result of rolling two balanced dice is 7 and the least probable results are 2 and 12. Looking at areas (and, hence, probabilities) for combined outcomes or events, we can also see a symmetry such that the probability of obtaining a value less than 7 is the same as the probability of obtaining a value greater than 7. More specific information on the probabilities is obtained by referring to the right-hand (probability) scale. ▲

Notation for Numerical Outcomes

While considering probabilities for trials with numerical outcomes, we will introduce some more mathematical notation. Rather than expressing the event "the result is a 7" with a capital letter, we use the expression "$x = 7$," where x represents the numerical result of the trial. The same expression will also be used for description of future events—in this case for "the result will be a 7."

EXAMPLE 4-44 $P(x = 7) = 1/6$ is read "the probability is $1/6$ that the result is (will be) a 7." Similarly, we write $P(x \leqslant 3) = 3/36 = 1/12$ instead of "the probability is $1/12$ that the result will be less than or equal to 3." ▲

The inequality sign can be modified to " $<$," indicating that the left side is less than (and not equal to) the right side.

EXAMPLE 4-45 As illustrations of these modifications, some probability statements on rolling two dice are: $P(x < 5) = 1/6$ (in words, the probability is $1/6$ that x is less than 5 where x is the sum for two dice), and (as we noted in Example 4-43) from the symmetry, $P(x < 7) = P(x > 7)$. As with our general limits on $P(E)$, we can use these signs twice in one expression; thus, $P(6 \leqslant x < 9) = 4/9$ is read "the probability is $4/9$ that the result is (will be) greater than, or equal to, 6 and less than 9." ▲

The general considerations outlined in these examples are summarized as follows:

> For trials with numerical outcomes we use mathematical expressions rather than capital letters to denote events. The probability that the outcome is (or will be) the number a is written
>
> $$P(x = a)$$
>
> The probability that the outcome is (will be) less than or equal to the number a is written
>
> $$P(x \leqslant a)$$
>
> The probability that the outcome is (will be) less than the number a is written
>
> $$P(x < a)$$
>
> With double use of signs, the probability that the outcome is (will be) between the numbers a and b inclusive is written
>
> $$P(a \leqslant x \leqslant b)$$

When we are interested only in probabilities, we use only one vertical scale—the probability scale—and we place it on the left.

Outcomes in a Continuum— Probability Density

In some cases we may have a very large number of possible outcomes. Considering such disparate possibilities as holes or very dense packing, the number of fibers in a cross section of material may be any number ranging from zero to several million. In another instance, even if reaction times can differ by no more than one full second from shortest to longest, measuring to the nearest millisecond produces 1000 possible outcomes. In the latter instance we have an example of a situation in which there are theoretically infinitely many outcomes. Theoretically, the time could be any one of the infinite values in the continuum of numbers from the shortest time to the longest. The number of possibilities has only been limited to 1000 because of the accuracy of the measuring device. In statistical analyses we will consider structures with infinitely many outcomes, either as correct theoretical models in themselves, or as reasonable approximations to reality. The corresponding smooth curve approximation for probabilities is similar to the smooth curve approximation to a histogram that was shown for frequencies in Figure 2-16. Our smooth curve for probabilities should, however, have the restriction that the total area between it and the horizontal scale must equal the total probability (i.e., 1).

> For a trial with very many outcomes, the probability graph may be approximated by a smooth curve as shown in Figure 4-7. For theoretical models with infinitely many outcomes in a continuum, the smooth curve is used as an exact representation. As with the other probability graphs, area is probability.

EXAMPLE 4-46 For the trial illustrated in Figure 4-7, the probability of obtaining a result x with a value between b and c inclusive [i.e., $P(b \leqslant x \leqslant c)$] is the area of the shaded section. ▲

General probability impressions are, again, available visually from relevant areas.

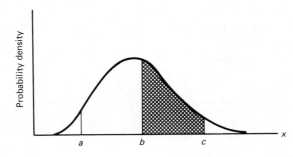

Figure 4-7 Continuous probability graph.

EXAMPLE 4-47 For the trial illustrated in Figure 4-7, the intervals of values with relatively high probability are those between *a* and *c*. Intervals of values with relatively low probability are those below *a* and above *c*. In a long run of trials, we should expect to see the majority of the trials result in values between *a* and *c* with only a few below *a* or above *c*. Values between *a* and *c* are "usual" values; other values are "rare" or "unusual," with those below *a* being "extremely" or "unusually" small and those above *c* being "extremely" or "unusually" large.

▲

Another point should be noted with regard to labeling continuous probability graphs. Although probability is *indicated by heights*, it is actually *measured by area*. It was appropriate to label the vertical scale "probability" in Figures 4-3 and 4-4 because the outcomes in these cases were whole numbers and the rectangles each had a base of 1, thus producing areas equal to heights. For outcomes that occur in a continuum, however, heights will not equal areas and thus will not equal probabilities.

> In probability graphs, heights, and hence the vertical scale, indicate which intervals have greater probability, but the actual probabilities must be calculated from areas. For this reason we should label the vertical scale differently—the usual terminology is to label the height as a *probability density* rather than probability.

Because probability is derived from area, *for outcomes in a continuum, we can only find probabilities for intervals*. If we try to find P($x = a$ exactly), we must find the area above this one point. Because a single point produces a base equal to 0, the area and hence the probability for a single exact point must also be 0. We can, however, find P($a \leqslant x \leqslant b$) because we now have an interval that produces a base of width $b - a$ to be used in the area calculation. Because each exact point in a continuum has probability 0, it does not matter whether intervals include or exclude the end points. We thus have the result that *for values chosen from a continuum*, P($a \leqslant x \leqslant b$), P($a < x < b$), P($a \leqslant x < b$), and P($a < x \leqslant b$) are all equal.

EXAMPLE 4-48 The weight of a compound that can be precipitated from a liter of solution can take any value between 0 and 0.10 gram. The amount precipitated is the result of a random trial with probabilities as indicated in Figure 4-8.

We can see that the graph is suitable as a probability graph. It does not cross below the horizontal scale and it has a total area of 1. [The area is found as the area of the full triangle—one-half the base times the height, that is, $(1/2) \times 0.10 \times 20 = 1$.]

The heights themselves cannot be probabilities since, to produce a total area of 1, the height of the triangle is 20, a number that exceeds 1 and thus cannot be a probability. The heights are probability densities as indicated on the vertical scale. The use of areas leads us to statements on probabilities about the precipitated weight *x*.

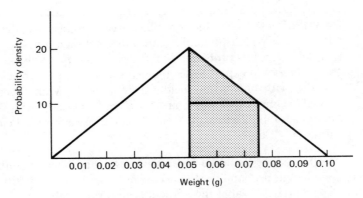

Figure 4-8 Probability graph for the weight of a precipitate.

We can see visually that the distribution is symmetric about 0.05 and, as a result, $P(x \leqslant 0.05) = P(0.05 \leqslant x) = 1/2$.

By partitioning the shaded region into a rectangle and triangle and finding their areas, we can calculate the shaded area to be $0.25 + 0.125 = 0.375$. We thus determine that $P(0.05 \leqslant x \leqslant 0.075) = 0.375$. We could have also excluded the end values from the interval to have $P(0.05 < x < 0.075) = 0.375$. Other probabilities are found in the same manner. ▲

PRACTICE PROBLEMS

P4-15 When a balanced coin is tossed five times, there are 32 equally likely outcomes such as *H T T H T*, denoting heads, then tails, then tails, then heads, then tails. The numbers of outcomes producing each of the total possible numbers of heads are:

Number of heads	Number of outcomes
0	1
1	5
2	10
3	10
4	5
5	1

(a) Draw a histogram/probability graph for the number of heads in five tosses of a coin.
(b) From the graph what can you see about $P(x \leqslant 2)$, where x is the number of heads?
(c) How does $P(x \leqslant 1)$ compare to $P(4 \leqslant x)$ (give a visual impression only)?

P4-16 A researcher has found that for intermediate-level subjects, the time to recognize a Roman numeral between one and ten and then to write down the corresponding Arabic numeral varies randomly from 1/2 second to 3/2 second, with probabilities as illustrated in Figure P4-1.

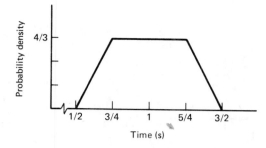

Figure P4-1 Probability graph for recognition times.

(a) Verify that the graph in Figure P4-1 is appropriate as a probability graph.
(b) If x denotes the random time for an individual subject, find
   (1) $P(x \leqslant 3/4)$
   (2) $P(1 < x)$
   (3) $P(3/4 < x < 5/4)$
   (4) $P(x < 2/3)$
   (5) $P(5/6 \leqslant x \leqslant 7/6)$

EXERCISES

4-30 When three dice are thrown together, there are 216 possible outcomes, producing totals from 3 to 18 with the following numbers of outcomes.

Total on dice	Number of outcomes
3	1
4	3
5	6
6	10
7	15
8	21
9	25
10	27
11	27
12	25
13	21
14	15
15	10
16	6
17	3
18	1

(a) Draw a histogram/probability graph for the results of throwing three dice.
(b) Without actually calculating probabilities, what can you determine from the graph about $P(x \leqslant 10)$?
(c) What can you see from the graph about $P(x \leqslant 5)$ as compared to $P(x \geqslant 16)$?
(d) Do totals of 3 or 18 appear from the graph to be usual values or rare values?

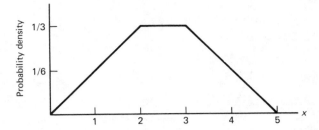

Figure 4-9 Probability graph.

4-31 (a) Verify that the graph in Figure 4-9 is appropriate as a probability graph.

(b) If a number is to be drawn at random from a continuum with probabilities as indicated by the graph, what is the probability that the number will be greater than 2?

4-32 When spun, a spinner is just as likely to stop at any given point on a circle as it is at any other. The circle has a circumference equal to 1 and any stopping point is designated by its distance measured clockwise around the circle from a fixed reference point.

(a) Draw a probability graph to illustrate probabilities for the outcomes resulting from spinning the spinner.

(b) Using the graph, and letting x denote the outcome, find

  (1) $P(x \leqslant 0.7)$

  (2) $P(0.4 \leqslant x \leqslant 0.6)$

  (3) $P(x \leqslant 0.95)$

4-33 If outcomes are picked from a continuum according to a random process, such that $P(x \leqslant 3) = 0.45$ and $P(4 \leqslant x) = 0.40$, find

(a) $P(x < 3)$

(b) $P(4 < x)$

(c) $P(3 \leqslant x \leqslant 4)$

4-34 Figure 4-10 is intended to be a probability graph. It is to provide a simple, yet reasonable, approximation to reality in determining probabilities on blood flow (milliliters per minute).

(a) Verify that the graph is appropriate as a probability graph.

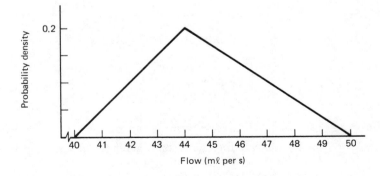

Figure 4-10 Probability graph for blood flow.

(b) Assuming that the graph is a reasonable approximation to reality and that x represents blood flow, find

(1) $P(x \leqslant 43.0)$
(2) $P(47.0 \leqslant x)$
(3) $P(43.0 < x < 47.0)$
(4) $P(42.5 \leqslant x \leqslant 47.5)$

SOLUTIONS TO PRACTICE PROBLEMS

P4-15 (a) The histogram/probability graph is illustrated in Figure P4-2.

(b) $P(x \leqslant 2)$ is seen from the symmetry to be $1/2$.

(c) They are equal.

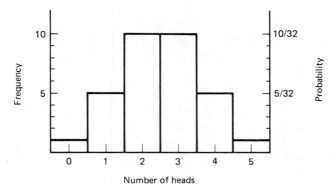

Figure P4-2 Histogram/probability graph for five tosses of a coin.

Number of heads

P4-16 (a) The graph does not cross below the horizontal scale. From $1/2$ to $3/4$, the graph produces a triangle with base $1/4$ and height $4/3$. From $3/4$ to $5/4$, it produces a rectangle with base $1/2$ and height $4/3$. From $5/4$ to $3/2$, it produces another triangle with base $1/4$ and height $4/3$. The total area is $[(1/2) \times (1/4) \times (4/3)] + [(1/2) \times (4/3)] + [(1/2) \times (1/4) \times (4/3)] = (1/6) + (2/3) + (1/6) = 1$, as required.

(b) (1) From the first triangle in part (a), we find $P(x \leqslant 3/4) = 1/6$.

(2) By symmetry, we see that $P(1 < x) = 1/2$.

(3) From the rectangle in part (a), we find $P(3/4 < x < 5/4) = 2/3$.

(4) From $1/2$ to $2/3$ the graph makes a triangle with base $(2/3) - (1/2) = 1/6$. Since the distance from $1/2$ to $2/3$ is $(1/6) \div (1/4) = 2/3$ of the distance across the first triangle in part (a), the corresponding height is $2/3$ of the total height or $(2/3) \times (4/3) = 8/9$. The area of the triangle from $1/2$ to $2/3$ is then $(1/2) \times (1/6) \times (8/9) = 2/27$. We then have $P(x < 2/3) = 2/27$.

(5) From $5/6$ to $7/6$, the graph makes a rectangle with height $4/3$, base $1/3$, and area $4/9$. We then have $P(5/6 \leqslant x \leqslant 7/6) = 4/9$.

4-6 SAMPLING DISTRIBUTIONS

In this section we introduce some of the intuitive concepts of sampling distributions. The general results that are stated will not be proven. We simply look at the behavior under different circumstances of two statistics from some small populations. The patterns that will develop from these examples will serve to introduce the concepts and to justify intuitively the general results of sampling distributions.

Throughout this section we will assume that, unless otherwise stated, samples are *random samples*. Specifically, samples are chosen at random so that each sample has an equal chance of being selected. We will further assume that sample members are chosen *without replacement*; that is, after we choose an individual item, we do not return it to the population for a possible repeat selection in the same sample. Finally, we will not distinguish between samples with regard to the order of selection. For example, the sample obtained by selecting the third then the seventh then the fifth members of a population will be considered to be the same as that obtained by selecting the seventh then the fifth then the third.

If we distinguish between samples with regard to order, the number of possible samples will increase; however, the probabilities relating to the statistics of interest to us will not change. Accordingly, the general results illustrated in this section will continue to be valid.

EXAMPLE 4-49   Consider a population of ten members having values: 3, 5, 4, 7, 3, 5, 5, 7, 6, and 5. Figure 4-11 illustrates a histogram for this population. A probability scale is included as well. This scale gives the probability of obtaining any given value when drawing a single member, at random, from this population. For example, $P(x = 7) = 2/10 = 1/5$.

We consider two parameters from this population, the mean and the variance (or standard deviation). They are $\mu = 5$ and $\sigma^2 = 1.8$ ($\sigma = 1.34$). For the variance, we could have also found the alternative population variance $S^2 = 2$.   ▲

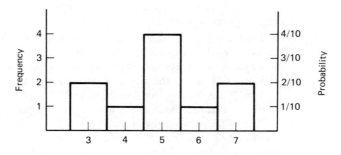

Figure 4-11   Histogram/probability graph for the population.

Sampling Distribution for the Sample Mean

Suppose that, as a sample, we draw two members at random from a population. As noted previously, in drawing each sample, we will not concern ourselves with the order in which the two members of the population are drawn. We will be concerned only with which two members we have. We may then consider that we have selected the two simultaneously.

EXAMPLE 4-50 With this sampling procedure, there are 45 equally likely possible samples of two members that we can draw from the population in Example 4-49: the first and the second members of the population, the first and the third, the first and the fourth,..., the ninth and the tenth. If we let x_1 represent the value of the first member in the sample and x_2 the value of the second, we can use (x_1, x_2) as the general representation of any sample.

Suppose that, for any sample, we are interested in one statistic, the mean $\bar{x} = (x_1 + x_2)/2$. The 45 possible samples and corresponding sample means are as follows:

Sample (x_1, x_2)	Mean $\bar{x}$	Sample (x_1, x_2)	Mean $\bar{x}$	Sample (x_1, x_2)	Mean $\bar{x}$
(3, 5)	4	(5, 6)	5.5	(3, 5)	4
(3, 4)	3.5	(5, 5)	5	(3, 5)	4
(3, 7)	5	(4, 7)	5.5	(3, 7)	5
(3, 3)	3	(4, 3)	3.5	(3, 6)	4.5
(3, 5)	4	(4, 5)	4.5	(3, 5)	4
(3, 5)	4	(4, 5)	4.5	(5, 5)	5
(3, 7)	5	(4, 7)	5.5	(5, 7)	6
(3, 6)	4.5	(4, 6)	5	(5, 6)	5.5
(3, 5)	4	(4, 5)	4.5	(5, 5)	5
(5, 4)	4.5	(7, 3)	5	(5, 7)	6
(5, 7)	6	(7, 5)	6	(5, 6)	5.5
(5, 3)	4	(7, 5)	6	(5, 5)	5
(5, 5)	5	(7, 7)	7	(7, 6)	6.5
(5, 5)	5	(7, 6)	6.5	(7, 5)	6
(5, 7)	6	(7, 5)	6	(6, 5)	5.5

The 45 means form a new population, the population of all possible means of samples of size $n = 2$ from the original population. Figure 4-12 is a histogram/probability graph for this new population, which is referred to as the *sampling distribution* for the mean of samples of size $n = 2$. ▲

When random samples of n members are taken from any population, the new population of all possible sample means $\bar{x}$ is referred to as *the sampling distribution for the mean of samples of size n*.

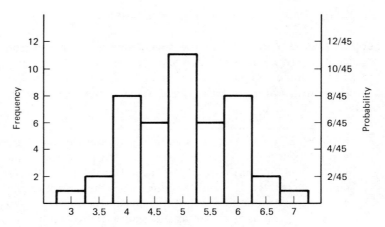

Figure 4-12 Histogram/probability graph for means of samples of size 2.

When we are interested in the sample means only, we can interpret each sample in two ways—as a sample of two members from the original population, or as a sample of one member from the population (sampling distribution) of sample means.

EXAMPLE 4-51 In Example 4-50, there are eight samples of two members of the original population with a mean of $\bar{x} = 4$. In the population of means (of samples of size 2), the value 4 has a frequency of 8. With either interpretation, $P(\bar{x} = 4) = 8/45$. ▲

Parameters for Sampling Distributions

We introduce notation for parameters of the sampling distribution as follows:

> The mean, variance, and standard deviation of this new population of $\bar{x}$'s are denoted by $\mu_{\bar{x}}$, $\sigma_{\bar{x}}^2$, and $\sigma_{\bar{x}}$.

EXAMPLE 4-52 For the sampling distribution in Example 4-50 we have a population of $N = 45$ $\bar{x}$'s which produce a sum $\sum \bar{x} = 225$ and a sum of squares $\sum \bar{x}^2 = 1161$. The parameters of interest are

$$\mu_{\bar{x}} = \frac{\sum \bar{x}}{N} = \frac{225}{45} = 5$$

$$\sigma_{\bar{x}}^2 = \frac{\sum \bar{x}^2}{N} - \mu_{\bar{x}}^2 = \frac{1161}{45} - 25 = 0.8$$

$$\sigma_{\bar{x}} = \sqrt{0.8} = 0.89$$ ▲

Suppose now that we consider taking random samples of size $n = 3$ from the original population. Again we consider only which population members are in the sample and not the order obtained. If we are again interested only in $\bar{x}$'s, we will notice that the change produces some similarities and some differences when compared to samples of size 2.

EXAMPLE 4-53 If we take samples of size $n = 3$ from the population in Example 4-49, there are 120, equally likely, possible samples: the first, second, and third members of the original population; the first, second, and fourth; ...; the eighth, ninth, and tenth. Using (x_1, x_2, x_3) as the general representation for the three values in the sample, the 120 samples are: $(3, 5, 4), (3, 5, 7), ..., (7, 6, 5)$. For each sample, we find the sample mean as $\bar{x} = (x_1 + x_2 + x_3)/3$. The 120 sample means corresponding to the 120 samples produce a new population. Figure 4-13 is a histogram/probability graph for this population and illustrates the sampling distribution for the mean of samples of size 3.

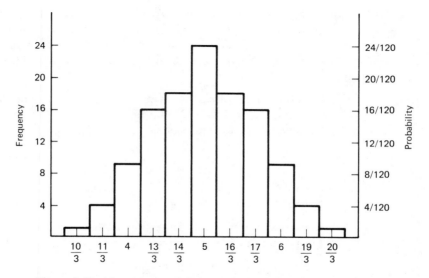

Figure 4-13 Histogram/probability graph for means of samples of size 3.

Again, when interested in the sample mean only, we can interpret the sample as a sample of three members of the original population or as a sample of one member of the population (sampling distribution) of sample means. The mean, variance, and standard deviation of this population of 120 $\bar{x}$'s are found to be $\mu_{\bar{x}} = 5$, $\sigma_{\bar{x}}^2 = 0.47$, and $\sigma_{\bar{x}} = 0.68$. ▲

Mean of Sample Means

Besides illustrating what is meant by a sampling distribution, the examples above also illustrate some general properties of the sampling distribution of the sample mean $\bar{x}$. Note that, in the examples, $\mu = 5$, $\mu_{\bar{x}} = 5$ for samples with $n = 2$, and $\mu_{\bar{x}} = 5$ for samples with $n = 3$ as well.

Regardless of the sample size, for random samples, the mean of the sampling distribution for the sample mean is always the population mean

$$\mu_{\bar{x}} = \mu$$

Since the mean $\mu_{\bar{x}}$ of the sampling distribution for the sample mean $\bar{x}$ is equal to the mean μ of the original population, the sample mean is said to be equal to the population mean "on the average." It is not biased toward values greater than μ nor values less than μ and is said to be an *unbiased estimator* of the population mean. We will thus use this statistic $\bar{x}$ to estimate the parameter μ in later inference procedures.

Variance of Sample Means and Effect of Sample Size

If we compare the variances for the original population and for the sampling distributions for the mean in the examples above, we see that $\sigma_{\bar{x}}^2$, the variance of the sample means, is smaller than σ^2, the variance of the original population. Furthermore, $\sigma_{\bar{x}}^2$ becomes smaller as n, the sample size, becomes larger. These facts result from the following:

For random samples chosen without replacement from a finite population of size N, the variance of the sampling distribution for the mean of samples of size n is

$$\sigma_{\bar{x}}^2 = \frac{\sigma^2}{n} \times \frac{N - n}{N - 1}$$

where $(N - n)/(N - 1)$ is a *finite population correction* factor.

As N increases to infinity, the finite population correction factors increase toward 1 and will be close to 1 if N is large and if n is small relative to N.

For random samples from a very large population, the variance of the sampling distribution for the mean of samples of size n approaches

$$\sigma_{\bar{x}}^2 = \frac{\sigma^2}{n}$$

as N increases.

The result $\sigma_{\bar{x}}^2 = \sigma^2/n$ applies for random samples of size n from an infinite population or for random samples of size n taken *with replacement* from a finite population.

Since larger sample sizes produce smaller variances, the sampling distributions based on larger values of n will have more frequency (probability) clustered near the mean $\mu_{\bar{x}} = \mu$ and, therefore, with larger n, we have a higher probability of obtaining a sample mean $\bar{x}$ close to the population mean μ. Accordingly, we choose sample sizes as large as are practical.

EXAMPLE 4-54 With the examples above, if $n = 2$, the probability that $\bar{x}$ differs from μ by less than 1 is $P(\mu - 1 < \bar{x} < \mu + 1) = P(4 < \bar{x} < 6) = 23/45 = 0.51$; whereas, if $n = 3$, $P(\mu - 1 < \bar{x} < \mu + 1) = 92/120 = 0.77$. ▲

Further illustrations of the properties described above, as well as others, can be obtained by considering samples from two slightly larger populations, one symmetric and the other skewed.

EXAMPLE 4-55 Consider first a symmetric population consisting of the 20 values: 1, 2, 2, 3, 4, 4, 4, 5, 5, 5, 5, 5, 5, 6, 6, 6, 7, 8, 8, and 9. Parameters of this population are: $\mu = 5$, $\sigma^2 = 4.1$, and $S^2 = 4.316$. If we take samples of size 3, we will find 1140 possible samples and the corresponding population of sample means produces $\mu_{\bar{x}} = 5$, which is μ, and $\sigma_{\bar{x}}^2 = 1.223$, which is equal to

$$\frac{\sigma^2}{n} \times \frac{N - n}{N - 1} = \left(\frac{4.1}{3}\right) \times \frac{20 - 3}{20 - 1} = 1.223$$

If we take samples of size 5, we will find 15,504 possible samples and the corresponding population of sample means produces $\mu_{\bar{x}} = 5$, which is μ, and $\sigma_{\bar{x}}^2 = 0.647$, which is $(\sigma^2/n) \times [(N - n)/(N - 1)]$, where $N = 20$ and $n = 3$.

Consider next a skewed distribution consisting of the 20 values: 1, 2, 2, 3, 3, 3, 3, 4, 4, 4, 4, 4, 5, 5, 5, 6, 6, 7, 8, and 9. Parameters for this population are $\mu = 4.4$, $\sigma^2 = 3.94$, and $S^2 = 4.147$. The 1140 samples of size 3 produce sample means with $\mu_{\bar{x}} = 4.4$, which is again μ, and $\sigma_{\bar{x}}^2 = 1.175$, which is again $(\sigma^2/n) \times (N - n)/(N - 1)$, where $N = 20$ and $n = 3$. The 15,504 samples of size 5 produce sample means with $\mu_{\bar{x}} = 4.4$ (again μ) and $\sigma_{\bar{x}}^2 = 0.622$ (again $(\sigma^2/n) \times [(N - n)/(N - 1)]$, where $N = 20$ and $n = 5$).

The two populations, together with the sampling distributions for $\bar{x}$ for $n = 3$ and $n = 5$, are illustrated in Figures 4-14 and 4-15 with profiles of frequency/probability histograms. Smooth curve approximations are also shown for the sampling distributions. ▲

Figures 4-14 and 4-15 demonstrate general sampling results:

The sampling distribution for $\bar{x}$ becomes more compact and concentrates more on μ as the sample size n increases. Graphs for the sampling distribution are very close to smooth curves which, even for skewed populations, approach a symmetric bell-shaped form as n increases.

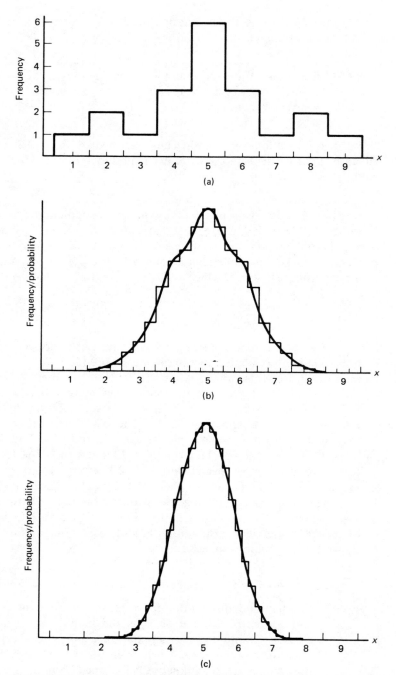

Figure 4-14(a) Original symmetric population. (b) Sampling distribution for the mean of samples of size 3 from the symmetric population. (c) Sampling distribution for the mean of samples of size 5 from the symmetric population.

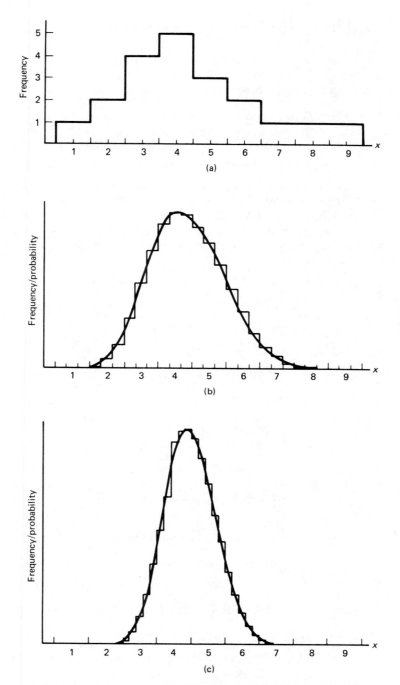

Figure 4-15(a) Original skewed population. (b) Sampling distribution for the mean of samples of size 3 from the skewed population. (c) Sampling distribution for the mean of samples of size 5 from the skewed population.

Sampling Distribution for the Sample Variance

In many instances of sampling, we will be interested in a second statistic, the sample variance s^2. As with the sample mean, we consider the sample variances from all possible samples and these form a new population.

> When random samples of n members are taken from any population, the new population of all possible sample variances s^2 is referred to as *the sampling distribution for the variance of samples of size n.*

EXAMPLE 4-56 If we consider the 45 samples of size 2 in Example 4-50 and the 45 corresponding sample variances instead of the sample means, we obtain a new population or sampling distribution for all possible values of s^2. The 45 sample variances for the samples as listed in Example 4-50 are:

2.0	0.5	0.5	0.0	2.0
0.5	2.0	0.5	0.5	0.5
8.0	2.0	0.5	2.0	0.0
0.0	0.0	4.5	2.0	2.0
2.0	0.0	2.0	2.0	0.5
2.0	2.0	0.5	8.0	0.0
8.0	0.5	8.0	4.5	0.5
4.5	0.0	2.0	2.0	2.0
2.0	4.5	2.0	0.0	0.5

This new population of $N = 45$ s^2's has sum $\Sigma s^2 = 90$ and sum of squares $\Sigma (s^2)^2 = 408$. The mean and variance of this population are

$$\mu_{s^2} = \frac{\Sigma s^2}{N} = \frac{90}{45} = 2.0 \quad \text{(which is the value for } S^2 \text{ of the original population)}$$

$$\sigma_{s^2}^2 = \frac{\Sigma (s^2)^2}{N} - \mu_{s^2}^2 = \frac{408}{45} - 4.0 = 5.07$$

This sampling distribution is illustrated in Figure 4-16(a). ▲

EXAMPLE 4-57 If we consider the sample variances for samples of size 3 as discussed in Example 4-53, we obtain the sampling distribution illustrated in Figure 4-16(b). This distribution has mean $\mu_{s^2} = 2$ (which is again the value of S^2 for the original population) and variance $\sigma_{s^2}^2 = 2.04$. ▲

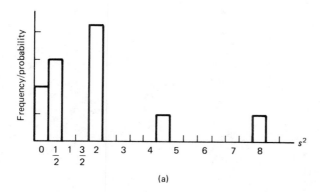

(a)

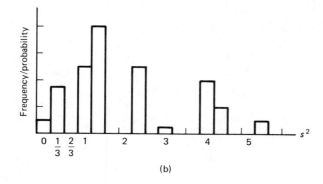

(b)

Figure 4-16(a) Sampling distribution for s^2 when $n = 2$. (b) Sampling distribution for s^2 when $n = 3$.

EXAMPLE 4-58 The sample variances from the 1140 samples of size 3 from the first population in Example 4-55 produce the sampling distribution shown in Figure 4-17(a) with mean $\mu_{s^2} = 4.136$ (which is the value of S^2 for the original population) and variance $\sigma_{s^2}^2 = 32.952$.

The 15,504 samples of size 5 produce the sampling distribution for s^2 shown in Figure 4-17(b) with mean $\mu_{s^2} = 4.136$ (again S^2) and variance $\sigma_{s^2}^2 = 6.294$. ▲

Mean of Sample Variances

The preceding examples exhibit properties which are again general results.

> When sampling at random without replacement from finite populations, the mean of the sampling distribution for the sample variance is $\mu_{s^2} = S^2$.

For infinite populations, there is only one variance, σ^2. For very large populations for which $(N - 1)/N$ is almost 1, σ^2 and S^2 are almost identical since

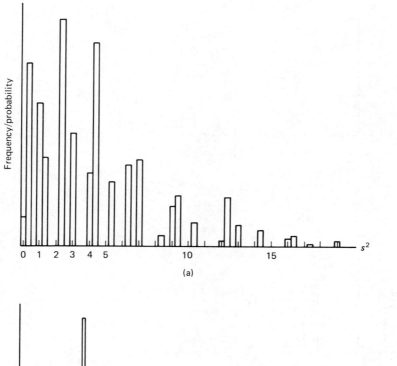

(a)

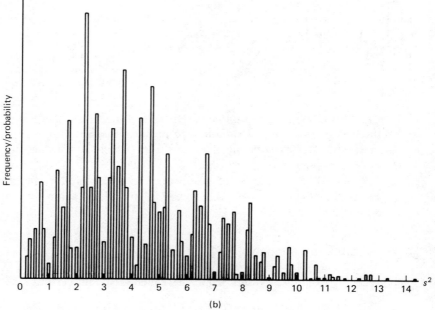

(b)

Figure 4-17(a) Sampling distribution for s^2 when $n = 3$. (b) Sampling distribution for s^2 when $n = 5$.

they are related through the equation $\sigma^2 = S^2 \times (N-1)/N$. In these cases we modify the general result.

> For random samples from very large (or infinite) populations, the mean of the sampling distribution for the sample variance is
>
> $$\mu_{s^2} = \sigma^2$$

This result also holds for random samples taken *with replacement* from finite populations.

Although the general expression for $\sigma_{s^2}^2$ is much too complicated for us to consider, the preceding examples do provide illustrations of a pattern that holds true in general.

> As the sample size is increased, the variance of the sampling distribution for the sample variance decreases.

Other Statistics

Statistics other than the sample mean and variance produce different sampling distributions. As we encounter other statistics of interest, we will consider their sampling distributions.

We will make use of these distributions in inference procedures. From the probabilities involved in the sampling distributions for a statistic, we may determine, for example, what values are likely to occur. We can also determine how confident we may be that a given known statistic is within a particular distance of a corresponding unknown parameter, and we may then make inferences about the unknown parameter accordingly.

PRACTICE PROBLEMS

P4-17 The remaining numbers of capsules in each of eight bottles are respectively: 9, 7, 8, 6, 9, 8, 9, and 10. Consider this collection of bottles to be a small population.
(a) Calculate μ, σ^2, and S^2 for the corresponding population of numbers of capsules.
(b) List all possible samples of size 2. (There are 28 such samples on the assumption that order is not important and that the members of the sample are chosen without replacement.) For each sample, calculate $\bar{x}$ and s^2.
(c) Calculate $\mu_{\bar{x}}$, $\sigma_{\bar{x}}^2$, and μ_{s^2} for the sampling distributions generated by the samples in part (b) and confirm that they are related to μ, σ^2 and S^2 as they should be.

P4-18 In each of the 365 days of one calendar year, the number of telephone calls placed between two communities was recorded. The resulting 365 numbers produced a population for that year with a mean $\mu = 6506$ and a standard deviation $\sigma = 1128$. The records of n days are to be selected as a sample and, for this sample, two statistics, $\bar{x}$ and s^2, are to be calculated.

(a) What is the mean for each of the sampling distributions for possible values of the two statistics?

(b) What is the variance of the sampling distribution for $\bar{x}$

   (1) if $n = 10$?

   (2) if $n = 25$?

4-35 The weights in grams of a population of 1000 specimens of experimental material produced mean $\mu = 150.00$ and variance $\sigma^2 = 625.00$ or $S^2 = 625.63$. In the experimentation process, samples of the specimens are to be taken for analysis.

(a) If all possible samples of size 30 were considered and the means of these samples calculated, what would be the mean $\mu_{\bar{x}}$ and variance $\sigma_{\bar{x}}^2$ of all of these $\bar{x}$'s?

(b) If the sample variances were calculated for the samples in part (a), what would be the mean μ_{s^2} of all of these s^2's?

4-36 Repeat Exercise 4-35 for samples of size 100.

4-37 Suppose that an infinitely large population has a variance of $\sigma^2 = 315$.

(a) What is the variance of the sampling distribution for the mean of samples of size 25?

(b) What is the mean of the sampling distribution for the variance of samples of size 25?

4-38 The numbers of previous conditioning sessions experienced by a population of eight laboratory rats consists of one 1, three 2's, two 3's, one 4, and one 5.

(a) Draw a histogram/probability graph for the numbers of sessions experienced by this population.

(b) Find μ, σ^2, and S^2 for the population.

(c) If all possible samples of size 2 are taken, what should be the mean $\mu_{\bar{x}}$ and variance $\sigma_{\bar{x}}^2$ for the sample means and the mean μ_{s^2} for the sample variances?

(d) Repeat part (b) for samples of size 3.

4-39 (a) Enumerate all of the samples in Exercise 4-38(c) (there are 28) and calculate $\bar{x}$ and s^2 for each sample. Verify that $\mu_{\bar{x}}$, $\sigma_{\bar{x}}^2$, and μ_{s^2} are what they should be by calculating them directly from the sampling distributions for $\bar{x}$ and s^2. Draw histograms/probability graphs for the sampling distributions.

(b) Enumerate all of the samples in Exercise 4-38(d) (there are 56) and, for these samples, repeat the procedures in part (a).

4-40 A titration apparatus is such that the (hypothetical) population of all of the (infinitely many) possible determinations that might be made on a single compound has a mean μ equal to the true compound level and a variance $\sigma^2 = 0.050$. For an individual compound, n sample titrations are to be performed and the sample mean $\bar{x}$ is to be used as an estimate of μ.

(a) What is the variance of the sampling distribution for $\bar{x}$ if

   (1) $n = 2$?

   (2) $n = 5$?

   (3) $n = 10$?

(b) How large must n be to reduce the variance of the sampling distribution to 0.002?

P4-17 (a) For this population we have $N = 8$, $\Sigma x = 66$, and $\Sigma x^2 = 556$. We then calculate $\mu = 66/8 = 8.25$, $\sigma^2 = 556/8 - 8.25^2 = 1.4375$, and $S^2 = (556 - 66^2/8)/7 = 1.6429$.

(b) We can systematically obtain all 28 samples of size 2 by pairing each number in the population with each other number that follows it in the list. We then find each possible pair (x_1, x_2) and the corresponding $\bar{x}$ and s^2 as follows, where $\bar{x} = (x_1 + x_2)/2$ and s^2 can be found quite simply from the defining formula as

$$\frac{(x_1 - \bar{x})^2 + (x_2 - \bar{x})^2}{2 - 1} = (x_1 - \bar{x})^2 + (x_2 - \bar{x})^2$$

(x_1, x_2)	$\bar{x}$	s^2	(x_1, x_2)	$\bar{x}$	s^2
(9, 7)	8	2.0	(8, 9)	8.5	0.5
(9, 8)	8.5	0.5	(8, 8)	8	0.0
(9, 6)	7.5	4.5	(8, 9)	8.5	0.5
(9, 9)	9	0.0	(8, 10)	9	2.0
(9, 8)	8.5	0.5	(6, 9)	7.5	4.5
(9, 9)	9	0.0	(6, 8)	7	2.0
(9, 10)	9.5	0.5	(6, 9)	7.5	4.5
(7, 8)	7.5	0.5	(6, 10)	8	8.0
(7, 6)	6.5	0.5	(9, 8)	8.5	0.5
(7, 9)	8	2.0	(9, 9)	9	0.0
(7, 8)	7.5	0.5	(9, 10)	9.5	0.5
(7, 9)	8	2.0	(8, 9)	8.5	0.5
(7, 10)	8.5	4.5	(8, 10)	9	2.0
(8, 6)	7	2.0	(9, 10)	9.5	0.5

(c) Since $\Sigma \bar{x} = 231$ and $\Sigma \bar{x}^2 = 1923$, we find that $\mu_{\bar{x}} = 231/28 = 8.25 = \mu$, as it should, and $\sigma_{\bar{x}}^2 = 1923/28 - (8.25)^2 = 0.6161$, which is equal to $(\sigma^2/n) \times [(N - n)/(N - 1)] = (1.4375/2) \times [(8 - 2)/(8 - 1)] = 0.6161$, as it should be. Since $\Sigma s^2 = 46$, we find that $\mu_{s^2} = 46/28 = 1.6429 = S^2$, as it should.

P4-18 (a) The means for the sampling distributions for $\bar{x}$ and s^2 are $\mu_{\bar{x}} = \mu = 6506$ and $\mu_{s^2} = S^2 = \sigma^2 \times N/N - 1 = 1128^2 \times 365/364 = 1,275,879.56$.

(b) Since $\sigma_{\bar{x}}^2 = \dfrac{\sigma^2}{n} \times \dfrac{N - n}{N - 1}$, we have

(1) $\sigma_{\bar{x}}^2 = (1128^2/10) \times (355/364) = 124,092.4$ if $n = 10$

(2) $\sigma_{\bar{x}}^2 = (1128^2/25) \times (340/364) = 47,539.62$ if $n = 25$.

4-7   CHAPTER REVIEW

4-41 In general, P(A *or* B) will equal _____.

4-42 If P(E) = 0.60, the odds against E are

(a) 3 to 2   (c) 2 to 3

(b) 10 to 6   (d) 6 to 1

4-43 For continuous data, $P(x < 3) = P(x \leqslant 3)$.
 (a) True
 (b) False
 (c) Depends on the particular data

4-44 If we take all possible samples of a given size from a population, the variance of the means of these samples will be
 (a) greater than the variance of the population
 (b) equal to the variance of the population
 (c) less than the variance of the population
 (d) there is no such thing as a variance of means

4-45 If A, B, and C are mutually exclusive, $P(A\ or\ B\ or\ C)$ will equal _____.

4-46 The sum of individual probabilities is always 1 for events that are ___ and ___.

4-47 In a probability graph, probability is represented by _____.

4-48 In words, $P(A|B)$ means _____.

4-49 In general, $P(A\ and\ B)$ will equal _____.

4-50 If we calculate probabilities such that A has a 30% chance of winning a championship, B has twice A's chance, and C has twice B's chance, we should
 (a) give odds of 10 to 3 against A
 (b) give odds of 12 to 3 for C
 (c) give odds of 3 to 2 for B
 (d) recalculate our probabilities

4-51 If A and B are mutually exclusive, $P(A\ and\ B) = P(A|B)$.
 (a) True
 (b) False

4-52 If A, B, and C are independent, $P(A\ and\ B\ and\ C)$ will equal _____.

4-53 Suppose that one card is to be drawn at random from a well-shuffled ordinary deck of 52 playing cards. What is the probability that the card will be
 (a) a "face" card (jack through ace)?
 (b) a red card?
 (c) a "face" card and a red card?
 (d) a "face" card or a red card?
 (e) neither a "face" card nor a red card?

4-54 What, if anything, is wrong with each of the following statements?
 (a) A, B, C, and D are events for a random trial such that A, B, and C are mutually exclusive and exhaustive and $P(D) = P(A\ and\ D) + P(B\ and\ D) + P(C\ and\ D)$.
 (b) A and B are events for a trial such that $P(A) = 1/5$, $P(B) = 2/3$, and $P(A\ or\ B) = 14/15$.
 (c) A, B, and C are mutually exclusive and exhaustive events such that $P(A\ or\ B) = 0.70$, $P(A\ or\ C) = 0.75$, and $P(A) = 0.30$.
 (d) E is an event such that $P(E) = 59/53$.
 (e) A and B are exhaustive events such that $P(A) = 0.60$ and $P(B) = 0.75$.

4-55 A, B, and C are three exhaustive events that can occur singly or in pairs, but not all together. Each pair has probability $1/8$ of occurring. The odds against A are 3 to 1, and the odds against B are 3 to 5. Find the odds against C.

4-56 Suppose that a player is dealt two cards from an ordinary deck. A third card is to be dealt to her and if it has a value between the values of the other two, she will win.

(a) (1) If the first two cards are 7, Q, what is the probability that she will win?
(2) Repeat part (1) with 8, J.
(b) Repeat part (a) assuming that in the course of play two 2's, one 3, two 4's, one 5, one 6, two 7's, two 9's, one J, one Q, one K, and three A's have been previously removed from the deck.

4-57 Past records indicate that 60% of complaints against doctors are settled on the basis of an initial investigation by the College of Physicians and Surgeons, 32% are settled by a review board, and 8% are referred to a disciplinary hearing. Claimants are generally satisfied in 84% of cases settled by initial investigation, 56% of those settled by review board, and only 30% of those referred to a disciplinary hearing. On the basis of past records, what is the probability that a complaint will be

(a) settled by a review board or referred to a disciplinary hearing?
(b) settled in an initial investigation or to the claimant's satisfaction?
(c) settled by a review board and to the claimant's satisfaction?
(d) referred to a disciplinary hearing but not settled to the claimant's satisfaction?

4-58 If 65% of respondents to an election survey are committed in their planned votes, what are the odds against an individual respondent's being uncommitted?

4-59 The membership of a single parents' association is 63% women and 37% men. Of the women, 58% are divorced, 27% are widowed, and 15% were never married. Of the men, 60% are divorced, 39% widowed, and 1% never married.

(a) What is the probability that a member chosen from this club will be
(1) a divorced woman?
(2) a man who has never been married?
(3) widowed?
(b) What is the probability that a divorced member is a man?
(c) What is the probability that a member who has never been married is a woman?

4-60 Surveys indicate the following analysis of families traveling on the weekend on a main highway throughout a vacation area.

Classification	Percentage
Area resident	12
Domestic tourist owning property	34
Domestic tourist renting accommodation	26
Foreign tourist owning property	4
Foreign tourist renting accommodation	10
Domestic traveler not stopping	11
Foreign traveler not stopping	3

What is the probability that a family traveling on the highway will be
(a) foreign?
(b) tourist renting accommodation?
(c) not stopping?
(d) area resident or property owner?

4-61 Suppose that a missile is subject to random effects after being fired such that it can strike the target or miss by a distance of up to 50 m. The distance between the target and the landing point is subject to the triangular probability graph illustrated in Figure 4-18.

(a) Confirm that this graph satisfies the required area conditions for a probability graph.

(b) Find the probability that the missile misses the target by a distance greater than 30 m.

(c) Find the probability that the missile lands at a distance of 20 m or less from the target.

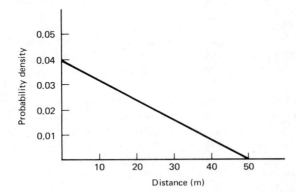

Figure 4-18 Probability density for missile distance.

4-62 Past experience indicates that 93% of students who consistently do the assigned work through the year obtain a passing grade, whereas only 30% who do not do the work obtain a passing grade.

(a) If a student is picked at random from a class in which 90% did the work, what is the probability that he or she will be one of the students to obtain a failing grade?

(b) If a student selected at random from a class in which 90% did the work consistently, obtained a failing grade, what is the probability that she or he did not do the work?

4-63 An investigator has been told that the number of service personnel on call in nonbusiness hours differs for the seven days of the week but is, on the average, four. The investigator plans to check the claim by choosing two days from the next seven as a sample. The numbers of personnel who will be on call through the seven days of the next week form a small population that consists of the following members: 3, 6, 3, 4, 4, 5, 3.

(a) (1) Find the mean μ to confirm the claim.

(2) Find the variances σ^2 and S^2.

(b) (1) Enumerate all possible samples of size 2 (there are 21) and, for each, find the sample mean.

(2) Using the results of part (b) (1), find the mean and variance for the means of samples of size 2.

(3) If a sample of size 2 is to be drawn from the original population, find $P(|\bar{x} - \mu| \geqslant 1.0)$, that is, find the probability that the sample mean will differ from μ (above or below) by at least 1.0.

(c) Repeat part (b) for all possible samples of size 3 (there are 35).

4-64 By our convention, μ is the symbol for _____.

4-65 In a probability graph the area under the curve between -0.7 and 0.3 represents _____.

4-66 If a member of a population has a score that is 1 standard deviation above the mean and everybody's score is increased by 10, the member's new score will be
(a) 1 standard deviation plus 10 above the new mean
(b) 1 standard deviation minus 10 above the new mean
(c) 1 standard deviation above the new mean

4-67 If A, B, and C are exhaustive, $P(A$ or B or $C)$ will equal _____.

4-68 To give a population a variance of 1, all of the values in the population must be divided by _____.

4-69 If $P(A) = 0.39$, $P(B) = 0.21$, and $P(C) = 0.28$, then A, B, and C
(a) may be mutually exclusive and exhaustive
(b) must be mutually exclusive
(c) must be exhaustive
(d) none of the above

4-70 A standard deviation of -3000
(a) indicates that the data are very spread out
(b) indicates that the data are all negative
(c) is a mistake
(d) none of the above

4-71 The total area under a probability graph is always _____.

4-72 If $\mu_{\bar{x}}$ is the mean of all $\bar{x}$'s from samples of size n from a population with mean μ and variance σ^2, then
(a) $\mu_{\bar{x}} = 0$
(b) $\mu_{\bar{x}} = \sigma/\sqrt{n}$
(c) $\mu_{\bar{x}} = \mu$
(d) $\mu_{\bar{x}} =$ depends on n

4-73 A cumulative "or more" distribution is based on boundaries or _____ limits.

4-74 If two events are mutually exclusive,
(a) they are independent
(b) they may be independent
(c) they are not independent
(d) the sum of their probabilities must always equal 1

4-75 If A has value 67 in a population with mean 50 and variance 64, and B has value 73 in a population with mean 54 and variance 100,
(a) A is more extreme
(b) B is more extreme
(c) A and B cannot be compared

4-76 The ages of the members of a club form a population that has a mean of 60 and a variance of 144. If a member of this club is selected at random, what can you say about the probability that this member's age will be between 42 and 78 inclusive? (*Hint*: Use Chebyshev's theorem.)

4-77 Suppose that two yellow dice, one red die, and one green die are concealed in a bag. One is drawn at random and rolled. The score obtained is equal to that showing on the die if it is yellow, one more than that on the die if it is red, and two more than that on the die if it is green.

(a) Plot a histogram of the number of ways of obtaining a score of 1, 2, 3, 4, 5, 6, 7, or 8.

(b) Find the mean score from this frequency distribution.

(c) Find the probability of obtaining
    (1) a score less than 5
    (2) a score of 6
    (3) a score equal to that showing on the die
    (4) an even score

4-78 The faculty of a particular university consists of 123 full professors, 247 associate professors, 186 assistant professors, and 28 lecturers. The mean salary for full professors is \$39,730. For associate professors, assistant professors, and lecturers, the mean salaries are \$29,080, \$23,305, and \$18,720. Find the mean salary of all members of faculty.

4-79 Suppose that four east–west streets A, B, C, and D intersect with four north–south streets W, X, Y, and Z and that the numbers of accidents recorded in a given year at the intersections are as indicated in the diagram.

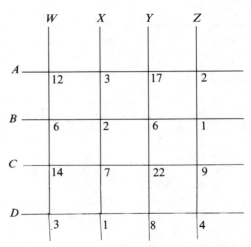

(a) If the record of one accident is selected at random, what is the probability that it occurred at an intersection along street A?

(b) What is the probability that it occurred at an intersection along street B or street X?

4-80 The following data represent the weights in grams of the iron content in a number of water samples from a spring:

374	329	388	305	327	309	309	240
322	311	323	363	260	307	256	288
358	288	379	310	301	355	252	319
243	345	361	332	369	247	358	265
403	323	329	340	311	277	329	358

(a) Find the mean and median weights.

(b) Find the range and the variance.

(c) Form a frequency distribution for these data.

(d) For this distribution, give the class boundaries, class limits, and class marks.

(e) Draw a frequency polygon of the distribution.

(f) Construct an "or more" cumulative distribution from this frequency distribution.

4-81 Suppose that after studying the four teams A, B, C, and D in a tournament, a fan believes that the odds are 2 to 1 against A as the winner of the tournament, 4 to 1 against B, 4 to 1 against C, and 5 to 1 against D.

(a) Do the odds satisfy the rules for probability?

(b) If the fan believes that the odds are correct for A and D and that B and C have equal chances of winning, what odds should he or she assign to B and to C?

(c) If C is eliminated from the tournament and the probabilities for A, B, and D remain in the same proportion to each other as in part (b), what are the new odds against winning for each of these three teams?

4-82 (a) (1) If a population has a variance of 100, what (using Chebyshev's theorem) is the minimum proportion of the population that is within 15 of the mean?

(2) If we take one value at random from the population in part (1), what is the minimum probability that it will be within 15 of the mean?

(b) (1) If we consider samples of size 4 from the population in part (a), what is the variance of the sampling distribution for the sample means?

(2) If we choose one sample of 4 and calculate the sample mean, what is the minimum probability that it will be within 15 of the mean?

(c) Repeat part (b) for samples of size 16.

4-83 The following figures represent the numbers of work stoppages affecting the various regions of the United States in 1973.

New England	264	East South Central	424
Middle Atlantic	1211	West South Central	205
East North Central	1436	Mountain	166
West North Central	349	Pacific	468
South Atlantic	1165		

Place the regions in order according to the numbers of work stoppages and, using this ordering, illustrate the data with a bar graph.

4-84 In order to investigate the mean length of time that employees of a corporation have been employed by the corporation, a sample of the lengths of times for a few employees is to be considered. The mean of this sample is to be used as an indicator of the mean for the corporation.

(a) What is the mean of all possible values of the sample means?

(b) If the corporation has 3500 employees whose lengths of time produce a variance $\sigma^2 = 50$, what will be the variance of the means of all possible samples of size 25?

(c) If the corporation is as indicated in part (b), what is the minimum probability that the sample mean will differ from the corporation mean by at most 4?

PROBABILITY DISTRIBUTIONS AND FURTHER SAMPLING DISTRIBUTIONS

5-1 INTRODUCTION

In Chapter 4 the basic ideas and notation of probability were introduced together with a preliminary discussion of sampling distributions. In order to make more use of probability, we must study further the methods for finding probabilities for various situations. To do so, we must consider general methods for determining the numbers of outcomes producing the events. We develop such methods in this chapter and use them to introduce a new concept, a probability function, which is a general expression used to determine probabilities for special types of trials with numerical outcomes. Using the ideas of probability functions, we also study more results for sampling distributions.

5-2 ENUMERATING EVENTS

Tree Diagrams

In the probability examples considered in Sections 4-2, 4-3, and 4-4, the numbers of outcomes and events were enumerated quite easily. In more complicated examples, a systematic approach will be required. In some cases, complication arises because an experiment or trial consists of a number of "subexperiments" or "subtrials."

One method of enumeration is to use a *tree diagram*, which is formed by constructing a "tree" consisting of sets of "branches" such that each subtrial of a trial has a number of branches equal to the number of possible outcomes.

EXAMPLE 5-1 An experiment consists of tossing three coins and we wish to enumerate all of the possible outcomes for this experiment. We consider the possible outcomes of the three subexperiments—the individual tosses of the three coins—and we develop a tree as follows. The first coin can result in heads (*H*) or tails (*T*), producing two branches, as shown in Figure 5-1. For each of these two results, the second coin can produce two outcomes, *H* or *T*, producing two branches from each of the previous two branch ends. We now have four branch ends. For each of these, the third coin produces two branches to produce, finally, the complete (horizontal) tree. The eight paths through the branches give the eight possible outcomes: *HHH, HHT, HTH, HTT, THH, THT, TTH,* and *TTT*. ▲

As illustrated in the following examples, branch paths do not always end at the same stage of the branching process, and numbers of branches may vary.

EXAMPLE 5-2 Alexander, denoted as *A*, and Bradshaw, denoted as *B*, are involved in a contest such that the first to win two games wins the contest. We wish to determine the numbers of possible two-game and three-game contests, assuming that there can be no tie games. We let each game be a subexperiment and, labeling branches according to the winners of the games, we produce the tree diagram in Figure 5-2. A square indicates Alexander as the contest winner and a circle indicates Bradshaw as the contest winner.

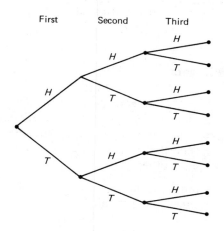

Figure 5-1 Tree diagram for three coins.

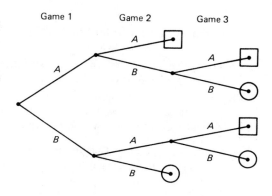

Figure 5-2 Tree diagram for the contest.

The six paths through the branches indicate the six possible contests and provide the following information: two of the contests consist of two games; four consist of three games; three of the contests produce Alexander as the winner; three produce Bradshaw as the winner. ▲

EXAMPLE 5-3 The three salespersons in a department are required to share responsibilities for aisles *I*, *II*, and *III*. They may be spread over different aisles or all congregated in one aisle. The area supervisor is not concerned with who is in which aisle but is interested in the number of ways the salespersons may be dispersed among the aisles as regards how many are in each aisle at any given time. The tree diagram in Figure 5-3 illustrates that there are ten possible ways of having the three salespersons spread across aisles *I*, *II*, and *III*. The branch labels indicate the numbers of salespersons in the aisles. ▲

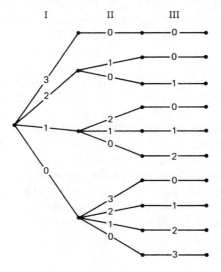

Figure 5-3 Tree diagram for salespersons in aisles.

General Rules for Independent Subexperiments

Although tree diagrams provide a systematic way to enumerate possibilities, they can become very complicated. In cases involving independent subexperiments, simple mathematical expressions are available to produce the numbers of outcomes of experiments. *Independent* subexperiments are such that the outcome in one subexperiment has no effect on the possible outcomes of others.

> If an experiment consists of k independent subexperiments, each with n possible outcomes, the total number of possible outcomes is n^k.

EXAMPLE 5-4 With three coins, we had three subexperiments, each with two possible outcomes, and, by doubling the number of branches at each stage, we obtained $2 \times 2 \times 2 = 2^3 = 8$ possible outcomes. Similarly, a tree diagram for four dice would produce $6 \times 6 \times 6 \times 6 = 6^4 = 1296$ paths, corresponding to the number of possible outcomes. ▲

> If the numbers of possible outcomes of k subexperiments are $n_1, n_2, \ldots, n_k$, the total number of outcomes is $n_1 \times n_2 \times \cdots \times n_k$.

EXAMPLE 5-5 If we toss two coins, roll a die, and draw a card from a deck of 52 cards, the total number of possible outcomes is $2 \times 2 \times 6 \times 52 = 1248$.

▲

Permutations

In some cases it is appropriate to consider all the possible orders of a set of objects.

EXAMPLE 5-6 Suppose that we shuffle part of an ordinary deck of cards—only the 13 spades—and that we wish to determine the number of different ways that the cards can be arranged in order as a result of the shuffle. There are 13 possible outcomes for the first card in the arrangement. For each possible first card, there are 12 cards remaining as possible outcomes for the second card. For each of these $13 \times 12 = 156$ selections of the first two cards, there are 11 possible outcomes for the third card; and so on. Continuing this process to produce a multiplication such as found in the counting rule above, we obtain the total number of ordered arrangements as $13 \times 12 \times 11 \times \cdots \times 3 \times 2 \times 1 = 6{,}227{,}020{,}800$. This number, written 13!, is referred to as *13 factorial* and is equal to the number of *permutations* of 13 objects. ▲

> The number of different ordered arrangements or *permutations* of n distinct objects is *n factorial* $= n! = n \times (n - 1) \times (n - 2) \times \cdots \times 3 \times 2 \times 1$. (*Note*: $0! = 1$.)

EXAMPLE 5-7 A psychologist is planning an experiment in which each subject must try each of four different stimulus–response schemes. In order to balance influences of order, the psychologist wishes to have one subject for each possible order in which the schemes can be presented. The number of subjects required is then the total number of possible scheme orders or $4! = 4 \times 3 \times 2 \times 1 = 24$.

▲

Sometimes it will be appropriate to consider selecting and ordering only some of the items available.

EXAMPLE 5-8 Suppose that we wish to determine the number of possible ordered arrangements that can result from drawing five cards from a full deck of 52 cards. There are 52 possible outcomes for the first card, then 51 for the second, $\ldots$, then 48 for the fifth. The total number of different arrangements or permutations is $52 \times 51 \times 50 \times 49 \times 48 = 311{,}875{,}200$. ▲

If we select x distinct objects from a total collection of n distinct objects, the total number of different permutations is calculated as $n \times (n - 1) \times (n - 2) \times \cdots \times (n - x + 1)$.

If we multiply this expression by $(n - x)!/(n - x)! = 1$, we still have the same value but it may be written now as

$$\frac{n \times (n - 1) \times \cdots \times (n - x + 1) \times (n - x) \times (n - x - 1) \times \cdots \times 3 \times 2 \times 1}{(n - x) \times (n - x - 1) \times \cdots \times 3 \times 2 \times 1}$$

$$= \frac{n!}{(n - x)!}$$

producing a more compact expression.

The number of permutations of x objects selected from a total of n distinct objects may be expressed as

$$\frac{n!}{(n - x)!}.$$

This second expression is more compact and will be used to develop other general expressions; however, actual calculations will involve the first expression unless tables of factorials are available.

EXAMPLE 5-9 A geography instructor is going to select and read 4 field trip reports from a total of 20 submitted. The impression that the instructor will hold after reading the reports depends on which reports are selected and the order in which they are read. Setting $n = 20$ and $x = 4$ in the second expression, the number of ways that the reports can be selected and read is expressed as $20!/16!$. Both $20!$ and $16!$ are very large numbers requiring rounding. If we round to six digits, we have $20! = 2.43290 \times 10^{18}$ and $16! = 2.09228 \times 10^{13}$, producing $20!/16! = 1.16280 \times 10^4 = 116,280$ ways in which the reports may be selected and read. Using the first expression, we obtain $20 \times 19 \times 18 \times 17 = 116,280$. ▲

Combinations

When we select only some of the items from the total available (such as when we choose a sample from a population), we will not always be concerned with the order of selection as the geography instructor was with the reports.

EXAMPLE 5-10 Let us consider drawing cards again. Normally, in games involving cards from a deck, we are interested only in the particular sets or *combinations* of cards and not in the order in which they were drawn. We might, therefore, wish to determine the number of different sets or combinations of five cards that can be obtained regardless of order. Suppose that there are N such combinations. We will arrive at the value of N by considering the number of distinct orders, for which we already have a solution. Each of these sets or combinations consists of five cards. Because, as noted earlier, five items produce 5! orders, each distinct combination of five cards produces 5! orders. Each of the N combinations differs from all of the others and, accordingly, produces 5! orders which differ from those produced by the other combinations. With 5! permutations from each of N combinations, we have a total of $N \times 5!$ permutations. Our previous permutations calculation procedure tells us that there are 52!/47! permutations; thus, we must have the result that $N \times 5! = 52!/47!$. Solving this equation, we find that $N = 52!/(5! \times 47!) = 2,598,960$.

▲

This result can be extended to the general case to produce the rule for combinations.

> The number of ways that x objects can be selected from n distinct objects without consideration of order is called the number of *combinations* of x objects selected from a total of n distinct objects and is equal to
>
> $$\frac{n!}{x!(n - x)!}$$
>
> This value is often written as $\binom{n}{x}$ and referred to as n *choose* x. (*Note:* $\binom{n}{0} = 1$.)

EXAMPLE 5-11 If, in Example 5-9, the geography instructor's impression depends only on which four reports are to be read and not on the order of reading, then the number of ways that the reports can be selected for reading may be expressed as

$$\binom{20}{4} = \frac{20!}{4! \times 16!}$$

▲

When calculating a numerical value for n choose x, a useful approach is to consider the larger of the two factorials in the denominator and to consider the factors in its expanded form, canceling out the corresponding factors in the expanded form of the factorial in the numerator.

EXAMPLE 5-12 In Example 5-11, the larger factor in the denominator is $16! = 16 \times 15 \times 14 \times \cdots \times 2 \times 1$. Each of the factors in this multiplication is present in the numerator:

$$20! = 20 \times 19 \times 18 \times 17 \times 16 \times 15 \times \cdots \times 2 \times 1$$
$$= (20 \times 19 \times 18 \times 17) \times (16 \times 15 \times \cdots \times 2 \times 1)$$
$$= (20 \times 19 \times 18 \times 17) \times 16!$$

We may thus cancel all the factors in 16! out of the numerator to leave only

$20 \times 19 \times 18 \times 17$ in the numerator and $4! = 4 \times 3 \times 2 \times 1$ in the denominator. We then have

$$\binom{20}{4} = \frac{20!}{4!16!} = \frac{20 \times 19 \times 18 \times 17}{4 \times 3 \times 2 \times 1} = 4845 \qquad \blacktriangle$$

EXAMPLE 5-13 For the drawing of five cards in Example 5-10, 47! is the larger value in the denominator of $\binom{52}{5} = 52!/(5!47!)$. We thus consider canceling 47! with the factors $47 \times 46 \times 45 \times \cdots \times 3 \times 2 \times 1$ from 52!, leaving $52 \times 51 \times 50 \times 49 \times 48$ in the numerator and the smaller factorial, $5! = 5 \times 4 \times 3 \times 2 \times 1$ in the denominator. We can then calculate $\binom{52}{5}$ more easily as $(52 \times 51 \times 50 \times 49 \times 48)/(5 \times 4 \times 3 \times 2 \times 1)$ which can be simplified further by dividing the 5 into 50, the 4 into 52, the 3 into 51, and the 2 into 48, to produce $13 \times 17 \times 10 \times 49 \times 24 = 2,598,960$. $\qquad \blacktriangle$

PRACTICE PROBLEMS

P5-1 A seminar leader must conduct five seminars over the next three days and can conduct no more than three per day. Use a tree diagram to determine how many ways there are of arranging the numbers of seminars to be conducted on each of the three days.

P5-2 On each of three nights in a weekend, there are four types of entertainment available: going to a concert, play, movie, or night club. Each night there is a different concert, play, movie, or night club act. With a selection of one type of entertainment each night, how many different ways can entertainment for the weekend be arranged?

P5-3 There are available to a laboratory six litters of guinea pigs, consisting of 5, 6, 4, 4, 6, and 3 guinea pigs. How many ways can an experimental group of guinea pigs be formed if one is chosen from each litter?

P5-4 There are five archeological teams to be assigned to five new projects. How many ways can they be assigned?

P5-5 A committee chairman must consult any 4 of the 12 other members on a committee to seek advice. In each consultation the chairman will inform the member of any previous consultations. As a result, the advice received depends not only on the members consulted, but also on the order of consultation. How many different ways can the advice be received?

P5-6 In order to assess the nature of evening meals offered on a 1-week vacation package, a sample of three of the seven supper menus is to be considered. How many ways can the sample be obtained?

EXERCISES

5-1 Two teams are involved in a championship series such that the first team to win three games (there are no ties) wins the championship.
(a) Set up a tree diagram to indicate all possible series variations. How many variations are there?

(b) How many variations involve complete alternation (i.e., no team wins two games in a row)?

(c) How many variations involve one of the teams winning three consecutive games?

(d) How many variations involve exactly four games?

5-2 A production schedule for a 5-day period is to be set up to complete four projects. The first project requires two full days, and, once started, must be completed. The others each require one full day. The second project must be completed before the third or fourth may be started; otherwise, the projects can be done in any order. Use a tree diagram to determine the number of possible production schedules.

5-3 How many possible outcomes are there for the toss of the two dice and three coins in Exercise 4-19?

5-4 An identification code involves a letter followed by two digits followed by another letter. If letter and digit repeats are allowed, how many possible codes are there?

5-5 Six trees—an ash, a birch, an elm, a maple, an oak, and a walnut—are to be planted along the side of a drive. In how many different orders can the trees be planted?

5-6 Three prizes—first, second, and third—are to be allocated to three individuals selected from a group of 150. How many ways can the prizes be allocated?

5-7 If the three prizes in Exercise 5-6 are all the same, so that order of allocation does not matter, how many ways can the prizes be allocated?

5-8 A researcher plans to study the effects of each of three different treatment processes. Fifteen experimental units are available and three are to be chosen—one for each process. In how many ways can the treatments be applied to the units?

5-9 If, in Exercise 5-8, only one treatment is to be studied, but it is to be tried with each of four units, how many ways can the treatment be applied to the units?

5-10 In the discussion on sampling distributions in Section 4-6, it was stated that, without concern for order of selection (i.e., considering combinations only), there are 45 possible samples of size 2 and 120 possible samples of size 3 from a population of ten members. It was also stated that there are 1140 possible samples of size 3 and 15,504 of size 5 from a population of size 20. Verify all of these statements.

5-11 A league rules committee is formed by selecting one member from each of the six teams in the league. If each team has ten members, how many ways can the committee be formed?

5-12 To complete preparation for an experimental study, a researcher must consult a number of readings over the next five days. The readings consist of ten abstracts, six research papers, two monographs, and one technical manual. Each of the monographs requires a full day and each can only be borrowed for one day. For consistency in approach to the other readings, the researcher must not change the types of readings within a half-day but may change from morning to afternoon. In each half-day, the researcher can read five abstracts, two research papers, or the manual. How many ways can the five days be organized with regard to the numbers of each type of readings done on each of the days?

5-13 Enough water is left in a sample to analyze the content of any two of lead, iron, dissolved carbon, chlorides, fluorides, and sodium. How many ways can the two contents be selected for analysis?

5-14 In an employment attitude survey, respondents listed, in order of importance, the following four reasons for selecting their jobs: flexibility of working hours, level of income, opportunity for promotion, and sense of accomplishment. How many possible response orders are there?

5-15 In order to study the effects of a carbon dioxide atmosphere versus an ordinary atmosphere on the diffusive resistance of wax on the leaves of five different types of plants in each of high light, low light, and darkness, a researcher wishes to have one plant of each type with each pairing of light level and atmosphere. How many plants are required?

5-16 Management policy for a courtyard park is to be assessed by soliciting the opinions of some of the families in the residential complex. The sample of families is to consist of families in three of the 15 apartments facing the court, two of the 10 apartments not facing the court, four of the 20 townhouses facing the court, and two of the 10 townhouses not facing the court. How many ways can the sample be formed?

5-17 It is necessary to organize five different workshops in four full days of a conference. Each workshop can be arranged for a morning, afternoon, or evening session. There need not be a workshop each day. Because of the differences among the workshops, it is necessary to consider which workshop, if any, is assigned in any given session.
(a) How many ways can the five different workshops be assigned to the different sessions?
(b) How many ways can the workshops be assigned if there are no evening sessions?

SOLUTIONS TO PRACTICE PROBLEMS

P5-1 There are 12 ways of arranging the seminars, as shown in Figure P5-1.

P5-2 The nights may be represented as $k = 3$ subtrials for each of which the entertainment possibilities represent $n = 4$ outcomes. We then have $n^k = 4^3 = 64$ ways of arranging entertainment.

P5-3 The choices from the litters represent $k = 6$ subtrials with $n_1 = 5$, $n_2 = 6$, $n_3 = 4$, $n_4 = 4$, $n_5 = 6$, and $n_6 = 3$ outcomes, respectively. The experimental group can thus be formed in $5 \times 6 \times 4 \times 4 \times 6 \times 3 = 5 \times 6^2 \times 4^2 \times 3 = 8640$ ways.

P5-4 The number of assignments is the number of orders or permutations of the five teams, which is $5! = 5 \times 4 \times 3 \times 2 \times 1 = 120$.

P5-5 Since the chairman must consider order, the number of ways of receiving advice is equal to the number of permutations of $x = 4$ selections from $n = 12$ or $n!/(n-x)! = 12!/8! = 12 \times 11 \times 10 \times 9 = 11,880$.

P5-6 It is sufficient to consider only which menus have been selected, not the order obtained. The number of possible samples is thus the number of combinations of $x = 3$ selected from $n = 7$ or $\binom{n}{x} = \dfrac{n!}{x!(n-x)!} = \dfrac{7!}{3!4!} = \dfrac{7 \times 6 \times 5}{3 \times 2 \times 1} = 35$.

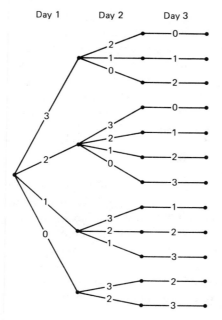

Figure P5-1 Numbers of seminars on each of 3 days.

5-3 PROBABILITY FUNCTIONS

Binomial Probability Function

With the development in Section 5-2 of the mathematical expressions for determining numbers of outcomes, we can now produce mathematical expressions for calculating probabilities. We first consider a mathematical formula that may be used in cases with several independent repetitions of the same trial.

EXAMPLE 5-14 Consider drawing cards at random from a well-shuffled deck of 52 playing cards. Unlike the usual circumstance of keeping cards as they are drawn, suppose that we are to draw five cards in succession and that each time that we draw a card, we observe it, replace it, and reshuffle the deck prior to the next draw. We wish to determine the probability that we will obtain two kings (and three nonkings).

One possible way to obtain two kings is to obtain the sequence $KKNNN$, where K denotes king and N denotes nonking. The probability that the first card will be a king is $4/52 = 1/13$. Since we replace this card and reshuffle the deck, the second draw is independent of the first and the probability that the second card will be a king is also $1/13$. Similarly, for each of the other draws, the probability of a nonking is $12/13$. The probability of the sequence is thus $(1/13) \times (1/13) \times (12/13) \times (12/13) \times (12/13) = (1/13)^2 \times (12/13)^3$.

Another way to obtain two kings is to obtain the sequence $KNNKN$ which, by similar reasoning, has probability $(1/13) \times (12/13) \times (12/13) \times$

$(1/13) \times (12/13) = (1/13)^2(12/13)^3$. In fact, any sequence with two K's and three N's will have the same probability $(1/13)^2 \times (12/13)^3$. If there are M distinct (mutually exclusive) sequences, the overall probability of obtaining two K's and three N's (in any order) is the sum of the individual sequence probabilities or $M(1/13)^2(12/13)^3$.

There is a distinct sequence for each choice of two positions in the sequence that are to be K's and, since there are five positions to choose from, the number of distinct sequences M is $\binom{5}{2}$, producing the probability

$$\binom{5}{2}(1/13)^2(12/13)^3 = 0.0465. \qquad \blacktriangle$$

The ideas in this example can be extended to the general case. Suppose that we have n independent trials instead of five; that, in individual trials, the probability of obtaining the outcome of interest (which we will denote as *success* or S) is p instead of $1/13$; that the probability of obtaining the complementary outcome (or event) (which we will denote as *failure* or F) is $1 - p$ instead of $12/13$; and that we are interested in the probability of x successes (and $n - x$ failures) instead of two (and three). There are $\binom{n}{x}$ sequences with x S's and $n - x$ F's, each with probability $p^x(1 - p)^{n-x}$; thus, the overall probability of obtaining x S's and $n - x$ F's is $\binom{n}{x}p^x(1 - p)^{n-x}$.

EXAMPLE 5-15 Suppose that we are to roll a balanced die and that we are interested in the probability of obtaining a number less than 3 on four of ten rolls. In this case we have $n = 10$, $x = 4$. Since two of the six faces on the die produce a number less than 3, we have $p = 1/3$ and $1 - p = 2/3$. The required probability is thus $\binom{10}{4}(1/3)^4(2/3)^6 = 0.228$. $\qquad \blacktriangle$

For any fixed n and p the probability of successes is found by substituting the value for x in the expression. This expression is an example of a *function* of x, denoted as $f(x)$, and we write $f(x) = \binom{n}{x}p^x(1 - p)^{n-x}$.

> A *function* is a rule or expression that assigns a value to one quantity on the basis of the value of another. The expression $y = f(x)$ indicates that the value of the quantity y is obtained by applying the rule to the value of the quantity x.

EXAMPLE 5-16 In an analysis of plant heights in trampled areas, the average maximum height of plants, denoted as y, was found to be related to the compressed bulk of the soil, denoted as x. The estimated relationship indicated that the average maximum height was equal to 53.5 minus 34.4 times the bulk. We may express this relationship as a function $y = f(x)$, where $f(x) = 53.5 - 34.4x$. $\qquad \blacktriangle$

Although it is usual to use x and y to represent the quantities and f to denote the function, other letters may be used as well.

EXAMPLE 5-17 The area A of a circle is determined from the value of the radius r and is equal to πr^2. The area is a function of the radius r and we may write A as $A(r)$, where the function is expressed as $A(r) = \pi r^2$. ▲

The function that we developed through Example 5-14 provides an expression for calculating probabilities for n independent repetitions of the same trial. This function is called the *binomial probability function*.

> The *binomial probability function* is used for any "experiment" consisting of n independent trials such that each of the outcomes of each individual trial falls into only one of two categories, which are denoted as *success* or *failure* and which have fixed probabilities p and $1 - p$, respectively. In such cases, the probability of obtaining x successes (and $n - x$ failures) in n trials is equal to
>
> $$f(x) = \begin{cases} \binom{n}{x} p^x (1 - p)^{n-x} & \text{for } x = 0, 1, 2, \ldots, n \\ 0 & \text{otherwise} \end{cases}$$

Note that, in expressing the function, we indicate the values of x for which the expression is valid. With six trials, for example, we cannot obtain ten successes. When a set of x's, such as "$x = 0, 1, 2, \ldots, n$" is indicated, the added condition "$= 0$ *otherwise*" is understood and need not always be stated.

EXAMPLE 5-18 In each of eight apartment complexes 25% of the apartments are modernized and 75% are dated. An inspector is about to conduct a survey on the basis of a sample formed by taking one apartment at random from each complex and we are interested in the probability that the sample will also be in the proportions 25% modernized and 75% dated. We may consider the sampling procedure to consist of $n = 8$ independent trials, each trial being the selection of an apartment from a complex. We may classify "modernized" as success and "dated" as failure. Since 25% of 8 is 2, we are interested in $x = 2$ successes and $n - x = 6$ failures. With 25% of the apartments in each complex being modernized the probabilities of success and failure for each trial are $p = 0.25$ and $1 - p = 0.75$. The required probability for two successes in eight trials is then $f(2) = \binom{8}{2} \times 0.25^2 \times 0.75^6 = 0.311$. ▲

For some values of n and p, values of the binomial probability function need not be calculated but may be obtained from Table I in Appendix B.

EXAMPLE 5-19 In an assessment of whether gamma-ray determinations of snow-water equivalents tend to be generally higher than gravimetric determinations, a researcher plans to use both methods in each of ten locations. The researcher assumes that it is just as likely for the gamma-ray determination to be the higher or the lower of the two and asks: "If this is so, what is the probability that the gamma-ray determination will be higher in eight or more of the ten locations?" Let "the gamma-ray determination is higher" be considered success. We then have $n = 10$ trials (the comparison of the two determinations at each of the ten locations is a trial) with probability $p = 0.5$ of success in each trial. The required probability is the probability of 8 or 9 or 10 successes and, since these are mutually exclusive, is found from the sum of the three probabilities $f(8)$, $f(9)$, and $f(10)$. The case $n = 10$ and $p = 0.5$ is included in Table I. We find the case $n = 10$ and then look in the column for $p = 0.5$. In the three rows for $x = 8$, 9, and 10, respectively, we find the values of $f(8)$, $f(9)$, and $f(10)$ to be 0.044, 0.010, and 0.001. The answer to the researcher's questions is then the sum $0.044 + 0.010 + 0.001 = 0.055$. ▲

Sampling With and Without Replacement

There are two ways of sampling from a population, and theoretical probabilities relating to the possible samples differ for the two sampling procedures.

EXAMPLE 5-20 Suppose in Example 5-14, concerned with the number of kings in five cards drawn from a deck, that cards are not replaced after they are drawn. The probability of a king will change from trial to trial depending on the results of previous trials. If we already have one king and two nonkings from the first three draws, the probability of another king on the fourth draw is 3/49, not 4/52. We must determine another rule for this situation. We consider all possible combinations of five cards that we can draw—there are $\binom{52}{5}$ combinations. There are $\binom{4}{2}$ ways of choosing two kings from the four kings in the deck, and for each of these choices, there are $\binom{48}{3}$ ways of drawing three nonkings from the 48 nonkings in the deck. The total number of ways of drawing two kings and three nonkings then becomes $\binom{4}{2} \times \binom{48}{3}$. The probability of drawing two kings and three nonkings when drawing without replacement is thus equal to $\left[\binom{4}{2} \times \binom{48}{3} \right] \Big/ \binom{52}{5} = 0.0399$. Note that this differs from the probability 0.0465 that we calculated in Example 5-14 for drawing two kings and three nonkings when drawing with replacement. ▲

When a sample is obtained by selecting members from a population such that each member selected is returned to the population prior to the selection of another so that any individual member may be selected more than once, the procedure is referred to as *sampling with replacement*. When selected members are not returned to the population so that any individual member may be only selected once, the procedure is referred to as *sampling without replacement*.

If the members of the population are classified into two categories, success and failure, then, using sampling with replacement as in the first procedure for drawing cards, each selection is independent of the others and probabilities may be determined by the binomial probability function.

> When *sampling with replacement* from a population of N members consisting of S "successes" and $N - S$ "failures," the probability of obtaining x successes in a random sample of size n is calculated by using the *binomial probability function* with $p = S/N$.
>
> When *sampling without replacement*, the probability is calculated by using a function called the *hypergeometric probability function*, which is defined below.

Hypergeometric Probability Function

By generalizing the cards example to consider the probability of x kings in a sample of n cards chosen without replacement from a deck of N cards including S kings, we may develop the expression for the hypergeometric probability function.

> If n items are selected at random without replacement from a population of N items consisting of S successes and $N - S$ failures, the probability of obtaining x successes (and $n - x$ failures) is found from the *hypergeometric probability function*,
>
> $$f(x) = \frac{\binom{S}{x} \times \binom{N - S}{n - x}}{\binom{N}{n}}$$
>
> where x cannot exceed S or n and where $n - x$ cannot exceed $N - S$.

EXAMPLE 5-21   We wish to draw a random sample of the attendance records of 5 of the employees in an office with a staff of 36 employees consisting of 32 women and 4 men. We sample without replacement since we do not wish to study any records twice. To find the probability that the sample will include the records of 3 women, we use the hypergeometric probability function with $N = 36$, $S = 32$, $N - S = 4$, $n = 5$, and $x = 3$. The probability is then equal to

$$f(x) = f(3) = \frac{\binom{32}{3} \times \binom{4}{2}}{\binom{36}{5}} = \frac{4960 \times 6}{376,992} = 0.079 \qquad \blacktriangle$$

Binomial Approximation for the Hypergeometric Function

If we have a very large finite collection such that S and $N - S$ are both much larger than the sample size n, then, even though we draw samples without replacement, the changes from trial to trial in the probabilities of success and failure may be negligible. The probabilities will remain fairly constant if we do not materially change the proportions in the population. As a rule of thumb, if x, the number of successes in the sample, does not exceed 5% of S, the number of successes in the population, and $n - x$, the number of sample failures, does not exceed 5% of $N - S$, the number of population failures, then the probability of success on each trial will remain fairly constant and approximately equal to S/N. The sampling conditions will then closely match those for the binomial with $p = S/N$.

> If $x \leqslant 0.5S$ and $n - x \leqslant 0.05(N - S)$, we may approximate the hypergeometric probability function with the binomial probability function using $p = S/N$.

EXAMPLE 5-22 We have 10,000 seeds, of which 90%, or 9000, will germinate, and we choose 10 at random to plant. If we use the hypergeometric probability function to find the probability that 6 of the 10 will germinate, we obtain

$$f(6) = \frac{\binom{9000}{6} \times \binom{1000}{4}}{\binom{10,000}{10}} = 0.01112$$

Noting that $x = 6 < 450 = 0.05S$ and $n - x = 4 < 50 = 0.05(N - S)$, we may use the binomial probability function with $p = S/N = 9000/10,000 = 0.9$ as an approximation to obtain $f(6) = \binom{10}{6}(0.9)^6(0.1)^4 = 0.01116$. The binomial approximation, in this case, is very close to the exact hypergeometric probability and is easier to calculate. In fact, when rounded to three decimals, both probabilities equal 0.011, which may be found without calculation as the entry in Table I corresponding to $n = 10, p = 0.9$, and $x = 6$. ▲

Poisson Probability Function

We will consider one other probability function, the *Poisson probability function*. The Poisson probability function is applied in cases in which events of a particular type occur completely at random in time or space at a fixed mean rate. The mean rate is denoted by λ (the Greek lowercase letter lambda) and is the average number of occurrences per unit of time, area, volume, and so on.

The *Poisson probability function* is

$$f(x) = \frac{\lambda^x e^{-\lambda}}{x!} \qquad for \ x = 0, 1, 2, \ldots$$

where e is Euler's number and is equal to 2.71828 to five places of decimals (values of $e^{-\lambda}$ may be obtained from Table II of Appendix B) and where λ is the mean rate of occurrence. The value of $f(x)$ is the probability of obtaining x occurrences in a particular unit of time, area, volume, and so on.

The Poisson probability function is often used for events that occur completely at random in time such that an event is just as likely to occur at one time point as at any other. It is applied in such cases as traffic analysis problems, demand on inventory problems, and arrival of customers at a service facility.

EXAMPLE 5-23 Telephone calls arrive at random at a switchboard at a mean rate of 2.5 calls per minute. We wish to know the probability that two calls will arrive in a particular 1-minute period. In this case one time unit is 1 minute and the mean rate per unit is $\lambda = 2.5$. The probability that the number of occurrences (calls) will be $x = 2$ is $f(2) = (2.5)^2 e^{-2.5}/2! = 6.25 \times 0.0821/(2 \times 1) = 0.257$. ▲

When events occur completely at random in space or a plane area without any tendency to cluster in a particular region, the Poisson probability function may be used to determine the probability of a particular number of occurrences in any given area or volume. Again the function is developed in terms of the mean rate of occurrence.

In some cases the mean rate may have to be adjusted to conform to the particular unit of time, area, or volume being considered.

EXAMPLE 5-24 Minor imperfections in material occur completely at random at a mean rate of 0.5 per square foot. We wish to know the probability that fewer than three imperfections will be present in 1 square yard of material. The unit of area of interest is 1 square yard, which is 9 square feet. The mean rate 0.5 per square foot must be adjusted to produce an appropriate mean rate $\lambda = 9 \times 0.5 = 4.5$ per square yard. To find the probability of fewer than three imperfections, we find $P(x < 3) = f(0) + f(1) + f(2)$ using the Poisson probability function with $\lambda = 4.5$. The probability is thus

$$\frac{4.5^0 e^{-4.5}}{0!} + \frac{4.5^1 e^{-4.5}}{1!} + \frac{4.5^2 e^{-4.5}}{2!} = e^{-4.5}(1 + 4.5 + 10.125)$$

From Table II we find that $e^{-4.5} = 0.011$ and the probability becomes $0.011 \times 15.625 = 0.172$. ▲

Poisson Approximation for the Binomial Function

The Poisson probability function may also be used as an approximation to the binomial probability function when n is large and p is small (a rule of thumb is $n \geqslant 20$

and $p \leqslant 0.05$ or $n \geqslant 100$ and $np \leqslant 10$). In such cases we may set $\lambda = np$ and calculate approximate binomial probabilities with the Poisson probability function.

EXAMPLE 5-25 If we heat 250 samples in glass beakers and the probability of breakage is 0.012 for each beaker, we have a binomial model with $n = 250$, $p = 0.012$. The probability that we will break $x = 2$ beakers is found as $f(x) = f(2) = \binom{250}{2}(0.012)^2(0.988)^{248} = 0.2245$. Since n is large and p is small, we may obtain an approximate probability by using the Poisson probability function with $\lambda = np = 250 \times 0.012 = 3$. We then have the simpler calculation $f(2) = 3^2 e^{-3}/2! = 9 \times 0.0498/2 = 0.2241$. ▲

Random Variables

In the cases considered so far in this section, we have been interested in trials that produce numerical outcomes, such as the number of heads in repeated tosses of a coin or the number of flaws in a segment of material. When a trial results in outcomes x that are numerical, we say that x is a *variable* taking different values at random and we refer to x as a *random variable*.

> If a trial or experiment has outcomes that are the numerical values of a variable x, we refer to x as a *random variable*.

EXAMPLE 5-26 In Example 5-25, x, the number of broken beakers in $n = 250$ sample heatings, is a random variable. ▲

Probabilities for the various possible values of a random variable are generally expressed through probability functions.

General Probability Functions

The binomial, hypergeometric, and Poisson probability functions are only three examples of probability functions. There are many other functions that are also probability functions. If any function produces a set of numbers that follow the rules developed previously for probabilities, it can be a probability function; otherwise, it cannot.

EXAMPLE 5-27 As will be illustrated in Example 5-28, the function

$$f(x) = \frac{x - 5}{15} \qquad for \ x = 6, 7, 8, 9, 10$$

is a probability function; however, the functions

$$g(x) = \frac{x^2 - 2}{20} \qquad \textit{for } x = 0, 1, 2, 3, 4$$

and

$$h(x) = \frac{3x + 5}{25} \qquad \textit{for } x = 1, 2, 3, 4, 5$$

are not.

For $g(x)$, the probability that $x = 0$ is found to be $g(0) = (0^2 - 2)/20 = -2/20$. Since this is negative, it cannot be a probability. The function $g(x)$ produces values that are not probabilities and thus is not a probability function.

For $h(x)$, consider $P(x = 1$ or $x = 2$ or $x = 3$ or $x = 4$ or $x = 5)$. Since these events are exhaustive, this probability should be 1. Since these events are mutually exclusive, this probability should also equal $P(x = 1) + P(x = 2) + P(x = 3) + P(x = 4) + P(x = 5) = h(1) + h(2) + h(3) + h(4) + h(5) = \sum_{x=1}^{5} h(x)$, but $\sum_{x=1}^{5} h(x) = 8/25 + 11/25 + 14/25 + 17/25 + 20/25 = 70/25 = 2.8$. Since 2.8 exceeds 1, it cannot be a probability. We have another function that cannot be a probability function. ▲

In Example 5-27, we have illustrated the two *required properties of a probability function*:

> To be a probability function, $f(x)$ must satisfy the two requirements: $0 \leqslant f(x)$ for all x, $\sum f(x) = 1$.

EXAMPLE 5-28 The function $f(x) = (x - 5)/15$ for $x = 6, 7, 8, 9$, and 10 is a probability function since it satisfies both requirements. Direct calculation produces the following values for $f(x)$ corresponding to each value of x:

x	$f(x) = (x - 5)/15$
6	1/15
7	2/15
8	3/15
9	4/15
10	5/15
Any other value	0
	$\overline{15/15}$

None of the values for $f(x)$ is negative and the sum of the values is $15/15 = 1$. ▲

We might have placed a further restriction on $f(x)$, namely that we must have $f(x) \leqslant 1$ for all x, since probabilities cannot exceed 1. This restriction is redundant, however, since a sum of nonnegative numbers would exceed 1 if any of the individual numbers exceeded 1.

Parameters for Probability Functions

With the development of probability functions, we introduce a new definition for a parameter. For a general population we have defined a parameter as *any* numerical measurement or characteristic. We are more restrictive in the case of probability functions.

> For a *probability distribution*, which is another name for a probability function, we define *parameters* as the constants that appear in the expression for the probability function $f(x)$ and characterize the particular distributions. The binomial distribution is characterized by the values of n and p; thus, the *binomial distribution* has two parameters, n and p. Similarly, the *hypergeometric distribution* has three parameters, N, S, and n. The *Poisson distribution* has one parameter, λ.

PRACTICE PROBLEMS

P5-7 It has been speculated that there is an 85% chance of receiving relief from nasal congestion by directing steam up the nostrils. If the speculation is correct and the steam treatment is used for a random sample of 15 people suffering nasal congestion, what is the probability that
(a) 13 will receive relief?
(b) at least 13 will receive relief?

P5-8 A biologist has 24 smallmouth bass and 6 largemouth bass in a tank and intends to study the gills of a sample of five fish caught at random from the tank. The biologist can only keep one fish out of the tank at a time and must return it to the tank before selecting another. What is the probability that the sampling procedure will produce a smallmouth bass three times and a largemouth bass twice?

P5-9 If the biologist in Practice Problem P5-8 has a small holding tank for the sample fish and does not replace a fish into the main tank before selecting another, what is the probability that the sample will include three smallmouth bass and two largemouth?

P5-10 Of 1800 pensioners, 630 are 75 years of age or older. If a random sample is to include 12 of the pensioners, what is the probability that
(a) five of them will be 75 years of age or older?
(b) fewer than three of them will be 75 years of age or older?

P5-11 Sporadic distortions of signals sent from a transmitter occur completely at random at a mean rate of 1.0 per minute. For 4-minute messages the distortions thus occur at a mean rate of 4.0 per message. What is the probability that the number of distortions in a 4-minute message will be
(a) (1) 2?
   (2) 1?
   (3) 0?
(b) 3 or more?

P5-12 It is claimed that each individual soft-drink bottle of a particular size and shape has only 1 chance in 1 million of exploding. If half a million such bottles are in circulation, what is the probability that
(a) none of them will explode?
(b) one of them will explode?
(c) two of them will explode?
(d) at least two of them will explode?

P5-13 A carnival game is such that a player may win nothing or a prize worth 5, 10, 15, 20, or 25 dollars. The corresponding probabilities for the possible values of winning are found from the function $f(x) = (30 - x)^2/2275$ for $x = 0, 5, 10, 15, 20,$ and 25.
(a) Verify that $f(x)$ is a probability function.
(b) Determine the probability of winning nothing.
(c) Determine the probability of winning a prize worth at least $20.

EXERCISES

5-18 Several shapes are drawn on a page and each is labeled with one of four colors. A child who cannot read has four crayons—one of each color—and is told to select five of the shapes and color them. An unlimited number of repeats is allowed (e.g., the child could color all shapes the same). Suppose that each choice of crayon is an independent random choice.
(a) What is the probability that the child will color all of the shapes correctly according to the labels?
(b) What is the probability that the child will color exactly three of the shapes correctly according to the labels?
(c) What is the probability that the child will color none of the shapes correctly according to the labels?
(d) What is the probability that the child will color at least three of the shapes correctly according to the labels?

5-19 A multiple-choice test consists of ten questions, each with three possible answers—one correct and two incorrect. Suppose that each of the questions is answered by a random selection of one of the possible answers.
(a) What is the probability that all of the answers chosen will be correct?
(b) What is the probability that none of the answers chosen will be correct?
(c) What is the probability that exactly three of the answers chosen will be correct?
(d) What is the probability that at least half of the answers chosen will be correct?

5-20 If seven representatives are to be selected at random from a group of 12 men and 16 women, what is the probability that the representatives will consist of 2 men and 5 women?

5-21 A birthday cake is cut into 12 pieces, 4 of which contain a hidden prize. If 4 pieces are chosen at random, what is the probability that 2 of them will contain a prize?

5-22 Eight heating dishes are to be chosen at random from a shipment of indistinguishable dishes, 20% of which are defective as a result of improper shipping.

What is the probability that two of the eight items chosen will be defective if the total number of items in the batch is

(a) 30?

(b) 50?

(c) several thousand?

5-23 A service counter is such that customers arrive at random at a mean rate of 8.0 per hour.

(a) What is the mean rate of arrivals per 15-minute period?

(b) If the counter is left unattended for 15 minutes, what is the probability that two customers will arrive and not receive service?

(c) If the counter is left unattended for 15 minutes, what is the probability that at most two customers will arrive and not receive service?

5-24 A particular characteristic is so rare that it is present in only 0.3% of individuals. In a random collection of 500 individuals, what is the probability that the characteristic will be present in

(a) two of the individuals?

(b) at most two of the individuals?

5-25 Suppose that $f(x)$ is a function whose values are given by

$$f(x) = \begin{cases} \dfrac{x}{55} & \text{for } x = 1, 2, 3, \ldots, 10 \\ 0 & \text{otherwise} \end{cases}$$

Show that $f(x)$ is a probability function.

5-26 If probabilities are as given by the function in Exercise 5-25, find

(a) $P(x = 4)$

(b) $P(x = 8)$

(c) $P(x \leqslant 3)$

(d) $P(2 < x < 7)$

(e) $P(x \leqslant 8)$

5-27 If 8 people are chosen at random from a group of 50 in the diplomatic corps of whom 15 majored in political science, what is the probability that the chosen group will include

(a) three who majored in political science?

(b) six who majored in political science?

(c) none who majored in political science?

5-28 After considering the sampling distribution of the mean, a team of geography students has determined that with each individual sample there is a 90% chance that the mean grain size in the sample will differ from the mean grain size in the riverbed population by at most 0.5 millimeter (mm). If each of the ten students in the team takes a sample, what is the probability that

(a) all of the sample means will be within 0.5 mm of the population mean?

(b) nine of the sample means will be within 0.5 mm of the population mean?

(c) at least eight of the sample means will be within 0.5 mm of the population mean?

5-29 If the incidence of a tumor is about 2 per 100,000 population, what is the probability of a tumor in two people in a sample selected at random from a population of several millions if the number of people in the sample is

(a) 2? (c) 10,000?

(b) 100? (d) 100,000?

5-30 To provide a holiday for employees when a statutory holiday falls on a Saturday, an employer may provide a holiday on the Friday or Monday. Of the 800 employees, 480 prefer Monday. If the personnel manager takes a random sample of ten employees, what is the probability that the majority of those in the sample will prefer Monday?

5-31 One-third of the three dozen subjects in an experiment are to receive a placebo instead of treatment. A balanced die is to be rolled and if it shows "5" or "6," a placebo is administered to the first subject. The same procedure is followed with each subsequent subject until one-third of all subjects have received the placebo or two-thirds have received the treatment. Of the first 12 subjects in the study, what is the probability that
(a) all will receive the placebo?
(b) none will receive the placebo?
(c) four will receive the placebo?

5-32 An art collection includes 15 paintings that are genuine originals and 5 that are forgeries. If five paintings are chosen for inspection, what is the probability that
(a) all of them will be forgeries?
(b) none of them will be forgeries?
(c) two of them will be forgeries?
(d) at least three of them will be forgeries?

5-33 Suppose that the probability of success on each trial is p.
(a) What is the probability of $n - 1$ successes on $x - 1$ trials (note the reversal of roles of x and n)?
(b) What is the probability of success on the xth trial?
(c) What is the expression for the probability function that will produce the probability that it will be necessary to have exactly x trials for exactly n successes where x may equal $n, n + 1, n + 2, \ldots$?

5-34 To make a commercial, a team will tape interviews with a number of people chosen at random from a population. If 30% of the population produce the desired reaction, what is the probability that, in order to obtain the desired reaction from four people, the team will have to tape
(a) four interviews?
(b) six interviews?
(c) eight or fewer interviews?
(d) at least ten interviews?

SOLUTIONS TO PRACTICE PROBLEMS

P5-7 Assuming that the probability of relief is 0.85 for each person, we have $n = 15$ repeated trials with a probability of $p = 0.85$ of success on each and the solutions are

(a) $f(13) = \begin{pmatrix} 15 \\ 13 \end{pmatrix} \times 0.85^{13} \times 0.15^2 = 105 \times 0.85^{13} \times 0.15^2 = 0.2856$

(b) $f(13) + f(14) + f(15) = \left\{ \begin{pmatrix} 15 \\ 13 \end{pmatrix} 0.85^{13} \times 15^2 \right\} + \left\{ \begin{pmatrix} 15 \\ 14 \end{pmatrix} \times 0.85^{14} \times 0.15^1 \right\} +$

$\left\{ \begin{pmatrix} 15 \\ 15 \end{pmatrix} \times 0.85^{15} \times 0.15^0 \right\}$

$= (105 \times 0.85^{13} \times 0.15^2) + (15 \times 0.85^{14} \times 0.15) + 0.85^{15}$

$= 0.2856 + 0.2312 + 0.0874 = 0.6042$

P5-8 Letting "success" denote smallmouth bass, we have $N = 30$, $S = 24$. We have sampling with replacement with $n = 5$ and $p = S/N = 0.80$. The solution is found as $f(3) = \binom{5}{3} \times 0.8^3 \times 0.2^2 = 0.2048$, or is found in Table I for $n = 5$, $p = 0.80$, and $x = 3$ as 0.205.

P5-9 We now have sampling without replacement and the solution is found as

$$\frac{\binom{24}{3} \times \binom{6}{2}}{\binom{30}{5}} = \frac{2024 \times 15}{142,506} = 0.2130$$

P5-10 Letting "success" denote 75 years or older, we have $N = 1800$ and $S = 630$ and $n = 12$.
(a) The solution is found exactly with the hypergeometric probability function as

$$\frac{\binom{630}{5} \times \binom{1170}{7}}{\binom{1800}{12}} = 0.2045$$

or from the binomial approximation with $p = 630/1800 = 0.35$, as $\binom{12}{5} \times 0.35^5 \times 0.65^7 = 0.2039$.

(b) Using the binomial approximation, the solution is

$$\left\{\binom{12}{0} \times 0.35^0 \times 0.65^{12}\right\} + \left\{\binom{12}{1} \times 0.35^1 \times 0.65^{11}\right\}$$

$$+ \left\{\binom{12}{2} \times 35^2 \times 0.65^{10}\right\}$$

$$= 0.65^{12} + (12 \times 0.35^2 \times 0.65^{11}) + (66 \times 0.35^2 \times 0.65^{10})$$

$$= 0.1513$$

P5-11 For random occurrences with a mean rate of 4.0, we use the Poisson probability function with $\lambda = 4.0$. From Table II we find that $e^{-4} = 0.018$ and the solutions are
(a) (1) $f(2) = 4^2 \times e^{-4}/2! = 16 \times 0.018/2 = 0.144$
    (2) $f(1) = 4^1 \times e^{-4}/1! = 4 \times 0.018/1 = 0.072$
    (3) $f(0) = 4^0 \times e^{-4}/0! = 1 \times 0.018/1 = 0.018$
(b) $1 - (f(0) + f(1) + f(2)) = 1 - 0.234 = 0.766$

P5-12 We have a binomial application with $p = 0.000001$ and $n = 500,000$ and we find approximate probabilities with the Poisson probability function with $\lambda = np = 0.5$. Using $e^{-0.5} = 0.607$ from Table II, we find the solutions as
(a) $f(0) = 0.607$
(b) $f(1) = 0.304$
(c) $f(2) = 0.076$
(d) $1 - [f(0) + f(1)] = 0.089$

P5-13 Values of x and $f(x)$ are as indicated in the table

Value x	$f(x) = (30 - x)^2/2275$
0	900/2275
5	625/2275
10	400/2275
15	225/2275
20	100/2275
25	25/2275
	2275/2275

(a) Since each value of $f(x)$ exceeds 0 and since $\Sigma f(x) = 1$, the function is a probability function.

(b) $f(0) = 900/2275 = 36/91$.

(c) $f(20) + f(25) = 125/2275 = 5/91$.

5-4 EXPECTATION

In the description of data, we developed summary values to represent "typical" or "average" values. In considering random trials, we can also develop summary values to represent typical or average results of such trials.

EXAMPLE 5-29 Suppose that we have a ticket in a small lottery to be operated in the following manner. There are 100 tickets numbered 1 to 100 and the numbers 1 to 100 are to be ordered at random in a list. The holder of the ticket with the same number as the first number in the list will win a first prize of $1000. As well, the holder of the ticket corresponding to the second number on the list will win a second prize of $250, and each of the holders of the tickets corresponding to the next five numbers will win a third prize of $50. No other prizes will be awarded.

The purchase of a ticket in this lottery is a random trial with outcomes and corresponding probabilities as indicated in the following table:

Prize	Number of tickets	Probability
$1000	1	1/100
250	1	1/100
50	5	5/100
0	93	93/100

In this lottery, with 100 tickets, a total of $1500 is to be paid in prizes. The average (mean) prize per ticket is, therefore, $15. We may consider a typical or average result to be the winning of $15. The term "typical" is more appropriate for a value similar to a mode and we will thus interpret "average" to be "mean" (this conforms to the method of obtaining the value $15—the total amount of money divided by the total number of tickets).

The value $15 is said to be the *mean* or the *expectation* of the random trial. The reason for using the term "mean" is indicated above. The use of the term "expectation" is explained below.

The expectation of the trial is the result that we *expect to obtain on the average*. We do not expect to win $15 on any individual trial—in fact, we know that we will never win $15 on any single operation of the lottery since there is no prize of $15. When we consider many repetitions of the lottery, however, we expect to win $1000 about once in 100 times or about 1% of the time. Similarly, we expect to win $250 about once in 100 times or about 1% of the time, $50 about five times in 100 or 5% of the time, and $0 about 93 times in 100 or 93% of the time. What we expect to win on the average is thus

$$\frac{(1 \times \$1000) + (1 \times \$250) + (5 \times \$50) + (93 \times \$0)}{100} = \$15$$

This value is a weighted mean, with the weights being the frequencies of occurrence of the individual values $1000, $250, $50, and $0.

This expression can also be written as $[(1/100) \times \$1000] + [(1/100) \times \$250] + [(5/100) \times \$50] + [(93/100) \times \$0] = \$15$. Again, we have a weighted mean; this time the weights are the relative frequencies of occurrence of the individual values. Note that we have apparently not divided by the total weight. In fact, the total weight must be 1 since the weights were relative frequencies, and it is not necessary to indicate division by 1. ▲

Expectation

The relative frequencies are, of course, the *probabilities* for the individual values; therefore, to obtain the expectation, we can multiply all of the individual values by their respective probabilities and then sum these products. It is this form of calculation that we use as the definition of *expectation*.

> If the outcomes of a trial consist of n mutually exclusive and exhaustive events $E_1, E_2, \ldots, E_n$ with individual values $V(E_1), \ldots, V(E_n)$ and probabilities $P(E_1), \ldots, P(E_n)$, then the *expectation*, or the *expected value*, or the *mean*, of the random trial is defined as
>
> $$\sum_{i=1}^{n} V(E_i) \times P(E_i)$$

EXAMPLE 5-30 We have 144 indistinguishable cans of four types of beverages, W, X, Y, and Z. We know that 12 of the cans contain the first beverage, which is 6% sugar; 48 of the cans contain the second beverage, which is 3% sugar; 72 of the cans contain the third beverage, which is 1% sugar; and 12 cans contain the fourth beverage, which is sugar free. If we consider the value of interest to be the sugar content, we have $V(W) = 6\%$, $V(X) = 3\%$, $V(Y) = 1\%$, $V(Z) = 0\%$, and $P(W) = 12/144 = 1/12$, $P(X) = 4/12$, $P(Y) = 6/12$, and $P(Z) = 1/12$. Now, if we choose one can at random, the expected value of the sugar content is $(6\% \times \frac{1}{12}) + (3\% \times \frac{4}{12}) + (1\% \times \frac{6}{12}) + (0\% \times \frac{1}{12}) = 24\%/12 = 2\%$. ▲

Fair Games

The concept of expectation may also be used to define a "fair game."

> A game involving the possible winning or losing of something of value is classified as a *fair game* if the *net expectation* is zero.
>
> If the game involves a *fixed fee* and possible winnings, the *net expectation* is the expected value of the winnings minus the fixed fee.

EXAMPLE 5-31 For example, in the lottery in Example 5-29 with an expected prize of $15, if the price per ticket or entry fee is T, the net expectation is $\$(15 - T)$. The lottery is a fair game if the price per ticket is $15. ▲

> If the game involves *possible winnings and possible losses*, the *net expectation* is the expected value of the winnings minus the expected value of the losses, or just the ordinary expectation if losses are classified as negative winnings.

EXAMPLE 5-32 Suppose that we are to draw a card at random from a well-shuffled deck of ordinary playing cards. If we draw an ace, our opponent pays us $26. If we draw a king, our opponent pays us $13. If we draw a 2, 3, or 4, we pay our opponent $13. If we draw any other card, there is no exchange of money. The expected value of our winnings is $[\$26 \times (1/13)] + [\$13 \times (1/13)] = \$3$. The expected value of our losses is $(3/13) \times 13 = \$3$. The net expectation is thus $0 and the game is fair. Alternatively, considering losses as negative winnings, the net expectation is the ordinary expectation or $[\$26 \times (1/13)] + [\$13 \times (1/13)] + [\$ - 13 \times (3/13)] = \$2 + \$1 - \$3 = \$0$, as before. ▲

> In a game of chance with an opponent such that the amount that we might lose is our stake or bet and the amount that we might win is our opponent's stake or bet, the *game is fair if the bets are in the same proportion as the odds of winning.*

EXAMPLE 5-33 If the odds in a game are 2 to 1 in favor of our opponent, then, for the game to be fair, our opponent should bet $2 for each $1 that we bet. If the odds are 3 to 2 in our favor, then, for the game to be fair, we should bet $1.50 for each $1 that our opponent bets. ▲

Mean for a Probability Function

For a trial or experiment with numerical outcomes x for which the probabilities are represented by a probability function $f(x)$, the value of an event is the number x itself, and the probability is $f(x)$; thus, the expectation is $\Sigma xf(x)$. The expectation in such cases is denoted by $E(x)$. Since a probability function or distribution may be interpreted as representing the probabilities for a *population* of possible values of a random variable x, the expectation may be interpreted as a population mean and is given the same symbol, μ.

We define the *mean*, or *expectation*, for a *probability function $f(x)$* to be

$$\mu = E(x) = \Sigma xf(x)$$

where the sum is over all possible values of x.

EXAMPLE 5-34 For the probability function in Example 5-28, $f(x) = (x - 5)/15$ for $x = 6, 7, 8, 9, 10$; thus, the mean is found as

$$\mu = E(x) = \sum_{x=6}^{10} \frac{x(x-5)}{15} = \frac{1}{15} \sum_{x=6}^{10} x(x - 5)$$

$$= \frac{1}{15} [(6 \times 1) + (7 \times 2) + (8 \times 3) + (9 \times 4) + (10 \times 5)]$$

$$= \frac{130}{15} = 8.67$$ ▲

Rules for Expectation

The following rules for expectations are useful for calculating values other than $E(x)$:

If x and y are random variables, and c is a constant-valued variable,

$$E(c) = c$$
$$E(cx) = cE(x)$$
$$E(x + c) = E(x) + c$$
$$E(x + y) = E(x) + E(y)$$

The first rule should be intuitively obvious. If an experiment always produces a constant value c, the average result must be c. The second and third rules are the results of similar arguments. If we multiply the outcome of a trial by a constant factor c or add to the outcome a constant c on every trial, the net average result must be adjusted accordingly. Finally, the last rule is explained by the argument that if we "pool" the results obtained from two experiments through addition, the average result is the same as that obtained by adding the average results from each of the experiments.

We may also find the expected value of a function of a random variable. (Here we mean a general function, not specifically a probability function.)

EXAMPLE 5-35 For the probability function in Example 5-28, we might wish to find the mean value of x^2. We find that x^2 takes the value of $36 = 6^2$ when $x = 6$; thus, the value 6^2 for x^2 has probability $f(6)$. Similarly, $x^2 = 7^2$ has probability $f(7)$, and so on, and the mean value for x^2 is

$$6^2 f(6) + 7^2 f(7) + 8^2 f(8) + 9^2 f(9) + 10^2 f(10)$$

$$= \sum_{x=6}^{10} x^2 f(x) = \frac{1}{15} \sum_{x=6}^{10} x^2 (x - 5) = \frac{1150}{15} = 76.67. \ \blacktriangle$$

The example illustrates the general result $E(x^2) = \sum x^2 f(x)$.

There is a more general rule for the expectation of any function $g(x)$ of a random variable x, and it is supported by arguments similar to those above.

> For any function $g(x)$ of a random variable x with probability function $f(x)$, the expectation of $g(x)$ is
>
> $$E(g(x)) = \sum g(x) f(x)$$

Variance and Standard Deviation for a Probability Function

In particular, we consider the mean value of the function $g(x) = (x - \mu)^2$, the square of the difference between the value x and the mean $\mu = E(x)$. For a population, the analogous value is the variance σ^2, which was found as the average squared deviation $\sum (x - \mu)^2 / N$. We use the same symbol for the variance for a probability function or distribution.

> We define the *variance for a probability function $f(x)$* to be
>
> $$\sigma^2 = E\big((x - \mu)^2\big) = \sum (x - \mu)^2 f(x)$$
>
> and the *standard deviation* to be
>
> $$\sigma = \sqrt{E\big((x - \mu)^2\big)}$$

Some algebraic manipulation using the rules of expectation will produce an alternative *calculating expression*,

$$\sigma^2 = E\big((x - \mu)^2\big) = E(x^2) - [E(x)]^2$$

EXAMPLE 5-36 For the probability function in Example 5-28, we have

$$\sigma^2 = E\left(\left(x - \frac{130}{15}\right)^2\right) = \sum_{x=6}^{10} \left(x - \frac{130}{15}\right)^2 \frac{x-5}{15}$$

$$= \sum_{x=6}^{10} \left(\frac{15x - 130}{15}\right)^2 \frac{x-5}{15} = \frac{1}{3375} \sum_{x=6}^{10} (15x - 130)^2 (x - 5)$$

$$= \frac{1}{3375}\big[(40^2 \times 1) + (25^2 \times 2) + (10^2 \times 3) + (5^2 \times 4) + (20^2 \times 5)\big]$$

$$= \frac{5250}{3375} = 1.56$$

or since we previously calculated $E(x) = 130/15$ and $E(x^2) = 1150/15$, we may find the variance more simply as

$$\sigma^2 = E(x^2) - [E(x)]^2 = \frac{1150}{15} - \left(\frac{130}{15}\right)^2$$

$$= \frac{350}{225} = 1.56$$

From either calculation, the standard deviation is

$$\sigma = \sqrt{1.56} = 1.25 \qquad \blacktriangle$$

Expectations for Particular Probability Functions

Applying these rules to the special distributions of Section 5-3, the means and variances are as follows:

For the binomial distribution,

$$\mu = np$$

$$\sigma^2 = np(1 - p)$$

In the long-run frequency interpretation of probability we expect the proportion of successes to be very close to the probability p. We might then intuitively expect that on the average this proportion of n trials would result in success producing an average number of successes as np. This argument provides an intuitive justification for the mean of the binomial probability function.

EXAMPLE 5-37 If there is a 60% chance that an individual rat will, after previous exposure, choose a tube with a sucrose solution instead of one with a water solution, and if an experiment consists of such a trial with each of 15 rats, the results of the experiment have probabilities determined by the binomial probability function with $n = 15$ and $p = 0.6$. In such experiments, we can expect that, on the average, $\mu = np = 15 \times 0.6 = 9$ rats of the 15 will choose the sucrose. The variance for the corresponding probability function is $np(1 - p) = 15 \times 0.6 \times 0.4 = 3.6$.

▲

For the hypergeometric distribution, in which the random variable x is the number of successes in the sample,

$$\mu = \frac{nS}{N}$$

$$\sigma^2 = n \times \frac{S}{N} \times \frac{N - S}{N} \times \frac{N - n}{N - 1}$$

If we set $p = S/N$, we obtain

$$\mu = np$$

$$\sigma^2 = np(1 - p) \frac{N - n}{N - 1}$$

EXAMPLE 5-38 Femurs from primates have been packaged in packs of 20, of which 12 are from arboreal specimens and 8 from terrestrial specimens. A random sample of five femurs is chosen from one package for analysis. Arbitrarily denoting arboreal as success, we apply the hypergeometric probability function with $N = 20$, $S = 12$, and $n = 5$. The expected number of arboreal femurs in such samples is $np = 5 \times 0.6 = 3$, where $p = S/N = 0.6$. The corresponding variance is $np(1 - p)(N - n)/(N - 1) = 5 \times 0.6 \times 0.4 \times 15/19 = 0.947$.

▲

Using the expressions based on $p = S/N$, we note that the hypergeometric probability function has the same expression for the mean as the binomial and almost the same expression for the variance as the binomial.

The variance for the hypergeometric differs from that for the binomial by an added factor $(N - n)/(N - 1)$, which is the same as the *finite population factor* considered in Section 4-6.

This factor may again be considered as a finite population factor because the binomial represents sampling with replacement, which is the same as sampling from an infinite (and hence unchanging) population, and the hypergeometric represents sampling without replacement from a finite collection or population.

The variance of the hypergeometric probability function will be very close to that of the binomial with $p = S/N$ if the factor $(N - n)/(N - 1)$ is close to 1. This factor will be close to 1 in cases in which the binomial is used to approximate the hypergeometric.

The mean and variance for the *Poisson distribution* are

$$\mu = \lambda$$

$$\sigma^2 = \lambda$$

Considering these expectations for the Poisson, we note the appropriateness of referring to the parameter λ as the mean rate.

P5-14 Depending on various contingencies, an insurance company will, in any year, pay claims against an individual policy of $1000, $750, $500, $250, or $0. From past records, the company has established the following probabilities for each claim.

Claim	Probability
$1000	0.001
750	0.002
500	0.005
250	0.010
0	0.982

(a) What is the expected claim for such a policy?

(b) If the company charges a premium of $10 for the policy, what is the expected gross profit (premium minus claim)?

(c) If expenses relating to a policy amount to $1.25 plus 10% of any claims, what is the expected expense per policy?

(d) What is the expected net profit from 10,000 such policies?

P5-15 You and an opponent are to wager on the roll of a pair of balanced dice such that if the dice show "7" your opponent will pay you $5; otherwise, you will pay your opponent $1. Is the game fair?

P5-16 What is the expected number of people who will receive relief from nasal congestion in Practice Problem P5-7?

P5-17 If there is a 25% chance that any individual truck tire will last for more than twice the anticipated number of miles, what are the mean and variance of the number that will do so in a sample of 12 such tires?

P5-18 A woman judge is to preside over hearings in ten cases chosen at random from 40 cases of which eight involve charges against women. For this judge's hearings, what are the mean and variance of the possible numbers of cases involving charges against women?

P5-19 What are the mean and variance of the possible numbers of passenger cars passing through a toll booth in a 15-minute period if such cars arrive randomly at a rate of 10 per period?

5-35 Past records indicate that 30% of all candidates succeed at an examination on the first try; 40% fail on the first try but succeed on the second; 20% fail on the first two tries but succeed on the third; and the remaining 10% finally succeed on the fourth try. If a candidate is to be selected at random and examined, what is the expected number of tries that will be required to achieve success?

5-36 If two balanced dice are tossed together, what is the expected value of the sum of the two numbers showing on the dice?

5-37 Suppose that it is believed that a given investment opportunity is such that there is a 5% chance of receiving a profit of 50% of the amount invested, a 10% chance of a 40% profit, a 15% chance of a 30% profit, a 25% chance of a 20% profit, a 20% chance of a 10% profit, a 10% chance of breaking even, a 10% chance of a 10% loss, and a 5% chance of a 20% loss. What is the net expected percentage return on the amount invested?

5-38 Suppose that eight cards—a 5, two 7's, a 9, two jacks, a queen, and an ace—are known to be out of play in a game with an ordinary deck of playing cards. A player is then dealt two cards, a 6 and a king, from the remainder of the deck and is to be dealt a third card. If the third card is between the first two (i.e., 7 through queen inclusive), the player wins; otherwise, the dealer wins. The player and dealer place bets such that if the player wins, the dealer pays $4, but if the dealer wins, the player pays $3. Is this game "fair"?

5-39 Find the mean and variance for the probability function given in Exercise 5-25.

5-40 In Exercise 5-18, what is the mean number of shapes that the child will color correctly according to the labels?

5-41 In Exercise 5-19, what are the mean and the standard deviation of the number of correct answers on the test?

5-42 In Exercise 5-20, how many men can be expected to be selected on the average in choosing a group of seven representatives?

5-43 Suppose that a survey is to be conducted by choosing, at random, 150 families from a community of 3000 families, 70% of whom are homeowners and 30% of whom rent their homes. What are the mean and the standard deviation of the number of homeowners in the sample of 150 families?

5-44 Repeat Exercise 5-43 assuming that the number of families in the community is sufficiently large that the finite population effect can be ignored: that is, such that the binomial rather than the hypergeometric model can be used.

5-45 The probability of making a faulty diagnosis with a chemical analysis is 0.001.
   (a) What are the mean and variance of the number of faulty diagnoses in
      (1) 10 analyses?
      (2) 100 analyses?
      (3) 10,000 analyses?
   (b) What are the mean and variance of the approximating Poisson distribution for 10,000 analyses?

5-46 Of 150 mentally deficient adults subject to supervision by a social agency, 60 are residents in institutional facilities. If a random sample of 15 cases is taken from the 150, what are the mean and variance of the number that involve institutional residents?

5-47 What are the mean and variance of the number of accidents on a highway in a 3-hour snowstorm if accidents in such snowstorms occur randomly at a mean rate of five per hour?

5-48 If 8% of the employable population in a community of 5000 employable people are unemployed, what are the mean and variance of the number unemployed in random samples of 100 from the employable people of this population?

5-49 What are the mean and variance of the number of unemployed people in random samples of 10,000 employable persons from a population of several millions with an unemployment rate of 10%?

5-50 A gaming establishment operates machines that players pay $0.25 to play. The possible prizes and their corresponding probabilities are as given in the table.

Prize	Probability
$10.00	0.002
5.00	0.010
2.50	0.020
1.00	0.100
0	0.868

If the establishment is open 24 hours a day and anticipates at least 100 plays a minute on its machines, at least how much can the establishment anticipate as a net average daily return from the machines?

SOLUTIONS TO PRACTICE PROBLEMS

P5-14 (a) The expected claim is found from the basic definition as (1000×0.001) + (750×0.002) + (500×0.005) + (250×0.010) + (0×0.982) = $7.50.

(b) To use the rule $E(x + c) = E(x) + c$, we consider the claims to be negative values for the company and the premium to be constant. We then have $E(x) = \$ - 7.50$, $c = \$10$, and the gross profit is $2.50.

(c) Since expenses are $1.25 + 10% of any claim x, the expenses on a claim are $1.25 + 0.1x$ and expected expenses are $E(1.25 + 0.1x) = E(1.25) + 0.1E(x) = \$1.25 + 0.1E(x) = \$2.00$.

(d) The cost to the company for any claim plus expenses is $1.25 + 1.1x$. The expected cost is thus $1.25 + 1.1E(x) = \$9.50$ or, alternatively, the expected claim plus the expected expenses: $7.50 + \$2.00 = \$9.50$. The net expected profit per claim is then $10.00 - \$9.50 = \$0.50$ and the net expected profit from 10,000 such policies is $10,000 \times \$0.50 = \$5000$.

P5-15 Since $P(x = 7) = 1/6$ for a pair of balanced dice, the odds against "7" are 5 to 1 and wagers of $5 to $1 produce a fair game. Alternatively, we might calculate the net expectation as $[\$5 \times (1/6)] + [(-\$1) \times (5/6)] = \$0$.

P5-16 With $n = 15$ and $p = 0.85$, the expected number is $np = 15 \times 0.85 = 12.75$.

P5-17 We have a binomial distribution with $n = 12$ and $p = 0.25$, producing the mean and variance as $\mu = np = 12 \times 0.25 = 3$ and $\sigma^2 = np(1 - p) = 12 \times 0.25 \times 0.75 = 2.25$.

P5-18 We have a hypergeometric distribution with $N = 40$, $S = 8$, and $n = 10$, producing $p = S/N = 0.2$ and mean and variance $\mu = np = 10 \times 0.2 = 2$ and $\sigma^2 = np(1 - p)(N - n)/(N - 1) = 10 \times 0.2 \times 0.8 \times 30/39 = 1.23$.

P5-19 Random arrivals at rate 10 produce the Poisson distribution with parameter $\lambda = 10$. The mean and variance are then $\mu = \lambda = 10$ and $\sigma^2 = \lambda = 10$.

5-5 CONTINUOUS VARIABLES—THE NORMAL DISTRIBUTION

The probability functions and summation expressions for expectations developed in previous sections applied to discrete variables. For random variables such as measurements that can take any values in a continuum, we refer to the distribution as a continuous distribution and we modify the previous results.

In these cases, the functions used in determining probabilities are referred to as *probability density functions*, because as noted with graphs in Section 4-5, these functions do not measure actual probability at a point but, instead, measure *density* of probability. The functions are generally represented in graphical form by smooth curves and, as illustrated in Figure 4-7, probabilities are found as areas across intervals under the curve.

One of the most important continuous distributions is the *normal distribution*. We consider this very important distribution in this section and introduce other continuous distributions in later chapters as the need arrives.

The normal distribution is represented by the probability density function

$$f(x) = \frac{1}{\sqrt{2\pi}\sigma} e^{-(x-\mu)^2/2\sigma^2} \qquad \text{for } -\infty < x < \infty$$

This function has two parameters, μ and σ. The graph of the function (illustrated in Figure 5-4) is a bell-shaped curve. It is symmetric about μ and its degree of spread is determined by the value of σ.

The parameters μ and σ are the mean and standard deviation, respectively, for the normal distribution. As illustrated in Figure 5-4, the normal distribution is a symmetric unimodal distribution and μ is also the median and the mode.

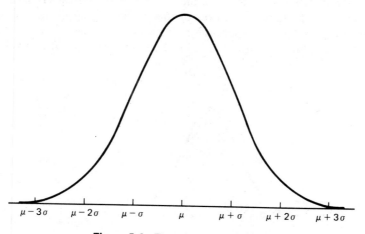

Figure 5-4   The normal distribution.

For all normal distributions, the probability is distributed as shown in Figure 5-4. For any fixed number k, $P(\mu - k\sigma \leqslant x \leqslant \mu + k\sigma)$ is thus the same for every normal distribution. In particular, if $k = 1.5$, this probability is 0.8664. We thus have the result that, for any normal distribution, 86.64% of the distribution will be within 1.5 standard deviations of the mean.

It is interesting to compare this result with the statement of Chebyshev's theorem in Section 3.3. According to that theorem, *for any distribution* (not just any normal), *at least* $100[1 - (1/1.5^2)]\% = 55.56\%$ of the distribution will be within 1.5 standard deviations of the mean.

Similar comparisons for selected values of k are as follows (see Exercise 5-63):

k	Interval	Minimum probability by Chebyshev's theorem	Actual probability for normal distributions
1.0	$\mu - \sigma$ to $\mu + \sigma$	0.0	0.6826
1.5	$\mu - 1.5\sigma$ to $\mu + 1.5\sigma$	0.5556	0.8664
2.0	$\mu - 2\sigma$ to $\mu + 2\sigma$	0.7500	0.9544
2.5	$\mu - 2.5\sigma$ to $\mu + 2.5\sigma$	0.8400	0.9876
3.0	$\mu - 3\sigma$ to $\mu + 3\sigma$	0.8889	0.9974
3.5	$\mu - 3.5\sigma$ to $\mu + 3.5\sigma$	0.9184	0.9995
4.0	$\mu - 4\sigma$ to $\mu + 4\sigma$	0.9375	0.9999

Standard Normal Distribution

Probabilities are given in Table IIIa of Appendix B for the normal distribution with $\mu = 0$ and $\sigma = 1$. This distribution is referred to as the standard normal distribution. A variable with the standard normal distribution is usually designated as z.

Table IIIa is such that, for any particular numerical value of z, the corresponding value in the table is the area of the shaded region illustrated in Figure 5-5. The entry in the table corresponding to a particular value z is thus the probability that the standard normal variable will be less than (or equal to) that value z.

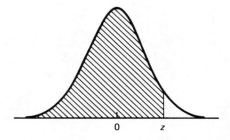

Figure 5-5 Illustration of Table IIIa.

EXAMPLE 5-39 If we enter the table with a z value of 1.35, we find the entry in the table (row 1.3, column 0.05, because $1.35 = 1.3 + 0.05$) to be 0.9115. There is thus a probability of 0.9115 that a variable z, with the standard normal distribution, will take a value less than 1.35; that is, $P(z \leqslant 1.35) = 0.9115$, as illustrated in Figure 5-6.

▲

We can find other types of probabilities from this first type of entry, by considering properties of the distribution. As the distribution is symmetric, the mean 0 is also the median and $P(z \leqslant 0) = 0.5000$. Combining this probability with a table entry, we can find the probability between 0 and positive values for z.

EXAMPLE 5-40 $P(0 \leqslant z \leqslant 1.35) = 0.9115 - 0.5000 = 0.4115$, as illustrated by the varied shading in part of Figure 5-7.

▲

Because of the symmetry of the distribution, the part of the curve below 0 is a mirror reflection of the part above 0. As a result, probabilities between 0 and negative values of z can be found by considering the reflections of the probabilities for the corresponding positive values of z.

EXAMPLE 5-41 By reversing Figure 5-7, we thus find that $P(-0.135 \leqslant z \leqslant 0) = 0.4115$.

▲

Combining these symmetric results, we can find probabilities for symmetric intervals about 0.

EXAMPLE 5-42 From the results above relating to the numerical value 1.35, we find that $P(-0.135 \leqslant z \leqslant 1.35) = 2 \times P(0 \leqslant z \leqslant 1.35) = 2 \times 0.4115 = 0.8230$, which we may also write as $P(|z| \leqslant 1.35) = 2 \times 0.4115 = 0.8230$.

▲

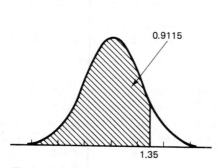

Figure 5-6 $P(z \leqslant 1.35)$.

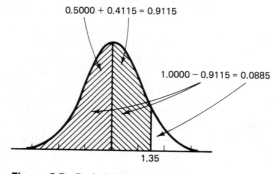

Figure 5-7 Probabilities related to $z = 1.35$.

We also know that the total area is the total probability and thus equal to 1. Combining this result with a table entry, we can find the probability that a standard normal will exceed a given numerical value.

EXAMPLE 5-43 Again from the entry corresponding to the numerical value 1.35, we find that $P(z \geqslant 1.35) = 1.0000 - 0.9115 = 0.0885$, as illustrated in the unshaded part of Figure 5-7. ▲

By taking the appropriate difference of values in Table IIIa, we can find the probability that z is between any two numerical values (inclusive or exclusive because, as noted in Chapter 4, for continuous variables, the probability at an exact point is 0).

EXAMPLE 5-44

$$P(-1.87 \leqslant z \leqslant -0.59) = P(z \leqslant -0.59) - P(z < -1.87)$$
$$= 0.2776 - 0.0307 = 0.2469$$
$$P(-2.50 < z < 1.50) = P(z < 1.50) - P(z \leqslant -2.50)$$
$$= 0.9332 - 0.0062 = 0.9270$$
$$P(0.25 \leqslant z \leqslant 1.75) = P(z < 1.75) - P(z < 0.25)$$
$$= 0.9599 - 0.5987 = 0.3612$$

▲

Normal Distributions Other Than the Standard

Rather than having many tables for various normal distributions, we transform nonstandard normal variables to standard normals. Probabilities can then be found from Table IIIa.

> If x is normal (i.e., has a normal distribution) with mean μ and standard deviation σ, then $z = (x - \mu)/\sigma$ is normal with mean 0 and standard deviation 1.

Because this is the form of adjustment that produces a standardized value or standard score as developed in Section 3-6, z is a standard normal variable.

> To find probabilities relating to any value for a nonstandard normal x, we find the probabilities relating to the corresponding value for z.

EXAMPLE 5-45 Measurements on the concentration of nonbiodegradable material in a waterway are assumed to follow a normal distribution with a standard deviation of $\sigma = 10.0$ and mean μ equal to the actual concentration level. It is believed that the mean μ is equal to 65.0. A value of 84.3 has been measured and following the researcher's reaction that "surely the value should have been less than 84.3," we wish to calculate an appropriate probability to assess this reaction. To do so, we consider a value x chosen at random from a normal distribution with mean 65.0 and standard deviation 10.0, and then we determine the probability that x will be less than 84.3 as expected by the researcher. We note that $x < 84.3$ when $(x - 65.0) < (84.3 - 65.0)$, and that this condition is true when $(x - 65.0)/10.00 < (84.3 - 65.0)/10.0 = 1.93$. Since $\mu = 65.0$ and $\sigma = 10.0$, $(x - 65.0)/10.0 = (x - \mu)/\sigma = z$; thus, $x < 84.3$ corresponds to $z < 1.93$ and

$$P(x < 84.3) = P\left(\frac{x - 65.0}{10.0} < \frac{84.3 - 65.0}{10.0}\right) = P(z < 1.93)$$
$$= 0.9732 \quad \text{from Table IIIa}$$

Note that this high probability tends to support the researcher's reaction that the value should be less than 84.3. In fact, there is a 97.32% chance that any value would be less than 84.3 and only a 2.68% chance of obtaining a value of 84.3 or larger.

The shaded area in the upper part of Figure 5.8 illustrates the desired probability for x. The corresponding z values and probability (shaded area) are illustrated in the lower part of Figure 5.8. ▲

EXAMPLE 5-46 As another example of a probability for x from the same distribution as that assumed in Example 5-45, let us consider the probability of

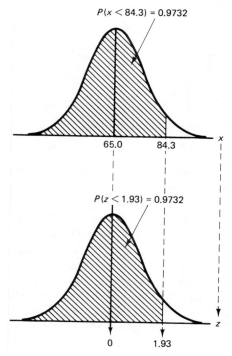

$P(x < 84.3) = 0.9732$

$P(z < 1.93) = 0.9732$

Figure 5-8 Illustration of $P(x < 84.3)$.

obtaining a value between 60 and 75. The required probability is

$$P(60.0 < x < 75.0) = P\left(\frac{60.0 - 65.0}{10.0} < \frac{x - 65.0}{10.0} < \frac{75.0 - 65.0}{10.0} \right)$$
$$= P(-0.50 < z < 1.00)$$

From Table IIIa we find that $P(-0.50 < z < 1.00) = P(z < 1.00) - P(z \leqslant -0.50) = 0.8413 - 0.3085 = 0.5328$. The required probability is then $P(60.0 < x < 75.0) = 0.5328$, as illustrated in Figure 5-9. ▲

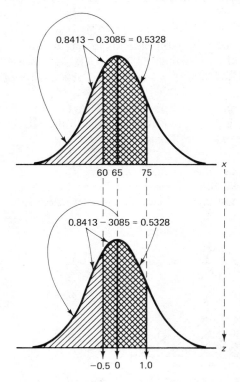

Figure 5-9 Illustration of $P(60.0 < x < 75.0)$.

Theoretically, the normal distribution applies to variables taking values in the whole continuum from $-\infty$ to ∞; however, practically, it is also useful as a reasonable approximation to reality in many other cases, such as the positive continuum from 0 to ∞, or an infinite collection of discrete values such as all the positive whole numbers, or a very large finite population.

EXAMPLE 5-47 Scores achieved on a psychological test have been found to be distributed in an appropriately bell-shaped symmetric distribution with mean 105 and standard deviation 15 so that, even though the population of all achieved scores is not continuous, it may be well represented by the smooth curve of a normal distribution. The appropriate normal distribution is the one with $\mu = 105$ and $\sigma = 15$. If, for instance, we then wish to find the probability that any individual selected at random will achieve a score greater than 130, we find the probability as $P(130 < x) = P[(130 - 105)/15 < (x - 105)/15] = P(1.67 < z) = 1.0000 - 0.9525 = 0.0475$ from Table IIIa because $(x - 105)/15 = (x - \mu)/\sigma = z$ and $(130 - 105)/15 = 1.67$. ▲

Normal Approximation for the Binomial Probability Function

The normal distribution can also be used to find approximate probabilities for the binomial distribution if n is large and p is neither too large nor too small.

> The normal approximation for the binomial distribution is used if np and $n(1 - p)$ are both sufficiently large. A usual rule of thumb is to use the approximation if both np and $n(1 - p)$ exceed 5.

We use the normal distribution that is closest to the binomial we are approximating in the sense of having the same mean and standard deviation.

> To approximate the binomial distribution with given n and p with a normal distribution, we use the normal with $\mu = np$ and $\sigma^2 = np(1 - p)$.

The graphical interpretation of this approximation is shown in Figure 5-10.

Figure 5-10 illustrates part of a probability graph for a binomial distribution. Each rectangle has base equal to 1 and height equal to the binomial probability. The area of the rectangle is then the binomial probability. Superimposed on the graph is an approximating normal curve. The area under the normal curve between $x - 0.5$

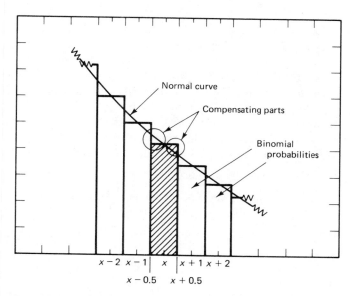

Figure 5-10   Illustration of normal approximation for the binomial.

and $x + 0.5$ (shaded) is approximately equal to the area of the rectangle which is the binomial probability for x. We thus see that the normal probability for the interval $x - 0.5$ to $x + 0.5$ is approximately equal to the binomial probability for the single value x. The addition and subtraction of 0.5 to produce an interval is called a *continuity correction*.

If we wish to find a binomial probability for a range of x values, we combine their corresponding intervals (after introduction of the continuity correction) to produce a single interval for the normal probability.

If we let P_B denote a binomial probability and P_N a normal probability and if we continue to use z to represent a standard normal random variable, normal approximations for binomial probabilities can be found from the following (in which the sign " $\doteq$ " is read "is approximately equal to").

> If x is a random variable having the binomial distribution with parameters n and p such that np and $n(1 - p)$ both exceed 5, then for any particular x value x_0,
>
> $$P_B(x = x_0) \doteq P_N(x_0 - 0.5 \leqslant x \leqslant x_0 + 0.5)$$
>
> $$= P\left(\frac{x_0 - 0.5 - np}{\sqrt{np(1 - p)}} \leqslant z \leqslant \frac{x_0 + 0.5 - np}{\sqrt{np(1 - p)}} \right)$$
>
> $$P_B(x < x_0) \doteq P_N(x \leqslant x_0 - 0.5) = P\left(z \leqslant \frac{x_0 - 0.5 - np}{\sqrt{np(1 - p)}} \right)$$
>
> $$P_B(x \leqslant x_0) \doteq P_N(x \leqslant x_0 + 0.5) = P\left(z \leqslant \frac{x_0 + 0.5 - np}{\sqrt{np(1 - p)}} \right)$$

EXAMPLE 5-48 Suppose that we are to choose 100 fuel pellets at random from a batch of several thousand (i.e., a large enough batch that the finite population correction factor may be ignored), of which 12% are defective. We wish to determine the probability that the number of defective pellets in the sample will be between 10 and 14 inclusive. Since the batch contains several thousand pellets, the probabilities will remain fairly constant from draw to draw, and we may use the binomial distribution with $n = 100$ and $p = 0.12$. We may, then, calculate the required probability from the binomial distribution as

$$f(10) + \cdots + f(14) = \binom{100}{10}(0.12)^{10}(0.88)^{90}$$

$$+ \cdots + \binom{100}{14}(0.12)^{14}(0.88)^{86} = 0.5583$$

Figure 5-11 shows a graph for the binomial distribution with $n = 100$ and $p = 0.12$ and has superimposed on it the normal distribution with mean $\mu = np = 12$ and variance $\sigma^2 = np(1 - p) = 10.56$ (and standard deviation

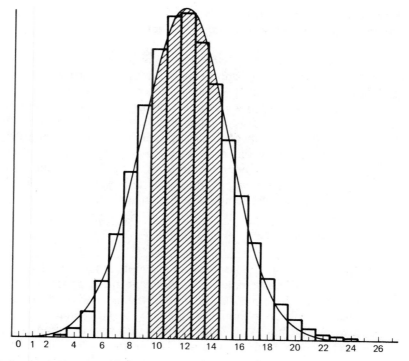

Figure 5-11 Binomial $n = 100$, $p = 0.12$, and normal $\mu = np = 12$, $\sigma^2 = np(1 - p) = 10.56$.

$\sigma = \sqrt{10.56} = 3.25$). Even though the binomial is slightly skewed, the symmetric normal provides a good approximation [$np > 5$ and $n(1 - p) > 5$] and we can use area under the normal curve to approximate the area for the graph of the binomial probability function.

If we use the normal approximation, the calculation is much simpler than that with the binomial probability function. Correcting for the continuity, we convert $f(10) + \cdots + f(14) = P(x = 10 \; or \; x = 11 \; or \cdots or \; x = 14)$ to $P(9.5 \leqslant x \leqslant 10.5 \; or \cdots or \; 13.5 \leqslant x \leqslant 14.5) = P(9.5 \leqslant x \leqslant 14.5)$, which we calculate as a normal probability, as shown in the shaded part of Figure 5-11. We thus calculate

$$P(9.5 \leqslant x \leqslant 14.5) = P\left(\frac{9.5 - 12}{3.25} \leqslant \frac{x - 12}{3.25} \leqslant \frac{14.5 - 12}{3.25} \right)$$
$$= P(-0.77 \leqslant z \leqslant 0.77)$$
$$= 0.5588 \qquad \text{which is very close to } 0.5583$$

Similarly, we can find other probabilities. The probability of fewer than ten defective pellets (i.e., nine defective pellets or fewer) is calculated from the normal approximation as $P_B(x \leqslant 9) \doteq P_N(x \leqslant 9.5) = P(z \leqslant -0.77) = 0.2206$.

▲

Normal Approximation
for the Hypergeometric Probability Function

For a sufficiently large value of N, the normal distribution may also approximate the hypergeometric distribution if S and $N - S$ are both reasonably large (i.e., if as previously discussed, the hypergeometric distribution could be approximated by a binomial distribution and that binomial could be approximated by a normal distribution). In this case, setting $p = S/N$ as usual, we use the normal approximation as with the binomial with $\mu = np$ but, employing the finite population correction, we have the standard deviation as

$$\sigma = \sqrt{np(1-p)\frac{N-n}{N-1}}$$

EXAMPLE 5-49   If the batch in Example 5-48 consists of $N = 1800$ pellets and we choose a sample of $n = 100$, the finite population correction factor $(N - n)/(N - 1) = 1700/1799 = 0.9450$ is employed. If 12% or 216 are defective, the approximate probability that a sample of 100 will contain between 10 and 14 defective pellets inclusive is found using a continuity correction just as above with $\mu = 12$ but now with

$$\sigma = np\sqrt{(1-p)(N-n)/(N-1)} = \sqrt{9.979} = 3.16$$

to produce a normal probability,

$$P(9.5 \leqslant x \leqslant 14.5) = P\left(\frac{9.5 - 12}{3.16} \leqslant \frac{x - 12}{3.16} \leqslant \frac{14.5 - 12}{3.16}\right)$$
$$= P(-0.79 \leqslant z \leqslant 0.79) = 2 \times 0.2852 = 0.5704$$

This value is very close to the exact hypergeometric probability which is .5715.

▲

Normal Approximation for the Poisson
Probability Function

The normal distribution may also be used to find approximate probabilities for the Poisson distribution. In this approximation the parameter λ should be at least 25 and the appropriate parameters for the normal are $\mu = \lambda$ and $\sigma = \sqrt{\lambda}$.

EXAMPLE 5-50   Vehicles pass a service station at random at a mean rate of 1 per minute and we wish to know the probability that more than 75 will pass the point in 1 hour. Since the mean rate is 1 per minute, the mean rate per hour is 60 and we use this as λ. We then introduce the continuity correction to change $P(75 < x)$ for the Poisson to $P(75.5 \leqslant x)$ for the normal. Using $\mu = \lambda = 60$ and $\sigma = \sqrt{\lambda} = \sqrt{60} = 7.75$, we produce a normal probability

$$P(75.5 \leqslant x) = P\left(\frac{75.5 - 60}{7.75} \leqslant \frac{x - 60}{7.75}\right) = P(2.00 \leqslant z)$$
$$= 1.0000 - 0.9772 = 0.0228 \qquad \blacktriangle$$

PRACTICE PROBLEMS

P5-20 If z is chosen at random from the standard normal distribution, find
(a) $P(z < 1.65)$
(b) $P(z < -0.83)$
(c) $P(-1.50 \leqslant z \leqslant 0)$
(d) $P(0 < z \leqslant 2.30)$
(e) $P(-2.00 \leqslant z \leqslant 2.00)$
(f) $P(1.53 < z)$
(g) $P(-0.67 < z)$
(h) $P(-1.86 \leqslant z \leqslant 0.54)$
(i) $(-2.25 \leqslant z \leqslant -1.25)$
(j) $P(0.10 < z \leqslant 1.55)$

P5-21 The amount of zinc oxide in an ointment varies from preparation to preparation such that the amount of zinc oxide in a gram of ointment is a random quantity following a normal distribution with a mean of 235.0 milligrams (mg) and a standard deviation of 0.7 mg. If 1 gram of the ointment is randomly selected and x denotes the amount of zinc oxide, find
(a) $P(x < 234.0)$
(b) $P(234.5 \leqslant x \leqslant 235.5)$
(c) $P(237.0 \leqslant x)$
(d) $P(233.3 < x < 236.8)$

P5-22 If the pollen counts for a species of flower vary randomly in a manner well represented by a normal distribution with mean 1000 and standard deviation 80, find the probability that an individual pollen count will be
(a) greater than 1200
(b) less than 775
(c) between 800 and 1100
(d) between 840 and 1160

P5-23 In 150 rolls of a pair of balanced dice, what is the probability that the number of doubles (i.e., two 1's, two 2's, etc.) will be
(a) between 20 and 30 inclusive?
(b) less than 15?
(c) more than 33?

P5-24 A random sample of 200 adults is to be taken in a municipality to investigate the proportion of the adult population who support a proposed change in the bylaws. If 35% of the adult population favor the change and the municipality is a

large metropolitan region, what is the probability that the number in the sample favoring the change will be
(a) at most 55?
(b) between 60 and 80 inclusive?
(c) at least 85?

P5-25 Repeat Practice Problem P5-24 for a village with an adult population of 2500.

P5-26 Summer storms of a sufficient intensity to overflow a storm sewer system occur randomly at a mean rate of three per year. If the storm sewer system is left unimproved for 10 years, what is the probability that in that 10 years, the number of such overflows will be
(a) at least 20?
(b) at most 45?
(c) between 25 and 35 inclusive?

EXERCISES

5-51 Suppose that a value z is to be selected at random from a standard normal distribution. Find
(a) $P(0 \leqslant z \leqslant 1.27)$
(b) $P(-1.39 \leqslant z \leqslant 0)$
(c) $P(-2.42 \leqslant z)$
(d) $P(-0.83 \leqslant z \leqslant 1.65)$
(e) $P(z \geqslant 3.02)$

5-52 Suppose that a value x is to be selected at random from a normal distribution with mean $\mu = 18.65$ and standard deviation $\sigma = 1.25$. Find
(a) $P(x \geqslant 16.24)$
(b) $P(18.00 \leqslant x \leqslant 20.00)$
(c) $P(x \geqslant 20.48)$.

5-53 According to research records, the thickness of an ice surface under given conditions varies about a mean thickness of 15.40 cm with a standard deviation of 2.38 cm. Suppose that the distribution of thickness values may reasonably be assumed to be normal. For any particular point of the surface selected at random, find the probability that the thickness of the ice will be
(a) less than 10 cm
(b) between 12.70 and 17.78 cm
(c) greater than 20.32 cm

5-54 Suppose that scores achieved on a social awareness test follow a distribution that is reasonably normal with a mean of 60.5 and a standard deviation of 10.4.
(a) Find the probability that an individual selected at random will have a score less than 40; hence, determine the percentage of all individuals with scores less than 40.
(b) Find the probability that an individual selected at random will have a score greater than 85; hence, determine the percentage of all individuals with scores greater than 85.

5-55 A balanced coin is to be tossed n times. Using the normal approximation, find the probability that between 48% and 52% of the tosses will be heads; that is, find the probability that the number of heads will be between $0.48n$ and $0.52n$ inclusive if

(a) $n = 100$

(b) $n = 2500$

(c) $n = 5000$

(d) $n = 10,000$

5-56 Find the probability that, in the survey in Exercise 5-43, at least 120 families in the sample will be homeowners.

5-57 Repeat Exercise 5-56, referring to Exercise 5-44 instead of 5-43.

5-58 A switchboard receives calls at random at a mean rate of five per minute. Find the probability that, in a 3-hour period, the number of calls will exceed 975.

5-59 Depths of water wells drilled in an area follow a normal distribution with a mean of 21.5 meters (m) and a standard deviation of 3.2 m. If x denotes the depth of an individual well, find

(a) $P(x < 15.0)$

(b) $P(20 \leqslant x \leqslant 25)$

(c) $P(30 < x)$

5-60 If material flaws occur randomly at a mean rate of 0.2 per square meter, what is the probability that the number of flaws in 5000 square meters of material will be

(a) between 975 and 1025 inclusive?

(b) greater than 980?

(c) greater than 1030?

5-61 If there is a 20% chance that an individual subject will respond to a hypnotic suggestion and suggestion is attempted with 150 subjects selected at random, what is the probability that the number who respond will be

(a) greater than 35?

(b) 16 or fewer?

(c) between 20 and 40 inclusive?

5-62 For a value z chosen at random from a standard normal, find

(a) $P(|z| \leqslant 1.96)$

(b) $P(1.28 < |z|)$

(c) $P(z < -1.65)$

5-63 (a) If a value is chosen at random from a normal distribution, find the probability that it will differ from the mean by at most

(1) 1.5 standard deviations

(2) 2 standard deviations

(3) 3 standard deviations

(b) Compare the results in part (a) to the lower bounds on these probabilities resulting from Chebyshev's theorem.

5-64 A random sample of 300 of the 200,000 subscribers to a magazine is to be taken in a survey to study the degree to which readers agree with editorial comment. What is the probability that the sample proportion will differ from the population proportion by at most 0.05 if the population proportion is

(a) 0.50?

(b) 0.60?

(c) 0.75?

5-65 Repeat Exercise 5-64 if the magazine is very specialized, with a limited list of 4200 subscribers.

P5-20 (a) $P(z < 1.65) = 0.9505$, directly from Table IIIa.

(b) $P(z < -0.83) = 0.2033$, directly from Table IIIa.

(c) $P(-1.50 \leqslant z \leqslant 0) = 0.5000 - P(z \leqslant -1.5) = 0.5000 - 0.0668 = 0.4332$.

(d) $P(0 < z \leqslant 2.30) = P(z \leqslant 2.30) - 0.5000 = 0.9893 - 0.5000 = 0.4893$.

(e) $P(-2.00 \leqslant z \leqslant 2.00) = 2[P(z \leqslant 2.00) - 0.5000] = 2(0.9772 - 0.5000) = 0.9544$.

(f) $P(1.53 < z) = 1.0000 - P(z < 1.53) = 1.0000 - 0.9370 = 0.0630$.

(g) $P(-0.67 < z) = 1.0000 - P(z < -0.67) = 1.0000 - 0.2514 = 0.7486$.

(h) $P(-1.86 \leqslant z < 0.54) = P(z < 0.54) - P(z < -1.86) = 0.7054 - 0.0314 = 0.6740$.

(i) $P(-2.25 \leqslant z \leqslant -1.25) = P(z \leqslant -1.25) - P(z < -2.25) = 0.1056 - 0.0122 = 0.0934$.

(j) $P(0.10 < z \leqslant 1.55) = P(z \leqslant 1.55) - P(z \leqslant 0.10) = 0.9394 - 0.5398 = 0.3996$.

P5-21 We have $\mu = 235.0$ and $\sigma = 0.7$, so that $(x - 235.0)/0.7 = (x - \mu)/\sigma = z$.

(a) $P(x < 234.0) = P\left(\dfrac{x - 235.0}{0.7} < \dfrac{234.0 - 235.0}{0.7} \right)$
$= P(z < -1.43) = 0.0764$

(b) $P(234.5 \leqslant x \leqslant 235.5) = P(-0.71 < z < 0.71) = 0.5222$

(c) $P(237.0 \leqslant x) = P(2.86 \leqslant z) = 0.0021$

(d) $P(233.3 < x < 236.8) = P(-2.43 < z < 2.57) = 0.9874$

P5-22 With $\mu = 1000$ and $\sigma = 80$, $(x - 1000)/80 = z$, where x is the individual pollen count.

(a) $P(1200 < x) = P\left(\dfrac{1200 - 1000}{80} < \dfrac{x - 1000}{80} \right)$
$= P(2.50 < z) = 0.0062$

(b) $P(x < 775) = P(z < -2.81) = 0.0025$

(c) $P(800 < x < 1100) = P(-2.50 < z < 1.25) = 0.8882$

(d) $P(840 < x < 1160) = P(-2.00 < z < 2.00) = 0.9544$

P5-23 Of the 36 possible outcomes, 6 produce doubles. We then have $n = 150$ trials with probability $p = 1/6$ of success on each. We approximate the binomial with the normal with $\mu = np = 25$ and $\sigma = \sqrt{np(1 - p)} = 4.56$. Employing the continuity correction, we obtain the following where a subscript B indicates binomial probability and a subscript N indicates normal probability.

(a) $P_B(20 \leqslant x \leqslant 30) \doteq P_N(19.5 \leqslant x \leqslant 30.5)$
$= P\left(\dfrac{19.5 - 25}{4.56} \leqslant \dfrac{x - 25}{4.56} \leqslant \dfrac{30.5 - 25}{4.56} \right)$
$= P(-1.21 \leqslant z \leqslant 1.21) = 0.7738$

(b) $P_B(x < 15) \doteq P_N(x \leqslant 14.5) = P(z \leqslant -2.30) = 0.0107$

(c) $P_B(33 < x) \doteq P_N(33.5 \leqslant x) = P(1.86 \leqslant z) = 0.0314$

P5-24 Assuming that the metropolitan region is sufficiently large that the finite population correction factor will be close enough to 1 to be omitted, we find approximate probabilities from the normal distribution with $\mu = np = 200 \times 0.35 = 70$ and $\sigma = \sqrt{np(1 - p)} = 6.75$. Employing the continuity correction and using a subscript N to denote the approximating normal probability, we find that

(a) $P(x \leqslant 55) \doteq P_N(x \leqslant 55.5) = P\left(\dfrac{x - 70}{6.75} \leqslant \dfrac{55.5 - 70}{6.75} \right)$
$= P(z \leqslant -2.15) = 0.0158$

(b) $P(60 \leqslant x \leqslant 80) \doteq P_N(59.5 \leqslant x \leqslant 80.5)$
$$= P(-1.56 \leqslant z \leqslant 1.56) = 0.8812$$
(c) $P(85 \leqslant x) \doteq P_N(84.5 \leqslant x) = P(2.15 \leqslant z) = 0.0158$

P5-25 If the population is only of size $N = 2500$, the finite population correction factor $(N - n)/(N - 1) = 2300/2499 = 0.920$ is employed. The standard deviation changes from 6.75 to $\sqrt{np(1 - p)(N - n)/(N - 1)} = 6.47$. Substituting 6.47 for 6.75 in each case produces the final probabilities in the solutions as
(a) $P(z \leqslant -2.24) = 0.0125$
(b) $P(-1.62 \leqslant z \leqslant 1.62) = 0.8948$
(c) $P(2.24 \leqslant z) = 0.0125$

P5-26 If the mean rate is three per year, then for a 10-year period the mean rate is $\lambda = 30$. The Poisson distribution for this experiment is approximated by the normal with $\mu = \lambda = 30$ and $\sigma = \sqrt{\lambda} = 5.48$. Again, employing the continuity correction and letting a subscript N denote the approximating normal probability, we find that
(a) $P(20 \leqslant x) \doteq P_N(19.5 \leqslant x) = P\left(\dfrac{19.5 - 30}{5.48} \leqslant \dfrac{x - 30}{5.48} \right)$
$$= P(-1.92 \leqslant z) = 0.9726$$
(b) $P(x \leqslant 45) \doteq P_N(x \leqslant 45.5) = P(z \leqslant 2.83) = 0.9977$
(c) $P(25 \leqslant x \leqslant 35) \doteq P_N(24.5 \leqslant x \leqslant 35.5)$
$$= P(-1.00 \leqslant z \leqslant 1.00) = 0.6826$$

5-6   MORE ON SAMPLING DISTRIBUTIONS

In Section 4-6 we considered examples of the sampling distribution for the mean of random samples from a small population. The results of those examples indicated some general results which were stated then and which we recall and extend further in this section. The first result related to the mean of the sampling distribution.

Mean of the Sample Mean

For *random* samples of size n from a population with mean μ and standard deviation σ, the sampling distribution for the sample mean has mean *equal to the population mean*
$$\mu_{\bar{x}} = \mu$$

Standard Deviation of the Sample Mean

The second general result was that the standard deviation of the sampling distribution for the mean decreases as the sample size increases. Two variations of the general result were noted.

For random samples of size n taken without replacement from a finite population of size N, the standard deviation of the sample mean is
$$\sigma_{\bar{x}} = \frac{\sigma}{\sqrt{n}} \sqrt{\frac{N - n}{N - 1}}$$

For a very large N, the finite population correction factor $(N - n)/(N - 1)$ is close enough to 1 to be omitted.

> For random samples of size n taken from a very large or infinite population, the standard deviation of the sample mean is
>
> $$\sigma_{\bar{x}} = \frac{\sigma}{\sqrt{n}}$$

Laws of Large Numbers

Since the standard deviation decreases as n increases, the probability tends to be concentrated closer to μ and the probability that $\bar{x}$ will be close to μ increases as n becomes large. This increase in probability is a result of one of the laws of probability known as the *laws of large numbers*.

> According to the *law of large numbers*, as the sample size increases, the probability that the mean of a random sample will be very close to the population mean approaches 1.

The result may be extended to statistics other than the sample mean. They too will have sampling distributions that tend to concentrate on their respective means as the sample size increases.

In particular, consider x/n, the proportion of successes in n independent repeats of the same random trial. Starting with the previous result that $E(x) = np$, where p is the probability of success and where n is the number of trials, the expectation of the sample proportion x/n is found from general expectation rules as $E(x/n) = E(x)/n = np/n = p$. The tendency of a sampling distribution to concentrate on its mean with increases in sample size indicates that the sampling distribution for the sample proportion tends to concentrate on p. This result leads to the statement of the law of large numbers as it applies to proportions.

> If the number of repeated independent trials of a given type is very large, there is a probability very close to 1 that the proportion of successes will be very close to the probability of success in an individual trial.

This result was illustrated in Exercise 5-55.

Normal Distribution for the Sample Mean

We can specify further the sampling distribution for the mean with the following results:

> If the original population of the x's is normal, the sampling distribution for $\bar{x}$, the mean for random samples of size n, will also be normal and will have mean μ, variance σ^2/n and standard deviation $\sigma/\sqrt{n}$.

EXAMPLE 5-51 After adjustment for usual reaction time to recognize a type of familiar symbol, times to recall the meaning of a coded symbol are normally distributed with mean $\mu = 200$ milliseconds (msec) and standard deviation $\sigma = 64$. We plan to take $n = 5$ sample times at random and wish to know the probability that the sample mean will differ from 200 by more then 10.

The means of such samples have mean $\mu_{\bar{x}} = \mu = 200$ and variance $\sigma_{\bar{x}}^2 = \sigma^2/n = 64/5 = 12.8$, so that $(\bar{x} - 200)/\sqrt{12.8} = (\bar{x} - \mu)/(\sigma/\sqrt{n}) = (\bar{x} - \mu_{\bar{x}})/\sigma_{\bar{x}} = z$. The probability that such a mean will differ from 200 by more than 10 is thus

$$
\begin{aligned}
P(10 < |\bar{x} - 200|) &= 1 - P(|\bar{x} - 200| \leqslant 10) \\
&= 1 - P(-10 \leqslant \bar{x} - 200 \leqslant 10) \\
&= 1 - P\left(\frac{-10}{\sqrt{12.8}} < \frac{\bar{x} - 200}{\sqrt{12.8}} < \frac{10}{\sqrt{12.8}}\right) \\
&= 1 - P(-2.80 < z < 2.80) \\
&= 1 - 0.9948 = 0.0052 \qquad \blacktriangle
\end{aligned}
$$

If the original population is not normal, the sampling distribution for the mean will still be approximately normal provided that n is sufficiently large. This latter result is due to the following very important theorem.

> The *central limit theorem* states that, for random samples of size n, the distribution of $(\bar{x} - \mu)/(\sigma/\sqrt{n})$ will become more and more like the standard normal distribution as n increases.

The nature of this result has already been illustrated graphically in Section 4-6, in which the sampling distributions for the mean of samples from two different nonnormal populations of size 20 are reasonably approximated by smooth curves which resemble the normal distribution even though the samples are only of size 5.

EXAMPLE 5-52 Bacterial counts follow a somewhat skewed distribution, with mean and standard deviation per specimen of $\mu = 300$ and $\sigma = 75$. We plan to take a random sample of 100 specimens and we wish to determine the probability that the sample mean will be greater than 305. Although the population is not necessarily normal, the sample size is large and we use the normal distribution with mean $\mu_{\bar{x}} = \mu = 300$ and standard deviation $\sigma_{\bar{x}} = \sigma/\sqrt{n} = 75/10 = 7.5$ as the sampling distribution for the mean. We then use the fact that $(\bar{x} - 300)/7.5 = (\bar{x} - \mu_{\bar{x}})/\sigma_{\bar{x}} = z$ to calculate $P(305 < \bar{x})$ as

$$
P\left(\frac{305 - 300}{7.5} < \frac{\bar{x} - 300}{7.5}\right) = P(0.67 < z) = 1.0000 - 0.7486 = 0.2514 \qquad \blacktriangle
$$

Finite Populations

Although sampling distributions in Section 4-6 resembled normal distributions, their variances were not of the form σ^2/n but rather $(\sigma^2/n) \times (N - n)/(N - 1)$. The general result for $\bar{x}$ must be modified when the finite population correction factor differs considerably from 1.

> When random samples of size n are taken without replacement from a finite population of size N, the sampling distribution for $\bar{x}$ will approach a normal distribution with mean μ and variance
>
> $$\frac{\sigma^2}{n} \times \frac{N - n}{N - 1}$$
>
> as N and n both become large.

If N is much larger than n, so that the finite population correction is close to 1, it may be ignored and the general result of the central limit theorem applies.

EXAMPLE 5-53 $N = 1000$ billing statements have final balances with a mean $\mu = \$7800$ and a standard deviation $\sigma = \$900$. An auditor is to take a random sample of $n = 100$ statements and calculate the mean balance and is interested in the probability that this sample mean will differ from the population mean by at most $200.

The sample mean will have a sampling distribution that is approximately normal with mean $\mu = 7800$ and standard deviation $(\sigma/\sqrt{n})$ $\times \sqrt{(N - n)/(N - 1)} = (900/10) \times \sqrt{900/999} = 85.4$. The required probability is then found as $P(|\bar{x} - 7800| \leqslant 200) = P(|(\bar{x} - 7800)/85.4| \leqslant 200/85.4) = P(|z| \leqslant 2.34) = 0.9808$. ▲

Probabilities on how close $\bar{x}$ will be to μ in terms of the standard deviation σ can be found on the basis of the sample size alone provided that the population is normal or the sample size is large enough to invoke the central limit theorem.

EXAMPLE 5-54 Suppose that we plan to take a random sample of size 64 from a population, not necessarily normal, with unknown mean and standard deviation μ and σ. We wish to find the probability that the sample mean will differ from the population mean by less than one-fourth of the standard deviation (i.e., $\sigma/4$). Assuming that $n = 64$ is large enough to rely on the central limit theorem (a usual rule of thumb is $n \geqslant 30$), we calculate this probability as

$$P\left(|\bar{x} - \mu| < \frac{\sigma}{4}\right) = P\left(-\frac{\sigma}{4} < \bar{x} - \mu < \frac{\sigma}{4}\right)$$

$$= P\left(\frac{-\sigma/4}{\sigma/8} < \frac{\bar{x} - \mu}{\sigma/8} < \frac{\sigma/4}{\sigma/8}\right) \qquad \left(\text{since } \frac{\sigma}{\sqrt{n}} = \frac{\sigma}{8}\right)$$

$$= P(-2 < z < 2) = 0.9544$$ ▲

Other Statistics

Sampling distributions for other statistics such as the sample variance will be developed in other chapters as the need arises. Other statistics follow the same general behavior as the sample mean. Their sampling distributions are spread around central values or measures of location and will tend to concentrate increasingly on these values as the sample size increases.

PRACTICE PROBLEMS

P5-27 Calves receiving implants shortly after birth show weights when shipped to market that exceed those for calves without implants. The increases follow a normal distribution with mean $\mu = 25$ lb and standard deviation $\sigma = 4$ lb. If 25 calves receive implants, what is the probability that the mean weight improvement will exceed 23 lb?

P5-28 If, when placed in close confinement for a fixed period of time, mice become aggressive such that the number of aggressive interactions is approximately normally distributed with mean $\mu = 35$ and standard deviation $\sigma = 7$, what is the probability that the mean number of aggressive interactions in 72 times periods will be
(a) less than 33?   (b) less than 37?

P5-29 If 2500 candidates write an aptitude test without time limit and the times that they require for the test have a mean $\mu = 61.3$ minutes and standard deviation $\sigma = 8.7$, what is the probability that the mean time of a random sample of 200 of these candidates will be
(a) more than 60.0 minutes?   (b) more than 63.0 minutes?

P5-30 What is the probability that the sample mean will differ from a population mean by at most one-tenth of the population standard deviation for a random sample of size
(a) 25?   (c) 400?
(b) 100?

P5-31 An ordinary deck of 52 playing cards is to be shuffled and then one card is to be selected at random, noted, and returned to the deck. What is the probability that the proportion of hearts will be between 0.24 and 0.26 if this process is repeated for a total of
(a) 300 times?   (c) 10,800 times?
(b) 4800 times?

EXERCISES

5-66 A random sample of size $n = 25$ is to be taken from a normal distribution with mean $\mu = 35.0$ and standard deviation $\sigma = 2.5$, and the sample mean $\bar{x}$ is to be found. Find
(a) $P(34.0 \leqslant \bar{x} \leqslant 36.0)$
(b) $P(33.5 \leqslant \bar{x})$
(c) $P(\bar{x} \geqslant 36.25)$

5-67 A random sample of 100 values is to be taken from a population (not necessarily normal) with mean 300 and standard deviation 25, and the sample mean $\bar{x}$ is to be calculated. Find

(a) $P(|\bar{x} - 300| > 5)$
(b) $P(296 \leqslant \bar{x} \leqslant 306)$
(c) $P(\bar{x} \geqslant 304)$
(d) $P(\bar{x} \leqslant 298)$

5-68 The weights of the contents of packages from a given packaging process have an approximate normal distribution with a mean of 2.268 kg and a standard deviation of 0.014 kg. If ten packages are selected at random, what is the probability that the mean weight of the packages will

(a) exceed 2.275 kg?
(b) be less than 2.260 kg?

5-69 Suppose that retention scores are known to exhibit a standard deviation of 15. If a sample of 100 retention scores is to be taken at random and the corresponding sample mean calculated, what is the probability that the mean of the sample will

(a) exceed the population mean by at least 3?
(b) be within 1 of the population mean?
(c) differ from the population mean by at least 2?

5-70 The numbers of persons accommodated in 1500 resort lodgings produced a mean of 800 and a standard deviation of 150.
(a) What is the probability that the mean of a random sample of 100 lodgings will be

(1) greater than 825?
(2) less than 775?
(3) between 750 and 830?
(b) Repeat part (a) for a sample of 150 lodgings.

5-71 Repeat Exercise 5-70 for a population of lodgings sufficiently large that the finite population correction may be ignored.

5-72 If there is a probability 0.001 that an individual lid will not seal properly, what is the probability that the percentage of improperly sealing lids will be less than 0.15% in a random collection of 25,000 lids?

5-73 If readings on the reducing sugar content of evergreen needles follow a normal distribution with standard deviation $\sigma = 5.0$ and mean μ equal to the actual sugar content, what is the probability that the mean of 25 readings will differ from the true content by at most 2.0?

5-74 Modifying tillage procedures to reduce runoff reduces soil erosion and increases retention of moisture and fertilizer, thus increasing crop yields. If individual yield improvements per acre show a mean of 4.2 bushels and a standard deviation of 1.3, what is the probability that modified tillage of 100 acres will produce a *total* improvement of more than 400 bushels?

5-75 If the numbers of days required before obtaining a hearing in 1200 cases produced a mean of 33.3 and a standard deviation of 6.2, what is the probability that the *total* waiting time for a random sample of 75 cases will have exceeded 2350 days?

P5-27 For samples of size $n = 25$, the sampling distribution for the sample mean has mean $\mu = 25$ and standard deviation $\sigma / \sqrt{n} = 4/\sqrt{25} = 0.8$. The required probability thus becomes $P(23 < \bar{x}) = P\left(\dfrac{23 - 25}{0.8} < \dfrac{\bar{x} - 25}{0.8}\right) = P(-2.5 < z) = 0.9938$.

P5-28 Since the sampling distribution for the mean of samples of size $n = 72$ has mean $\mu = 35$ and standard deviation $\sigma / \sqrt{n} = 7/\sqrt{72} = 0.82$, the required probabilities are

(a) $P(\bar{x} < 33) = P\left(\dfrac{\bar{x} - 35}{0.82} < \dfrac{33 - 35}{0.82}\right) = P(z < -2.44) = 0.0073$

(b) $P(\bar{x} < 37) = P(z < 2.44) = 0.9927$

P5-29 The sampling distribution for the mean of a sample of size $n = 200$ from the population of size $N = 2500$ will have mean $\mu = 61.3$ and standard deviation

$$\frac{\sigma}{\sqrt{n}} \times \sqrt{\frac{N - n}{N - 1}} = \frac{8.7}{\sqrt{200}} \times \sqrt{\frac{2300}{2499}} = 0.59.$$

The required probabilities are thus

(a) $P(60.0 < \bar{x}) = P\left(\dfrac{60.0 - 61.3}{0.59} < \dfrac{\bar{x} - 61.3}{0.59}\right)$
    $= P(-2.20 < z) = 0.9861$

(b) $P(63.0 < \bar{x}) = P(2.88 < z) = 0.0020$

P5-30 For a general n the probability is

$$P\left(|\bar{x} - \mu| < \frac{\sigma}{10}\right) = P\left(\frac{|\bar{x} - \mu|}{\sigma/\sqrt{n}} < \frac{\sigma/10}{\sigma/\sqrt{n}}\right) = P\left(|z| < \frac{\sqrt{n}}{10}\right)$$

Substituting 25,100 and 400 for n, we find the required probabilities as

(a) $P(|z| < 0.5) = 0.3830$

(b) $P(|z| < 1.0) = 0.6826$

(c) $P(|z| < 2.0) = 0.9544$

P5-31 For a general n the probability is $P(0.24 < x/n < 0.26) = P(0.24n < x < 0.26n)$. Assuming that the sample size is large enough to use the normal approximation and that the continuity correction will be negligible and noting that $p = 0.25$, we find the probability as

$$P\left(\frac{0.24n - 0.25n}{\sqrt{n \times 0.25 \times 0.75}} < \frac{x - 0.25n}{\sqrt{n \times 0.25 \times 0.75}} < \frac{0.26n - 0.25n}{\sqrt{n \times 0.25 \times 0.75}}\right)$$

$$= P\left(\frac{-0.01n}{\sqrt{3n/16}} < z < \frac{0.01n}{\sqrt{3n/16}}\right)$$

$$= P\left(|z| < 0.04\sqrt{n/3}\right)$$

Substituting $n = 300$, 4800, and 10,800, we find the required probabilities as

(a) $P(|z| < 0.4) = 0.3108$

(b) $P(|z| < 1.6) = 0.8904$

(c) $P(|z| < 2.4) = 0.9836$

5-76 For a probability function $f(x)$, the value $\Sigma\, xf(x)$ is called the _____.

5-77 The finite population correction factor applied to the variance of a sampling distribution for the mean
    (a) reduces the variance
    (b) increases the variance
    (c) may either increase or decrease the variance

5-78 The value $\binom{n}{x}$ (i.e., "n choose x") will equal the value $\binom{n}{n-x}$
    (a) always
    (b) never
    (c) only if $x = n/2$

5-79 For the central limit theorem to apply, the original population must have a normal distribution.
    (a) True
    (b) False

5-80 If we wish to calculate $P(x > 25)$ for a binomial distribution and we change to a normal, then, for the normal, we should calculate $P(x \geqslant \underline{\quad})$.

5-81 To be a probability function, $f(x)$ must satisfy
    (a) $-1 \leqslant f(x) \leqslant 1$
    (b) $0 \leqslant f(x)$ and $\Sigma\, f(x) = 1$
    (c) $0 \leqslant f(x) \leqslant 1$ and $\Sigma\, f(x) < \infty$
    (d) $\Sigma\, f(x) = 0$

5-82 If the value in the normal tables corresponding to $z = 1.16$ is 0.8770, then $P(-1.16 \leqslant z \leqslant 1.16)$ is equal to _____.

5-83 When the Poisson probability function is used to approximate the binomial, the value of the parameter λ is determined as _____.

5-84 The expectation of a probability function $f(x)$ is
    (a) always positive
    (b) $\Sigma f(x)$
    (c) $\Sigma xf(x)$
    (d) $\Sigma x/n$

5-85 The area under the graph of normal distribution between μ and ∞ is equal to __.

5-86 A planting bed is to include one small evergreen—a juniper, a yew, or small cedar; one large evergreen—a full-size cedar, a blue spruce, a fir, or a pine; and one flowering bush—a lilac, a snowball, a forsythia, a spiraea, or a rose. How many ways can the three plants be selected?

5-87 An urn contains five counters, three red and two blue. Five drawings are made without replacement.
    (a) Sketch a tree diagram to indicate all possible drawing orders (e.g., *BRRBR*) for the five draws. How many are there?
    (b) In how many orders does it take exactly two draws to obtain both blue counters? In how many does it take three? four? five?

5-88 (a) A player in a card game has received five cards and discarded two—the ace of hearts and 6 of clubs. He still has the 8, 9, and 10 of diamonds. What is the probability that two replacement cards will give the player five diamonds in a row?

(b) If the player has only received the three diamonds and four other players also have three cards each (all showing) consisting of the jack of diamonds, the 5 of diamonds, and ten nondiamonds, what is the probability that two additional cards will produce five diamonds?

(c) In each of parts (a) and (b), what is the probability that the player will end up with five diamonds not in a row?

5-89 A sample of eight saplings is to be chosen at random from a stock consisting of 15 of a hardy strain and 25 of a weaker strain.

(a) What is the probability that the eight saplings will all be hardy?

(b) What is the probability that the proportion of hardy saplings in the sample will be the same as that in the original stock?

5-90 Suppose that $f(x)$ is a function such that $f(x) = (5 - x^2)/15$ for $x = -2, -1, 0, 1, 2$ [and $f(x) = 0$ for all other values of x]. Show that $f(x)$ is a probability function and find the corresponding mean and variance.

5-91 Boat traffic arrives at the first lock in a canal system in a fully random pattern at a mean rate of five vessels per 15-minute period. The lock can accommodate at most ten vessels in any 15-minute period. In a particular 15-minute period, what is the probability that the lock will not be able to accommodate all of the arriving vessels?

5-92 (a) If 10% of the output of a machine is defective and one item is drawn at random from a batch of several thousand produced by this machine, what is the probability that it will be defective?

(b) (1) If five items are drawn from this batch, what is the probability that exactly one will be defective?

(2) What is the probability that at least one will be defective?

5-93 Suppose that someone will give us $4.00 each time that we roll a 1 with a balanced die. How much should we pay when we roll a 2, 3, 4, 5, or 6 to make this game fair?

5-94 If the probability that the value of a certain stock will remain the same is 0.46, the probabilities that its value will increase by $0.50 or $1.00 per share are, respectively, 0.17 and 0.23, and the probability that its value will decrease by $0.25 per share is 0.14, what is the expected gain per share?

5-95 A machine produces steel pins whose lengths are normally distributed with a mean of 1 inch (in.) and a standard deviation of 0.05 in.

(a) (1) What is the probability that a pin chosen at random will have a length between 1.01 and 1.04 in.?

(2) If 100 steel pins are drawn at random, how many of them would you expect to have lengths less than 1.02 in.?

(3) If the mean of the 100 pins in part (2) is taken, what is the probability that the mean will be between 0.99 and 1.01 in.?

(b) Repeat part (a) (2) and (3) if there are 1000 pins with $\mu = 1$ and $\sigma = 0.05$.

5-96 Suppose that the distance d required for a vehicle to stop under certain conditions has a normal distribution with mean 110 and variance 95. Find

(a) $P(d \geqslant 118)$

(b) $P(d \geqslant 103)$

(c) $P(105 \leqslant d \leqslant 115)$

5-97 A monkey has been trained to push an illuminated button whenever a number flashes on a screen. There are ten buttons labeled 0, 1, 2,..., 9 and the numbers flashed on the screen are random selections of these numbers. If the monkey has no tendency to push the same button as the number, but pushes a button at random, what is the probability that the monkey will push the same button as the number flashed

(a) in at most 2 of 10 trials?

(b) in more than 6 of 20 trials?

(c) in more than 50 of 400 trials?

5-98 A fishing tank at a sporting show contains 36 speckled trout and 24 brown trout.

(a) How many ways can six speckled trout be selected without replacement?

(b) How many ways can four brown trout be selected without replacement?

(c) What is the expected number of speckled trout in a sample of ten caught at random without replacement if each individual trout is just as likely to be caught as any other?

(d) What is the probability that, of ten caught at random without replacement, six will be speckled trout?

5-99 Of 2000 prescriptions filled by a pharmacist, 1200 are subject to refill. If a sample of 200 prescriptions is selected at random, what is the probability that between 110 and 135 of them (inclusive) will be subject to refill?

5-8 GENERAL REVIEW

5-100 A value of 80 would appear to be reasonable as the standard deviation for a set of data with lowest value 73.05 and highest value 126.15.

(a) True   (b) False

5-101 If we have a population with mean μ and variance σ^2 and we form a new population by changing each x to $(x - \mu)/\sigma$, the new mean and variance will be _____ and _____ .

5-102 If the mean is subtracted from all the values of a symmetric population, the new median will be _____.

5-103 If A and B are mutually exclusive, P(A *and* B) will equal

(a) P(A) + P(B)

(b) a number greater than 0 and less than P(A) + P(B)

(c) 1.0

(d) 0.0

5-104 A parameter is a characteristic of a _____ or a _____ .

5-105 If a few classes in a frequency distribution have class intervals that are half those for the other classes, then, in a histogram, the heights for these few classes should be

(a) twice the frequencies   (c) equal to the frequencies

(b) half the frequencies

5-106 The values P(A) = 0.49, P(B) = 0.65 are possible

(a) if A, B are mutually exclusive

(b) if A, B are mutually exclusive and exhaustive

(c) if A, B are exhaustive

(d) never

5-107 If 100 is subtracted from all the values in a set of data, all these differences are squared, and the squares are added up, the expression for the final result is _____.

5-108 If A, B, and C are exhaustive events, $P(A \text{ or } B \text{ or } C)$ is _____.

5-109 The area under a probability graph between 1.5 and 2.7 represents _____.

5-110 Counters marked 1, 2, and 3 are placed in a bag, and one is withdrawn and replaced. If the operation is repeated three times, what is the probability of obtaining a total of 6?

5-111 (a) If 10% of the population are hunters, and 0.1% of hunters use a bow and arrow and a member of the population is selected at random, what are the odds against that member's being a hunter who uses a bow and arrow?

(b) If the population in part (a) consists of several hundred millions, what is the probability that a survey of 20,000 will include two or fewer hunters who use bows and arrows?

5-112 (a) If 25% of a certain type of seed fails to germinate and ten such seeds picked at random from several thousand are planted, find the probability that at most three will fail to germinate.

(b) If 400 seeds are planted, find the probability that at most 20% fail to germinate.

5-113 Suppose that $f(x) = (15 - x)/15$ for $x = 10, 11, 12, 13, 14 = 0$ otherwise.

(a) Show that $f(x)$ is a probability function.

(b) Find the expectation of x.

5-114 A project consists of five individual tasks. One, two, or three tasks must be performed each day until the project is finished.

(a) Draw a tree diagram to illustrate all of the possible work schedules (do not differentiate between individual tasks—consider only the number of tasks performed each day).

(b) Considering all possible schedules, what is the mean number of days required for the project? What is the mode? What is the median?

(c) Draw a frequency histogram for the number of days required to complete the project.

(d) If a schedule is drawn at random and each schedule is equally likely to be drawn, what is the probability that more than 3 days will be required to complete the project?

5-115 Suppose that a particular advertising lottery is such that anyone can enter (once only) by mailing in a free coupon, and a draw will take place from the coupons received. There is a first prize of value $500 and five consolation prizes of value $25 each. If it costs 35 cents for a stamp and envelope to mail the coupon, is it worthwhile (in the sense of mathematical expectation) to enter the lottery if we may assume that at least 2000 coupons will be mailed in?

5-116 A record of 100 years shows the following numbers of years in which a river flooded, 0, 1, 2, 3, 4, or 5 times:

Number of floods	Number of years
0	24
1	35
2	24
3	12
4	4
5	1

(a) Illustrate these results with a histogram.

(b) What are the mean, the mode, and the variance of the number of floods per year?

(c) Suppose that the number of floods per year follows a Poisson distribution with parameter λ, and that there is enough equipment available for the next year to withstand 2 floods. Using the mean from part (b) as the parameter, λ, what is the probability that the equipment will prove to be insufficient?

5-117 Suppose that, of 100 indistinguishable nuclear fuel bundles, 3 have defective radiation shields. Suppose that a preliminary screening process is such that if a defective shield is tested, there is a 95% chance that it will be detected and a 5% chance that it will be falsely classified as nondefective; and if a nondefective shield is tested, there is a 90% chance that it will be properly identified and a 10% chance that it will be falsely classified as defective.

(a) If one of the bundles is picked at random, what is the probability that it will have a defective shield?

(b) If one is picked at random, and the shield is tested, what is the probability that it will be classified as defective?

(c) If one is picked at random and the shield is tested and, as a result, classified as defective, what is the probability that it is actually defective?

(d) If one is picked at random and the shield is tested, and, as a result, classified as nondefective, what is the probability that it is nondefective?

5-118 If 3000 shipments have weights with a mean $\mu = 500$ kg and standard deviation $\sigma = 20$ kg, what is the probability that the total weight in a random selection of 250 shipments will exceed 124,500 kg?

5-119 In November 1979, gasoline prices were found to be as given in the following table, indicating total cost per gallon and how much of the total cost represented taxes.

Country	Total cost (U.S. dollars/ U.S. gal)	Tax
Canada (Montreal)	$0.71	$0.21
United States (New York State)	0.97	0.17
United Kingdom	1.76	0.86
West Germany	2.16	0.89
France	2.46	1.55
Italy	2.55	1.80

(a) Illustrate these data with a bar graph such that the bars represent total cost per gallon. Shade an appropriate portion of each bar to represent tax.

(b) Draw a bar graph of the percentages of total costs that represent tax.

5-120 In a game of chance involving the rolling of two dice, a roll of "7" is ignored and bets are made on whether the roll will produce a number less than 7 or greater than 7. The player with the dice has rolled a very large number of consecutive numbers less than 7 and is about to roll the dice again. Bearing in mind the law of large numbers as it relates to proportions, should you bet on a number less than 7 or against a number less than 7?

ESTIMATING

6-1 INTRODUCTION

Estimating unknown population values from the information in sample data is one of the major applications of statistical methodology. In this chapter we consider the estimation of population parameters on the basis of the values of sample statistics.

It is important not only to choose appropriate statistics but also to be able to assess the reliability of such statistics as estimating quantities. To make the appropriate choice and to perform the assessment, we study the behavior of possible statistics by considering their sampling distributions.

6-2 ESTIMATING PARAMETERS

EXAMPLE 6-1   Suppose that the manufacturer of a new pesticide wishes to know what proportion of insects present in an enclosed space will be killed in a given time after a particular quantity of the pesticide is introduced into the space. The value of interest is the overall proportion of all insects that are susceptible to the pesticide under the given conditions. This value can also be regarded as the probability that any insect chosen at random will be susceptible.

The pesticide is to be tried on a random sample of 1000 insects in an experiment. The possible results of the experiment can be represented by a binomial distribution in which the number of trials n is 1000, the probability of success p (the probability that an individual insect will be killed) is unknown, and the number of insects killed is x. Considering the sampling distribution for x, we note that, on the average, x will be equal to np (in this case, $1000p$) the mean of the binomial distribution, and that $x/1000$ (the sample proportion) will thus be equal to the unknown value p on the average. Because it is on the average the correct value, we choose $x/1000$ as the appropriate statistic to estimate the parameter p and we refer to it as an unbiased estimator. ▲

Unbiased Estimators

> If a statistic has a sampling distribution with a mean
> equal to a parameter of interest, that statistic is said to
> provide an *unbiased estimator* of the parameter.

The property of unbiasedness is often desired of statistics and, in many problems of estimation, statistics such as the sample proportion are chosen, in part, because they provide unbiased estimates for the parameters of interest. Two other statistics that we have considered that are unbiased estimators are the sample mean and the sample variance. As we have noted earlier from their sampling distributions, they are unbiased estimators of the population mean and variance, respectively.

Minimum Variance Unbiased Estimators

Unbiasedness is not the only desirable property of a statistic. If we wish to estimate the parameter λ of a Poisson distribution, we have a choice of two unbiased statistics, the sample mean and the sample variance, since the mean and the variance of the distribution are both equal to the value of the parameter. It turns out that the sampling distribution of the mean $\bar{x}$ has smaller variance than the sampling distribution of the variance s^2; thus, besides being equal to λ on the average, the sample mean also has higher probability of being close to λ in individual samples. For this reason, we choose the sample mean instead of the sample variance as the appropriate statistic.

This result illustrates another desirable property of a statistic—that its sampling distribution has a variance no larger than that for the distribution for any other unbiased statistic. Such a statistic is not only said to be unbiased, it is said to be *minimum variance unbiased*. There are other properties that might also be considered; however, we will not develop any further properties. The concepts of unbiased and minimum variance estimators serve as sufficient examples of aspects of statistics to be considered in the choice of an estimate.

6-3 EVALUATING ESTIMATES AND SAMPLE SIZES

The example with regard to the Poisson distribution illustrated a method of assessing a statistic in terms of probability. We choose estimates with small variance in order to have a high probability of obtaining an estimate that is close to the true parameter value. In evaluating a statistic, we may try to determine the probability that it will be within a particular distance of the true value of the parameter being estimated. Determining such a probability requires the use of a sampling distribution.

EXAMPLE 6-2 A psychometrist plans to assess a random sample of 25 members of a large group with a pattern recognition testing procedure that is known to produce test scores that have a normal distribution with a standard deviation of 16. The mean response of the 25 sample members is to be used to estimate the overall mean for the full group and it is desired that the "error" in the estimate be at most 5. It is of interest to determine the probability that the estimate will differ from true mean value by at most 5.

From Chapter 5 we know that the sample mean $\bar{x}$ will have a sampling distribution that is normal with mean $\mu_{\bar{x}}$ equal to the true mean value μ of the population and with standard deviation $\sigma_{\bar{x}} = \sigma/\sqrt{n}$.

We may thus use the sampling distribution for the mean of a sample from a normal population to calculate

$$P(|\bar{x} - \mu| \leqslant 5) = P(-5 \leqslant \bar{x} - \mu \leqslant 5)$$

$$= \left(\frac{-5}{\sigma/\sqrt{n}} \leqslant \frac{\bar{x} - \mu}{\sigma/\sqrt{n}} \leqslant \frac{5}{\sigma/\sqrt{n}} \right)$$

$$= P(-1.56 \leqslant z \leqslant 1.56) \qquad (\text{since } \sigma = 16, n = 25)$$

$$= 0.8812$$

The psychometrist thus has an 88% chance of obtaining an estimate within 5 of the true value if the mean of a sample of size 25 is to be used as the estimate. ▲

Sample Size for Estimating a Mean

If the probability that an estimator will be sufficiently close to the true parameter value is not high enough, increasing the sample size may be necessary.

EXAMPLE 6-3 Suppose that the psychometrist in Example 6-2 is not satisfied with the probability obtained in the evaluation of the estimate, but would rather have an estimating procedure that provides a 98% chance of obtaining an estimate within 5 of the true value. An increase in the probability can be obtained if the variance σ^2/n of the estimate is decreased; this can be achieved by increasing the sample size n.

Let n be the required sample size. We then rewrite the probability statement as $P(-5/(16/\sqrt{n}) \leqslant z \leqslant 5/(16/\sqrt{n})) = 0.98$ as desired. In this probability statement, we have two numerical z-values which are symmetric about 0 and include the central 98% of the standard normal distribution as illustrated in Figure 6-1. As illustrated in Figure 6-1, only 1% of the distribution is above $5/(16/\sqrt{n})$. In other words $5/(16/\sqrt{n})$ must be the 99th

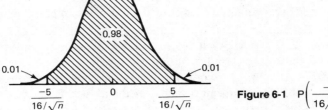

Figure 6-1 $P\left(\dfrac{-5}{16/\sqrt{n}} \leqslant z \leqslant \dfrac{5}{16/\sqrt{n}}\right).$

percentile of the standard normal distribution. As illustrated below, the 99th percentile of the standard normal distribution is equal to 2.326.

We then have the result $5/(16/\sqrt{n}) = 2.326$ which produces $n = (16 \times 2.326/5)^2 = 55.40$. As it is not possible to take a sample of size 55.4, the sample size should be rounded *up* to 56. A sample size 55 is a reduction from 55.4 and involves a decrease in probability. With a sample size of 56, there is a probability slightly more than 0.98 that the sample mean will be within 5 of the true population mean. The minimum sample size meeting these requirements is thus 56. ▲

Percentiles of the Standard Normal

Example 6-3 illustrates only one of several procedures that require the use of percentiles of the standard normal distribution.

> Percentiles of the standard normal distribution are denoted by a subscripted z, the subscript being equal to the probability above the percentile. The $(1 - \alpha) \times 100th$ *percentile* is denoted as z_α for any α (the Greek lowercase letter alpha) between 0 and 1.

Percentiles of the standard normal distribution are found in Table IIIb of Appendix B, illustrated in Fig. 6-2(a).

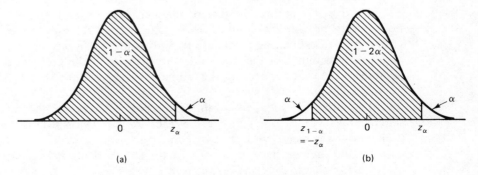

Figure 6-2 Illustration of Table IIIb.

EXAMPLE 6-4 The 99th percentile is the 0.99×100th or $(1 - 0.01) \times 100$th percentile. It has probability 0.01 above it (i.e., in the upper portion of the distribution) and is denoted as $z_{0.01}$. From Table IIIb we find $z_{0.01} = 2.326$, as used in Example 6-3.
▲

> Because of the symmetry of the standard normal distribution, lower percentiles of the standard normal are negatives of upper percentiles; that is, $z_{1-\alpha} = -z_\alpha$, as illustrated in Figure 6-2(b).

EXAMPLE 6-5

$$z_{0.95} = -1.645 = -z_{0.05}$$
▲

General Sample Size for Estimating a Mean

The general solution for sample size determination is found by reconsidering Example 6-3 in more general terms. In general, we designate the maximum tolerable error as E (instead of 5), the standard deviation as σ (instead of 16), and the desired probability as $1 - \alpha$ (instead of 0.98). This produces $P(-E/(\sigma/\sqrt{n}) < (\bar{x} - \mu)/(\sigma/\sqrt{n}) < E/(\sigma/\sqrt{n})) = 1 - \alpha$ which leads to $P(z > E/(\sigma/\sqrt{n})) = \alpha/2$ (instead of 0.01); thus, $E/(\sigma/\sqrt{n}) = z_{\alpha/2}$, where, for any numerical value of $\alpha/2$, $z_{\alpha/2}$ is found in Table IIIb in the row labeled with the numerical value of $\alpha/2$. We then find the required sample size by solving for n and rounding up to the next integer with the following result:

> To obtain a probability of at least $1 - \alpha$ that a random sample will have mean $\bar{x}$ that differs from the mean μ of a normal distribution by at most E, the sample size n should be at least
>
> $$n = \left(\frac{\sigma z_{\alpha/2}}{E} \right)^2$$
>
> where σ is the population standard deviation.

Also, if we are assessing the mean of a large sample or if we are calculating a sample size that we expect to be large, then, by the central limit theorem introduced in Chapter 5, we may use the foregoing procedures *even if the population is not normal.*

EXAMPLE 6-6 A chemist uses an analysis procedure for measuring concentrations which produces readings that exhibit a standard deviation of 0.04 and a mean equal to the true concentration level. The mean $\bar{x}$ of a number of sample readings is to be used to estimate μ. It is desired that there be a 95% chance of obtaining an estimate within 0.01 of the true mean. In this case, $\sigma = 0.04$ and $E = 0.01$. Also, $1 - \alpha = 0.95$; thus, $\alpha = 0.05$ and $\alpha/2 = 0.025$. The value $z_{\alpha/2}$ is thus $z_{0.025}$ and is found to be 1.960 from Table IIIb. Finally, the required sample size is found as $(\sigma z_{\alpha/2}/E)^2 = [(0.04 \times 1.96)/0.01]^2 = 61.47$, which should be increased to 62.
▲

Sample Size for Estimating a Proportion or Percentage

The techniques described above can also be used in the estimation of proportions or percentages if the (anticipated) sample size is sufficiently large to justify approximating the corresponding binomial or hypergeometric distribution with the normal distribution.

EXAMPLE 6-7 In a study on the effects of a steam treatment for nasal congestion, it is of interest to know the minimum sample size necessary to have a 90% chance that the proportion of cold sufferers in a sample who obtain relief from nasal congestion within a short time after the steam treatment will differ from the population proportion by at most 0.07. ▲

Very Large or Infinite Populations

For very large or infinite populations, the number of successes x may be considered as having a binomial distribution with mean np and standard deviation $\sqrt{np(1-p)}$. This binomial is approximated by the normal with the same mean and standard deviation.

We have already noted that an appropriate estimate of the population proportion p is the sample proportion x/n. Using our rules with regard to the adjustment of data and expectations, we divide the mean and standard deviation of x by n to find the mean and standard deviation of x/n. Our procedure will then be based on the distribution for x/n.

> The sample proportion x/n may be considered to have a sampling distribution that is approximately normal with a mean of p and a standard deviation of $\sqrt{p(1-p)/n}$.

Suppose now that we plan to estimate p with x/n and we wish the error in the estimate to be at most E in magnitude. We want our estimate to satisfy the condition $|x/n - p| \leqslant E$. As with estimating a mean, we cannot guarantee the result but we may specify a desired minimal probability, say $1 - \alpha$, that the condition will hold true.

EXAMPLE 6-8 In Example 6-7 we wish to have a probability of $1 - \alpha = 0.90$ that the proportion of cold sufferers in the sample who obtain relief will differ from the true population proportion p by at most $E = 0.07$. We thus wish to find the smallest n which will satisfy the probability statement $P(|x/n - p| \leqslant 0.07) = .90$. ▲

The result that we wish to achieve may be written as a general probability statement $P(|x/n - p| \leqslant E) = 1 - \alpha$ which may be rewritten as $P(-E \leqslant x/n - p$

$\leqslant E) = 1 - \alpha$ and then further adjusted to

$$P\left(\frac{-E}{\sqrt{p(1-p)/n}} \leqslant \frac{x/n - p}{\sqrt{p(1-p)/n}} \leqslant \frac{E}{\sqrt{p(1-p)/n}} \right) = 1 - \alpha$$

or

$$P\left(\frac{-E}{\sqrt{p(1-p)/n}} \leqslant z \leqslant \frac{E}{\sqrt{p(1-p)/n}} \right) = 1 - \alpha$$

By the same arguments as those used for finding n to estimate μ, we know that the final probability statement will be satisfied if $E/\sqrt{p(1-p)/n} = z_{\alpha/2}$. Solving this expression for n, we find the required sample size as $n = (z_{\alpha/2}/E)^2 \times p(1-p)$.

Unfortunately, this expression involves the unknown value p. We must base our solution on a value of $p(1-p)$ that will produce a satisfactory n for all possible values of p. To be safe, we base our solution on the value of $p(1-p)$ that produces the largest sample size—that is, the value for which $p(1-p)$ is a maximum. The maximum value of $p(1-p)$ is $1/4$ and occurs when $p = 1/2$. If we have no prior knowledge on p, we replace $p(1-p)$ with $1/4$ and then calculate the sample size n.

> In order to have a probability of at least $1 - \alpha$ that a random sample proportion x/n will differ from a population proportion p by at most E, then, with no prior knowledge on p, the sample size should be at least
>
> $$n = \frac{\left(z_{\alpha/2}/E\right)^2}{4}$$

EXAMPLE 6-9 In Examples 6-7 and 6-8, we have been given no prior knowledge on the possible value for p. We then use the "safe" sample size based on $E = 0.07$ and $1 - \alpha = 0.90$, which produces $\alpha = 0.10$, $\alpha/2 = 0.05$, and $z_{\alpha/2} = z_{0.05} = 1.645$. The sample size must be at least $(z_{\alpha/2}/E)^2/4 = (1.645/0.07)^2/4 = 138.1$. A sample size of at least 139 is required for the study. ▲

If p is different from $1/2$, this value for n will be larger than necessary and, if possible, we should try to reduce it. The required sample size decreases as the difference between p and $1/2$ increases. If, for example, we know that p cannot exceed 0.3, then 0.3 is as close as p can come to $1/2$ and we base our solution on $p(1-p) = 0.3 \times 0.7 = 0.21$ to produce $n = (z_{\alpha/2}/E)^2 \times 0.21$. If, as another example, we know that p cannot be less than 0.8, then 0.8 is as close as p can come to $1/2$ and we use $0.8 \times 0.2 = 0.16$ and $n = (z_{\alpha/2}/E)^2 \times 0.16$.

> If prior knowledge indicates that p differs from $1/2$, the required minimum sample size to obtain a probability of at least $1 - \alpha$ that x/n will differ from p by at most E is
>
> $$n = \left(\frac{z_{\alpha/2}}{E} \right)^2 \times p^*(1 - p^*)$$
>
> where p^* is the closest value to $1/2$ that p may take.

EXAMPLE 6-10 In the nasal congestion example, if it can be reasonably assumed that the proportion of the population who would obtain relief is at least 0.70, p^* is set equal to 0.70 and the new minimum sample size is calculated as $n = (1.645/0.07)^2 \times 0.7 \times 0.3 = 116.0$. A sample size of at least 116 would be required. ▲

Since percentages are just proportions multiplied by 100, we may find sample sizes to produce adequate estimates of percentages by converting the requirements on the percentage estimate to requirements on the corresponding proportion and solving for n.

EXAMPLE 6-11 In order to estimate the percentage of a given type of plant that will bear fruit within 2 years of transplanting, a random sample of plants is to be studied and the percentage of successful plants is to be used as the estimate. It is a requirement of the study that there be a 95% chance that the estimate will be within four percentage points of the true value. The requirements are reexpressed as requirements on the estimate as a proportion and then the sample size is determined on that basis. The population and sample percentages are converted to proportions by dividing by 100 and the error limit, four percentage points, is adjusted accordingly to an error limit of $4/100$ or 0.04. To use the general solution for n, we note that, in our example, $1 - \alpha = 0.95$ (since we want a 95% chance of achieving the desired accuracy) and $z_{\alpha/2} = z_{0.025} = 1.960$. We have the error limit as $E = 0.04$ and the solution for the required sample size is $n = (1.96/0.04)^2/4 = 600.25$. Although this should theoretically be taken up to 601, a sample size of 600 might be more convenient and would produce almost the required probability. Furthermore, since this is a maximum value and required only if p is $1/2$, we can justify using the more convenient value for n.

Considering the last comment further, suppose that we can reasonably assume that p will not be near $1/2$. Suppose that we can assume that at most 35% of all such plants will meet the requirements. We may then assume that $p \leqslant 0.35$. Now, since p may be assumed to be no closer to $1/2$ than 0.35, we may use $0.35 \times 0.65 = 0.2275$ rather than $1/4$ as the value for $p^*(1 - p^*)$ and produce a new solution for the sample size as $n = (1.960/0.04)^2 \times 0.2275 = 546.2$. The minimum required sample size now is 547. ▲

Finite Populations

If the population is not very large, the hypergeometric model applies and the results found above must be modified to incorporate the finite population correction. We will consider only the case in which the population and (anticipated) sample size are still large enough to use the normal approximation to the hypergeometric distribution. The previous procedure for very large or infinite populations may be repeated with the introduction of the finite population correction factor to change the variance of the approximating normal for x from $np(1 - p)$ to $np(1 - p)(N -$

$n)/(N-1)$. Again dividing x by n to produce the sample proportion, we now produce an approximate distribution for x/n as a normal with mean p and standard deviation $\sqrt{[p(1-p)/n] \times (N-n)/(N-1)}$. Setting up the normal approximation as before, we find that to determine the sample size we must solve $E/\sqrt{[p(1-p)/n] \times (N-n)/(n-1)} = z_{\alpha/2}$ for n with the following result:

In order to obtain a probability of at least $1 - \alpha$ that a random sample proportion x/n will differ from a population proportion p by at most E for a population of size N when sampling without replacement, the sample size should be at least

$$n = \frac{Nz_{\alpha/2}^2}{z_{\alpha/2}^2 + (N-1)E^2/[p^*(1-p^*)]}$$

where p^* is the closest value to $1/2$ that p may take. If no prior information on p is available, p^* is set equal to $1/2$.

EXAMPLE 6-12 An official wishes to estimate the percentage of 4000 applicants for employment in the federal civil service who are fluent in a foreign language. The estimate is to be based on a sample percentage and is to be such that there will be a 90% chance of obtaining an estimate that is within six percentage points of the true value. The estimate is first found as a proportion and then changed to a percentage by multiplying by 100. In terms of a proportion, the tolerable error (six percentage points) is 0.06, the desired probability, $(1 - p)$, is 0.90, and $z_{\alpha/2}$ is $z_{0.05} = 1.645$. A safe sample size is found by using $1/4$ for $p^*(1 - p^*)$ and is $n = \dfrac{4000 \times 1.645^2}{1.645^2 + \left(3999 \times \dfrac{0.06^2}{1/4}\right)} =$

179.5 which is rounded up to 180.

If there is reason to believe that at most 25% of the 4000 applicants are fluent in a foreign language, then the sample size is larger than required. Considering that p is no closer to $1/2$ than $p^* = .25$, it is reasonable to substitute $p^*(1 - p^*) = 0.25 \times 0.75 = 0.1875$ and to use a sample size of $[4000 \times 1.645^2]/[1.645^2 + (3999 \times 0.06^2/0.1875)] = 136.2$ which is rounded up to 137.

It is interesting to note that without the finite population correction, these sample sizes would have been $(1.645/0.06)^2/4 = 187.9$ producing 188 or $(1645/0.06)^2 \times 0.25 \times 0.75 = 140.9$ producing 141. ▲

There are many other methods for evaluating potential errors in estimates and for determining sample sizes. These examples are sufficient, however, to illustrate the nature of such procedures.

Random Samples

The calculation of the appropriate sample size is not all that is required in deciding what sample to take. We must also have an appropriate procedure for determining which members of the population are included in the sample. The previous sample size calculations and the inference procedures that are to follow are all based on the assumptions of particular sampling distributions for the statistics employed in the analyses. These assumptions are valid only if the sampling procedures involve random selection of population members. The selection must be such that the probability that an individual population value will be included in the sample is the same as that indicated by the assumed underlying probability distribution.

> To ensure the validity of statistical procedures, samples chosen for analysis must be *random samples*. Unless otherwise stated, *all samples* in this book are assumed to be *random samples* taken without replacement.

P6-1 Environmental assessments involve counts of the number of particles of dust and other pollutants in the air. If repeated measurements with a particle counter are normally distributed about the actual particle content level with a standard deviation of 25 micrograms per cubic meter ($\mu g/m^3$), how many readings must be taken to have a 95% chance that the sample mean $\bar{x}$ will differ from the true level by at most 5 $\mu g/m^3$?

P6-2 (a) If a sample of members of a large national church denomination is to be taken to estimate the proportion of members who favor a specific church policy, how large a sample must be taken to provide a 90% chance that the estimate will be within five percentage points of the true proportion?

(b) How large must the sample be for the survey in part (a) if it can be safely assumed that at least two-thirds of the members favor the policy?

P6-3 (a) A researcher investigating an organization employing 3500 employees plans to conduct a survey to estimate the percentage of employees in favor of a change in the benefits provided by the organization. The researcher is quite confident that the proportion exceeds 75% but would like to be able to estimate the value more closely and would in fact like to have an estimate from the survey sample that will have a 98% chance of being within four percentage points of the true value. What is the minimum number of employees that should be included in the sample?

(b) How large should the sample be for the survey in part (a) if the researcher has no prior information on the proportion?

6-1 A researcher knows that measurements on the amount of a pollutant dissolved in a fixed quantity of liquid will follow a normal distribution with a mean μ equal to the "true" average level and a standard deviation σ equal to 2.50. If the researcher plans to use the mean $\bar{x}$ of a random sample of measurements to estimate the true average μ, how many measurements must be taken for there to be a 95% chance that the error in the estimate will be at most 0.50?

6-2 A bottle-filling process is known to fill bottles with a variable quantity of liquid (not to overflowing) such that the standard deviation of the volumes in bottles filled by the process is 0.0050 liters; that is, 5.0 milliliters (ml). How many bottles must be sampled in order to have a probability 0.98 that the mean volume in the sample will be within 1.0 ml of the mean volume produced by the process?

6-3 Suppose that a study is to be conducted to determine the percentage of cases in which a particular remedy will provide a successful cure for a disease. This unknown percentage is to be estimated with the corresponding percentage in a sample of cases.

(a) How many cases should be studied in order to have a 90% chance that the estimate will be within four percentage points of the unknown percentage?

(b) If it may be assumed that the unknown percentage is at least 70%, how many cases must be studied to have a 90% chance that the estimate will be within four percentage points of the unknown value?

6-4 Suppose that a survey is to be conducted to determine what percentage of housing units in a community meet a particular set of standards. The estimate of the community percentage is to be the sample percentage. If it can be assumed that at least 80% of the housing units in the community meet the standards, and if there are about 10,000 units in the community, how many units should be included in the sample to obtain a 95% chance that the error in the estimate will not exceed three percentage points?

6-5 A research laboratory is involved in several different types of assessments with measurement procedures that produce different variances depending on the materials being assessed. A common relative standard is obtained by insisting that each assessment involve a sufficient number of measurements so that there is a 95% chance that the sample mean will be within one-half of the standard deviation of the true mean level of the material being assessed. How many measurements should be taken in each case?

6-6 A sample survey is to be conducted to estimate the percentage of the 2500 extension students who visit a university library more than twice a week. It is desired that there be a probability of at least 0.90 that the sample percentage will be within two percentage points of the true percentage, and a probability of at least 0.95 that the sample percentage will be within four percentage points.

(a) Which of the requirements is more restrictive (i.e., requires a larger sample size)?

(b) What is the smallest sample size that will satisfy both requirements.

6-7 If the weights of individual bones exhibit a standard deviation of 7.5 g, how many bones must be weighed to produce a probability of at least 0.98 that the sample mean will differ from the mean weight of all such bones by at most 1.0 g?

6-8 How many individuals must be included in a survey to estimate the proportion of the population who have traveled abroad in the past year if it is a requirement of the survey that there be a 95% chance that the sample percentage will be within three percentage points of the actual population percentage?

6-9 Complete the solution of $E/\sqrt{[p(1-p)/n] \times (N-n)/(N-1)} = z_{\alpha/2}$ to verify the solution for n given in the text of this section for the case of estimating a proportion for a finite population.

P6-1 Since $1 - \alpha = 0.95$, we have $\alpha = 0.05$, $\alpha/2 = 0.025$, and $z_{\alpha/2} = z_{0.25} = 1.960$. Substituting this value together with $\sigma = 25$ and $E = 5$ in the solution for n, we find the minimum sample size as $(25 \times 1.960/5)^2 = 96.04$. Theoretically, a sample size of at least 97 is required.

P6-2 (a) In terms of proportions, the tolerable error of 5 percentage points is $E = 0.05$. Since $1 - \alpha = 0.90$, $z_{\alpha/2} = z_{0.05} = 1.645$. The minimum sample size is then $(1.645/0.05)^2/4 = 270.6$ or 271.

    (b) With the condition that the proportion is at least 2/3, we set $p^* = 2/3$ and the minimum sample size is $(1.645/0.05)^2 \times (2/3) \times (1/3) = 240.5$ or 241.

P6-3 (a) Since we have a finite population of size $E = 3500$, we use the sample size incorporating the finite population correction. We have $1 - \alpha = 0.98$, producing $z_{\alpha/2} = z_{0.01} = 2.326$, an error of four percentage points or, in terms of proportions, $E = 0.04$, and the assumption that p is no closer to 1/2 than $p^* = 0.75$. The required sample size is thus found as

$$\frac{3500 \times 2.326^2}{2.326^2 + \left[3499 \times 0.04^2/(0.75 \times 0.25)\right]} = 536.9 \quad \text{or} \quad 537$$

    (b) With no prior knowledge on p, we use $p^* = 0.5$ to produce the required sample size as

$$\frac{3500 \times 2.326^2}{2.326^2 + \left[3499 \times 0.04^2/(0.5 \times 0.5)\right]} = 681.1 \quad \text{or} \quad 682$$

6-4 INTERVAL ESTIMATES FOR POPULATION MEANS

EXAMPLE 6-13 Suppose that the psychometrist in Example 6-3 does decide to use the calculated size of $n = 56$. Then, prior to the sampling, there is a 98% chance that the sample mean will be within 5 of the true population mean. Suppose further that a sample so taken produces a sample mean of 62. The psychometrist does not know whether this value actually is within 5 of the true mean because the true mean is unknown. On the basis of the high presample probability, however, it is reasonable to be confident that the specific value, 62, is within 5 of the true mean. In fact, the degree of confidence can be measured in a manner similar to the measurement of probability.

Since the psychometrist had a presample 98% chance of obtaining a value of $\bar{x}$ within 5 of the true mean μ, she can be 98% confident that the sample value, 62, is within 5 of the true mean, or equivalently, that the true mean is within 5 of the specific value 62, that is, in the interval from 57 to 67. Since it has a related degree of confidence of 98%, this interval is called a 98% *confidence interval for the mean*.

We might consider a further interpretation of the confidence that can be attached to such an interval estimate through the long-run frequency approach to probability. If we employ the procedure as above, then, from the sampling distribution for $\bar{x}$, we know that there is a 98% chance that $\bar{x}$ will be within 5 of the true value of μ in the single sample. We might therefore expect that, if we repeated the procedure a great many times, $\bar{x}$ would be within 5 of the true value of μ 98% of the time and would fail to be within 5 of μ the other 2% of the time. In other words, we may anticipate that the interval $\bar{x} - 5$ to $\bar{x} + 5$ would include μ 98% of the time and fail to include μ the other 2% of the time.

Since we may anticipate that the interval $\bar{x} - 5$ to $\bar{x} + 5$ would include μ 98% of the time in repeated applications, we may be 98% confident that, in the single case with $\bar{x} = 62$, the interval 57 to 67 does include μ. ▲

Confidence Intervals for the Mean

Generalizing the procedure of Example 6-13, we may produce the general expressions for interval estimates or confidence intervals for the mean.

Population Variance Known

In general, for any samples from normal populations or for large samples from other populations, $(\bar{x} - \mu)/(\sigma/\sqrt{n})$ has a standard normal distribution and (using symmetry) we have $P(-z_{\alpha/2} \leqslant (\bar{x} - \mu)/(\sigma/\sqrt{n}) \leqslant z_{\alpha/2}) = 1 - \alpha$. Multiplying through by $\sigma/\sqrt{n}$ and then adding μ throughout, we obtain the result that $P(\mu - z_{\alpha/2}\sigma/\sqrt{n} < \bar{x} < \mu + z_{\alpha/2}\sigma/\sqrt{n}) = 1 - \alpha$. We therefore have a presample probability of $1 - \alpha$ that $\bar{x}$ will be within $z_{\alpha/2}\sigma/\sqrt{n}$ of μ and a postsample confidence of $100 \times (1 - \alpha)\%$ that μ is within $z_{\alpha/2}\sigma/\sqrt{n}$ of $\bar{x}$.

This result leads to the following confidence interval for the mean of a normal population. For *large samples*, this confidence interval may also be used for *nonnormal populations*.

A $100 \times (1 - \alpha)\%$ *confidence interval for the population mean is*

$$\bar{x} - \frac{z_{\alpha/2}\sigma}{\sqrt{n}} \text{ to } x + \frac{z_{\alpha/2}\sigma}{\sqrt{n}}$$

which is sometimes written as $\bar{x} \pm z_{\alpha/2}\sigma/\sqrt{n}$, where "$\pm$" is read "plus or minus" and indicates that the end points of the interval are obtained by subtracting $z_{\alpha/2}\sigma/\sqrt{n}$ from $\bar{x}$ and adding $z_{\alpha/2}\sigma/\sqrt{n}$ to $\bar{x}$.

EXAMPLE 6-14   In order to determine whether a rainfall a few days after the eruption of Mount St. Helens on May 18, 1980, had cleared enough volcanic ash from the air to permit reasonably safe business operations in a Spokane, Washington, service station, the station operator consulted a particle counter which could be assumed to produce particle counts approximately normally distributed with a standard deviation of $\sigma = 25$ $\mu g/m^3$. A random sample of $n = 15$ readings in a short period of time produced a mean of $\bar{x} = 532.4$. On the basis of these data, the operator wished to determine a 95% confidence interval for the true mean particle count. Setting $1 - \alpha = 0.95$ so that $z_{\alpha/2} = z_{0.025} = 1.960$, the operator could calculate the 95% confidence interval as $\bar{x} \pm 1.960\sigma/\sqrt{n}$, which is $532.4 \pm 1.960 \times 25/\sqrt{15} = 532.4 \pm 12.7$ or 519.7 to 545.1. The operator could be 95% confident that the mean particle content of the air about the service station was between 519.7 and 545.1 $\mu g/m^3$. ▲

In the confidence interval, the sample mean $\bar{x}$ provides the location of the interval. The width $(\bar{x} + z_{\alpha/2}\sigma/\sqrt{n}) - (\bar{x} - z_{\alpha/2}\sigma/\sqrt{n}) = 2z_{\alpha/2}\sigma/\sqrt{n}$ is determined by the standard deviation, the degree of confidence, and the sample size. As we might intuitively expect, with a known population variance and a fixed degree of confidence, an increase in the sample size thus produces a decrease in the width of the interval.

EXAMPLE 6-15 If the service station operator in Example 6-14 obtained the sample mean $\bar{x} = 532.4$ from $n = 30$ rather than $n = 15$ readings, the width of the confidence interval would have been decreased and the interval would have been $532.4 \pm 1.96 \times 25/\sqrt{30} = 532.4 \pm 8.9$ or 523.5 to 541.3. The width of the interval would have been $2 \times 8.9 = 17.8$ instead of $2 \times 12.7 = 25.4$. ▲

Population Variance Unknown — Sample Size Large

The expression for the confidence interval involves σ, the standard deviation of the population. In many experiments, however, the population standard deviation is not known. In these cases it is necessary to substitute an estimated value for the standard deviation. The value to be substituted is s, the value of the sample standard deviation. If the sample size is reasonably large (a usual rule of thumb is $n > 30$), s may be expected to be close enough to σ that the confidence interval may be used with no adjustment other than this substitution.

A *large sample confidence interval for the population mean* μ when σ is unknown is

$$\bar{x} \pm \frac{z_{\alpha/2}s}{\sqrt{n}}$$

EXAMPLE 6-16 Suppose that light bulbs of a particular kind have lifetimes with an unknown mean and an unknown standard deviation. It is of interest to find a 99% confidence interval for the mean lifetime of light bulbs of this kind. A sample of 100 bulbs produced a set of lifetimes with a mean of 987.3 and a standard deviation of 64.1 hours. Since $1 - \alpha = 0.01$, $z_{\alpha/2} = z_{0.005} = 2.576$ and the confidence interval is $987.3 \pm 2.576 \times 64.1/10$ or 987.3 ± 16.5. We can be 99% confident that the true mean lifetime of such bulbs is between 970.8 and 1003.8 hours. ▲

We do not have as much control over the width of the interval *if the population variance is unknown* as we had when the variance was known. If the interval is wider than we wish, we might consider sampling further to increase the sample size; however, it is possible that further sampling would produce a larger sample standard deviation that might offset the effect of the increase in the sample size.

EXAMPLE 6-17   If we sampled 50 more bulbs in Example 6-16 to produce a total sample size of 150, we might obtain a final mean of 991.4 and a standard deviation of 79.2, producing a 99% confidence interval for the mean as $991.4 \pm 2.576 \times 79.2/\sqrt{150}$ or 991.4 ± 16.7, which is slightly wider than before even though the sample size is larger. ▲

Population Variance Unknown — Sample Size Small

The probability of s being considerably different from σ is larger for smaller samples and, for such samples, it is not sufficient to substitute s for σ without also modifying the confidence interval further. The modification involves the sampling distribution for a new statistic, $t = \dfrac{\bar{x} - \mu}{s/\sqrt{n}}$.

> For random samples of size n from a normal distribution with mean μ, the statistic $t = \dfrac{\bar{x} - \mu}{s/\sqrt{n}}$ has a sampling distribution referred to as the *t-distribution with $n - 1$ degrees of freedom*, where n is the size of the sample.

The t-distribution is similar to the standard normal distribution. It is symmetric about a mean of 0 but has a larger variance than the normal. Intuitively, we can explain the larger variance as being due, in part, to the fact that t reflects two sources of variation: the variation in $\bar{x}$ and the variation in s.

> Percentiles of the t-distribution are given in Table IV of Appendix B. The $(1 - \alpha) \times 100$th percentile of the t-distribution with f degrees of freedom is found in the row labeled with the numerical value for f and the column labeled t_α, where the subscript is the numerical value for α. This percentile is denoted as
> $$t_{\alpha, f}$$

Table IV is illustrated in Figure 6-3. Note that, as indicated in Figure 6-3, we will on occasion simply write t_α.

EXAMPLE 6-18   The 95th percentile of the t-distribution with $f = 13$ degrees of freedom is found in Table IV in the row labeled 13 and the column labeled $t_{0.050}$ and is equal to $t_{0.050, 13} = 1.771$. ▲

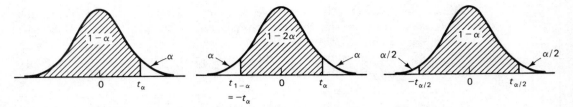

Figure 6-3 Illustration of Table IV.

The t-distribution approaches the standard normal as f increases, and as a rule of thumb, if f is at least 30, we might consider the statistic t to have the standard normal distribution. Reading the entries in any column, we see that, as the degrees of freedom increase, the percentiles become closer to the mean 0 (the distribution becomes more compact and its variance decreases) until they finally become approximately equal to the standard normal value, which appears in the last row. Figure 6-4 illustrates the standard normal distribution as well as t-distributions with 4 and with 16 degrees of freedom.

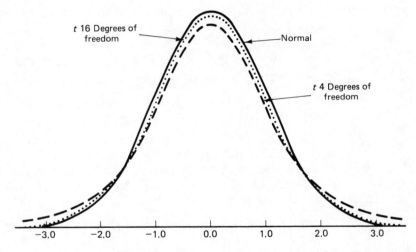

Figure 6-4 The normal distribution = ———; the t-distribution with 4 degrees of freedom = — —; the t-distribution with 16 degrees of freedom = ...

Because of the symmetry of the t-distribution, lower percentiles are found as negatives of upper percentiles,

$$t_{1-\alpha, f} = -t_{\alpha, f}$$

EXAMPLE 6-19 Just as $t_{0.05}$ "cuts off" the upper 5% of the distribution, $-t_{0.05}$ cuts off the lower 5%; that is, the area (probability) under the curve to the right of $t_{0.05}$ is 0.05 and the area to the left of $-t_{0.05}$ is also 0.05. The area between the two is thus 0.90. For a sample of size 14, we thus have the probability statement $P(-t_{0.05, 13} \leqslant (\bar{x} - \mu)/(s/\sqrt{14}) \leqslant t_{0.05, 13}) = 0.90$, or, using the value 1.771 from Table IV, $P(-1.771 \leqslant (\bar{x} - \mu)/(s/\sqrt{14}) \leqslant 1.771) = 0.90$. ▲

In the general case with a sample size n, we have $f = n - 1$ and $P(-t_{\alpha/2, n-1} \leqslant \dfrac{\bar{x} - \mu}{s/\sqrt{n}} \leqslant t_{\alpha/2, n-1}) = 1 - \alpha$. If we adjust this probability statement as we did for the standard normal, we may produce a new confidence interval.

A $100 \times (1 - \alpha)\%$ *small-sample* confidence interval for the *mean* of a *normal population* when σ is unknown is

$$\bar{x} \pm \frac{t_{\alpha/2, n-1}s}{\sqrt{n}}$$

EXAMPLE 6-20 A pain killer administered to a population of subjects suffering pain can be assumed to produce a population of times until experiencing relief that is approximately normal. A random sample of 18 subjects produced a mean of 15.4 minutes and a standard deviation of 3.2 minutes. Finding $t_{0.025, 17}$ from Table IV as 2.110, we calculate a 95% confidence interval for the "true" mean time until relief as $15.4 \pm 2.110 \times 3.2/\sqrt{18}$ or 15.4 ± 1.6. We can be 95% confident that the true mean time until relief after administration of this pain killer is in the interval from 13.8 to 17.0 minutes. ▲

We should be cautious in using the interval for small samples from nonnormal populations because the t-distribution does require that the original population be normal (or at least approximately normal).

Although the central limit theorem justifies the use of the large-sample confidence interval for nonnormal populations, in the case of small samples the confidence intervals are only valid for populations that are (approximately) normal.

PRACTICE PROBLEMS

P6-4 A laboratory procedure for assessing soluble sugar in plants is such that in the range 100 to 200 milligrams (mg) of glucose per gram of dry weight, repeat measurements will follow a normal distribution with mean μ equal to the true sucrose level and standard deviation $\sigma = 3.0$ mg/g dry weight. If five measurements on a carnation section produce a mean of $\bar{x} = 146.35$ mg of glucose per gram of dry weight, find a 98% confidence interval for the true mean glucose level.

P6-5 After being deprived of sleep, 72 subjects were assessed for the reduction in their reflexes from levels obtained after adequate sleep. The 72 subjects produced reductions with a mean of $\bar{x} = 26.2\%$ and a standard deviation of $s = 4.3\%$.

Determine a 90% confidence interval for the mean reduction that can generally be expected after sleep deprivation.

P6-6 On the basis of relationships between basicranial structure and upper respiratory systems in extant primates, studies of fossils help in the reconstruction of respiratory systems of extinct species. In one assessment on a number of hominoid fossils the distances in centimeters from the front to the back of the teeth were found to be 5.39, 6.56, 5.47, 5.78, 5.50, 6.42, 5.85, and 5.07. On the basis of this sample, find a 95% confidence interval for the mean distance in this species. What assumption is made in using this procedure?

EXERCISES

6-10 Suppose that the researcher in Exercise 6-1 takes only 40 measurements and the sample mean is 15.83. Find a 90% confidence interval for the true average level.

6-11 To assess an individual's score on an attitude test, it is necessary to know, among other things, the mean of the scores that would be achieved by the population of all subjects. Suppose that a sample of 50 subjects produced scores with a mean of 52.1 and a standard deviation of 10.4. Determine a 95% confidence interval for the mean of the population.

6-12 Because of natural fluctuations in a measurement process, repeat assessments on a chemical aspect of blood fluctuate according to a normal distribution with a mean μ equal to the actual value in the blood and standard deviation $\sigma = 0.020$. A sample of ten assessments on a given blood sample produced the values 1.013, 1.009, 0.991, 0.983, 1.041, 1.015, 1.003, 1.020, 0.972, and 0.959. Find a 98% confidence interval for the actual value in the blood.

6-13 Repeat Exercise 6-12 assuming that σ, the standard deviation of the measurement process, is unknown.

6-14 Suppose that the crop potential of a given area is to be assessed on the basis of the yields from 20 sample plots that produced the yield values 142, 68, 41, 108, 132, 105, 98, 63, 54, 124, 70, 85, 65, 93, 100, 136, 146, 78, 81, and 123. Assuming that yields follow a normal distribution, find an interval of values as an estimate for the true mean yield. Determine this interval in such a way that you can be 95% confident that it does include the true mean.

6-15 A sample of 75 hourly outputs for light punch press operators produced a mean of 602.7 units and a standard deviation of 20.3. On the basis of these sample data, construct an interval estimate for the actual mean hourly output such that you can be 98% confident that the actual mean is included in the interval.

6-16 If the procedure for determining snow-water equivalents produces a normal distribution of readings with a standard deviation of $\sigma = 7.5$ mm, how many sample readings must be taken to produce a 95% confidence interval for the true mean μ with a width 5.0 mm?

6-17 If a random sample of ten values from a normal population produces a mean of 113.65 and a standard deviation of 18.32, what can you say about the value of the population mean so that you can be 95% confident that what you say is true?

6-18 Prior to taking a random sample of a given size from a normal population with a known standard deviation, a researcher determines that the resulting confidence interval for the mean would be twice as wide as desired. How should the sample size be adjusted?

P6-4 With $1 - \alpha = 0.98$, $z_{\alpha/2} = z_{0.01} = 2.326$ and the confidence interval is $\bar{x} \pm z_{\alpha/2}\sigma/\sqrt{n} = 146.35 \pm 2.326 \times 3.0/\sqrt{5} = 146.35 \pm 3.12$ or 143.23 to 149.47.

P6-5 Using the large-sample procedure, the confidence interval is found as $\bar{x} \pm z_{\alpha/2}s/\sqrt{n} = 26.2 \pm 1.645 \times 4.3/\sqrt{72} = 26.2 \pm 0.8$ or 25.4 to 27.0.

P6-6 The small-sample confidence interval for the case that σ is unknown is required. The sample data produce $n = 8$, $\bar{x} = 5.755$, and $s = 0.5139$. From Table IV, $t_{\alpha/2, n-1} = t_{0.025, 7} = 2.365$. The confidence interval is then $\bar{x} \pm t_{\alpha/2, n-1}s/\sqrt{n} = 5.755 \pm 2.365 \times 0.5139/\sqrt{8} = 5.755 \pm 0.430$ or 5.325 to 6.185. In using this procedure it is assumed that the distribution of measurements is normal.

6-5 CONFIDENCE INTERVALS FOR POPULATION VARIANCES AND STANDARD DEVIATIONS

Confidence intervals for the variance or standard deviation of a *normal distribution* can also be developed from the sampling distribution of an appropriate statistic. The sampling distribution for the sample variance s^2 was considered for a particular set of examples in Section 4-6 and, as noted there and again in Section 6-2 when sampling from infinite populations such as the normal distribution, the sampling distribution for s^2 has a mean equal to the population variance σ^2. This result provides some justification for using s^2 as an estimate for σ^2. To develop an interval estimate for σ^2 we consider the sampling distribution of a statistic that is a modification of s^2.

Small Samples

> If s^2 is the variance of a random sample of size n from a *normal population* with variance σ^2, the statistic $(n - 1)s^2/\sigma^2$ has a sampling distribution referred to as the χ^2-*distribution* (chi-square distribution) *with $n - 1$ degrees of freedom.* (χ is the Greek lowercase letter chi, pronounced "ky").

Note that this statistic is just a result of rescaling s^2. The sampling distribution of s^2 was seen in Section 4-6 to be positively skewed and, as we might expect, the χ^2-distribution is also positively skewed.

Using the previously developed form of a subscript notation for z_α and $t_{\alpha, n-1}$, we produce notation for percentiles of the χ^2-distribution.

> Percentiles of the χ^2-distribution are given in Table V of Appendix B and illustrated in Figure 6-5. The $(1 - \alpha) \times 100$th percentile of the χ^2-distribution with f degrees of freedom is denoted as

$$\chi^2_{\alpha, f}$$

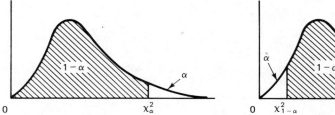

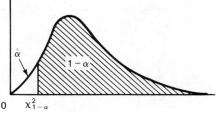

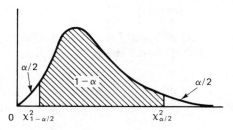

Figure 6-5 Illustration of Table V.

As with the t we will on occasion simply write χ_α^2.

Since $(n-1)s^2/\sigma^2$ must always be positive, we cannot find a value $-\chi_\alpha^2$ analogous to $-t_\alpha$ or $-z_\alpha$. Lower percentiles must thus be expressed in the form $\chi_{1-\alpha, f}^2$.

A confidence interval for the variance or standard deviation is obtained by considering the sampling distribution of $(n-1)s^2/\sigma^2$. Using the previously stated result that this statistic follows the χ^2-distribution with $f = n-1$ degrees of freedom, and noting that, by definition of percentiles, a proportion $1-\alpha$ of the χ^2-distribution is found between $\chi_{1-\alpha/2}^2$ and $\chi_{\alpha/2}^2$, we have the general probability statement $P(\chi_{1-\alpha/2, n-1}^2 \leqslant (n-1)s^2/\sigma^2 \leqslant \chi_{\alpha/2, n-1}^2) = 1-\alpha$, where s^2 is calculated from a sample of size n. After some algebraic manipulation of this probability statement, we produce the following result:

A $100 \times (1-\alpha)\%$ *confidence interval for the variance* σ^2 of a *normal* population is

$$\frac{(n-1)s^2}{\chi_{\alpha/2, n-1}^2} \quad \text{to} \quad \frac{(n-1)s^2}{\chi_{1-\alpha/2, n-1}^2}$$

Taking square roots produces the *confidence interval for the standard deviation*.

EXAMPLE 6-21   For Example 6-20, $n = 18$ and $s^2 = 10.24$. Suppose that we want a 98% confidence interval for σ, the population standard deviation. Since $1-\alpha = 0.98$, $\alpha/2 = 0.01$ and $1-\alpha/2 = 0.99$. From Table V we find $\chi_{0.01, 17}^2 = 33.409$ and $\chi_{0.99, 17}^2 = 6.408$. The 98% confidence interval for σ^2 is then

$17 \times 10.24/33.409$ to $17 \times 10.24/6.408$ or 5.21 to 27.17. Taking square roots, the confidence interval for σ is $\sqrt{5.21}$ to $\sqrt{27.17}$ or 2.3 to 5.2. ▲

Table V will be useful for samples of size 31 or less since the degrees of freedom are given up to 30. For sample sizes exceeding 31, large-sample results will provide an approximation.

Large Samples

For large values of f, the cube root of a χ^2 statistic divided by f has a distribution that is approximately normal with mean $1 - 2/9f$ and variance $2/9f$.

Large-sample confidence intervals are thus obtained by replacing χ^2 percentiles with the appropriate approximate percentiles.

For large values of f,

$$\chi^2_{\alpha, f} \doteq f\left(1 - 2/9f + z_\alpha\sqrt{\frac{2}{9f}}\right)^3$$

We may use this result to reexpress the confidence intervals for large samples.

An approximate $100 \times (1 - \alpha)\%$ *large-sample confidence interval for the variance* of a normal distribution is

$$\frac{s^2}{\left[1 - 2/9(n - 1) + z_{\alpha/2}\sqrt{2/9(n - 1)}\right]^3}$$

to

$$\frac{s^2}{\left[1 - 2/9(n - 1) - z_{\alpha/2}\sqrt{2/9(n - 1)}\right]^3}$$

The approximate *confidence interval for the standard deviation* is found by taking square roots.

EXAMPLE 6-22   In a study to determine the extent of natural variability in the capabilities of laboratory mice, four dozen mice were tested and their test scores produced a variance of 138.6. Considering $n = 48$ to be a large sample, we use approximate confidence interval procedures. For an approximate 95% confidence interval we use $z_{\alpha/2} = z_{0.025} = 1.960$ to find the interval as

$$\frac{138.6}{\left(1 - \frac{2}{9 \times 47} + 1.960\sqrt{\frac{2}{9 \times 47}}\right)^3}$$

to

$$\frac{138.6}{\left(1 - \dfrac{2}{9 \times 47} - 1.960\sqrt{\dfrac{2}{9 \times 47}}\,\right)^{3}}$$

or 96.05 to 217.53. Taking square roots of these values, we find an approximate 95% confidence interval for the standard deviation as 9.8 to 14.7. ▲

PRACTICE PROBLEMS

P6-7 Calculate 90% confidence intervals for the variance and standard deviation of the concentrations of artifical food coloring in different lots if a sample of 12 lots produced the following values: 0.013, 0.016, 0.009, 0.018, 0.014, 0.010, 0.013, 0.014, 0.012, 0.014, 0.015, and 0.010. What assumption is made in the use of this procedure?

P6-8 To assess the consistency of measurements taken by a lab technician trainee, a supervisor asked the trainee to measure 60 samples. To avoid a bias against considerable variability, the supervisor did not inform the trainee that the samples were all from the same lot and should produce very consistent measurements. Find a 95% confidence interval for the variance of measurements that would be produced by this trainee on any single lot if the sample of 60 measurements produced a variance of 4.90.

EXERCISES

6-19 Determine a 95% confidence interval for the variance of (the population of) scores received on a psychological test if a normal distribution can be assumed and a sample of 25 scores produced a sample variance of 549.8.

6-20 Suppose that the standard deviation σ for the measurement process of Exercise 6-12 is unknown. Using the data in that exercise, determine a 98% confidence interval for σ.

6-21 Determine a 90% confidence interval for the standard deviation of snow depths across a region at a particular time if the depths at 50 sampling points produced a sample standard deviation of 7.8 cm.

6-22 Using the data from Practice Problem P6-6, calculate a 95% confidence interval for the standard deviation of distances from the front to the back of the teeth in this species.

6-23 The numbers of trials to reach criterion for 50 inbred rats displayed a sample standard deviation of 8.3. Estimate the population standard deviation for such rats with an interval such that you can be 98% confident that the interval does include the population value.

6-24 Complete the algebraic manipulation to demonstrate that $P(\chi^2_{1-\alpha/2,\,n-1} \leqslant (n-1)s^2/\sigma^2 \leqslant \chi^2_{\alpha/2,\,n-1}) = 1 - \alpha$ does lead to the confidence interval given for σ^2.

P6-7 To avoid six places of decimals in squared values, the data are all multiplied by 1000. The new data produce $\Sigma x = 158$ and $\Sigma x^2 = 2156$. From Table V we find that $\chi^2_{0.95, 11} = 4.575$ and $\chi^2_{0.05, 11} = 19.675$. Writing $(n - 1)s^2$ as $\Sigma x^2 - (\Sigma x)^2/n = 75.667$, we calculate the confidence interval for σ^2 for the adjusted data as $75.667/19.675$ to $75.667/4.575$ or 3.85 to 16.54. Taking square roots, we find the confidence interval for the standard deviation as 1.96 to 4.07. Dividing by 1000^2 and 1000, respectively, we convert these confidence intervals to the confidence intervals for the original data as 3.85×10^{-6} to 16.54×10^{-6} for the variance, and 0.00196 to 0.00407 for the standard deviation. A normal distribution is assumed.

P6-8 The large-sample confidence interval is required and is found to be

$$\frac{4.90}{\left(1 - \dfrac{2}{9 \times 59} + 1.960\sqrt{\dfrac{2}{9 \times 59}}\right)^3}$$

to

$$\frac{4.90}{\left(1 - \dfrac{2}{9 \times 59} - 1.960\sqrt{\dfrac{2}{9 \times 59}}\right)^3}$$

or 3.52 to 7.29.

6-6 CONFIDENCE INTERVALS FOR A POPULATION PROPORTION OR PERCENTAGE

As indicated before, when we wish to estimate a population proportion p, the sample proportion x/n is an appropriate statistic to consider. In the estimation of percentages, we merely find the corresponding proportion estimate and multiply by 100. As with the mean and variance, interval estimates may also be determined for a proportion and hence for a percentage.

Very Large or Infinite Populations

Let us first consider cases in which the population is infinite or sufficiently large, so that even though we sample without replacement, the binomial model may be assumed.

Confidence Intervals from Charts

After determining the sample proportion x/n, no further calculation is necessary when appropriate tables are available.

Table VI of Appendix B consists of two charts that give 95% and 99% confidence intervals for the population proportion for various sample sizes and sample proportions.

The curved bands correspond to sample size and the vertical lines correspond to the sample proportion. The horizontal lines correspond to the end points of the confidence interval.

If the sample proportion is less than 0.5, x/n is indicated on the bottom scale. The two points where the vertical line above the value for x/n crosses the two bands corresponding to the value of n produce the end points of the confidence interval. The end points are read off the left-hand scale after taking horizontal lines from the crossing points to the left-hand scale.

If the sample proportion is greater than 0.5, it is indicated on the top scale and the end points of the confidence interval are found on the right-hand scale. Figure 6-6 provides a general illustration of the chart.

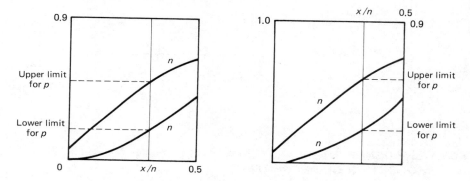

Figure 6-6 Illustration of Table VI.

If there are no bands corresponding to the sample size for a particular sample, we must use interpolation; that is, we must estimate the approximate location of each band between the bands for the closest sample sizes given.

As n increases, the bands become closer together, indicating that, as we might expect, larger sample sizes produce smaller confidence intervals. With small n and with large or small x/n (generally corresponding to large or small p), the confidence intervals are not symmetric about x/n. (By "large" or "small" we mean close to 1 or 0.) These intervals are not symmetric because in these cases, the binomial model is not a very symmetric model.

EXAMPLE 6-23 In a random sample of 60 employees of a large corporation, 19 were found to be opposed to a proposed change in working hours. On the basis of this sample with $x/n = 19/60 = 0.32$ and $n = 60$, we refer to Table VIa for a 95% confidence interval for the proportion of employees in the

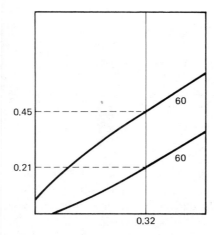

0.45

60

0.21

60

0.32

Figure 6-7 Confidence interval for employee survey.

corporation opposed to the change. The interval is 0.21 to 0.45, as illustrated in Figure 6-7. ▲

Confidence Intervals Based on the Normal Approximation

If a table such as Table VI is not available, approximate confidence intervals for proportions can be obtained by extending the ideas involved in determining sample sizes for estimating proportions and in finding confidence intervals of the mean. In determining sample sizes, we used the result that the sample proportion x/n (for large samples) has a distribution that is approximately normal with mean p and variance $p(1 - p)/n$. We then have the probability statement $P(-z_{\alpha/2} \leq \dfrac{x/n - p}{\sqrt{p(1 - p)/n}} \leq z_{\alpha/2}) = 1 - \alpha$. By the same development as that used for the mean, we might first attempt to produce a confidence interval as $x/n \pm z_{\alpha/2}\sqrt{p(1 - p)/n}$. Unfortunately, this expression involves the unknown value p itself.

To overcome this problem, we must square each of the three parts of the inequality inside the probability statement and then solve the resulting quadratic inequalities to produce the following result.

A $100 \times (1 - \alpha)\%$ *large-sample* confidence interval for a proportion p is

$$\dfrac{\dfrac{x}{n} + \dfrac{z_{\alpha/2}^2}{2n} \pm z_{\alpha/2}\sqrt{\dfrac{\dfrac{x}{n}\left(1 - \dfrac{x}{n}\right)}{n} + \dfrac{z_{\alpha/2}^2}{4n^2}}}{1 + \dfrac{z_{\alpha/2}^2}{n}}$$

EXAMPLE 6-24 The results of the employee survey produce a 95% large sample confidence interval as

$$\frac{\dfrac{19}{60} + \dfrac{1.96^2}{120} \pm 1.96 \sqrt{\dfrac{(19/60) \times (41/60)}{60} + \dfrac{1.96^2}{4 \times 3600}}}{1 + \dfrac{1.96^2}{60}}$$

or 0.21 to 0.44. ▲

As n becomes very large, the parts of the confidence interval involving $z_{\alpha/2}^2/n$ become negligible and may be omitted. The resulting expression is the same as that which would be obtained by replacing p with x/n in the expression resulting from the first attempt to form a confidence interval.

> A *very large sample* approximate $100 \times (1 - \alpha)\%$ confidence interval for a proportion p is
>
> $$\frac{x}{n} \pm z_{\alpha/2} \sqrt{\frac{(x/n)(1 - x/n)}{n}}$$

EXAMPLE 6-25 In a survey of 750 adult members of a community, 450 respondents expressed a belief that a recreation director should be appointed. On the basis of these results, a very large sample approximate 90% confidence interval for the proportion of all adults in the community favoring the appointment of a director is $0.60 \pm 1.645\sqrt{0.60 \times 0.40/750} = 0.60 \pm 0.03$ or 0.57 to 0.63. ▲

Finite Populations

If the population is finite and the sample very large, the hypergeometric distribution should be used as the underlying model and the finite population correction factor should be employed. For such a case, the last approximate confidence interval above is modified simply through the inclusion of the finite population correction.

> An approximate $100 \times (1 - \alpha)\%$ confidence interval for a proportion p of a *finite population of size N* on the basis of a very large sample is
>
> $$\frac{x}{n} \pm z_{\alpha/2} \sqrt{\frac{(x/n)(1 - x/n)}{n}} \times \frac{N - n}{N - 1}$$

EXAMPLE 6-26 If the adult community in Example 6-25 includes only $N = 7500$ adults, the confidence interval is modified to $0.600 \pm 1.645\sqrt{\dfrac{0.60 \times 0.40}{750} \times \dfrac{6750}{7499}} = 0.600 \pm 0.028$ or 0.572 to 0.628. ▲

P6-9 Of 200 cases sampled at random in a survey on legal suits, 82 involved companies bringing suits against individuals. Find a 99% confidence interval for the proportion of all legal suits that involve a company suing an individual
(a) using charts
(b) using the large-sample expression based on the normal approximation

P6-10 In a survey of cases of duodenal ulcers, bleeding occurred in 130 of 1000 cases sampled. Determine a 95% confidence interval for the proportion of all duodenal ulcers that bleed.
(a) using charts
(b) using the very large sample approximation

P6-11 In a collective study by a few colleges, 600 of the 6000 freshmen were selected at random and tested for proficiency in English. Of these 600 freshmen, 270 failed the test. Calculate a 98% confidence interval for the proportion of all the freshmen who lack proficiency in English.

6-25 In a survey of 100 children exposed to television advertising of a particular toy, 58 chose that toy over another not advertised on television. Using Table VI, determine a 99% confidence interval for the proportion of all children who would, in the same circumstances, choose the advertised toy.

6-26 Without using Table VI, determine an approximate 99% confidence interval for Exercise 6-25.

6-27 Suppose that 500 students at a college with a total enrollment of 4500 were surveyed and, of these 500 students, 348 were in favor of a modification to the fee structure for the use of the athletic facilities. Determine an approximate 95% confidence interval for the proportion of the full student body in favor of the modification.

6-28 When a supposedly balanced die was rolled 24 times, it showed "1" on 12 of the rolls. Find a 95% confidence interval for the probability that this die will show "1" when rolled.

6-29 When contacted by telephone, 85 of 250 members of an organization indicated that they would wait for a second renewal notice before paying their dues. Each member who does not pay upon receipt of the first notice costs the organization an additional 50 cents. Find a 95% confidence interval for the amount of money that the organization must budget to cover these additional costs if there are 2000 members.

6-30 Estimate the percentage of Indian homes on reserves that would be classified as poor if, in a sample of 200 homes, 52 were found to be poor. Estimate the percentage with an interval determined in such a way that you can be 98% certain it does include the true value.

6-31 Determine a 90% interval estimate for the percentage of air travelers who would rather travel by train if 215 of 1000 air travelers included in a survey expressed a preference for the train.

6-32 Determine a 95% confidence interval for the number of forgeries in a collection of 2000 paintings if a sample inspection of 250 revealed 50 forgeries.

6-33 Changing the inequalities to equalities and then squaring the terms in the probability statement that leads to the approximate confidence intervals for a proportion produces $z_{\alpha/2}^2 = (x/n - p)^2/[p(1 - p)/n]$. Show that the two solutions for p from this equation are the limits of the large-sample confidence interval for p.

SOLUTIONS TO PRACTICE PROBLEMS

P6-9 (a) The data produce $x/n = 0.41$ and, for $n = 200$, the chart in Table VIb produces a 99% confidence interval as 0.32 to 0.50.

(b) The large-sample confidence interval produces

$$\frac{0.41 + 2.576^2/400 \pm 2.576\sqrt{0.41 \times 0.59/200 + 2.576^2/160{,}000}}{1 + 2.576^2/200}$$

$$= \frac{0.4266 \pm 0.0911}{1.0332}$$

or 0.325 to 0.501.

P6-10 (a) With $n = 1000$ and $x/n = 0.13$, Table VIa produces a 95% confidence interval as 0.11 to 0.15.

(b) The very large sample confidence interval produces $0.13 \pm 1.96\sqrt{0.13 \times 0.87/1000} = 0.13 \pm 0.02$ or 0.11 to 0.15.

P6-11 Using the very large sample confidence interval with the finite population correction factor, we find the 98% confidence interval as $0.45 \pm 2.326\sqrt{(0.45 \times 0.55/600) \times (5400/5999)} = 0.45 \pm 0.04$ or 0.41 to 0.49.

6-7 CHAPTER REVIEW

6-34 If we construct an interval of values from a sample in such a manner that we would expect the resulting interval to include the true percentage of a population in a given category in 19 of 20 samples, we call this interval a ___ % ___ for a _____ .

6-35 The area under the normal curve to the left of $z_{\alpha/2}$ is _____.

6-36 The t-distribution is such that $t_{1-\alpha} = -t_{\alpha}$.
(a) True
(b) False

6-37 The width of a confidence interval for the mean of a normal population with known variance is constant from sample to sample if the size of the sample and degree of confidence are fixed.
(a) True
(b) False
(c) Independent of the variance

6-38 An unbiased estimator for a parameter is
(a) always equal to the parameter in value
(b) always close to the parameter in value
(c) equal to the parameter in value on the average

6-39 If x is drawn at random from a χ^2-distribution with 14 degrees of freedom, $P(\chi^2_{0.95, 14} \leqslant x < \chi^2_{0.05, 14}) = $ _____ .

6-40 For any fixed tolerable error E, the probability that the sample mean will be within E of the population mean can be increased
(a) always
(b) never
(c) by changing the sample size

6-41 A chemist is interested in determining the sulfur content of a given organic chemical. The experiment performed to do so is such that the measurement of the sulfur content obtained can be considered to be a value taken on by a random variable x. Studies on substances having a known sulfur content have shown that x can be considered to be normally distributed with a mean equal to the true sulfur content (expressed as percent by weight) and a standard deviation of 0.5 (expressed as percent by weight). To improve the estimate of the sulfur content, the experiment is repeated on several different samples. How many times should the experiment be repeated if it is desired that the probability that the estimate deviates from the true sulfur content by more than 0.1 (percent by weight) is not greater than 0.05.

6-42 Suppose that the following 16 laboratory measurements were recorded on a physical constant.

2.9962	2.9965	2.9964	2.9962	2.9963	2.9967	2.9966	2.9962
2.9964	2.9963	2.9956	2.9962	2.9964	2.9965	2.9966	2.9969

(a) Find a 95% confidence interval for the true value of the physical constant.
(b) What assumptions are used to find the confidence interval in part (a)?
(c) If the true value is a constant, why take more than one measurement? Why are the measurements not all equal?

6-43 (a) Response times to a given stimulus are known to have a population standard deviation of 1.75. Find an interval of values that you can be 95% confident will include a true mean response time if a sample of 15 times produced the following results.

30.5	36.1	32.3	33.0	35.3	33.9	33.7	31.2
34.3	31.8	33.5	34.9	37.2	32.6	33.2	

(b) Suppose that the population standard deviation is not known in part (a).
(1) Find an interval of values as required in part (a).
(2) Find an interval of values that you can be 90% confident includes the true standard deviation.

6-44 A survey is to be conducted in a given area to determine the percentage of senior high school students who drink alcoholic beverages. It is desired that there be at least a 90% chance that the estimate will be within three percentage points of the true value.

(a) With no prior knowledge of the true percentage, how large must the sample of students in the survey be?

(b) How large must the sample be if it is strongly suspected that the percentage is at least 75%? (Assume, for simplicity, that all students sampled in the survey will respond, that they will respond honestly, and that the finite population effects may be ignored.)

6-45 Repeat Exercise 6-44 with the additional information that there are about 8000 senior high school students in the region being studied.

6-46 Find a 95% confidence interval for the standard deviation of gas consumptions in individual cars of a particular model if gas consumptions for a sample of 50 such cars produced a standard deviation of 1.48 liters per 100 kilometers.

6-47 In a study of different possible treatments for a particular disease, 75 cases refused any of the treatments and, of these, 16 failed to recover. Find a 95% confidence interval for the true mortality rate (proportion of failures) in untreated cases of this disease.

6-48 Sample assessments on the basic psychiatric rating scale for 10 subjects with enlarged ventricles produced a mean and standard deviation of 51.7 and 21.2. Estimate the mean rating in such cases with an interval determined in such a way that you can be 98% confident that it includes the "true" value.

6-49 How many television viewers must be included in a survey if there is to be a 95% chance that the percentage of those in the sample watching a given television network is within four percentage points of the population value?

6-50 Find a 90% confidence interval for the total amount of contaminant in 4000 units if the amounts in a sample of 300 units produced a mean and standard deviation of 0.50 and 0.11 microgram.

6-8 GENERAL REVIEW

6-51 If two sets of data have mean 50 and range 100, they must be identical sets of data.
(a) True
(b) False

6-52 If we change a set of data by subtracting 0.55 from each value and then multiplying these values by 1000 and if the resulting values produce a confidence interval for the variance as 4.86 to 10.73, a confidence interval for the variance based on the original values would be
(a) 0.00486 to 0.01073
(b) 0.55486 to 0.56073
(c) 4.86×10^{-6} to 10.73×10^{-6}
(d) 0.5500486 to 0.5501073

6-53 When the finite population correction is included, the width of a confidence interval
(a) increases
(b) is unchanged
(c) decreases

6-54 The result of an experiment that is "expected" to occur (i.e., the expectation)
  (a) may never occur
  (b) must occur a given proportion of times
  (c) always occurs
  (d) none of these

6-55 If two different samples of the same size are taken from a normal population with a known variance, the two resulting 95% confidence intervals for the mean will be of the same length.
  (a) True
  (b) False
  (c) Not necessarily

6-56 The 90th percentile of the t-distribution with f degrees of freedom is
  (a) smaller than
  (b) equal to
  (c) greater than the 90th percentile of the standard normal distribution.

6-57 The sampling distribution for s^2 is
  (a) negatively skewed
  (b) symmetric
  (c) positively skewed

6-58 The variable $(\bar{x} - \mu)/(\sigma/\sqrt{n})$ has mean and variance, respectively, of _____ and _____ .

6-59 The classes 0 to 5, 10 to 15, 20 to 25, 30 to 35, 40 to 45, 50 to 55, 60 to 65 appear unsuitable for a frequency distribution because _____.

6-60 If the sample size is increased, the width of a confidence interval for the mean of a normal population with unknown variance will decrease.
  (a) True
  (b) False
  (c) Not necessarily

6-61 (a) If a certain type of link is such that 5% of a large batch of several thousand will break under stress and 10 of these links are picked at random, find the probability that more than one link will break.
  (b) If 200 are picked, find the probability that at most 12 will break.
  (c) Repeat part (a) assuming that there are only 60 links in the batch.

6-62 (a) Verify that the graph in Figure 6-8 is appropriate as a probability graph.
  (b) If we are to draw a number at random from a continuum with probabilities as indicated by the graph, what is the probability that the number will be less than 1?

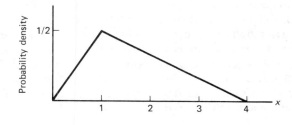

Figure 6-8 Probability graph.

6-63 The following is a sample of heating costs (in dollars) for 51 homes in a particular area.

306	301	309	250	305	281	279	254	210	248	285
275	216	200	282	294	302	282	270	298	278	249
295	274	267	221	240	312	301	300	232	321	288
295	310	273	265	241	266	304	234	250	265	305
293	294	288	268	315	326	283				

(a) Find the mean and median costs.

(b) Find the range and standard deviation.

(c) Form a frequency distribution for this sample using equal-width classes with 200 to 209 as the first class. For each class, give the class limits, class marks, and class boundaries.

(d) Sketch a frequency polygon for this sample.

(e) Form an "or more" cumulative distribution for this sample.

(f) According to Chebyshev's theorem, how many of the individual costs should be within $45 of the mean? How many of the actual costs are within $45 of the mean?

6-64 If $f(x) = (3x - 10)^2/1095$ for $x = 5, 6, 7, 8, 9, 10$, show that $f(x)$ is a probability function and find the corresponding mean and variance.

6-65 Suppose that each question on a multiple-choice quiz with four possible answers per question is answered as follows: A card is drawn from an ordinary deck: if it is a spade, the first answer is chosen; if it is a heart, the second is chosen; if it is a diamond, the third is chosen; and if it is a club, the fourth is chosen. If there are ten questions, what is the probability of fewer than three correct answers?

6-66 A librarian wishes to estimate the *number* of volumes in a given collection that have been signed out more than twice in the last year. The collection includes 3500 volumes. If the estimate is to be based on a sample of volumes and the librarian wants to have a 95% chance that the estimated number will be within 125 of the actual number, how many volumes should be included in the sample

(a) if the librarian has no prior knowledge?

(b) if the librarian believes very strongly that at most 1400 of the volumes have been signed out more than twice?

6-67 In a traffic accident study conducted over a 1-month period, an urban area was divided into six different regions. In the first region, 10 intersections were studied and they produced a mean of 6.3 accidents per intersection. In the second region, 12 intersections produced a mean of 3.25. In the third, fourth, fifth, and sixth regions, 16, 8, 10, and 14 intersections, respectively, produced means of 0.75, 1.25, 0.1, and 0.5, respectively. Considering all of the regions, what was the mean number of accidents per intersection during the month?

6-68 Of 144 lighters, 132 are acceptable and will light 95% of the time. The other 12 are faulty and will light only 60% of the time. One lighter chosen at random has been tried twice and was found to light both times. What is the probability that this lighter will light in at least eight of the next ten tries?

6-69 Find the value of α in each of the following.

(a) $z_\alpha = 1.960$

(b) $z_\alpha = -1.645$

(c) $z_\alpha = 1.00$

(d) $-z_{\alpha/2} = -2.50$

(e) $z_{\alpha/2} = 0.44$

(f) $t_{\alpha, 15} = 1.753$

(g) $t_{\alpha/2, 10} = 3.169$

(h) $-t_{\alpha/2, 8} = -2.306$

(i) $\chi^2_{\alpha, 12} = 21.026$

(j) $\chi^2_{\alpha, 24} = 10.856$

6-70 In a sample of 150 members of a particular age group, 43 indicated that they used marijuana. Estimate the proportion of the full age group that uses marijuana. Give the estimate in the form of an interval such that you can be 90% confident that the true proportion is in the interval.

6-71 Suppose that a small population consists of the following members: 1, 3, 2, 2, 1, 4, 1.

(a) Find the mean and variance of the population.

(b) Enumerate all possible samples of size 2 and, for each, find the sample mean. Sample without replacement.

(c) Using the results of part (b), find the mean and variance for the means of samples of size 2.

(d) If a random sample of size 2 is to be drawn from the population, what is the probability that the mean of that sample will be greater than 2.5?

6-72 A sample of 15 measurements of a particular attribute of a single blood sample was taken with a new measuring device and the resulting measurements were:

$$12.15 \quad 12.51 \quad 12.43 \quad 12.20 \quad 12.63 \quad 12.39 \quad 12.47 \quad 12.12$$

$$12.55 \quad 12.17 \quad 12.38 \quad 12.49 \quad 12.30 \quad 12.33 \quad 12.16$$

(a) Estimate the variance of measurements displayed by this device.

(b) Calculate a 98% confidence interval for the standard deviation.

(c) Assuming that measurements taken with this device are equal to the true level of the attribute on the average, calculate a 95% confidence interval for the true level of this particular blood sample.

(d) What distribution assumption is made in parts (b) and (c)?

6-73 Suppose that five shipments are to be assigned to two different carriers in such a way that all shipments are carried and neither carrier carries more than three shipments. How many ways can the shipments be assigned?

6-74 If a 90% confidence interval for the population mean is to be calculated from each of 20 samples, what is the probability that at least 18 of the intervals will include the population mean?

6-75 Use a dollar-sign pictograph to illustrate the following world expenditures in 1977:

Military expenditures	$434 billion
Education expenditures	$374 billion
Health expenditures	$186 billion

6-76 Individual rats produce maze scores for each variation of a particular maze structure that are normally distributed with a standard deviation $\sigma = 4.8$. How many sample maze scores must be obtained to estimate the true mean maze score μ if it is required that there be a 95% chance that the sample mean will differ from μ by no more than 1.0.

TESTING HYPOTHESES

7-1 INTRODUCTION

In Chapter 6 we developed techniques for estimating parameter values with a single value or with an interval estimate. The estimation techniques are exploratory in nature and help to answer general questions such as: What is the value of the population mean? In some investigations, we may be interested in one particular value for a parameter and we may require techniques to assess the plausibility of this value. In these cases we will require confirmation techniques to help answer more particular questions such as: Is 50.0 seconds the mean time required by upper-year students to recognize obscured patterns?

A statement specifying that a parameter has a particular value is referred to as a *hypothesis* and, when we assess the plausibility of that statement, we are said to be testing the hypothesis. The general procedures for performing such assessments are referred to as *hypothesis-testing* procedures. There are many considerations involved in hypothesis testing. We attempt to integrate them all by developing the general ideas in the next section on the basis of one illustration. The one illustration will involve investigating the mean μ of a normal population when the population variance σ^2 is known. The decision rules resulting from this development are summarized in Section 7-3 together with procedures for testing hypotheses about population means in other situations. Sections 7-4 and 7-5 include procedures for testing hypotheses about variances and proportions.

EXAMPLE 7-1 In previous experiments, a psychologist chose subjects from freshman classes and asked them to try to identify an obscured pattern on a display card. For each subject, the psychologist recorded the time required to identify the pattern correctly. Collecting all of the results from a very large number of freshmen over several years, the psychologist noted that the response times were normally distributed with a mean of 50.0 seconds and a standard deviation of 4.0 seconds. The psychologist now wishes to compare the responses for upper-level students in psychology classes with those of the freshmen. It is possible that, after taking a number of psychology courses, subjects will sharpen reactions to generally reduce response times, will generally show no material change in response times, or because of the boredom of repetition, will generally increase response times. The psychologist plans to take a sample of 25 responses for upper-level students to investigate whether the mean response time for upper-level students is 50.0 or whether it is some other value. The appropriate procedure is to employ a hypothesis test in which the statement being assessed for plausibility is the statement: The mean response time is 50.0; that is, $\mu = 50$. ▲

Null Hypothesis and Alternative Hypothesis

In hypothesis testing we are usually trying to decide between two hypotheses, with one of the hypotheses being very specific about the value of a parameter or the relationship between a number of parameters and the other hypothesis providing a more general alternative statement. The first hypothesis, the specific statement that is being assessed for plausibility, is referred to as the *null hypothesis*. It is usually denoted as H_0. The term null hypothesis has developed from the many investigations in which there is a test of the specific hypothesis that a particular treatment has no effect.

The second hypothesis, usually a more general statement, is adopted as an alternative if the null hypothesis is found to be implausible. This second hypothesis is referred to as the *alternative hypothesis*. It is usually denoted as H_A.

> The primary statement being tested for plausibility in a hypothesis test is referred to as the *null hypothesis* and denoted as H_0. The statement to be adopted if the null hypothesis is found to be implausible is the *alternative hypothesis*, denoted as H_A.

EXAMPLE 7-2 In Example 7-1 the null hypothesis is "$H_0 : \mu = 50$." The alternative hypothesis is the statement that μ takes on some other value; that is, the alternative hypothesis is "$H_A : \mu \neq 50$." ▲

Test Statistic

The assessment of the plausibility of the null hypothesis is based on statistical evidence from sample data. In each investigation we try to choose one statistic as an appropriate indicator of the plausibility of the null hypothesis.

> The statistic selected to provide evidence about the null hypothesis is called the *test statistic*.

EXAMPLE 7-3 For the hypothesis test in Example 7-1, inferences on a population mean μ are required. An appropriate test statistic to consider from the sample of response times is the sample mean $\bar{x}$, since, as we know from previous chapters, the sample mean is a reasonable indicator of the true value of the population mean. ▲

Acceptance Region and Rejection or Critical Region — A Decision Rule

The decision on the plausibility of the null hypothesis is based on comparing the actual observed value of the test statistic to what we would expect to observe if the null hypothesis were true. If the actual result is inconsistent with what we would expect if the null hypothesis were true, we reject the null hypothesis; otherwise, we either accept it or (as discussed later) we reserve judgment.

EXAMPLE 7-4 In our pattern recognition example, suppose that it is reasonable to assume that even though the population mean μ may change, the population of response times will still have a standard deviation $\sigma = 4.0$. The sample mean $\bar{x}$ will thus have a sampling distribution that is a normal distribution with a mean $\mu_{\bar{x}}$ equal to the population mean response time μ and a standard deviation equal to $\sigma/\sqrt{n} = 4.0/\sqrt{n}$. If the null hypothesis is true, and if a sample of $n = 25$ response times is taken, the sampling distribution of $\bar{x}$ will then be a normal distribution with a mean of 50.0 and a standard deviation of $4.0/\sqrt{25} = 0.80$. Accordingly, if the null hypothesis were true, we would expect $\bar{x}$ to be "close" to 50.0 because its distribution would then be centered on the mean 50.0 and have a small standard deviation 0.80. In this case, if $\bar{x}$ is not close to 50.0, it will not meet our expectations. The null hypothesis will be rejected; the statement "$\mu = 50.0$" will be found to be implausible. ▲

In general, we consider the sampling distribution for the chosen test statistic to determine which of its values are inconsistent with the null hypothesis (are unlikely to occur if H_0 is true) but are consistent with the alternative hypothesis. These values provide evidence to contradict the null hypothesis and lead to a decision to reject H_0. The set of these values is referred to as the *rejection region* or *critical region*.

We also determine which values of the test statistic are consistent with the null hypothesis (are likely to occur if H_0 is true). These values provide evidence to support the null hypothesis and lead to a decision to accept H_0. They belong to a set of values called the *acceptance region*.

The acceptance region also includes any values of the test statistic that are unlikely to occur if H_0 is true but even less likely to occur if H_0 is false. (Such values may occur in one-sided tests, as illustrated later in Example 7-14). These values also lead to a decision to accept H_0.

The acceptance region is, in fact, complementary to the rejection region and consists of all possible values of the test statistic not included in the rejection region.

> Values of the test statistic that are inconsistent with the null hypothesis but are consistent with the alternative hypothesis form the *rejection region* or *critical region*. Values of the test statistic not included in the rejection region form the *acceptance region*. Conclusions on the null hypothesis are based on the following *decision rule*: If the value of the test statistic is included in the rejection region, reject H_0 and adopt H_A instead. If the value of the test statistic is included in the acceptance region, accept H_0 or reserve judgment.

EXAMPLE 7-5   Suppose, in Example 7-4, that close to 50.0 is defined as within 1.2 of 50.0; that is, a value close to 50.0 is a value between 48.8 and 51.2 inclusive. The rejection region consists of values not close to 50.0. It is formed of two parts, values of $\bar{x}$ for which $\bar{x} < 48.8$ and values of $\bar{x}$ such that $\bar{x} > 51.2$. The acceptance region consists of all values of $\bar{x}$ such that $48.8 \leqslant \bar{x} \leqslant 51.2$.

The decision rule is as follows: If $\bar{x} < 48.8$ or if $\bar{x} > 51.2$, reject H_0 and adopt H_A. If $48.8 \leqslant \bar{x} \leqslant 51.2$, accept H_0 or reserve judgment.   ▲

Type I and Type II Errors

As we did with estimating procedures in Chapter 6, we also consider an evaluation of a hypothesis-testing procedure. We first consider what might happen if the null hypothesis is, in fact, true. We may either obtain data that will lead to the correct decision to accept the true null hypothesis, or we might obtain data producing a value of the test statistic in the rejection region, thus leading to the error of rejecting the true null hypothesis.

> Rejecting a true null hypothesis is called a *type I error*.

The probability of making a type I error is a *conditional probability*—it is the probability of rejecting the null hypothesis given that it is true. This probability is usually denoted as α.

$$P(\text{type I error}) = P(\text{reject } H_0 \mid H_0 \text{ true}) = \alpha$$

Now we consider what might happen if the null hypothesis is, in fact, false. We may either obtain data that will lead to the correct decision to reject the null hypothesis, or we might obtain data producing a value of the test statistic in the acceptance region, thus leading to the error of accepting the false null hypothesis.

Accepting a false null hypothesis is called a *type II error*.

The probability of making a type II error is also a conditional probability, the probability of accepting the null hypothesis given that it is false. It is usually denoted as β (the Greek lowercase letter beta).

$$P(\text{type II error}) = P(\text{accept } H_0 \mid H_0 \text{ false}) = \beta$$

We summarize in the following table:

Decision	True Condition	
	H_0 true	H_0 false
Accept H_0	correct decision Probability $= 1 - \alpha$	type II error Probability $= \beta$
Reject H_0	type I error Probability $= \alpha$	correct decision Probability $= 1 - \beta$

We first illustrate the case in which H_0 is true and there is a possibility of a type I error.

EXAMPLE 7-6   We might assess the proposed decision procedure in Example 7-5 by evaluating α, the probability that it will result in an incorrect decision if the null hypothesis is, in fact, true.

It is easier to first find $1 - \alpha$, the probability of correctly accepting the true hypothesis, and to then subtract this value from 1 to produce α.

According to the decision rule, $H_0 : \mu = 50.0$ is to be accepted if $48.8 \leqslant \bar{x} \leqslant 51.2$; hence, $P(\text{accept } \mu = 50.0) = P(48.8 \leqslant \bar{x} \leqslant 51.2)$. Since $(\bar{x} - \mu)/(\sigma/\sqrt{n}) = z$, we may convert the probability to a standard normal probability

$$P\left(\frac{48.8 - \mu}{\sigma/\sqrt{n}} \leqslant \frac{\bar{x} - \mu}{\sigma/\sqrt{n}} \leqslant \frac{51.2 - \mu}{\sigma/\sqrt{n}} \right) = P\left(\frac{48.8 - \mu}{\sigma/\sqrt{n}} \leqslant z \leqslant \frac{51.2 - \mu}{\sigma/\sqrt{n}} \right)$$

Since a sample of size $n = 25$ is used and since $\sigma = 4.0$, then $\sigma/\sqrt{n} = 0.8$, as previously noted. Now, if the hypothesis is, in fact, true, then $\mu = 50.0$ and the probability statement becomes P(accept the null hypothesis given that $\mu = 50.0$)

$$= P\left(\frac{48.8 - 50.0}{0.8} \leqslant z \leqslant \frac{51.2 - 50.0}{0.8} \right)$$

$$= P(-1.5 \leqslant z \leqslant 1.5) = 0.8664$$

If the null hypothesis is true, the probability that the psychologist will make a correct decision is $1 - \alpha = 0.8664$. Considering the complement, we find that if the null hypothesis is true, the probability that the psychologist will make a type I error and reject H_0 is $\alpha = 1 - (1 - \alpha) = 1 - 0.8664 = 0.1336$. ▲

Level of Significance

It is usual in hypothesis testing to specify a desired α in advance and to determine the appropriate decision rule to achieve the desired α.

The value of α is generally chosen to be small, in part so that there will be a small probability of a type I error. With a small α, an experimenter can be quite confident that H_0 is false before rejecting it. A small α provides a stronger statement when the null hypothesis is rejected. If α is small, values of the test statistic that cause a rejection of the null hypothesis are very rare or *significantly different* from what we expect when the hypothesis is true. Such values of the test statistic, referred to as *significant* values, thus provide strong evidence to contradict the null hypothesis.

The value of α indicates, in a probability sense, how rare a sample result must be, that is, how significantly it must differ from what would be expected if the null hypothesis were true, before H_0 will be rejected. As a result, α is usually called the *level of significance*.

In a hypothesis test, the level of significance is

$$\alpha = P(\text{reject } H_0 \mid H_0 \text{ true})$$

It is very often expressed as a percentage.

EXAMPLE 7-7 Suppose that the psychologist in our illustration does not want a probability of a type I error as high as the value 0.1336 found in Example 7-6, but would prefer a probability as low as 0.05 (while maintaining the same sample size). The psychologist wants a level of significance of $\alpha = 0.05$ or 5%. To achieve this level, the decision rule must be changed so that $P(\text{reject } H_0 \mid H_0 \text{ true}) = 0.05$ or, equivalently, so that $P(\text{accept } H_0 \mid H_0 \text{ true}) = 1 - 0.05 = 0.95$. The decision rule was originally based on whether $\bar{x}$ is "close" to 50.0. As illustrated in Figure 7-1, the definition of "close" must be modified. The hypothesis was to be accepted if the difference between $\bar{x}$ and 50.0 did not exceed 1.2. To modify the definition of close, this maximum difference 1.2 must be replaced with a new value d which will produce $P(\text{accept}$ the null hypothesis when $\mu = 50.0) = 0.95$, that is, such that $P(\mid \bar{x} - 50.0 \mid \leqslant d \mid \mu = 50.0) = 0.95$. We can rewrite this as $P(-d \leqslant \bar{x} - 50.0 \leqslant d \mid \mu = 50.0) = 0.95$ or $P(-d/(\sigma/\sqrt{n}) \leqslant (\bar{x} - 50.0)/(\sigma/\sqrt{n}) \leqslant d/(\sigma/\sqrt{n}) \mid \mu = 50.0) = 0.95$. Since $50.0 = \mu$ when the null hypothesis is true, $(\bar{x} - 50)/(\sigma/\sqrt{n}) = z$, and since, as previously noted, $\sigma/\sqrt{n} = 0.8$, we can write the probability statement as $P(-d/0.8 \leqslant z \leqslant d/0.8) = 0.95$. From the use of percentiles in the development of the confidence interval procedures in Chapter 6 we know that $P(-z_{0.025} \leqslant z \leqslant z_{0.025}) = 0.95$. For the probability statement with d to be correct, d must therefore be such that $d/0.8 = z_{0.025}$ or

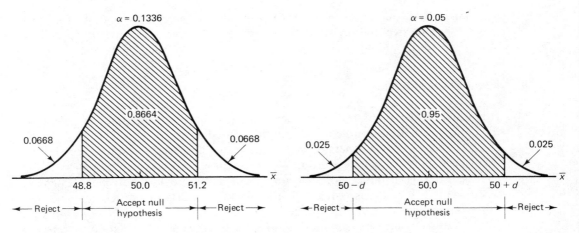

Figure 7-1 Acceptance and rejection regions.

1.960. As a result, d must equal $1.960 \times 0.8 = 1.57$. The psychologist will now accept the hypothesis if $|\bar{x} - 50.0| \leqslant 1.57$, that is, if $48.43 \leqslant \bar{x} \leqslant 51.57$.

The decision rule is modified to the following: Reject H_0 if $\bar{x} < 48.43$ or if $\bar{x} > 51.57$. Accept H_0 or reserve judgment if $48.43 \leqslant \bar{x} \leqslant 51.57$. ▲

A General Procedure

A general procedure (which will be repeated in Section 7-3) can be developed by considering the general analogies to the particular results of our response-time illustration. In the general case, we test the null hypothesis that μ is some specific value μ_0 (rather than 50.0). We base the procedure on a sample of size n (rather than 25) from a population that is assumed to have a normal distribution with a known standard deviation σ (rather than 4.0). To achieve a level of significance α (rather than 0.05), we have a decision rule to reject the null hypothesis if $\bar{x}$ differs from μ_0 by more than a critical distance d. The value of d must satisfy $P(|\bar{x} - \mu_0| \leqslant d \mid \mu = \mu_0) = 1 - \alpha$ [rather than $P(|\bar{x} - 50.0| \leqslant d \mid \mu = 50.0) = 0.95$]. It is found by equating $d/(\sigma/\sqrt{n})$ to $z_{\alpha/2}$ (rather than $d/0.8$ to $z_{0.025} = 1.960$). The appropriate value for d is thus $z_{\alpha/2} \times \sigma/\sqrt{n}$ (rather than $1.960 \times 0.8 = 1.57$). The acceptance region for the test statistic $\bar{x}$ then becomes $\mu_0 - z_{\alpha/2}\sigma/\sqrt{n} \leqslant \bar{x} \leqslant \mu_0 + z_{\alpha/2}\sigma/\sqrt{n}$ (rather than $48.43 \leqslant \bar{x} \leqslant 51.57$), producing the following decision rule:

> In order to test $H_0: \mu = \mu_0$ against $H_A: \mu \neq \mu_0$ with a level of significance α on the basis of the results of a sample of size n from a normal population with standard deviation σ, the decision rule is as follows: If $\bar{x} < \mu_0 - z_{\alpha/2}\sigma/\sqrt{n}$ or if $\bar{x} > \mu_0 + z_{\alpha/2}\sigma/\sqrt{n}$, reject H_0. If $\mu_0 - z_{\alpha/2}\sigma/\sqrt{n} \leqslant \bar{x} \leqslant \mu_0 + z_{\alpha/2}\sigma/\sqrt{n}$, accept H_0 or reserve judgment.

Statistical Proof — Burden of Proof

When we obtain a "significant" value for the test statistic, we may claim that we have sufficient evidence to declare that the null hypothesis is false. We say that such a result *proves the null hypothesis false*. This "proof," which we may refer to as *statistical proof*, is not absolute proof because, even though we may statistically "prove" the null hypothesis false, it is possible that the null hypothesis is, in fact, true and that a type I error has been made. Choosing a small α reduces the probability of such an error and strengthens the proof. This consideration of proof applies only when we reject the null hypothesis. When we obtain a result that leads to accepting the null hypothesis, we have enough evidence in support of the null hypothesis to *establish that it is plausible*; however, we cannot prove that it is true. In a sense, we accept the hypothesis "by default" since we have failed to prove it false. As a result, in a hypothesis test, the *alternative bears the burden of proof*.

We may consider a further aspect of interpreting the results of a hypothesis test through a consideration of conditional probabilities. The probability of accepting the null hypothesis given that it is true is P(accept H_0 | H_0 is true) and is equal to $1 - \alpha$. Similarly, P(reject H_0 | H_0 is true) $= \alpha$, the chosen level of significance.

Now, in general, $P(A | B)$ does not equal $P(B | A)$. It is not reasonable, therefore, to believe that P(H_0 is true | accept H_0) = P(accept H_0 | H_0 is true) = $1 - \alpha$. In particular, we should not, for example, say that there is a 95% chance that the null hypothesis is true if we accept it in a test with a 5% level of significance.

When we accept a null hypothesis, we do so not because it has a high probability of being true, but because we have failed to obtain sufficient evidence in the data to reject it. When we obtain a "significant" value of the test statistic and reject the null hypothesis, we say that we have sufficient evidence to declare that the hypotheses is false because there would be only a small probability (i.e., α) of obtaining such a significant value if the null hypothesis were true.

EXAMPLE 7-8 If the psychologist in our illustration obtains a sample mean of 47.2, then, according to the decision rule of Example 7-7, H_0 that $\mu = 50.0$ should be rejected. Because the value 47.2 is an unusual result for $\bar{x}$ when the null hypothesis is true, it provides proof that the null hypothesis is false. If, however, a sample mean of 49.0 is obtained, then, even though it leads to accepting the hypothesis, it does not provide proof that the hypothesis is true. In fact, the value 49.0 is closer to (and, hence, more consistent with) a true mean level of 48.5 or 49.5 than it is to the hypothesized value of 50.0. ▲

Reserving Judgment

Because accepting the null hypothesis is a form of "default" decision and involves the risk of a type II error, we may wish, *where possible*, to avoid accepting the null hypothesis. If an immediate decision is not necessary, we may wish to *reserve judgment* rather than accept the null hypothesis, especially if we obtain a value for the test statistic that is in the acceptance region but close to the rejection region.

Choice of Null Hypothesis and Alternative Hypothesis

Consideration of the preceding interpretations on significance and proof is necessary in choosing the null hypothesis and the alternative hypothesis. The choice is determined by the fact that the alternative bears the burden of proof. Experimenters must often set up a null hypothesis in the hope that it can be rejected in order to prove a claim. As a result an experimenter's *research hypothesis*—a hypothesis for which strong supporting evidence is desired—will often become the alternative hypothesis in a statistical test.

EXAMPLE 7-9 In our response-time illustration, in the hope of finding sufficient evidence to support the research hypothesis that upper-level students differ from freshmen with regard to mean response time, the psychologist has set up a test with the null hypothesis that upper-level students do not differ but have the same mean response time $\mu = 50.0$. The research hypothesis that $\mu \neq 50.0$ becomes the alternative hypothesis and bears the burden of proof in the test. ▲

Type II Errors — Operating Characteristic Curve

To this point we have considered α and $1 - \alpha$, the probabilities that apply when the null hypothesis is true. We now consider β, the probability of making a type II error and accepting a null hypothesis when it is, in fact, false.

EXAMPLE 7-10 Suppose that, in our illustration, the true mean response time is not 50.0 as hypothesized, but 47.5. According to the decision rule in Example 7-7, the psychologist will accept the false hypothesis that $\mu = 50.0$ if $48.43 \leqslant \bar{x} \leqslant 51.57$. With $\mu = 47.5$, $\bar{x}$ will have a sampling distribution that is a normal distribution with mean 47.5 and standard deviation $\sigma/\sqrt{n} = 0.8$, as before. As a result, $(\bar{x} - 47.5)/0.8 = z$ and we find the probability of a type II error when $\mu = 47.5$ as

$$\beta = P(\text{accept } H_0 \mid \mu = 47.5) = P(48.43 \leqslant \bar{x} \leqslant 51.57 \mid \mu = 47.5)$$

$$= P\left(\frac{48.43 - 47.5}{0.8} \leqslant \frac{\bar{x} - 47.5}{0.8} \leqslant \frac{51.57 - 47.5}{0.8} \Big| \mu = 47.5 \right)$$

$$= P(1.16 \leqslant z \leqslant 5.08) = 0.1230$$

If the true mean concentration level is 47.5, there is a probability of 0.1230 that the procedure will produce data that lead the psychologist to make a type II error and accept the false hypothesis that the mean level is 50.0.

Figure 7-2 illustrates the sampling distributions and probabilities of errors for the cases that $\mu = 50$ and $\mu = 47.5$. ▲

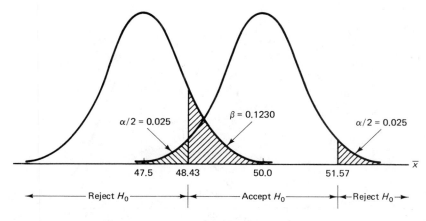

Figure 7-2 Type I and type II error probabilities.

Although there is only one α related to the hypothesis test, there are many β's. Each value of μ that is possible according to the alternative hypothesis produces a β.

EXAMPLE 7-11 In our response-time illustration, suppose that μ is neither 50.0 as given in the null hypothesis, nor 47.5 as considered in Example 7-10, but is instead 48.0. To accept H_0 is to commit a type II error and the probability of this is $\beta = P(\text{accept } H_0 \mid \mu = 48.0)$. We determine this β by repeating the calculation of β in Example 7-10 with $\mu = 48.0$ instead of 47.5 to find that the probability of accepting the (false) hypothesis (that $\mu = 50.0$) is $\beta = 0.2946$ if, in fact, $\mu = 48.0$.

Repeating the calculation with $\mu = 49.0$ instead produces $\beta = P(\text{accept } H_0 \mid \mu = 49.0) = 0.7611$. ▲

Since there are many possible values for β, it is usual to use a graphical presentation of the many β's as an indication of the behavior of a test.

> The *operating characteristic curve* for a hypothesis test is a graph illustrating $P(\text{accept } H_0)$ for each value of the parameter being investigated.

In the test of $H_0: \mu = \mu_0$ the operating characteristic curve is a plot of the values of $P(\text{accept } H_0: \mu = \mu_0 \mid \mu = \mu^*)$ for all possible values μ^* that μ might take, including μ_0.

EXAMPLE 7-12 Figure 7-3 illustrates the operating characteristic curve for the response-time illustration with decision rule as given in Example 7-7. ▲

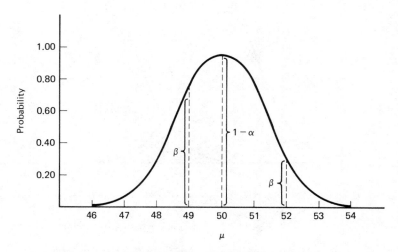

Figure 7-3 Probability of accepting $\mu = 50.0$ as a function of μ.

Note that the operating characteristic curve illustrates the probability of accepting the null hypothesis as a function of the true value of the parameter; thus, the height of the curve is equal to β for each value of the parameter except the hypothesized value. At this one point (where the null hypothesis is true) it is $1 - \alpha$, the probability of accepting a true null hypothesis.

A good test is one for which there is a high probability of accepting the null hypothesis when it is true, or nearly true, and a low probability of accepting the null hypothesis when the true value of the parameter is quite different from the hypothesized value. In Figure 7-4, μ_0 is the parameter value under the null hypothesis. The dashed curve is the operating characteristic curve for a relatively poor test and the solid curve is that for a relatively good test, both tests having the same level of significance.

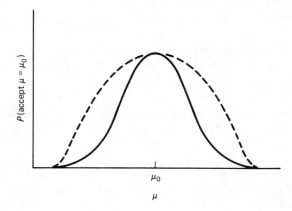

Figure 7-4 Operating characteristic curves.

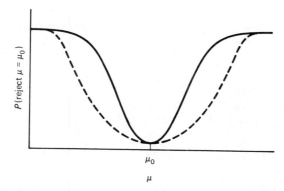

Figure 7-5 Power curves.

Power

Sometimes, instead of considering β, the probability of accepting a false hypothesis, we consider $1 - \beta$, the probability of rejecting a false hypothesis.

> The probability of rejecting a false hypothesis is referred to as the *power* of a test.

Figure 7-5 illustrates the power curves corresponding to Figure 7-4.

Sample Size for Specific β

If assessments of the power or operating characteristic curve of a test indicate the need for improvement, we might consider increasing the sample size. It would be possible to achieve increased power by expanding the rejection region, but to do so would increase α to a value greater than the chosen value. In fact, with a fixed sample size, a gain (or loss) in β is accompanied, in a particular test, with a loss (or gain) in α, and vice versa; accordingly, although we may wish to choose α very small, we must bear in mind (as indicated in Figure 7-6, in which μ_0 is the hypothesized

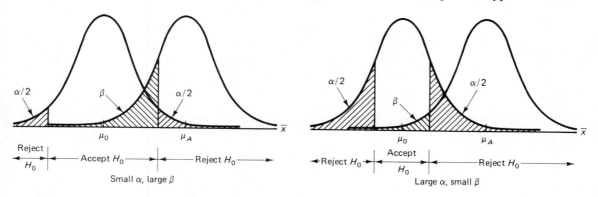

Figure 7-6 Gain (loss) in β produces loss (gain) in α.

value and μ_A an alternative value) that small values of α may produce large values of β. It is possible, however, to decrease β and, at the same time, maintain (or even decrease) α by choosing a larger sample size.

EXAMPLE 7-13 Suppose, in our illustration, that the psychologist does not believe that the test is powerful enough when $\mu = 48.0$, but would like the probability of rejecting the hypothesis in this case to be greater than the value of $1 - \beta = 1 - 0.2946 = 0.7054$ found from Example 7-11. Suppose that it is desirable to have β as high as 0.90.

To increase $1 - \beta$ without decreasing α, the psychologist must choose a new sample size n. If α is still 0.05, then, as before, the hypothesis will be accepted if $\bar{x}$ is sufficiently close to 50.0. The definition of "close" was within a distance d, where $d = z_{0.025}\sigma/\sqrt{n}$ was found as 1.960×0.8. With a new n but the same $\sigma = 4.0$, the new value for d becomes $z_{0.025} \times (4.0/\sqrt{n}) = 1.96 \times (4.0/\sqrt{n})$. The procedure now is to accept the hypothesis if $|\bar{x} - 50.0| \leqslant z_{0.025} \times (\sigma/\sqrt{n}) = 1.960 \times (4.0/\sqrt{n})$ that is, if $50.0 - (1.960 \times 4.0/\sqrt{n}) \leqslant \bar{x} \leqslant 50.0 + (1.960 \times 4.0/\sqrt{n})$. The psychologist would like the probability of this event to be $1 - 0.90 = 0.10$ when $\mu = 48.0$. To determine the appropriate value for n, the probability of the event is set equal to 0.10 and then to make use of standard normal tables, $\mu = 48.0$ is subtracted throughout and everything is divided by $\sigma/\sqrt{n} = 4.0/\sqrt{n}$ to produce

$$P\left(\frac{50.0 - \left[1.960 \times (4.0/\sqrt{n})\right] - 48.0}{4.0/\sqrt{n}} \leqslant \frac{\bar{x} - \mu}{\sigma/\sqrt{n}} \leqslant \frac{50.0 + \left[1.960 \times (4.0/\sqrt{n})\right] - 48.0}{4.0/\sqrt{n}} \right) = 0.10.$$

In this expression, σ has been replaced with its known value 4.0 in the left and right terms and μ has been replaced with the value being considered, 48.0, in both terms. The middle term is then a standard normal z.

Since the limits on $\bar{x}$ were centered around 50.0, which is above 48.0, it is reasonable to expect that the two numerical limits will represent values above 48.0 and that the larger of the two will, as illustrated in Figure 7-7, be far enough into the extreme upper tail of the distribution to have no effect on the probability. The lower term must, therefore, cut off the upper 10% of the distribution. The lower term in the standard normal probability statement must, therefore, cut off the upper 10% of the standard normal and must be $z_{0.10} = 1.282$. We then solve $(50.0 - [1.960 \times (4.0/\sqrt{n})] - 48.0)/(4.0/\sqrt{n}) = 1.282$ to find n. Subtracting the 48.0 from 50.0 to produce 2.0 and then multiplying the top and the bottom of the left-hand side by $\sqrt{n}/4.0$, we obtain $2.0\sqrt{n}/4.0 - 1.960 = 1.282$; thus, $\sqrt{n} = (1.282 + 1.960) \times (4.0/2.0) = 6.48$, and $n = (6.48)^2 = 42$. With this calculation, we thus find that the psychologist can maintain a level of significance of $\alpha = 0.05$ and reduce β to 0.10 when $\mu = 48.0$ if the sample size is increased to 42. With this new sample size, the critical distance d becomes $1.960 \times (\sigma/\sqrt{n}) = 1.960 \times (4.0/\sqrt{42}) = 1.21$ and the new acceptance region is $48.79 < \bar{x} < 51.21$. ▲

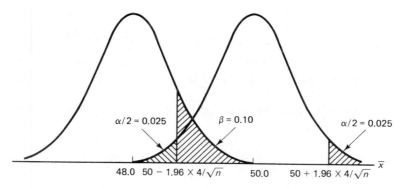

Figure 7-7 Determining n for fixed α, β.

The procedure leading to the solution of n can be generalized beyond this particular example. We let μ_0 denote the hypothesized mean (50.0 in the example), μ_A the alternative value being considered (48.0 in the example), σ the population standard deviation (4.0 in the example), and α and β the probabilities of type I and type II errors, respectively (0.05 and 0.10 in the example).

> In a test of the null hypothesis $H_0 : \mu = \mu_0$ against the alternative $H_A : \mu \neq \mu_0$, with level of significance α, when μ is an alternative value μ_A, the probability of a type II error will be at most β if the sample size is no smaller than
> $$n = \sigma^2 \left(\frac{z_{\alpha/2} + z_\beta}{\mu_0 - \mu_A} \right)^2$$
> where σ^2 is the variance of the population.

Note that as with sample-size calculations in Chapter 6, calculated values of n are rounded up to the next whole number.

Summarizing the General Procedure

In terms of our response-time illustration, we may outline the general procedure for testing hypotheses. First, we determine the statement that is to be the null hypothesis H_0 (in the illustration, H_0 is $\mu = 50.0$), and the statement that is to be the alternative hypothesis H_A (in the illustration, H_A is $\mu \neq 50.0$). We then select an appropriate test statistic ($\bar{x}$ in the illustration). We select the level of significance α and then, considering the sampling distribution for this statistic when the hypothesis is true, we find the appropriate (general) rejection region [in this case, the region consists of all $\bar{x}$'s such that $|\bar{x} - 50.0| > 1.96 \times (4.0/\sqrt{n}\,)$]. We choose the sample size n, perhaps on the basis of desired power for the test (in the illustration we ultimately chose $n = 42$), and finally, we calculate the numerical rejection region (in the example with $n = 42$ it is all $\bar{x}$'s such that $|x - 50.0| > 1.21$). On this basis we produce a *decision rule*, a rule that tells us what conclusion to draw for each possible result of the experiment or sample (in the illustration, the decision rule is to reject the null hypothesis if $\bar{x} < 48.79$ or if $\bar{x} > 51.21$, and to accept the null hypothesis or reserve judgment otherwise). After developing the decision rule, we may then conduct an experiment or take a sample and perform the test of the hypothesis with the resulting data.

We may then summarize the test procedure in a compact format, as illustrated in the following example.

EXAMPLE 7-14 The test procedure for our response-time illustration with $n = 42$ as determined in Example 7-13 may be summarized as follows:

$H_0 : \mu = 50.0$

$H_A : \mu \neq 50.0$

Level of significance: 0.05

Test statistic: $\bar{x}$

Sample size: 42

Decision rule: reject H_0 and adopt H_A if $\bar{x} < 48.79$ or if $\bar{x} > 51.21$; otherwise, accept H_0 or reserve judgment ▲

Two-Sided Tests and One-Sided Tests

The test we have just described is an example of a *two-sided test* because the alternative hypothesis includes values on both sides (above or below) of the hypothesized value μ_0, and the rejection region has two parts. The two parts of the rejection region consist of the two extreme tails of the sampling distribution for $\bar{x}$; thus, this test may also be referred to as a *two-tailed test*.

If the alternative hypothesis states specifically that the parameter is greater than the hypothesized value or states specifically that the parameter is less than the hypothesized value, the hypothesis test is said to be a *one-sided test* or a *one-tailed test*.

EXAMPLE 7-15 In the response-time illustration, suppose that the psychologist believes that upper-level students have generally quicker reactions than freshmen and produce a mean response time that is less than 50.0. The psychologist thus has a research hypothesis that μ is less than 50.0. The psychologist hopes to obtain sufficient evidence to support (i.e., statistically "prove") the claim that $\mu < 50.0$. Accordingly, this research hypothesis bears the burden of proof in a hypothesis test and becomes the alternative hypothesis.

To determine the appropriate null hypothesis, we consider the opposite statement—that upper-level students are no quicker than freshmen; that is, the mean response time for upper-level students is at least 50.0. This reasoning would lead the psychologist to test $H_0: \mu \geqslant 50.0$ against $H_A: \mu < 50.0$. The statement that $\mu \geqslant 50.0$ is a *composite hypothesis* that includes many values for μ. For the psychologist to set and use α, it is necessary for H_0 to specify a single value for μ. We consider the hypothesized value for μ which is closest to the set of values included in the alternative hypothesis and which thus produces the maximum probability of a type I error. If we consider a single hypothesized μ, there will be one single sampling distribution for the test statistic when the null hypothesis is true and α may be determined from this distribution. We thus consider the test of $H_0: \mu = 50.0$ versus $H_A: \mu < 50.0$. Since the alternative hypotheses consist of values that are all on one side of (below) the hypothesized value, this test is a one-sided or one-tailed test.

We will continue to consider $\bar{x}$ as an appropriate test statistic, as we are still interested in inferences on the population mean μ. The decision rule is developed as follows.

If the sample mean $\bar{x}$ is close to 50.0, the statement $\mu = 50.0$ is plausible and the null hypothesis may be accepted. If the resulting $\bar{x}$ is very much smaller than 50.0, the statement $\mu = 50.0$ may be rejected as implausible. Because such evidence indicates that $\mu < 50.0$, the more general composite hypothesis that $\mu \geqslant 50.0$ may also be rejected. If the resulting $\bar{x}$ is very much greater than 50.0, this result may be inconsistent with the statement that $\mu = 50.0$, but it is even less consistent with the statement $\mu < 50.0$ because it suggests instead that $\mu > 50.0$. It would be foolish to reject $\mu = 50.0$ in favor of $\mu < 50.0$ in this case; accordingly, $\mu = 50.0$ may be accepted. In effect, it is the more general composite null hypothesis $\mu \geqslant 50.0$ that is accepted. In this test, the psychologist accepts $H_0: \mu = 50.0$ (or $\mu \geqslant 50.0$) unless $\bar{x}$ is very much smaller than 50.0, where "very much smaller" is defined in the following:

The decision rule for this one-sided test is to reject H_0 if $\bar{x} < c$ and to accept H_0 or reserve judgment, otherwise, where c is a critical value chosen so that $P(\bar{x} < c \mid \mu = 50.0)$ will equal α, the level of significance.

Now with $n = 42$ as previously established and with σ still assumed to be 4.0, $\bar{x}$ is transformed to a standard normal with the change $z = (\bar{x} - \mu)/(\sigma/\sqrt{n}) = (\bar{x} - \mu)/(4.0/\sqrt{42})$. This result becomes $z = (\bar{x} - 50.0)/(4.0/\sqrt{42})$ if $H_0: \mu = 50.0$ is true. We may then write $P(\bar{x} < c \mid \mu = 50.0) = \alpha$ as $P\big((\bar{x} - 50.0)/(4.0/\sqrt{42}) < (c - 50.0)/(4.0/\sqrt{42}) \mid \mu = 50.0\big) = \alpha$ which then becomes $P\big(z < (c - 50.0)/(4.0/\sqrt{42})\big) = \alpha$. The value of c is chosen to satisfy this statement. Now if we continue to choose $\alpha = 0.05$, then c must satisfy the probability statement $P\big(z < (c - 50.0)/(4.0/\sqrt{42})\big) = 0.05$. For this to be true, $(c - 50.0)/(4.0/\sqrt{42})$ must be the 5th percentile of the standard normal distribution, which is $z_{0.95} = -z_{0.05} = -1.645$. We thus have $c = 50.0 - (1.645 \times 4.0/\sqrt{42}) = 49.0$. The test is summarized as:

$H_0: \mu = 50.0 \; (\mu \geqslant 50.0)$

$H_A: \mu < 50.0$

Level of significance: 0.05

Test statistic: $\bar{x}$

Sample size: 42

Decision rule: reject H_0 and adopt H_A if $\bar{x} < 49.0$; otherwise, accept H_0 or reserve judgment

▲

General Procedure

As with a two-sided test, the one-sided test can be generalized by repeating the development in the example with general analogies substituted for the specific numerical values to produce the following summary for the test of $H_0 : \mu = \mu_0$ (actually $\geqslant \mu_0$) against $H_A : \mu < \mu_0$.

$H_0 : \mu = \mu_0 \, (\mu \geqslant \mu_0)$

$H_A : \mu < \mu_0$

Level of significance: α

Test statistic: $\bar{x}$

Decision rule: reject H_0 and adopt H_A if $\bar{x} < \mu_0 - z_\alpha \sigma / \sqrt{n}$; otherwise, accept H_0 or reserve judgment

A similar development produces the following summary for the opposite one-sided test of $H_0 : \mu = \mu_0$ (actually $\mu \leqslant \mu_0$) against $H_A : \mu > \mu_0$.

$H_0 : \mu = \mu_0 \, (\mu \leqslant \mu_0)$

$H_A : \mu > \mu_0$

Level of significance: α

Test statistic: $\bar{x}$

Decision rule: reject H_0 and adopt H_A if $\bar{x} > \mu_0 + z_\alpha \sigma / \sqrt{n}$; otherwise, accept H_0 or reserve judgment

Operating Characteristic Curves for One-Sided Tests

Because of their different form, one-sided tests produce different operating characteristic curves or power graphs from those produced by two-sided tests. The symmetric two-sided tests produce symmetric operating characteristic curves. One-sided tests are not symmetric. As a result, as illustrated in the following example, they do not produce symmetric operating characteristic curves.

EXAMPLE 7-16 In Example 7-15 the psychologist may wish to know the probability of a type II error if, in fact, μ is actually equal to 48.5. Now

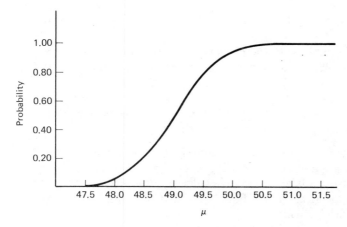

Figure 7-8 Probability of accepting $\mu = 50.0$ as a function of μ.

P(accept H_0) is P($\bar{x} \geqslant 49.0$), which may be written as P($(\bar{x} - \mu)/(\sigma/\sqrt{n}) \geqslant (49.0 - \mu)/(\sigma/\sqrt{n})$). Substituting the known value $\sigma = 4.0$, the chosen sample size $n = 42$, and the alternative value being considered, $\mu = 48.5$, then, given that $\mu = 48.5$, we have ($\bar{x} -$ 48.5)/(4.0/ $\sqrt{42}$) = ($\bar{x} - \mu$)/($\sigma/\sqrt{n}$) = z and the probability becomes P(accept $\mu = 50.0 \mid \mu = 48.5$) = P($z \geqslant (49.0 - 48.5)/(4.0/\sqrt{42})$) = P($z \geqslant$ 0.81) = 0.2090. If $\mu = 49.0$, the probability of making a type II error and accepting H_0 that $\mu = 50.0$ is $\beta = 0.2090$. The power, that is, the probability of rejecting the (false) hypothesis, is $1 - 0.2090 = 0.7910$.

The probability of a type II error can be found for other alternative values of μ. For $\mu = 48.0$ we may repeat the process, substituting 48.0 for 48.5 to find $\beta = 0.0526$. For $\mu = 47.5$, we find that $\beta = 0.0075$. Similarly, we may calculate P(accept H_0) for any other value of μ. With these values we may plot the operating characteristic curve as shown in Figure 7-8. ▲

Sample Size for Specific β in One-Sided Tests

Sample size modification can be used in one-sided tests as well as two-sided tests to improve power for a particular alternative.

EXAMPLE 7-17 Let us suppose that the psychologist is not fully satisfied with the procedure in Example 7-16, but would like to reduce β to 0.10 when $\mu = 48.5$ (while maintaining $\alpha = 0.05$). Again, a new sample size n must be determined.

Replacing 42 with the (to be determined) new value n, the critical value for $\bar{x}$ now becomes $c = 50.0 - (1.645 \times 4.0/\sqrt{n})$. Setting P(accept $\mu = 50.0 \mid \mu = 48.5$) equal to 0.10 as required and substituting the new n and c produces P($z \geqslant [50.0 - (1.645 \times 4.0/\sqrt{n}) - 48.5]/(4.0/\sqrt{n})$) = 0.10. For this probability statement to hold true, the term on the right must equal $z_{0.10} = 1.282$. Adjusting this term as before when solving for n in the two-sided test produces

$\sqrt{n} = (1.282 + 1.645) \times 4.0/1.5 = 7.81$ and the required sample size is $n = (7.81)^2 = 61$.

With the new sample size, the critical value for $\bar{x}$ becomes $c = 50.0 - (1.645 \times 4.0/\sqrt{61}) = 49.2$, and the new decision rule is to reject H_0 if $\bar{x} < 49.2$.

▲

The solution for n can be generalized as before. This time the subscript on z changes from $\alpha/2$ to α, and we have the following result:

> In a one-sided test of the null hypothesis that the mean of a normal population is μ_0, if the level of significance is α, then, when μ is an alternative value μ_A, the probability of a type II error will be at most β if the sample size is chosen no smaller than
>
> $$n = \sigma^2 \left(\frac{z_\alpha + z_\beta}{\mu_0 - \mu_A} \right)^2$$
>
> where σ^2 is the variance of the population.

In this section we have used one illustration to develop the general ideas of hypothesis testing. We have also produced the procedures that apply for tests on the mean of a normal population with a known variance. In the next section we repeat these test procedures with a modified test statistic and then produce the techniques to be used when the population standard deviation is unknown. In the following sections we develop the techniques for testing hypotheses about population proportions or percentages.

PRACTICE PROBLEMS

P7-1 A sample of size 16 is to be taken from a normal population with a standard deviation of 8. Give the decision rule for the hypothesis tests with the following null hypotheses and alternative hypotheses if the level of significance is 5%.
   (a) $H_0 : \mu = 30$, $H_A : \mu \neq 30$
   (b) $H_0 : \mu = 30$, $H_A : \mu < 30$
   (c) $H_0 : \mu = 30$, $H_A : \mu > 30$

P7-2 Repeat Practice Problem P7-1 if the level of significance is 1%.

P7-3 The amount of stabilizer in a milk drink is supposed to be 7.50 milliliter per liter (ml/ℓ). Individual assessments on 1 liter of the drink are known to be normally distributed with a mean μ equal to the actual stabilizer content (ml/ℓ) and a standard deviation σ of 0.30 ml/ℓ. A sample of assessments is to be made on a batch of the drink to determine whether the amount of stabilizer is as it should be or whether there is sufficient statistical evidence that it is not as it should be.
   (a) Set up an appropriate null hypothesis and alternative hypothesis for the determination.
   (b) What is the probability that the null hypothesis will be accepted when it is true if the decision rule is to declare the amount inappropriate whenever the mean of nine measurements exceeds 7.665 or is less than 7.335? What is the level of significance?

(c) What should be the rejection region in part (b) if the level of significance is to be 5%?

(d) If the actual amount is 7.80, what is the probability that the test with the rejection region in part (c) would produce a type II error?

(e) Repeat part (d) for actual amounts of 7.0, 7.1, 7.2, 7.3, 7.4, 7.6, 7.7, 7.9, and 8.0. Sketch the operating characteristic curve.

(f) How many individual assessments are required to have a probability of 0.05 of a type I error and a probability of 0.10 of a type II error when the actual amount is 7.75?

(g) Is there sufficient statistical evidence at the 5% level of significance to reject the null hypothesis if a sample of 16 assessments produced a mean of 7.68?

P7-4 Sources of variation such as operator variability are such that the number of units per hour processed at one stage of an assembly system are normally distributed with a mean of 300 and a standard deviation of 8. It is claimed that the process can be modified so that the mean number of units will be increased by more than 2% while the standard deviation will be unchanged. Because of the costs involved, the modification should not be introduced unless there is sufficient statistical evidence that the increase does exceed 2%. On the other hand, it is highly desirable that the process be modified if the increase is as much as 4%.

(a) Set up an appropriate hypothesis test for assessing a sample of hourly outputs with the modification to determine whether the modification should be introduced in general. Base the decision rule on the minimum sample size necessary to produce a probability of a type I error of at most 0.01 if the increase is no more than 2% and a probability of a type II error of at most 0.05 if the increase is at least 4%.

(b) Determine the probability of rejecting the null hypothesis for actual increases of $1, 1\frac{1}{2}, 2, 2\frac{1}{2}, 3, 3\frac{1}{2}, 4, 4\frac{1}{2}, 5$ and $5\frac{1}{2}$% and sketch the power curve.

(c) Should the modification be introduced if a sample of 28 hourly outputs produced a mean of

(1) 303.8?   (2) 308.2?   (3) 312.4?

EXERCISES

7-1 What are the null hypotheses and alternative hypotheses for the following?

(a) A study is to be conducted to determine whether the mean of a population is 100 or whether it can be proven to differ from this value.

(b) Sample results are to be studied to determine whether they provide statistical evidence that the mean particle count exceeds the maximum tolerable level of 875.

(c) A sample of maximum joint rotations is to be analyzed to see whether there is sufficient statistical evidence that the mean maximum rotation is less than 180°.

7-2 Suppose that individual scores on a test for success potential are known to follow a normal distribution with a standard deviation of 12. It is hypothesized that the mean of the distribution is 60 and this null hypothesis is to be tested on the basis of the sample mean of the scores of 20 individuals selected at random. The null hypothesis is to be rejected if the sample mean exceeds 65 or is less than 55.

(a) If the null hypothesis is true, what is the probability that it will be accepted?

(b) What is the level of significance for this testing procedure?

7-3 A process is known to produce values that follow a normal distribution with a standard deviation $\sigma = 0.10$. If the process is under control, the mean of the normal distribution is $\mu = 20.00$. The null hypothesis that the process is under control is to be tested against the alternative hypothesis that it is out of control (i.e., μ is not 20.00). The hypothesis is to be tested on the basis of the mean of a sample of ten values.

(a) Set up a testing procedure with a 5% level of significance.

(b) If $\mu = 20.08$, what is the probability that the testing procedure will result in a type II error?

(c) Repeat part (b) for $\mu = $ 19.88, 19.90, 19.92, 19.94, 19.96, 19.98, 20.02, 20.04, 20.06, 20.10, and 20.12.

(d) Sketch the operating characteristic curve for this testing procedure.

(e) What decision should be made if the mean of ten sample values is
  (1) 20.05?
  (2) 19.91?

7-4 Suppose, in Exercise 7-3, that there should only be a 10% chance of accepting the null hypothesis that $\mu = 20.00$ if μ is, in fact, 20.08.

(a) Determine a new sample size for the testing procedure.

(b) Using the new testing procedure, repeat Exercise 7-3, parts (b), (c), and (d).

7-5 Measurements on the amount of sulfur dioxide in a fixed quantity of acid rain follow a normal distribution with mean μ equal to the mean amount of sulfur dioxide and a standard deviation of 0.15. It is claimed that the mean amount μ exceeds 3.50.

(a) Set up the appropriate null hypothesis and alternative hypothesis to be used in determining whether the claim can be statistically proven.

(b) What sample size would be used if the test is to have a 1% level of significance and if there should only be a 5% chance of accepting the (false) null hypothesis if μ is actually 3.60?

(c) What is the decision rule for the hypothesis test in part (a) using a 1% level of significance and the sample size found in part (b).

(d) If a sample of the size determined in part (b) is taken and if the mean of the sample is 3.58, can the claim be considered to be statistically proven?

7-6 Determinations of the number of fibers in a cross section of plant material follow a normal distribution with a standard deviation of 15. The mean μ of the normal distribution is the actual mean number of fibers for such a cross section. It is believed that μ is 300 and the null hypothesis that $\mu = 300$ is to be accepted if the mean of a sample of 25 determinations is between 292.65 and 307.35 and is to be rejected otherwise.

(a) What is the level of significance for this hypothesis test?

(b) What is the probability of a type II error if μ is actually (1) 290? (2) 310? (3) 315?

SOLUTIONS TO PRACTICE PROBLEMS

P7-1 The decision in each case is made on the basis of the sample mean $\bar{x}$.

(a) The rejection region consists of the extreme 5% of the sampling distribution that would result for $\bar{x}$ if the null hypothesis were true. For the two-sided test the region consists of two symmetric parts $\bar{x} > 30 + (z_{0.025} \times \sigma/\sqrt{n}) = 30 + (1.960 \times 8/4) = 33.92$ and $\bar{x} < 30 - (1.960 \times 8/4) = 26.08$.

(b) For this one-sided test, $\bar{x}$ must be in the lower extreme 5% and the rejection region is $\bar{x} < 30 - (z_{0.05} \times \sigma/\sqrt{n}) = 30 - (1.645 \times 8/4) = 26.71$.

(c) For this one-sided test, the rejection region consists of the extreme upper 5% or $\bar{x} > 30 + (1.645 \times 8/4) = 33.29$.

P7-2 For the two-sided test, we replace $z_{0.025}$ with $z_{0.005} = 2.576$, and for the one-sided tests we replace $z_{0.05}$ with $z_{0.01} = 2.326$ to produce the following rejection regions:

(a) $\bar{x} < 30 - (2.576 \times 8/4) = 24.848$ or $\bar{x} > 30 + (2.576 \times 8/4) = 35.152$

(b) $\bar{x} < 30 - (2.326 \times 8/4) = 25.348$

(c) $\bar{x} > 30 + (2.326 \times 8/4) = 34.652$

P7-3 (a) We are attempting to determine whether there is sufficient evidence that the amount of stabilizer is not as it should be. This latter statement thus bears the burden of proof and is the alternative. The null hypothesis and alternative hypothesis thus become $H_0 : \mu = 7.50$ and $H_A : \mu \neq 7.50$.

(b) P(accept H_0) = P($7.335 \leqslant \bar{x} \leqslant 7.665$). With $n = 9$, we thus have the conditional probability

$$\text{P(accept } H_0 \mid H_0 \text{ true)} = \text{P}\left(\frac{7.335 - 7.50}{0.3/3} \leqslant \frac{\bar{x} - \mu}{\sigma/\sqrt{n}} \leqslant \frac{7.665 - 7.50}{0.3/3} \right)$$

$$= \text{P}(-1.65 \leqslant z \leqslant 1.65) = 0.9010$$

The level of significance is P(reject $H_0 \mid H_0$ true) $= 1 - 0.901 = 0.099$ or 9.9%.

(c) The two-sided rejection region consists of the two parts $\bar{x} < 7.50 - (z_{0.025} \times \sigma/\sqrt{n}) = 7.50 - (1.960 \times 0.3/3) = 7.304$ and $\bar{x} > 7.50 + (1.960 \times 0.3/3) = 7.696$.

(d) P(type II $\mid \mu = 7.80$) = P(accept $H_0 \mid \mu = 7.80$)
$$= \text{P}(7.304 \leqslant \bar{x} \leqslant 7.696 \mid \mu = 7.80)$$
$$= \text{P}\big((7.304 - 7.80)/(0.3/3) \leqslant (\bar{x} - \mu)/(\sigma/\sqrt{n}) \leqslant (7.696 - 7.80)/(0.3/3)\big)$$
$$= \text{P}(-4.96 \leqslant z \leqslant -1.04) = 0.1492$$

(e) Repeating part (d) with $\mu = 7.6$, 7.7, 7.9, and 8.0 produces the probabilities 0.8300, 0.4840, 0.0207, and 0.0012. Because of the symmetry of the two-sided test, the probabilities for $\mu = 7.0$, 7.1, 7.2, 7.3, and 7.4 are the same as those for $\mu = 8.0$, 7.9, 7.8, 7.7, and 7.6. Using these probabilities as well as the fact that P(accept $H_0 \mid \mu = 7.50$) $= 1 - \alpha = 0.95$, we plot the operating characteristic curve as shown in Figure P7-1.

(f) In order to have $\alpha = 0.05$ and to have $\beta = 0.10$ when $\mu = 7.75$, the sample size must be at least

$$n = \sigma^2 \left(\frac{z_{\alpha/2} + z_\beta}{\mu_0 - \mu_A} \right)^2 = \sigma^2 \left(\frac{z_{0.025} + z_{0.10}}{\mu_0 - \mu_A} \right)^2$$

$$= 0.09 \times \left(\frac{1.960 + 1.282}{7.50 - 7.75} \right)^2 = 15.1$$

The sample size should thus be 16.

(g) With $n = 16$, we reject the null hypothesis if $\bar{x} < 7.50 - (1.960 \times 0.3/4) = 7.353$ or if $\bar{x} > 7.50 + (1.960 \times 0.3/4) = 7.647$. Since $\bar{x} = 7.68$ exceeds 7.647, it is in the rejection region and we reject the null hypothesis. We thus have sufficient statistical evidence at the 5% level of significance to conclude that the amount of stabilizer is not as it should be.

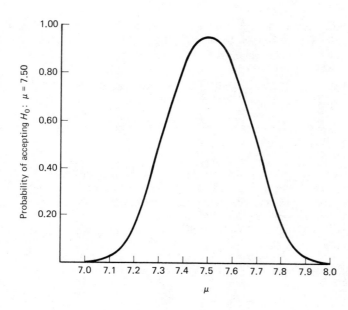

Figure P7-1 Probability of accepting $\mu = 7.50$ as a function of μ.

P7-4 (a) Since the process is not to be modified unless there is evidence of an increase in excess of 2%, the burden of proof is on the statement that the increase exceeds 2%, that is, that the mean exceeds $300 \times 1.02 = 306$. We thus have $H_0 : \mu \leqslant 306$ and $H_A : \mu > 306$ or $H_0 : \mu = 306$ versus $H_A : \mu > 306$. In order to have a level of significance $\alpha = 0.01$ in this one-sided test and to have a probability of a type II error of $\beta = 0.05$ when the increase is 4%, that is, when $\mu = 300 \times 1.04 = 312$, the sample size must be at least

$$n = \sigma^2 \left(\frac{z_\alpha + z_\beta}{\mu_0 - \mu_A} \right)^2 = 64 \times \left(\frac{2.326 + 1.645}{306 - 312} \right)^2 = 28$$

The decision rule is then to reject the null hypothesis if $\bar{x} > 306 + (2.326 \times 8/\sqrt{28}) = 309.5$.

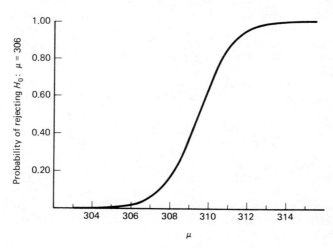

Figure P7-2 Operating characteristic curve.

(b) Through the choice of n to satisfy $\alpha = 0.01$ and $\beta = 0.05$ for an increase of 2%, we know that the probability of rejecting the null hypothesis for actual increases of 2% and 4% are 0.01 and 0.95. For an actual increase of 1%, μ becomes $300 \times 1.01 = 303$ and the probability of accepting the null hypothesis in this case is $P(\bar{x} \leqslant 309.5 \mid \mu = 303) = P\big(z \leqslant (309.5 - 303)\ / (8/\sqrt{28}) = 4.30\big) = 0.0000$. Similarly, actual increases of $1\frac{1}{2}$, $2\frac{1}{2}$, 3, $3\frac{1}{2}$, $4\frac{1}{2}$, and 5% produce probabilities of 0.0005, 0.0934, 0.3707, 0.7454, 0.9960, and 0.9999. Using these values, we sketch the power curve as shown in Figure P7-2.

(c) (1) Even without the hypothesis test, we should not introduce the modification since the sample increase is *less* than 2%.

(2) The sample increase exceeds 2%, but 308.2 is not large enough to be significant. The modification should not be introduced.

(3) The value 312.4 is in the rejection region. We should, therefore, reject the hypothesis and introduce the modification.

7-3   TESTING HYPOTHESES ABOUT A MEAN

Normal Population, Variance Known

In Section 7.2 we developed and illustrated a procedure for *testing hypotheses about the value of the mean of a normal population with a known variance*. We now modify the test statistic in the general procedure. In the general case, the hypothesized value is denoted as μ_0 and, in the two-sided test, the null hypothesis is rejected if $|\bar{x} - \mu_0| > z_{\alpha/2} \times \sigma/\sqrt{n}$. After dividing by $\sigma/\sqrt{n}$, this condition may be written $|(\bar{x} - \mu_0)/(\sigma/\sqrt{n})| > z_{\alpha/2}$. The value $(\bar{x} - \mu_0)/(\sigma/\sqrt{n})$ is a modification of the test statistic $\bar{x}$ and is itself a test statistic z. Values of $\bar{x}$ that are extremely large or small relative to μ_0 produce extremely large or small values of z. If the null hypothesis is true, the sampling distribution for z is the standard normal distribution. For a level of significance α, the significant values of z are those values in the extreme tails of the distribution that satisfy $z < -z_{\alpha/2}$ or $z > z_{\alpha/2}$; that is, those values that satisfy $|z| > z_{\alpha/2}$, as indicated above.

The test with modified test statistic is summarized in the general case in the following form illustrated in Figure 7-9(a).

$H_0 : \mu = \mu_0$

$H_A : \mu \neq \mu_0$

Level of significance: α

Sample size: n

Test statistic: $z = \dfrac{\bar{x} - \mu_0}{\sigma/\sqrt{n}}$

Decision rule: reject H_0 and adopt H_A if $|z| > z_{\alpha/2}$; otherwise, accept H_0 or reserve judgment

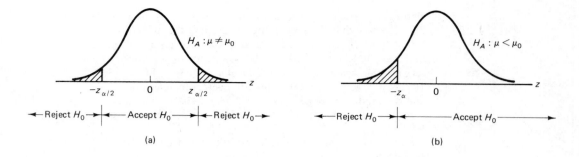

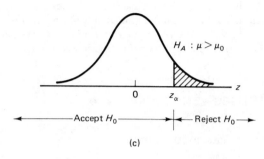

Figure 7-9 Testing H_0: $\mu = \mu_0$.

EXAMPLE 7-18 The two-sided test procedure summarized in Example 7-14 will have a modified test statistic $(\bar{x} - 50.0)/(4.0/\sqrt{42})$ because we had $\mu_0 = 50.0$, $\sigma = 4.0$, and $n = 42$ in the illustration. With $\alpha = 0.05$, $z_{\alpha/2}$ becomes $z_{0.025} = 1.960$. The test is then summarized as

$H_0 : \mu = 50.0$

$H_A : \mu \neq 50.0$

Level of significance: 0.05

Sample size: 42

Test statistic: $z = \dfrac{\bar{x} - 50.0}{4.0/\sqrt{42}}$

Decision rule: reject H_0 and adopt H_A if $|z| > 1.960$; otherwise, accept H_0 or reserve judgment ▲

For the one-sided test with alternative hypothesis $\mu < \mu_0$, the null hypothesis is rejected for (significantly) small values of $\bar{x}$ or z, that is, for values of z such that $z < -z_\alpha$. This test is written in the general case in the following form, illustrated in Figure 7-9(b).

$H_0 : \mu = \mu_0 \qquad (\mu \geq \mu_0)$

$H_A : \mu < \mu_0$

Level of significance: α

Sample size: n

Test statistic: $z = \dfrac{\bar{x} - \mu_0}{\sigma/\sqrt{n}}$

Decision rule: reject H_0 and adopt H_A if $z < -z_\alpha$; otherwise, accept H_0 or reserve judgment

---

EXAMPLE 7-19   Substituting $\mu_0 = 50.0$, $\sigma = 4.0$, $n = 61$, and using $-z_\alpha = -z_{0.05} = -1.645$ in the summary above, the test developed in Example 7-15, but with the sample size as determined in Example 7-17, is written as

$H_0 : \mu = 50.0 \qquad (\mu \geqslant 50.0)$

$H_A : \mu < 50.0$

Level of significance: 0.05

Sample size: 61

Test statistic: $z = \dfrac{\bar{x} - 50.0}{4.0/\sqrt{61}}$

Decision rule: reject H_0 and adopt H_A if $z < -1.645$; otherwise, accept H_0 or reserve judgment   ▲

For the one-sided test with alternative hypothesis $\mu > \mu_0$, the null hypothesis is rejected for (significantly) large values of $\bar{x}$ that correspond to (significantly) large values of z; that is, values of z such that $z > z_\alpha$. This test is illustrated in Figure 7-9(c) and is written in the general case as follows:

$H_0 : \mu = \mu_0 \qquad (\mu \leqslant \mu_0)$

$H_A : \mu > \mu_0$

Level of significance: α

Sample size: n

Test statistic: $z = \dfrac{\bar{x} - \mu_0}{\sigma/\sqrt{n}}$

Decision rule: reject H_0 and adopt H_A if $z > z_\alpha$; otherwise, accept H_0 or reserve judgment

---

EXAMPLE 7-20   Repeated use of a measurement process has indicated that in investigations on the amount of algae in a fixed quantity of water, weight determinations for the algae will be normally distributed with mean μ equal to the actual mean algae content and with standard deviation $\sigma = 2.50$ g. An environmentalist claims that, under given conditions, the mean algae weight will exceed 20.0 g. A sample of $n = 15$ measurements has produced a sample mean weight $\bar{x} = 21.8$ g. The environmentalist wants to know whether these data provide sufficient evidence, at the 1% level of significance, that the claim is true.

In this investigation, the burden of proof is on the claim. It is appropriate to employ a hypothesis test with alternative hypothesis $H_A : \mu > 20.0$. The corresponding null hypothesis is $H_0 : \mu = 20.0$ (or, more broadly, $H_0 : \mu \leqslant 20.0$). With σ assumed to be 2.50, with a sample size of $n = 15$, and with $\alpha = 0.01$ producing $z_\alpha = z_{0.01} = 2.326$, the test is summarized as:

$$H_0 : \mu = 20.0 \qquad (\mu \leqslant 20.0)$$
$$H_A : \mu > 20.0$$

Level of significance: 0.01

Sample size: 15

Test statistic: $z = \dfrac{\bar{x} - 20.0}{2.50/\sqrt{15}}$

Decision rule: reject H_0 and adopt H_A if $z > 2.326$; otherwise, accept H_0 or reserve judgment

Substituting the sample value $\bar{x} = 21.8$ into the test statistic produces $(21.8 - 20)/(2.50/\sqrt{15}) = 2.789$. Since $2.789 > 2.326$, we may reject H_0. At the 1% level of significance, there is sufficient evidence that the mean weight exceeds 20.0 g, as claimed. ▲

Nonnormal Populations

If the population is not normal, the testing procedures described above will still apply provided that *n is large* (*n* is often considered to be large if it is greater than 30) as a result of the central limit theorem. If the sample size is not large enough for the central limit theorem to apply, we must use distribution-free procedures, which are introduced in Chapter 12.

With large samples, we may overcome not only the problem of a nonnormal population, but also the problem of an unknown population variance. As with confidence intervals for the mean of a normal population with unknown variance, we substitute the sample standard deviation s for the unknown population value σ and rely again upon the results of the central limit theorem.

Population Variance Unknown — Sample Size Large

If the variance of the population is unknown, the testing procedures will still apply for normal or nonnormal populations provided that *n is large* (> 30) *and the test statistic is modified to* $z = (\bar{x} - \mu_0)/(s/\sqrt{n})$, where s is the standard deviation of the sample data.

EXAMPLE 7-21 The weights of a sample of 40 capsules are to be used to determine whether there is sufficient empirical evidence that capsules of this type have a mean weight greater than 10.00 g. There is no prior knowledge on the value of the population variance. Considering that the alternative hypothesis bears the burden of proof in a hypothesis test, it is appropriate to have $\mu > 10.00$ as the alternative (μ is the true mean weight); accordingly, the appropriate null hypothesis to test is $\mu = 10.00$.

Suppose that the evidence will be considered to be strong enough if the sample mean is so large that it would occur only with a probability of at most 0.01 if μ were 10.00 (or less than 10.00). The level of significance is, therefore, chosen to be 0.01 or 1%. Expressing the level of significance as a percentage, we may refer to the test as a 1% test.

The testing procedure is summarized as:

$H_0 : \mu = 10.00$ ($\mu \leqslant 10.00$)

$H_A : \mu > 10.00$

Level of significance: 0.01

Sample size: 40

Test statistic: $z = \dfrac{\bar{x} - 10.00}{s/\sqrt{40}}$

Decision rule: reject H_0 and adopt H_A if $z > 2.326$; otherwise, accept H_0 or reserve judgment

Suppose that the data produced $\bar{x} = 9.77$ and $s = 0.50$. Since 9.77 is less than 10.00, it cannot logically provide evidence that the true mean is greater than 10.00. We cannot reject the null hypothesis. We do not have sufficient evidence that the true mean weight is greater than 10.00 g.

Suppose instead that the data produced $\bar{x} = 10.12$ and $s = 0.50$. Since the sample mean in this case is greater than 10.00, it may provide sufficient evidence and we therefore proceed to calculate the test statistic $z = (10.12 - 10.00)/(0.50/\sqrt{40}) = 1.518$. This value does not exceed $z_{0.01} = 2.326$; therefore, we cannot reject the null hypothesis. The data do not provide sufficient evidence that the true mean weight is greater than 10.00 g. We must accept H_0 or reserve judgment.

If, instead, the data produced $\bar{x} = 10.24$ and $s = 0.50$, we calculate the test statistic as $z = (10.24 - 10.00)/(0.50/\sqrt{40}) = 3.036$. This value does exceed 2.326 and, in this case, we can reject the hypothesis. We do have sufficient evidence that the true mean weight of such capsules is greater than 10.00 g. ▲

Normal Population, Variance Unknown — Sample Size Small

If the population variance is unknown and n is small (at most 30), the test procedure must be modified further. In this case the population must be (*at least approximately*) *normal*. The test statistic is $t = (\bar{x} - \mu_0)/(s/\sqrt{n})$ and, if the hypothesis is true, the sampling distribution for t is the t-distribution with $n - 1$ degrees of freedom. For a two-sided test, the significant values of t are the extremely large or small values corresponding to values of $\bar{x}$ that are extremely large or small relative to μ_0. These values are such that $t > t_{\alpha/2, n-1}$ or $t < -t_{\alpha/2, n-1}$, that is, such that $|t| > t_{\alpha/2, n-1}$. For a one-sided test with alternative $\mu < \mu_0$, the hypothesis is rejected for (significantly) small values of t corresponding to small values of $\bar{x}$, that is, such that $t < -t_{\alpha, n-1}$. Similarly, for one-sided tests with alternative $\mu > \mu_0$, the hypothesis is rejected if $t > t_{\alpha, n-1}$. With z's changed to t's, Figures 7-9 illustrates these tests. Again, we may summarize the three general cases in the previous format.

The two-sided test is summarized as:

$H_0 : \mu = \mu_0$

$H_A : \mu \neq \mu_0$

Level of significance: α

Sample size: n

Test statistic: $t = \dfrac{\bar{x} - \mu_0}{s/\sqrt{n}}$

Decision rule: reject H_0 and adopt H_A if $|t| > t_{\alpha/2, n-1}$; otherwise, accept H_0 or reserve judgment

---

EXAMPLE 7-22 A plastics firm requires that the thickness of sheets of plastic from a given production have a mean of 0.1500. A sample of 18 measurements is available to determine whether the mean of the production meets the requirements. In assessing the mean thickness, we hypothesize that $\mu = 0.1500$ and we accept this null hypothesis unless the data provide sufficient evidence that the mean does not meet requirements.

Suppose that we choose a 5% test and that we must make a decision (we may not reserve judgment). The critical t value is $t_{\alpha/2, n-1} = t_{0.025, 17} = 2.110$ and the test procedure is:

$H_0 : \mu = 0.1500$

$H_A : \mu \neq 0.1500$

Level of significance: 0.05

Sample size: 18

Test statistic: $t = \dfrac{\bar{x} - 0.1500}{s/\sqrt{18}}$

Decision rule: reject H_0 and adopt H_A if $|t| > 2.110$; otherwise, accept H_0

Suppose that the sample produces $\bar{x} = 0.1496$ and $s = 0.0012$. The test statistic is $t = (0.1496 - 0.1500)/(0.0012/\sqrt{18}) = -1.414$; therefore, $|t| = 1.414$ is less than 2.110 and the null hypothesis cannot be rejected. Although the *sample mean* is different from the required value for the *population mean*, it may have resulted naturally by chance and is not sufficiently different to provide statistical evidence that the population mean is not the required value.

▲

The one-sided test with alternative hypothesis $\mu < \mu_0$ is summarized as:

$H_0 : \mu = \mu_0 \qquad (\mu \geqslant \mu_0)$

$H_A : \mu < \mu_0$

Level of significance: α

Sample size: n

Test statistic: $t = \dfrac{\bar{x} - \mu_0}{s/\sqrt{n}}$

Decision rule: reject H_0 and adopt H_A if $t < -t_{\alpha,\, n-1}$; otherwise, accept H_0 or reserve judgment

The one-sided test with alternative hypothesis $\mu > \mu_0$ is written as:

$H_0 : \mu = \mu_0 \qquad (\mu \leqslant \mu_0)$

$H_A : \mu > \mu_0$

Level of significance: α

Sample size: n

Test statistic: $t = \dfrac{\bar{x} - \mu_0}{s/\sqrt{n}}$

Decision rule: reject H_0 and adopt H_A if $t > t_{\alpha,\, n-1}$; otherwise, accept H_0 or reserve judgment

---

EXAMPLE 7-23 An anthropologist disputes the theory that a hominoid genera of primates display forearm rotations of 180°, but claims instead that the mean maximum rotation is much less—in fact, even less than 170°. A sample of maximum rotations for $n = 10$ subjects is available to investigate the claim that the mean rotation is less than 170°.

The appropriate procedure is to test $H_0 : \mu = 170$ versus $H_A : \mu < 170$. To convince colleagues, the anthropologist must have a very significant result and chooses a 1% level of significance. The test procedure then has the decision rule to reject the null hypothesis if $t < -t_{0.01, 9} = -2.821$, where $t = (\bar{x} - 170)/(s/\sqrt{10})$. Suppose that the sample rotations produce mean $\bar{x} = 157.2$ and standard deviation $s = 9.73$. The test statistic t is then $(157.2 - 170.0)/(9.73/\sqrt{10}) = -4.16$. Since $-4.16 < -2.821$, the null hypothesis may be rejected and the anthropologist has sufficient evidence at the 1% level that the mean maximum rotation is less than 170°, as claimed.   ▲

P7-5 Determine whether there is sufficient evidence at the 2% level that the mean of a normal population with standard deviation $\sigma = 10.0$ differs from 100.0 if a sample of 25 observations produced a mean $\bar{x} = 93.8$.

P7-6 One of the requirements of an ice cream parlor is that tubs of ice cream produce a mean of 84 scoops. Set up an appropriate 5% testing procedure to determine whether the attendants are obtaining the appropriate mean number of scoops if the numbers of scoops obtained from 72 tubs produced a mean $\bar{x} = 83.7$ and a standard deviation $s = 1.43$.

P7-7 In a study on time awareness, 12 subjects were asked to estimate a time duration under each of two conditions. One condition involved waiting in an empty room until summoned. The other condition involved being engaged in "busy" talk. In each case, the actual time was the same, and the two conditions were undergone in different weeks to avoid subject awareness of the equal time length. The experimenter performed the experiment to support the claim that the mean of the differences of waiting time estimate minus busy time estimate exceeds 5.0.

(a) Set up an appropriate 5% testing procedure for the experimenter's purpose.

(b) The 12 subjects' waiting time estimates exceeded their busy time estimates by amounts producing mean $\bar{x} = 7.8$ and standard deviation $s = 3.2$. Do these data provide sufficient evidence at the 5% level of significance to substantiate the experimenter's claim?

7-7 Performance scores on an intelligence test can be assumed to be normally distributed with a standard deviation of 10.0. Set up a 1% hypothesis test with a two-sided alternative hypothesis to test the null hypothesis that $\mu = 80.0$ on the basis of a sample of 20 scores and determine whether the null hypothesis should be accepted if the sample mean of the 20 scores is 73.1.

7-8 Repeat Exercise 7-7 if the population standard deviation cannot be assumed to be 10.0 but is unknown and if the sample standard deviation is 11.2.

7-9 A sample of 50 measurements on the percent of methane in gas from oxidation ponds produced a mean of 71.4% and a standard deviation of 10.3%. By setting up an appropriate null hypothesis and alternative hypothesis, determine whether these data provide sufficient evidence at the 5% level that the mean percent of methane is less than 75%.

7-10 Suppose that, after sampling 20 records at random, a sociologist finds the following durations (to the nearest tenth of a year) of marriages ending in divorce:

| 10.1 | 21.2 | 13.8 | 11.1 | 10.9 | 9.2 | 6.6 | 12.3 | 7.8 | 15.1 |
| 2.6 | 14.3 | 14.9 | 5.4 | 8.7 | 4.8 | 19.4 | 26.3 | 24.5 | 21.6 |

By setting up an appropriate null hypothesis and alternative hypothesis, determine whether these data provide proof, at the 5% level, that the mean duration of marriages ending in divorce in the population has decreased from an earlier value of 14.9. What distribution assumption is made in applying the hypothesis test?

7-11 Suppose that, at each of 15 sample points, erosion rates were measured for two different time periods and the differences of the two rates (second minus first) were

2.10	6.64	−1.94	−4.31	5.84	−12.11	−1.59	−5.59
9.25	−7.68	−5.39	3.05	8.48	0.97	−2.82	

Test the null hypothesis that there has been no difference over time on the average, that is, that, for the differences, $\mu = 0$, against the alternative hypothesis that there has been a difference. Use a 2% level of significance.

7-12 A designer claims that by smoothing out parts of a particular automobile body to reduce air resistance, the average fuel consumption can be reduced below 8.0 liters per 100 km. In an attempt to support the claim, the designer has obtained a sample of fuel consumption for 15 modified automobiles. Do the results of the sample provide sufficient evidence to support the claim at the 1% level of significance if the 15 sample values produced a mean of 7.4 and a standard deviation of 0.8?

7-13 It is claimed that an income assurance program can lead to annual hog production increases in excess of 15%. Suppose that 36 farms were covered by the program as an experimental group and these farms showed annual increases with a mean of 16% and a standard deviation of 3.2% while production of noncovered farms was unchanged.

(a) Would these results substantiate the claim
 (1) at the 5% level of significance?
 (2) at the 1% level?
(b) In coming to the conclusions in part (a), why is it important that production of noncovered farms was unchanged?

SOLUTIONS TO PRACTICE PROBLEMS

P7-5 It is necessary to test $H_0 : \mu = 100$ against $H_A : \mu \neq 100$ and to reject H_0 if $|z| > z_{\alpha/2} = z_{0.01} = 2.326$, where $z = (\bar{x} - \mu_0)/(\sigma/\sqrt{n}) = (\bar{x} - 100)/(10/5)$. Since $\bar{x} = 93.8$, we have $z = -3.1$ and $|z| = 3.1 > 2.326$. There is sufficient evidence that μ differs from 100.

P7-6 It is appropriate to test $H_0 : \mu = 84$ against $H_A : \mu \neq 84$ and to reject H_0 if $|z| > z_{\alpha/2} = z_{0.025} = 1.96$, where $z = (\bar{x} - \mu_0)/(s/\sqrt{n}) = (\bar{x} - 84)/(s/\sqrt{72})$. The data produce $|z| = (83.7 - 84)/(1.43/\sqrt{72}) = -1.78$; thus, $|z| < 1.96$ and H_0 may not be rejected. There is not enough evidence at the 5% level that the mean number of scoops differs from 84 and we may accept the null hypothesis that the mean number of scoops is 84 as required, or, since 1.78 is close to the rejection value 1.96, we may wish to reserve judgment.

P7-7 (a) It is appropriate to test $H_0 : \mu = 5.0$ against $H_A : \mu > 5.0$ and to reject H_0 if $t > t_{\alpha/2, n-1} = t_{0.05, 11} = 1.796$, where $t = (\bar{x} - \mu_0)/(s/\sqrt{n}) = (\bar{x} - 5.0)/(s/\sqrt{12})$.
(b) The data produce $t = (7.8 - 5.0)/(3.2/\sqrt{12}) = 3.03 > 1.796$ and H_0 may be rejected. The data do provide sufficient evidence to support the claim at the 5% level.

7-4 TESTING HYPOTHESES ABOUT A VARIANCE OR STANDARD DEVIATION

In some cases it will be of interest to investigate specific values for the variance, or equivalently, the standard deviation of a population. We will develop the procedures *for normal populations* by considering an example.

EXAMPLE 7-24 In Example 7-22 the firm may not only set requirements on the mean thickness, but may also require that the sheets of plastic be reasonably uniform in thickness. The uniformity of the thickness can be determined in terms of the standard deviation of the various thickness values. If the standard deviation is small, there is little variation in the individual values (i.e., the thickness is uniform). A large standard deviation indicates considerable variation of thickness in the sheets of plastic.

Suppose that requirements are such that the standard deviation should be at most 0.0010. A sample of $n = 18$ values produced a sample standard deviation of 0.0012, suggesting that perhaps the requirements have not been met but that the standard deviation is too large. To determine whether this "suggestion" may be regarded as sufficient evidence, it is appropriate to formulate a test of the hypothesis that the population standard deviation is 0.0010 with the alternative that it is greater than 0.0010.

Intuitively, we expect to reject the hypothesis if s, the sample standard deviation, is (significantly) greater than 0.0010. We recall from Section 6-5 that, *for a normal population*, $\chi^2 = (n - 1)s^2/\sigma^2$ has a χ^2-distribution with $n - 1$ degrees of freedom; thus, if the hypothesis is true, $\chi^2 = 17s^2/(0.0010)^2$ must have a χ^2-distribution with 17 degrees of freedom. It is appropriate, therefore, to use χ^2 as the test statistic. Since large values of s correspond to large values of χ^2, we reject the hypothesis if χ^2 is significantly large. By following the previously developed principles for hypothesis testing, we determine that this value is significantly large with level of significance α if it exceeds the critical value $\chi^2_{\alpha, n-1}$. If the level of significance is 5%, this critical value is $\chi^2_{0.05, 17} = 27.587$.

For the sample data, $\chi^2 = 17 \times (0.0012)^2/(0.0010)^2 = 24.48$. Although this approaches significance, it is not a significant value and the data do not provide sufficient evidence at the 5% level that the sheets of plastic fail to meet the uniformity requirements. We might accept the null hypothesis or, since we have come close to significance, we may reserve judgment pending further investigation. ▲

Small Samples

From this particular example, we can develop and summarize the general test procedures for use with small samples. (Small samples will be interpreted as samples for which n is 31 or less.)

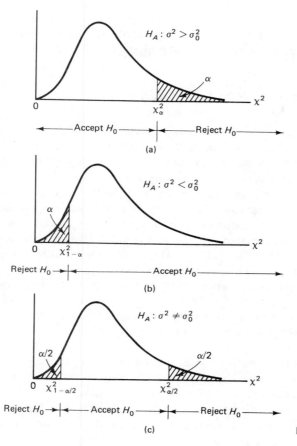

Figure 7-10 Testing $H_0: \sigma^2 = \sigma_0^2$.

For the example, the general case is a one-sided test with alternative $\sigma > \sigma_0$ (where σ_0 is the hypothesized value for the population standard deviation). This case is illustrated in Figure 7-10(a) and is summarized in terms of the standard deviation or variance as:

$H_0: \sigma = \sigma_0$ (i.e., $\sigma^2 = \sigma_0^2$)

$H_A: \sigma > \sigma_0$ (i.e., $\sigma^2 > \sigma_0^2$)

Level of significance: α

Sample size: n

Test statistic: $\chi^2 = \dfrac{(n-1)s^2}{\sigma_0^2}$

Decision rule: reject H_0 and adopt H_A if $\chi^2 > \chi^2_{\alpha,\, n-1}$; otherwise, accept H_0 or reserve judgment

For a one-sided test with alternative hypothesis $\sigma < \sigma_0$, it is appropriate to reject the null hypothesis for significantly small values of s, or equivalently s^2, which correspond to small values of χ^2, that is, the values of χ^2 in the lower tail which are less than $\chi^2_{1-\alpha}$. The general test procedure is illustrated in Figure 7-10(b) and is summarized as above but with the alternative hypothesis and decision rule restated as:

$H_A : \sigma < \sigma_0$; (i.e., $\sigma^2 < \sigma_0^2$)
Decision rule: reject H_0 and adopt H_A if $\chi^2 < \chi^2_{1-\alpha, n-1}$;
otherwise, accept H_0 or reserve judgment

EXAMPLE 7-25 It is required that water analysis procedures be sufficiently precise that measurements on the amount of fluoride in solution display a standard deviation that is very small. An analysis procedure will be considered to be suitable only if there is sufficient evidence at the 1% level on the basis of a sample of $n = 15$ measurements that the standard deviation is less than 0.2 milligram per liter. On the assumption of a normal distribution, the water analysis procedures are evaluated in a test of the null hypothesis $H_0 : \sigma = 0.2$, $H_A : \sigma < 0.2$ with $\alpha = 0.01$. The null hypothesis is rejected and the procedure considered sufficiently precise if $\chi^2 < \chi^2_{0.99, 14} = 4.660$, where $\chi^2 = 14s^2/(0.2)^2$.

In a sample evaluation, 15 measurements produced a standard deviation of 0.17. These results produce $\chi^2 = 14 \times 0.17^2/0.2^2 = 10.115$. Since 10.115 is not less than 4.660, we may not reject H_0. The standard deviation may be 0.2 or larger and the analysis procedure is not sufficiently precise. ▲

A two-sided test with alternative $\sigma \neq \sigma_0$ involves rejecting the hypothesis for unusually large or small values of s, that is, for significantly large or small values of χ^2. By considering the rejection region to be split into two parts of the χ^2-distribution representing the upper $\alpha/2$ and the lower $\alpha/2$, as illustrated in figure 7-10(c), the general two-sided test may be summarized as above with the alternative hypothesis and decision rule changed to:

$H_A : \sigma \neq \sigma_0$ (i.e., $\sigma^2 \neq \sigma_0^2$)
Decision rule: reject H_0 and adopt H_A if $\chi^2 < \chi^2_{1-\alpha/2, n-1}$ or if
$\chi^2 > \chi^2_{\alpha/2, n-1}$; otherwise, accept H_0 or reserve judgment

Large Samples

For samples of size greater than 31, the degrees of freedom will exceed those of Table V, and we must use the normal approximation to the χ^2 as considered in Section 6-5.

For convenience we restate the results for approximate values for χ^2_α and $\chi^2_{1-\alpha}$ as they would apply in the one-sided tests.

$$\chi^2_{\alpha, f} \doteq f\left(1 - \frac{2}{9f} + z_\alpha\sqrt{\frac{2}{9f}}\right)^3$$

$$\chi^2_{1-\alpha, f} \doteq f\left(1 - \frac{2}{9f} - z_\alpha\sqrt{\frac{2}{9f}}\right)^3$$

For the hypothesis tests, the degrees of freedom will be $f = n - 1$.

For the two-sided test, $\chi^2_{\alpha/2}$ and $\chi^2_{1-\alpha/2}$ are required. We obtain their approximations by replacing α with $\alpha/2$ as the subscripts on χ^2 and z in the expressions above.

EXAMPLE 7-26 A psychologist has changed to a new supplier of laboratory mice and, before using the mice for some experiments, must ensure that they are comparable to those previously used. In one measure of group activity it is necessary to determine whether the new mice will exhibit a standard deviation that differs from 4.5, the value produced by mice from the previous supplier. The psychologist investigates the standard deviation on the basis of a sample of four dozen mice. To do so, the psychologist tests $H_0 : \sigma = 4.5$ against $H_A : \sigma \neq 4.5$ with level of significance $\alpha = 0.05$. The null hypothesis is to be accepted and the mice considered equivalent to previous mice in this respect if the data produce s^2 such that the test statistic $\chi^2 = (n - 1)s^2/\sigma_0^2 = 47s^2/(4.5)^2$ falls in the acceptance region: $\chi^2_{0.975,47} \leqslant 47s^2/4.5^2 \leqslant \chi^2_{0.025,47}$, where $z_{0.025} = 1.96$ produces the approximations

$$\chi^2_{0.975,47} = 47\left[1 - 2/(9 \times 47) - 1.96\sqrt{2/(9 \times 47)}\,\right]^3 = 29.95$$

and

$$\chi^2_{0.025,47} = 47\left[1 - 2/(9 \times 47) + 1.96\sqrt{2/(9 \times 47)}\,\right]^3 = 67.82$$

The 48 mice produced data such that $s = 5.1$. The test statistic is $\chi^2 = 47 \times 5.1^2/4.5^2 = 60.37$, which is in the acceptance region. The new mice may be considered equivalent to the previous mice in this respect as H_0 may be accepted; however, we may wish to reserve judgment and sample further since 60.37 is so close to the rejection region. ▲

PRACTICE PROBLEMS

P7-8 It is of interest to know whether a change in diet will affect the standard deviation of weight gain rates displayed in cattle. Previously, the standard deviation was $\sigma = 0.10$ kg/day. A sample of 24 cattle with the new diet produced weight gain rates with a standard deviation of 0.14 kg/day. Do these data provide evidence of a difference at the 5% level?

P7-9 Suppose that it is claimed that altering the bus schedule to increase the number of buses in some time periods and decrease them in others will "smooth out" the numbers of passengers per bus changing at a transfer point by decreasing the standard deviation of the passenger counts below 12. If the scheduling is modified for a trial period and a sampling of 100 passenger counts produces a standard deviation of 11.13, is the claim substantiated at the 1% level?

Section 7-4   *Testing Hypotheses About a Variance or Standard Deviation*   **279**

7-14 In order to meet homogeneity requirements, maze run scores for a particular maze structure should exhibit a standard deviation of 2.5 or less for a particular strain of rat. Suppose that the scores of a sample of 15 such rats produced a standard deviation of 3.1. By setting up a hypothesis test with appropriate null hypothesis and alternative hypothesis, determine whether these results provide sufficient evidence at the 5% level of significance that the standard deviation of all such scores exceeds 2.5. What distribution assumption is made in applying this hypothesis test?

7-15 Suppose that past records show that water levels in a particular controlled waterway displayed a distribution that was (approximately) normal with a standard deviation of 0.75. A change in the control process has been implemented and it is claimed that this change would reduce the variation in water levels. Suppose that a sample of 14 measurements on water levels taken after the implementation of the new controls and at sufficiently spread time intervals as to be independent produced a standard deviation $s = 0.54$. On the basis of an appropriate hypothesis test, determine, at the 1% level, whether the standard deviation σ is, as claimed, less than 0.75.

7-16 Suppose that the value $s = 0.54$ in Exercise 7-15 resulted from 42 rather than 14 measurements. Repeat Exercise 7-15 with this new sample size.

7-17 A therapeutic program is intended to reduce variability of mood in patients experiencing extreme attitude differences. Patients of the type being considered for the program have previously displayed psychiatric rating scores displaying a normal distribution with a standard deviation of 20.00. Is there sufficient evidence at the 5% level that the program does reduce variability if a sample of 12 ratings produced after introduction of the program displayed a standard deviation of 11.2?

7-18 If a sample of six dozen packages shows a standard deviation in the weights of 58.3 g, does the sample result provide sufficient evidence at the 5% level to substantiate a consumer advocate's claim that variation is excessive, showing, in general, a standard deviation in excess of 50 g?

SOLUTIONS TO PRACTICE PROBLEMS

P7-8 In order to determine whether there is any change (increase or decrease), we use a two-sided test of $H_0 : \sigma = 0.10$ against $H_A : \sigma \neq 0.10$. With $n = 24$ and $\alpha = 0.05$, we reject H_0 if $\chi^2 < \chi^2_{0.975, 23} = 11.689$ or if $\chi^2 > \chi^2_{0.025, 23} = 38.076$, where $\chi^2 = 23s^2/0.10^2$. The data produce $\chi^2 = 23 \times 0.14^2/0.10^2 = 45.08$. Since $45.08 > 38.076$, H_0 may be rejected. At the 5% level, the data do indicate a difference.

P7-9 The claim that $\sigma < 12$ bears the burden of proof and becomes the alternative hypothesis producing the test as $H_0 : \sigma = 12$ against $H_A : \sigma < 12$. The null hypothesis is rejected at the 1% level if $\chi^2 < \chi^2_{0.99, 99} = 99[1 - 2/(9 \times 99) - 2.326\sqrt{2/(9 \times 99)}\,]^3 = 69.22$, where 2.326 is $z_{0.01}$. The data produce $\chi^2 = (n - 1)s^2/\sigma^2 = 99 \times 11.3^2/12^2 = 87.79$. Since 87.79 is not less than 69.22, H_0 may not be rejected and the claim has not been substantiated at the 1% level. ▲

Hypothesis tests for proportions or percentages are developed in the same manner as those for means and standard deviations. In the case of small samples, results are evaluated on the basis of binomial probabilities. The normal approximation to the binomial is used for large samples.

Small Samples

We will use an example to develop one-sided tests first since they are easier to develop than two-sided tests.

EXAMPLE 7-27 A small survey is to be conducted to study the effect of a large increase in the price of coffee. It is intended to determine whether the results of the survey provide sufficient evidence at the 5% level that, despite the price increase, more than half the people in a particular category drink coffee rather than any other beverage. The appropriate procedure is to test the null hypothesis that the proportion of the people who drink coffee rather than other beverages is 0.5, against the alternative hypothesis that the proportion is greater than 0.5.

In the survey, a sample of 16 people is to be chosen at random. The number of coffee drinkers in the sample may be considered to be a random variable x having the binomial distribution with the parameter p equal to the true proportion of people in the category who are coffee drinkers and with n equal to 16. If the null hypothesis is true, p is 0.5 and the expected value for x, the number of coffee drinkers in such samples, will equal $np = 16 \times 0.5 = 8$.

Intuitively, we determine that the null hypothesis should be rejected if x is large (significantly larger than 8). We then attempt to determine a critical value c such that, if the null hypothesis is true, $P(x > c) = \alpha$, the chosen level of significance. Since the binomial distribution is discrete, we may not be able to obtain a critical value producing the exact level of significance that we choose. Suppose, for example, that we choose a level of significance of 5%. If $c = 10$, we find that $P(x > c | H_0$ is true$) = P(x > 10 | p = 0.5) = \sum_{x=11}^{16} \binom{16}{x}(0.5)^x(0.5)^{16-x} = 0.105$. This probability is too large; therefore, $c = 10$ is not extreme enough. We thus try the next possible value $c = 11$ (since the binomial is not continuous, we can have no value between 10 and 11). In this case $P(x > c | H_0$ is true$) = P(x > 11 | p = 0.5) = \sum_{x=12}^{16} \binom{16}{x}(0.5)^x(0.5)^{16-x} = 0.038$. Since this probability is less than 0.05, the value $c = 11$ is sufficiently extreme; in fact, it is more extreme than necessary for the chosen level of significance. Because we cannot obtain a value of c between 10 and 11, we cannot have a level of significance between 0.105 and 0.038. We cannot obtain the chosen level of significance. In order for a result to be at least significant as that with the chosen level of significance, we select the more extreme value as the critical value. The test may then be summarized as:

$$H_0: p = 0.5$$

$$H_A: p > 0.5$$

Level of significance: 5%   (actual 3.8%)

Sample size: 16

Test statistic: x

Decision rule: reject H_0 and adopt H_A if $x > 11$;
otherwise, accept H_0 or reserve judgment   ▲

EXAMPLE 7-28 If, in Example 7-27, it is of interest instead to determine whether the sample results provide evidence that *less* than half of the people in the category drink coffee, the alternative is reversed. The data will now provide sufficient evidence with an unusually *small* number of coffee drinkers (i.e., with a value of x significantly less than 8). We must now determine a new value c such that if p is 0.5, then $P(x < c) = \alpha$. Again, we may be unable to obtain α exactly. We will choose the largest c that satisfies $P(x < c) \leqslant \alpha$.

Suppose that we choose $\alpha = 0.05$ again. With $n = 16$ and $p = 0.5$ (the hypothesized value), we must find the largest c that satisfies $P(x < c) = 0.05$. To determine this value, we calculate, successively, $P(x < 1) = P(x = 0)$, $P(x < 2) = P(x = 0) + P(x = 1)$, $P(x < 3) = P(x = 0) + P(x = 1) + P(x = 2)$, and so on, as indicated in the following table, until we have arrived at the correct value.

VALUES OF $P(x = c)$ AND $P(x < c)$ FOR THE BINOMIAL
WITH $n = 16$ AND $p = 0.5$

c	$P(x = c)$	$P(x < c)$
0	0.0000	—
1	0.0002	0.0000
2	0.0018	0.0002
3	0.0085	0.0020
4	0.0278	0.0105
5	0.0667	0.0383
6	0.1222	0.1050

We stop the table at 6 since $P(x < 6) = 0.105$, which exceeds $\alpha = 0.05$; that is, 6 is too large for c. We note that $P(x < 5) = 0.0383$, which does not exceed $\alpha = 0.05$; thus, 5 is the largest value for c for which $P(x < c|H_0$ is true) $\leqslant \alpha$. In this case, the test may be written as:

$$H_0: p = 0.5$$

$$H_A: p < 0.5$$

Level of significance: 5%   (actual 3.8%)

Sample size: 16

Test statistic: x

Decision rule: reject H_0 and adopt H_A if $x < 5$;
otherwise, accept H_0 or reserve judgment ▲

The test procedure is somewhat more complicated for a two-sided alternative because we must find upper and lower critical values so that the upper extreme has approximately the same probability as the lower extreme and such that the probabilities are as large as possible without allowing their sum to exceed the level of significance.

EXAMPLE 7-29 The percentage of a given plant strain meeting approved standards was 60%. New growing conditions may alter this up or down or have no appreciable effect. It is of interest to determine on the basis of a sample of 20 plants whether there is sufficient evidence at the 5% level of significance of a change. The original percentage, 60%, is interpreted as a proportion 0.6 and we hypothesize that $p = 0.6$ with alternative hypothesis $p \neq 0.6$. The null hypothesis is to be rejected if x, the number of plants in the sample meeting standards, is significantly larger or smaller than $np_0 = 20 \times 0.6 = 12$, the expected value under H_0. We must find two critical values c_1 and c_2 such that $P(x < c_1 | H_0 \text{ true}) \doteq P(x > c_2 | H_0 \text{ true})$ and such that $P(x < c_1 | H_0 \text{ true}) + P(x > c_2 | H_0 \text{ true}) \leqslant \alpha$.

In our case $\alpha = 0.05$ and if H_0 is true, we have binomial probabilities with $n = 20$ and $p = 0.6$. To determine the appropriate values for c_1 and c_2, we tabulate and accumulate probabilities again as in Example 7-28, but this time, we accumulate from both ends, as illustrated in the following table of probabilities for the hypothesized case (binomial with $n = 20$ and $p = 0.6$).

c_1	$P(x = c_1)$	$P(x < c_1)$	c_2	$P(x = c_2)$	$P(x > c_2)$
1	0.0000	—	20	0.0000	—
2	0.0000	0.0000	19	0.0005	0.0000
3	0.0000	0.0000	18	0.0031	0.0005
4	0.0003	0.0000	17	0.0123	0.0036
5	0.0013	0.0003	16	0.0350	0.0159
6	0.0049	0.0016	15	0.0746	0.0509
7	0.0146	0.0065			
8	0.0355	0.0211			
9	0.0710	0.0566			

From the table we note that $P(x < 8) = 0.0211$ and $P(x > 16) = 0.0159$. The sum of these two probabilities is 0.0370 and, as required, does not exceed 0.05. Expanding the rejection region to include 8 or 16 will produce a probability in either extreme in excess of the level of significance 0.05.

To obtain a level of significance as close as possible to (without exceeding) 0.05 and to split this probability into two approximately equal parts, it is appropriate to choose two critical values $c_1 = 8$ and $c_2 = 16$ and to reject the null hypothesis if $x < 8$ or if $x > 16$.

The test is summarized as:

$H_0 : p = 0.60$

$H_A : p \neq 0.60$

Level of significance: 5%   (actual 3.7%)

Sample size: 20

Test statistic: x

Decision rule: reject H_0 and adopt H_A if $x < 8$
or if $x > 16$; otherwise, accept or reserve judgment   ▲

As before, from the examples, we can develop and summarize three general test procedures. For the two-sided test, we have

$H_0 : p = p_0$

$H_A : p \neq p_0$

Level of significance: α (actual value $\leqslant \alpha$)

Sample size: n

Test statistic: x

Decision rule: reject H_0 and adopt H_A if $x < c_1$ or if $x > c_2$,
where c_1 and c_2 are such that $P(x < c_1 | H_0 \text{ true}) + P(x > c_2 | H_0 \text{ true})$
is as large as possible without exceeding α
and such that $P(x < c_1 | H_0 \text{ true})$
and $P(x > c_2 | H_0 \text{ true})$ are approximately equal;
otherwise, accept H_0 or reserve judgment

For the one-sided test with alternative $p > p_0$, we have H_A and the decision rule modified to produce

$H_A : p > p_0$

Decision rule: reject H_0 and adopt H_A if $x > c$
where c is such that $P(x > c | H_0 \text{ true})$
is as large as possible without exceeding α;
otherwise, accept H_0 or reserve judgment

For the one-sided test with alternative $p < p_0$, we have H_A and the decision rule modified to produce

$H_A : p < p_0$

Decision rule: reject H_0 and adopt H_A if $x < c$
where c is such that $P(x < c | H_0 \text{ true})$
is as large as possible without exceeding α;
otherwise, accept H_0 or reserve judgment

Exceeding the Chosen Level of Significance

Although the value of c is to be such that the probability of rejecting the hypothesis does not exceed the chosen level of significance, in some cases it may be preferable to exceed the chosen level if the value of c will produce a probability slightly in excess of the level but the next possible value would produce a probability very much less than the chosen level.

EXAMPLE 7-30 If 20 people are to be sampled in the coffee study in Example 7-27, a value of 13 for c will produce a probability of a type I error of 0.058, slightly larger than the chosen 0.05; whereas a value of 14 for c will produce a probability of 0.021, much less than the chosen value. It may, therefore, be more appropriate to choose $c = 13$. ▲

A Test Format That Does Not Require c

The determination of the critical value c or critical values c_1 and c_2 can be avoided in the foregoing tests by calculating the probability of obtaining a value as extreme as the observed result when the null hypothesis is true and then rejecting the null hypothesis if this probability is less than the level of significance.

Letting x_0 denote the actual observed number in the sample and letting x denote a variable with a binomial distribution with parameters equal to the sample size n and the hypothesized proportion p_0, the "reject" criteria in the decision rules become:

For $H_A : p \neq p_0$
reject H_0 and adopt H_A if $P(x \leqslant x_0 | H_0 \text{ true}) \leqslant \alpha/2$ or if $P(x \geqslant x_0 | H_0 \text{ true}) \leqslant \alpha/2$; otherwise, accept H_0 or reserve judgment

For $H_A : p > p_0$
reject H_0 and adopt H_A if $P(x \geqslant x_0 | H_0 \text{ true}) \leqslant \alpha$; otherwise, accept H_0 or reserve judgment

For $H_A : p < p_0$
reject H_0 and adopt H_A if $P(x \leqslant x_0 | H_0 \text{ true}) \leqslant \alpha$; otherwise, accept H_0 or reserve judgment

---

EXAMPLE 7-31 Suppose that, in Example 7-29, only 17 plants were sampled and, of these, 15 met the standards. As before we have $H_0 : p = 0.60$, $H_A : p \neq 0.60$, and $\alpha = 0.05$. The sample of size $n = 17$ has produced $x_0 = 15$. Now, for the binomial model with $n = 17$ and $p = 0.60$, we find that $P(x \geqslant 15) = 0.0123$. This is less than 0.025, which is $\alpha/2$, thus indicating that 15 is in the extreme upper 2.5% and is a significant value. (We use $\alpha/2$ since we have a two-sided test.) We can, therefore, reject the hypothesis and conclude that the proportion of plants meeting standards is not 0.60; that is, the percentage is not 60%.　▲

Large Samples

If n is sufficiently large to use the normal approximation for the binomial [i.e., if $np_0 > 5$ and $n(1 - p_0) > 5$], then, if the null hypothesis that $p = p_0$ is true, $z = (x - np_0)/\sqrt{np_0(1 - p_0)}$ will have a sampling distribution that is approximately standard normal. Unusually large or small values of x will produce unusually large or small values of z. We will use z as a test statistic after modifying x by the addition or subtraction of $\frac{1}{2}$ as a continuity correction.

The two-sided test may then be written as:

$H_0 : p = p_0$

$H_A : p \neq p_0$

Level of significance: α

Sample size: n

Test statistic: $z = \dfrac{x \pm \frac{1}{2} - np_0}{\sqrt{np_0(1 - p_0)}}$

Decision rule: reject H_0 and adopt H_A if $|z| > z_{\alpha/2}$; otherwise, accept H_0 or reserve judgment, where we use $-\frac{1}{2}$ as the continuity correction if $x > np_0$ and $+\frac{1}{2}$ as the correction if $x < np_0$

The one-sided tests are as above with the alternative, test statistic, and decision rule modified to

$H_A : p > p_0$

Test statistic: $z = \dfrac{x - \frac{1}{2} - np_0}{\sqrt{np_0(1 - p_0)}}$

Decision rule: reject H_0 and adopt H_A if $z > z_\alpha$; otherwise, accept H_0 or reserve judgment

and

$H_A : p < p_0$

Test statistic: $z = \dfrac{x + \frac{1}{2} - np_0}{\sqrt{np_0(1 - p_0)}}$

Decision rule: reject H_0 and adopt H_A if $z < -z_\alpha$; otherwise, accept H_0 or reserve judgment

EXAMPLE 7-32 Suppose that in Example 7-27, the survey on coffee drinkers is to involve a sample of 250 people. Both $np_0 = 250 \times 0.5 = 125$ and $n(1 - p_0) = 250 \times 0.5 = 125$ greatly exceed 5 and we may use the large-sample normal approximation. To still have a level of significance of 5%, we will reject the null hypothesis if $z > 1.645$, since $z_{0.05} = 1.645$. With $n = 250$ and $p_0 = 0.5$, the test statistic z is $(x - \frac{1}{2} - 125)/\sqrt{250 \times 0.5 \times 0.5}$.

Suppose that 145 of the 250 people in the sample turn out to be coffee drinkers; then $x = 145$ and $z = (145 - \frac{1}{2} - 125)/\sqrt{250 \times 0.5 \times 0.5} = 2.47$. Since this value exceeds 1.645, the null hypothesis may be rejected and, at the 5% level, the survey provides sufficient empirical evidence that, despite the price increase, over half the people in the particular category drink coffee rather than any other beverage. ▲

PRACTICE PROBLEMS

P7-10 A random "decision-maker" device is supposed to provide unbiased decisions on whether each subject in an experiment will receive the treatment being investigated or a placebo by generating a red or green signal with red and green each having probability $1/2$. The device is to be tested at the 5% level for evidence of any bias on the basis of n sample decisions. Set up a hypothesis test indicating the test statistic and the critical values for the test statistic
  (a) for $n = 12$
  (b) for $n = 100$

P7-11 An official claims to be unbiased in the selection of men and women to serve on panels. Is there sufficient evidence at the 5% level that the official actually has a preference for men if
  (a) in a sample of selections from 15 pairs of individuals (each pair consists of a man and woman who may be judged of equal capabilities and one individual is selected from each pair), the official selected 10 men and 5 women?
  (b) in a sample of selections from 150 pairs, the official selected 100 men and 50 women?

P7-12 Is there sufficient evidence at the 1% level that more than 30% of freshmen fail to meet mathematical standards if
  (a) 6 freshmen out of a sample of 13 failed to meet the standards?
  (b) 60 freshmen out of a sample of 130 failed to meet the standards?

EXERCISES

7-19 Determine critical values c for the test statistic x in each of the following. In each case indicate the actual value of α.
  (a) $H_0 : p = 0.5$ against $H_A : p \neq 0.5$ with $n = 14$ and $\alpha \doteq 0.05$
  (b) $H_0 : p = 0.6$ against $H_A : p > 0.6$ with $n = 12$ and $\alpha \leqslant 0.02$
  (c) $H_0 : p = 0.2$ against $H_A : p < 0.2$ with $n = 15$ and $\alpha \leqslant 0.05$

7-20 Is there sufficient evidence at the 5% level that the plaintiff wins more than 60% of legal suits if the plaintiff won in 70% of the cases in a survey of 20 cases?

7-21 Repeat Exercise 7-20 for a sample of 120 cases.

7-22 Suppose that a researcher claims that, without prior conditioning, girls would choose a particular "boy's" activity rather than an alternative "girl's" acitivty more than half the time. By setting up an appropriate hypothesis test, determine whether the claim can be substantiated at the 5% level if 10 girls of a sample of 15 did choose the "boy's" activity.

7-23 Repeat Exercise 7-22 if 100 girls of a sample of 150 chose the "boy's" activity.

7-24 It is believed that potential passengers would be equally split on preference of two modes of transportation. Is there sufficient evidence to reject the belief at the 5% level if 12 potential passengers of a sample of 18 preferred mode A over mode B?

7-25 Repeat Exercise 7-24 if 36 potential passengers of a sample of 54 preferred mode B over mode A.

7-26 (a) What is the probability that the null hypothesis will be rejected in Practice Problem P7-10(a) if the probability of a red is (1) 0.55? (2) 0.60? (3) 0.65?
(b) Sketch the power curve for the test in Practice Problem P7-10(a).

7-27 Sketch the power curve for the test in Practice Problem P7-10(b).

SOLUTIONS TO PRACTICE PROBLEMS

P7-10 In each case we are testing $H_0: p = \frac{1}{2}$ against $H_A: p \neq \frac{1}{2}$ where p may be chosen as P(red).
(a) Letting the test statistic be x, the number of reds generated, the rejection region consists of two parts, $x < c_1$ and $x > c_2$, where the critical values c_1 and c_2 are chosen so that $P(x < c_1|H_0 \text{ true}) + P(x > c_2|H_0 \text{ true}) \leq 0.05$. Using the binomial with $p = p_0 = \frac{1}{2}$ and $n = 12$, we develop the following table to find $c_1 = 3$, $c_2 = 9$, and the actual α is 0.038.

c_1	$P(x = c_1)$	$P(x < c_1)$	c_2	$P(x = c_2)$	$P(x > c_2)$
0	0.0002	—	12	0.0002	—
1	0.0029	0.0002	11	0.0029	0.0002
2	0.0161	0.0031	10	0.0161	0.0031
3	0.0537	0.0192	9	0.0537	0.0192
4	0.1208	0.0729	8	0.1208	0.0729

(b) In the large-sample case with $n = 100$, the test statistic is $z = (x \pm \frac{1}{2} - np_0)/\sqrt{np_0(1 - p_0)} = (x \pm \frac{1}{2} - 50)/5$, where we use $+\frac{1}{2}$ if $x < 50$ and $-\frac{1}{2}$ if $x > 50$. The rejection region is $|z| > z_{0.025} = 1.96$.

P7-11 We are testing $H_0: p = \frac{1}{2}$ against $H_A: p > \frac{1}{2}$.
(a) The data produce $n = 15$ and $x_0 = 10$, so that $P(x \geq x_0|H_0 \text{ true})$ is found from the binomial with $n = 15$ and $p = \frac{1}{2}$ as $P(x \geq 10) = 0.151$. Since 0.151 exceeds $\alpha = 0.05$, we may not reject the null hypothesis. There is not sufficient evidence of bias.
(b) In this case we have a large-sample result and the test statistic is $z = (x - \frac{1}{2} - np_0)/\sqrt{np(1 - p_0)}$. Numerically, the test statistic is $z = (100 - \frac{1}{2} - 75)/\sqrt{150 \times 0.5 \times 0.5} = 4.00$, which exceeds $z_{0.05} = 1.645$. We may reject the null hypothesis. There is sufficient evidence of bias in this case.

P7-12 The appropriate test is $H_0: p = 0.30$ against $H_A: p > 0.30$.

(a) With $n = 13$ and $x_0 = 6$, we have $P(x \geqslant x_0 | H_0 \text{ true}) = P(x \geqslant 6)$ for the binomial with $n = 13$ and $p = 0.30$. This probability is found to be 0.165, which exceeds $0.01 = \alpha$. We may not reject the null hypothesis. There is insufficient evidence to substantiate the claim.

(b) For this large-sample case, the test statistic is $z = (x - \frac{1}{2} - np_0)/\sqrt{np_0(1 - p_0)} = (60 - \frac{1}{2} - 39)/\sqrt{27.3} = 3.923$, which exceeds $2.326 = z_{0.01}$. In this case we may reject the null hypothesis and we do have sufficient evidence to substantiate the claim.

7-6 HYPOTHESIS TESTING AND CONFIDENCE INTERVALS AND p-LEVELS

Having developed the general testing procedures for population means, variances, and proportions, we will now consider, through examples, two further aspects of hypothesis testing: the relationship to confidence intervals, and the empirical significance or p-level of a test statistic.

Hypothesis Tests and Confidence Intervals

As might be expected, since the same sampling distribution for the relevant statistic is used in both cases, there is a relationship between a confidence interval for a given parameter and a test of a hypothesis on a value for that parameter. We will not develop the general relationship for all cases since an example should serve to indicate the general pattern that can be developed for each case.

Two-Sided Tests

EXAMPLE 7-33   As an illustration, we consider Example 7-22 of Section 7-3, involving the null hypothesis that the mean thickness of sheets of plastic in a given production is 0.1500. In that example, the hypothesis is accepted since $\bar{x}$ is within $t_{\alpha/2, n-1}s/\sqrt{n}$ of μ_0; that is, 0.1496 is within 0.0006 of 0.1500, and 0.1500 is thus a plausible value for μ.

If, instead, we calculate a 95% confidence interval for μ as $\bar{x} \pm t_{\alpha/2}s/\sqrt{n}$ or 0.1496 ± 0.0006, we can be 95% confident that μ is within 0.0006 of 0.1496, that is, in the interval from 0.1490 to 0.1502. Since 0.1500 is in this interval, it is one of the plausible values for μ.

In both cases, we obtain the same conclusion because we use the same criterion—whether or not the difference between $\bar{x}$ and the possible value for μ is less than $t_{\alpha/2}s/\sqrt{n}$. The confidence interval, however, provides more information than the hypothesis test. The test indicates only whether 0.1500 is a plausible value or not, whereas the confidence interval indicates all the values that are plausible—the values between 0.1490 and 0.1502. If, for example, the null hypothesis were that $\mu = 0.1505$, then, with the data as given, the null hypothesis could be rejected since 0.1505 is not in the confidence interval. The value $\mu = 0.1505$ is not within 0.0006 of $\bar{x} = 0.1496$; therefore, $\bar{x}$ is not within 0.0006 of 0.1505 and would thus be in the rejection region in a hypothesis test.

▲

Similar arguments for inferences on other parameters lead to the following general decision rule based on confidence intervals.

> For any parameter θ (the lowercase Greek letter theta) (θ may represent a mean μ, a standard deviation σ, a proportion p, or any other parameter of interest), a two-sided test of $H_0: \theta = \theta_0$ against $H_A: \theta \neq \theta_0$ with level of significance α may be written with the decision rule: Reject H_0 if θ_0 is not included in a $100(1 - \alpha)\%$ confidence interval for θ; accept H_0 or reserve judgment if θ_0 is included in the confidence interval.

One-Sided Tests

The confidence intervals that we have developed are two-sided and thus correspond only to two-sided tests. One-sided confidence intervals can be derived, however, that would correspond to one-sided hypothesis tests. The one-sided confidence interval would have a limit on only one side, the expression for the limit being the same as one of the limits in the two-sided interval with the subscript $\alpha/2$ or $1 - \alpha/2$ on the tabulated percentile changed to α or $1 - \alpha$, as appropriate.

Equivalently, a one-sided $100(1 - \alpha)\%$ confidence interval could be obtained by finding an ordinary two-sided $100(1 - 2\alpha)\%$ interval and then using only one of the two limits.

The appropriate resulting one-sided intervals will correspond to hypothesis tests with one-sided alternatives.

EXAMPLE 7-34   One of the requirements placed on dry food pellets is that they display a standard deviation of less than 0.02 g. Samples of 20 pellets are taken from successive lots and the lot is considered sufficiently uniform if there is statistical evidence at the 1% level that the standard deviation of the lot is less than 0.02 g. A sample of 20 pellets has produced $s = 0.012$.

The uniformity may be tested in a test of $H_0: \sigma = 0.02$ against $H_A: \sigma < 0.02$, in which H_0 is rejected if $\chi^2 = 19s^2/0.02^2 < \chi^2_{0.99, 19} = 7.633$. For the sample data, $\chi^2 = 19 \times 0.012^2/0.02^2 = 6.840 < 7.633$. We may therefore reject H_0 and conclude that there is sufficient evidence that $\sigma < 0.02$.

The uniformity may also be investigated with a confidence interval. Since we must have σ less than a specific value (namely 0.02), we want a one-sided interval with an upper limit. Using the upper limit of a confidence interval for a standard deviation and replacing $1 - \alpha/2 = 0.995$ with $1 - \alpha = 0.99$ in the subscript on the χ^2 percentile, we may be 99% confident that σ is less than

$$\sqrt{(n-1)s^2/\chi^2_{1-\alpha, n-1}} = \sqrt{19 \times s^2/\chi^2_{0.99, 19}} = \sqrt{19 \times 0.012^2/7.633} = 0.019.$$

Since we may be 99% confident that σ is less than 0.019, we may be at least 99% confident that σ is less than 0.02 and that the pellets are sufficiently uniform. ▲

If we are, in fact, interested only in a single value for a parameter, performing a hypothesis test is appropriate; otherwise, calculating a confidence interval may be preferred since we would then obtain information on many values.

The confidence interval approach to hypothesis testing helps to provide a further interpretation on the strength of an "accept" or "reject" statement. In Example 7-33, in which we accept the null hypothesis, we may be 95% confident that μ is in the interval 0.1490 to 0.1502, including the possibility that $\mu = 0.1500$. There is no reason to doubt that $\mu = 0.1500$; hence, we may not reject this value. On the other hand, we many not concentrate all of our confidence on this single value, but must spread it over the entire interval. In other words, we can accept the null hypothesis, but we cannot be 95% confident that it is true.

In Example 7-34, not only can we reject the null hypothesis, but as indicated in that example, we can be at least 99% confident that $\sigma < 0.02$; that is, we can be at least 99% confident that the null hypothesis is false.

> We found in Section 7-2 that we cannot, in general, find the *probability* that the hypothesis is true or false. We now find, however, that if we reject the hypothesis in a test with a level of significance α, we can have *confidence* of at least $100(1 - \alpha)\%$ that the null hypothesis is false.

p-Levels for Hypothesis Tests

The "strength" of a reject statement in a hypothesis test may be indicated as above through a confidence interval approach. It may also be indicated through the use of the empirical level of significance or *p-level* that a test statistic exhibits. As with the confidence interval approach, we will illustrate these ideas only through examples.

> The *empirical level of significance* is the level of significance for which a test statistic is just barely significant.

EXAMPLE 7-35 Suppose that we wish to test $H_0: p = 0.7$ against $H_A: p < 0.7$ on the basis of $n = 11$ trials and the statistic x_0. If we obtain $x_0 = 4$, we find that $P(x \leqslant x_0 | H_0 \text{ true}) = P(x \leqslant 4 | p = 0.7, n = 14)$, and the latter probability is 0.022. The test statistic would be significant at the 5% level or 3% level because 2.2% is less than either of these. The test statistic would not be significant at the 1% level, because 2.2% exceeds 1%, and so on. The test statistic is just barely significant at the 2.2% level. The empirical level of significance or the actual degree of significance indicated by this particular result is 2.2%. As a probability, this is referred to as the *p-level*. We indicate the *p*-level in this example by reporting "$p = 0.022$." ▲

In a hypothesis test, the empirical level of significance is referred to as the *p*-level. It measures how significant the observed sample test statistic actually is.

Unlike the level of significance α, which is set by the experimenter, the *p*-level is determined by the data.

In tests involving distributions such as the *t*-distribution or χ^2-distribution, for which we have only a selection of tabulated percentiles, we may not be able to ascertain the exact empirical level of significance as above but must be satisfied by considering the most extreme tabulated percentile exceeded by the test statistic. We then indicate, as a *p*-level, the corresponding upper limit on the empirical level of significance.

EXAMPLE 7-36   In the test on σ in Example 7-34, we saw that $\chi^2 = 6.480$ was significant at the 1% level. If we consult Table V of Appendix B we will also find that $\chi^2 = 6.840 < 6.844 = \chi^2_{0.995, 19}$ and is significant at the 0.5% level. It may also be significant at the 0.49% level, but as we do not know $\chi^2_{0.9951, 19}$, we cannot say for sure. We do know, however, that since $\chi^2 < \chi^2_{0.995, 19}$ it is barely significant for some α less than 0.005. We report this result as a *p*-level with the statement "χ^2 is significant with $p < 0.005$."   ▲

In two-sided tests, we determine how significantly large or small the test statistic is by finding P(test statistic $\geq$ observed value) or P(test statistic $\leq$ observed value). We then double this probability to produce the *p*-level because we must account for possible significance in two tails. (Recall that, in a two-tailed test, each part of the rejection region has probability $\alpha/2$, which is doubled to produce the level of significance α.)

EXAMPLE 7-37   In an investigation of mean concentration, a researcher wishes to test $H_0 : \mu = 10.0$ against $H_A : \mu \neq 10.0$ on the basis of 15 sample measurements producing $\bar{x} = 10.9$ and $s = 1.4$. The data produce test statistic $t = (\bar{x} - \mu_0)/(s/\sqrt{n}) = (10.9 - 10.0)/(1.4/\sqrt{15}) = 2.49$. This value falls between the tabulated values $t_{0.025, 14} = 2.145$ and $t_{0.01, 14} = 2.624$; thus, $P(t > 2.49)$ is between 0.025 and 0.01. An upper limit (from our tables) on P(test statistic > observed value) = $P(t > 2.49)$ is 0.025. We can specify no further since 0.01, the next value indicated in the tables, is less than the probability and cannot be an upper limit. To find an upper limit on the *p*-level, we double the upper limit on $P(t > 2.49)$ to produce $2 \times 0.025 = 0.05$. We may report the results as "$t = 2.49$ is significant in a two-sided test with $p < 0.05$."   ▲

PRACTICE PROBLEMS

P7-13 A sample of data from a normal distribution has resulted in a 95% confidence interval for the mean μ as 68.3 to 75.5. On the basis of these data, indicate which of the following null hypotheses should be accepted at the 5% level.

(a) $H_0 : \mu = 70.0$ against $H_A : \mu \neq 70.0$
(b) $H_0 : \mu = 65.0$ against $H_A : \mu \neq 65.0$
(c) $H_0 : \mu = 80.0$ against $H_A : \mu \neq 80.0$
(d) $H_0 : \mu = 75.0$ against $H_A : \mu \neq 75.0$

P7-14 In a test of $H_0 : \mu = 15.0$ against $H_A : \mu > 15.0$ for a normal population, the null hypothesis is rejected for significantly large values of $z = (\bar{x} - 15.0)/(\sigma/\sqrt{n})$. Suppose that $\sigma = 10$ and $n = 25$. Find the empirical level of significance corresponding to a sample mean $\bar{x}$ equal to
(a) 18.8
(b) 19.6
(c) 21.4

P7-15 In a study on short-term memory in monkeys, the numbers of trials (in hundreds) to achieve the criterion of 90% correct choices in color discrimination were obtained for $n = 10$ monkeys and they produced mean $\bar{x} = 1.46$ and standard deviation $s = 0.92$. Find an appropriate one-sided 95% confidence interval for the mean number of trials (in hundreds) and on the basis of the interval, test $H_0 : \mu = 1.00$ against $H_A : \mu > 1.00$.

P7-16 For the hypothesis test in Practice Problem P7-15, indicate the p-level if the numbers for $n = 10$ monkeys produced
(a) $\bar{x} = 1.63, s = 0.85$
(b) $\bar{x} = 1.52, s = 0.95$
(c) $\bar{x} = 1.81, s = 0.90$

7-28 If measurements on the diameters of several bearings produced a 99% confidence interval for the mean diameter of all such bearings as 0.983 to 1.009, should the null hypothesis that $\mu = 1.000$ be accepted or rejected in a 1% test with alternative $\mu \neq 1.000$?

7-29 Suppose that 15 measurements on the same compound with a particular measurement process produced a sample standard deviation of 0.18. Find a 95% confidence interval for the standard deviation of measurements determined by this process and, on the basis of this confidence interval, test the null hypothesis that $\sigma = 0.125$ in a two-sided 5% test.

7-30 An archeologist found that, of a sample of 40 pottery items, 33 had been made according to a particular process. Determine a 95% confidence interval for the proportion of all comparable items made by the process and using this interval, test the null hypothesis that 75% of all such items were made by the process. Use a two-sided alternative and 5% level of significance.

7-31 The mean lifetime for watch batteries must exceed 8800 hours.
(a) Assuming normality for lifetimes, find an appropriate *one-sided* 99% confidence interval for the mean lifetime if 15 batteries produced lifetimes with a mean of 8856 and standard deviation of 48.
(b) On the basis of the interval in part (a), test $H_0 : \mu = 8800$ against $H_A : \mu > 8800$ with level of significance $\alpha = 0.01$.

7-32 Students are required to practice titration techniques in order to produce reasonable uniformity. For a given test solution, repeat titrations should produce a standard deviation σ that is less than 0.50.

(a) Find an appropriate *one-sided* 95% confidence interval for σ for a given student if 12 sample titrations on the test solution produced a sample standard deviation $s = 0.28$.

(b) Use the interval in part (a) to test $H_0 : \sigma = 0.50$ against $H_A : \sigma < 0.50$ at the 5% level.

7-33 Find the empirical level of significance for the test result in Exercise 7-7.

7-34 Find the *p*-level for the test result in Exercise 7-8.

7-35 (a) Determine a 99% *one-sided* confidence interval corresponding to the hypothesis test in Exercise 7-12.

(b) Determine the *p*-level for the hypothesis test in Exercise 7-12.

7-36 Determine the *p*-level for the results of the hypothesis test in Exercise 7-17.

7-37 Indicate the general expressions for one-sided $100(1 - \alpha)\%$ confidence intervals for the mean of a normal population on the basis of samples of size n and assuming a known population standard deviation σ.

7-38 Repeat Exercise 7-37 for unknown σ and for $n < 30$.

7-39 What are the general expressions for one-sided $100(1 - \alpha)\%$ confidence intervals for the variance of a normal distribution?

7-40 Give a general expression for the empirical level of significance corresponding to an observed x_0 in the test of $H_0 : p = p_0$ against $H_A : p > p_0$ if the sample size n is too small to use the normal approximation.

SOLUTIONS TO PRACTICE PROBLEMS

P7-13 (a), (d). Since both 70.0 and 75.0 are included in the confidence interval, these null hypotheses may be accepted. Note, however, that since 75.0 is just barely within the interval we may wish to reserve judgment in part (d).

(b), (c). Since neither 80.0 nor 65.0 is in the interval, neither of these null hypotheses should be accepted.

P7-14 (a) Since $\bar{x} = 18.8$ produces $z = 1.9$, the empirical level of significance is $P(z \geqslant 1.9) = 0.0287$.

(b) Similarly, $\bar{x} = 19.6$ produces $z = 2.3$ and an empirical level 0.0107.

(c) Similarly, $\bar{x} = 21.4$ produces $z = 3.2$ and an empirical level 0.0007.

P7-15 For the alternative hypothesis $\mu > 1.00$, we must determine at least how large μ might be. We thus require the one-sided confidence interval with the lower limit $\bar{x} - t_{0.05, 9} s / \sqrt{10} = 1.46 - (1.833 \times 0.92 / \sqrt{10}) = 0.93$. We may be 95% confident that μ is at least 0.93. The interval includes values greater than 1.00, but it also includes 1.00 and values less than 1.00. The null hypothesis $\mu = 1.00$ is, therefore, acceptable and cannot be rejected.

P7-16 (a) The data produce $t = (\bar{x} - 1.00)/(s / \sqrt{10}) = (1.63 - 1.00)/(0.85 / \sqrt{10}) = 2.34$. Since t exceeds $t_{0.025, 9} = 2.262$ but does not exceed the next percentile $t_{0.01, 9} = 2.821$, we may conclude that the empirical level of significance is less than 0.025 but not less than 0.01. We thus have the *p*-level as $p < 0.025$.

(b) Similarly, $\bar{x} = 1.52$ and $s = 0.95$ produce $t = 1.73$ and $p < 0.10$.

(c) Similarly, $\bar{x} = 1.81$ and $s = 0.90$ produce $t = 2.85$ and $p < 0.01$.

7-41 If we accept the null hypothesis in a 5% test, the probability that the null hypothesis is true
(a) is 5%
(b) is 95%
(c) cannot be determined without more information

7-42 The level of significance α in a test of a hypothesis is
(a) always 5%
(b) sometimes 95%
(c) the probability of a type I error
(d) determined after the data have been collected

7-43 To test $H_0 : \mu = \mu_0$ against $H_A : \mu < \mu_0$ with test statistic $t = (\bar{x} - \mu_0)/(s/\sqrt{n})$ and level of significance α, the rejection region is _____.

7-44 The potential probabilities of a type I and type II error can be reduced simultaneously in a hypothesis test
(a) by choosing a larger rejection region
(b) by choosing a smaller rejection region
(c) by choosing a different sample size
(d) never

7-45 In a test of $H_0 : \sigma^2 = \sigma_0^2$ against $H_A : \sigma^2 > \sigma_0^2$ such that H_0 is rejected if $(n - 1)s^2/\sigma_0^2 > \chi^2_{0.05, n-1}$, the probability of a type I error is
(a) 0.05
(b) 0.10
(c) undetermined
(d) depends on σ_0^2

7-46 If sample data produce a 95% confidence interval for the population mean μ as 63.2 to 85.7, then in a two-sided hypothesis test with level of significance 5% based on these data, $H_0 : \mu = 65.0$
(a) may be accepted
(b) may be rejected
(c) may not be judged without further information

7-47 In a test of $H_0 : \mu = 0$ against $H_A : \mu > 0$, if the data produce test statistic $t = 1.93$ and if for this case $t_{0.05} = 1.729$, $t_{0.025} = 2.093$, and $t_{0.01} = 2.539$, the p-level is $p <$ _____.

7-48 A type I error in a test of a hypothesis is the acceptance of a false hypothesis.
(a) True
(b) False

7-49 In a 5% test of $H_0 : \mu = 35$ against $H_A : \mu > 35$ based on a normal population, if a sample of 100 produces $\bar{x} = 30$, $s_x = 10$, the hypothesis should be rejected.
(a) True
(b) False

7-50 According to specification requirements, certain computer parts must have a mean of 1.800. It may be assumed that the standard deviation of such parts is 0.050 and that they follow a normal distribution. A 5% hypothesis test is to be used to determine whether there is sufficient evidence that the parts do not meet specification requirements:

(a) Set up the test and determine the decision rule if a sample of 25 parts is to be assessed.

(b) Find the probability of accepting the (false) null hypothesis for each of the following values for the true mean: 1.750, 1.760, 1.770, 1.780, 1.790, 1.810, 1.820, 1.830, 1.840, and 1.850.

(c) Sketch the power function for this test.

7-51 Repeat Example 7-50 with a sample size of 75.

7-52 A new packaging process is under investigation to determine whether it will produce packaged weights which are more consistent (i.e., less variable) than those produced by the current process, which exhibits a standard deviation of 0.15. A sample of measurements from the new process produced the following values:

| 16.07 | 15.79 | 15.83 | 16.14 | 16.22 | 16.09 | 16.17 | 15.94 |
| 16.01 | 15.88 | 16.09 | 16.07 | 15.95 | 15.97 | 16.13 | |

Determine whether the data provide sufficient evidence at the 5% level that the new process is more consistent.

7-53 Determine an appropriate one-sided 95% confidence interval for the standard deviation in Exercise 7-52.

7-54 On the basis of the data in Exercise 7-52, find a 95% confidence interval for the mean weight and, using this interval, test $H_0 : \mu = 16.00$ against $H_A : \mu \neq 16.00$ with a 5% level of significance.

7-55 Suppose that in a follow-up study 10 years after the introduction of a municipal flag, 250 citizens were surveyed at random and of these, 26 still disapproved of the flag. Do the results of this survey provide sufficient empirical evidence to prove, at the 5% level, that the percentage of citizens disapproving of the flag has been reduced below 15%?

7-56 In a sample of 25 test cases, the following dosages of a drug required to produce a given response resulted:

1.07	0.79	0.83	1.14	1.22	1.09	1.17	1.10	1.26
1.10	1.04	1.17	0.94	0.86	1.19	1.01	1.12	
0.83	1.02	1.20	0.85	1.03	0.95	1.13	0.98	

Assuming a normal distribution, find a 95% confidence interval for the mean required dosage and, using this interval, test the hypothesis that the mean required dosage is 1.00 with a two-sided 5% test.

7-57 What is the expression for the minimum sample size required to have level of significance α in a test of $H_0 : \mu = \mu_0$ against $H_A : \mu \neq \mu_0$ if the probability of a type II error must be at most β when μ is actually a value μ_A such that $|\mu_0 - \mu_A| > k\sigma$ for some positive number k?

7-58 A professor claims that the first-year class in a proposed program will have to be offered in sections since the backgrounds, and hence capabilities, of incoming students will be so varied as to produce too great a variation in the final marks. If a sample of 25 students was selected and if these students produced scores on a general capabilities test that produced a standard deviation $s = 16.3$, find an appropriate one-sided 95% confidence interval indicating at least how large the standard deviation of such scores would be for incoming students in general.

7-59 (a) For the data in Exercise 7-58, test $H_0 : \sigma = 12.5$ against $H_A : \sigma > 12.5$ at the 5% level.

(b) Indicate the p-level for the test result in part (a).

7-60 (a) In hypothesis tests, as in confidence intervals, if very large samples are taken from finite populations, the variance of the sampling distribution for a statistic may have to be modified by the finite population correction factor. Apply the appropriate modification to the large sample test statistic for $H_0 : p = p_0$.

(b) In a survey of $n = 350$ employees of a total of $N = 2700$, $x = 283$ indicated that they liked the opportunity to work overtime. Do these data substantiate the claim, at the 1% level, that more than 75% of all the employees like the opportunity to work overtime?

7-8 GENERAL REVIEW

7-61 The value -10 appears reasonable for the standard deviation of a set of numbers that range from -50 to 0.
(a) True
(b) False

7-62 The appropriate mathematical model for describing the distribution of outcomes in a coin-tossing experiment is
(a) the normal curve
(b) the binomial distribution
(c) mathematical expectation

7-63 If a hypothesis test with $H_A : \sigma > \sigma_0$ produces the test statistic $\chi^2 = 23.6$, and if $\chi^2_{0.05} = 18.3$, $\chi^2_{0.01} = 23.2$, and $\chi^2_{0.005} = 25.2$, we report the p-level as $p < \underline{\hspace{1cm}}$.

7-64 If we wish to calculate $P(19 \leqslant x \leqslant 31)$ for a binomial and we convert to a normal, then for the normal, we should calculate $P(\underline{\hspace{1cm}} \leqslant x \leqslant \underline{\hspace{1cm}})$.

7-65 If we calculate a probability as 3.69, we conclude that $\underline{\hspace{2cm}}$.

7-66 If we can reject a hypothesis at an extreme level of significance such as 0.001, the hypothesis cannot possibly be true.
(a) True
(b) False

7-67 If A and B are mutually exclusive events, $P(A|B)$ is equal to
(a) $P(A) + P(B) - P(A \text{ and } B)$
(b) 0.0
(c) 0.1
(d) $P(A \text{ or } B)/P(B)$

7-68 If a sample from a normal population produces a 95% confidence interval for the mean as 13.7 to 16.8, then on the basis of this sample, the hypothesis $\mu = 17$ versus the alternative $\mu \neq 17$ should be rejected at the 5% level.
(a) True
(b) False
(c) Impossible to tell without further information

7-69 The area under a probability graph is used to measure $\underline{\hspace{2cm}}$.

7-70 Suppose that you are engaged in a game of chance with five other players and that everybody has an equal chance of winning any given round. The game is such that the first person to win two rounds wins the game. If the game has gone five rounds already and you have been one of the losers every time, what are the odds against your being one of the losers again in the sixth round? What is the probability that the game will end on the seventh round?

7-71 In a study on voting patterns, people of different national origins were presented with two fictitious candidates in a fictitious election. One of the candidates was of the same national origin as the person, the other was not. Of the 80 people involved in the study, 48 voted for a candidate of the same origin as themselves.
(a) Find a 95% confidence interval for the overall proportion of people who vote for a candidate of the same national origin.
(b) Using the results of part (a), test the hypothesis that this overall proportion is 0.5.

7-72 The following is a sample of days absent from work for 58 employees over a 1-year period.

0	8	5	0	1	12	0	7	2	21	7	6	2	9
1	0	6	1	2	23	2	19	11	13	3	16	6	8
3	7	7	2	2	6	22	15	1	16	7	1	2	5
1	1	1	8	15	18	0	20	9	3	7	17	4	7
0	5												

(a) Find the mean and median numbers of days absent from work.
(b) Find the range and standard deviation.
(c) Form a frequency distribution using 0 to 2 as the first class and, for each class, give the class mark, limits, and boundaries.
(d) Sketch a histogram for this sample.
(e) Form a "more than" cumulative distribution for the sample.

7-73 It is claimed that the lifetimes of light bulbs of a particular type have a normal distribution with mean 1000 and variance 8500.
(a) If the claim is true and a bulb is picked at random;
   (1) What is the probability that it will last at least 1150 hours?
   (2) What is the probability that it will last less than 900 hours?
(b) If the mean lifetime of a sample of 25 such bulbs is 960 hours, does this provide sufficient evidence, at the 5% level, that the overall mean is less than 1000 hours? (Assume that the variance is 8500, as claimed.)

7-74 A particular measuring device produces measurements that may be assumed to follow a normal distribution. It is necessary that the variance of these measurements should be at most 0.0225. The following data represent a sample of measurements (on the same object) using this device:

3.33	2.98	2.84	3.12	3.13	3.29	2.74	3.07	3.10
2.69	2.97	3.29	2.85	2.99	2.70	3.25	3.03	

(a) Determine whether the device meets the necessary requirements with an appropriate hypothesis test and a 5% level of significance.
(b) Find an appropriate one-sided 95% confidence interval for the true variance.

7-75 A subcommittee is to be formed by choosing 5 people at random from a committee of 20 people consisting of 12 elected members and 8 members appointed ex officio.

(a) What is the probability that the subcommittee will consist entirely of ex officio members of the committee?

(b) What is the probability that the membership of the subcommittee will be in the same proportions as that of the committee?

7-76 Consider a modification to a manufacturing process that is claimed to increase daily average productivity from the present average of 300 units. Suppose that it can be assumed that the standard deviation will remain at the present 50 units. Suppose that it is decided to test the claim at the 5% level of significance with a sample of 25 days:

(a) Set up the test so that the burden of proof is on the claim.

(b) If the hypothesis of the test is true, what error might we make when we come to our conclusion? What is the probability of this?

(c) If the increase is, in fact, 25 units, what error might we make? What is the probability of this?

(d) What sample size is required to reduce the probability of the second error to 10% without increasing the probability of the first?

7-77 The extent of acid rain in an area has been claimed to be such that the annual fall of sulfuric acid exceeds 300 milliliters per acre. Set up an appropriate hypothesis test to investigate the claim, and indicate the p-level that would result from the following data summaries based on samples of size $n = 15$. For results that do not even achieve significance for α as large as 0.10, report "not significant" instead of the p-level.

(a) $\bar{x} = 325$, $s = 35$

(b) $\bar{x} = 312$, $s = 40$

(c) $\bar{x} = 332$, $s = 36$

(d) $\bar{x} = 314$, $s = 29$

(e) $\bar{x} = 306$, $s = 27$

7-78 Units are obtained from four different suppliers with different rates of defective items. Percentages of total stock received from the suppliers and defective rates are as follows:

Supplier:	A	B	C	D
Percentage of full stock provided by supplier	35%	20%	15%	30%
Percentage of defects in supplier's stock	1%	8%	10%	5%

(a) If a unit is selected at random, what is the probability that it will be defective?

(b) If a unit selected at random is defective, what is the probability that it was supplied by C?

(c) If a unit selected at random is not defective, what is the probability that it was supplied by A?

(d) If a unit selected at random from a batch (provided by one supplier) is not defective, what is the probability that a second unit from the same batch will also be nondefective?

7-79 Before petitioning a municipal council to approve extended opening hours, a shopkeeper wishes to determine a reasonable estimate of the proportion of all regular customers who would actually shop during the extended hours. The estimate is to be the proportion found in a sample survey. If there are 3500 regular customers, if the shopkeeper believes very strongly that the proportion will not exceed 0.35, and if there is to be at least a 90% chance that the estimate will be within 0.04 of the true value, how many customers should be included in the survey?

7-80 There are 20 comparable manufacturing plants available to study as cases of plant shutdowns. The ownerships and markets served by the plants are as tabulated.

Market served	Ownership		
	Totally domestic	Totally foreign	Shared domestic/foreign
Domestic	3	1	1
Foreign	1	5	1
Both	2	2	4

Consider as a trial the random selection of a plant to be studied as a test case.
(a) (1) How many possible outcomes are there for this trial and what are these outcomes?
(2) List a possible set of events for this trial.
(b) For the plant chosen as a test case, determine
(1) the probability that it will be totally foreign owned
(2) the probability that it will serve both markets
(3) the chance that it will have partial or total domestic ownership
(4) the odds against its serving only the domesitc market

7-81 Individual experimental units of a given type have been found occasionally to be of no use at all, but in some cases have been found to be usable in a number of experiments. The percentage breakdown is as follows:

Number of times unit was useable	Percentage of units
0	5
1	30
2	40
3	20
4	5

How many experiments can we expect to be able to do with ten units selected at random?

7-82 If mistakes in recording survey data occur randomly at a mean rate of one mistake per 3000 figures, what is the probability that a data bank including 15,000 figures will have fewer than four mistakes?

COMPARING MEANS AND VARIANCES

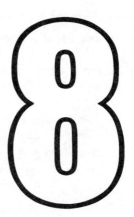

8-1   INTRODUCTION

In Chapter 7 we developed techniques for testing the values of a mean or a variance in a single population. In this chapter we consider techniques for comparing means and variances of two populations.

While developing the hypothesis-testing procedures, we will also make use of the relationships between hypothesis tests and confidence intervals established in Chapter 7 to develop methods for finding confidence intervals for comparing two means or two variances.

Since we have already developed the general concepts for these procedures in previous chapters, some of the results in this chapter are established briefly without as much intuitive argument and with examples following rather than preceding the general procedures.

In the development of hypothesis tests, the level of significance will be assumed to be α in all cases. The sizes of the first and second samples will be denoted as n_1 and n_2, respectively. *The statement of the decision rule will be abbreviated to an indication of the rejection region only.* It will be understood in every case, however, that we adopt H_A whenever we reject H_0 and that if the test statistic is not in the rejection region, we accept H_0 or reserve judgment.

As well, in the statement of confidence intervals, it will be understood that the intervals are $100(1 - \alpha)\%$ *two-sided* confidence intervals, unless otherwise stated.

8-2 COMPARING MEANS

Independent Samples

In the comparison of two population means, we will consider the empirical difference of two sample means as an indication of the actual difference, if any, between the population means. In the analysis we make use of the following results.

> If two independent random variables have distributions that are normal, their difference will have a distribution that is normal with a mean equal to the difference of the two original population means and a variance equal to the sum of the original population variances.

Variances Known

We first consider the case of independent samples from two normal distributions with known variances. The first population is normal with unknown mean μ_1 and known variance σ_1^2. The second is normal with unknown mean μ_2 and known variance σ_2^2. If we take samples of sizes n_1 and n_2, respectively, from these populations, the sampling distributions for the sample means will be normal with means μ_1 and μ_2, respectively, and variances σ_1^2/n_1 and σ_2^2/n_2, respectively; accordingly, by the general result given above, $\bar{x}_1 - \bar{x}_2$, the difference between the sample means, will have a normal distribution with mean $\mu_1 - \mu_2$ and variance $\sigma_1^2/n_1 + \sigma_2^2/n_2$. We thus have the following:

With normality and independence assumptions as above,

$$z = \frac{(\bar{x}_1 - \bar{x}_2) - (\mu_1 - \mu_2)}{\sqrt{\sigma_1^2/n_1 + \sigma_2^2/n_2}}$$

has a standard normal distribution. If $\mu_1 = \mu_2$,

$$z = \frac{\bar{x}_1 - \bar{x}_2}{\sqrt{\sigma_1^2/n_1 + \sigma_2^2/n_2}}$$

has a standard normal distribution.

Testing Equality of μ_1 and μ_2

To test the null hypothesis that $\mu_1 = \mu_2$ (i.e., that $\mu_1 - \mu_2 = 0$), it is appropriate to consider a test statistic $z = (\bar{x}_1 - \bar{x}_2)/\sqrt{\sigma_1^2/n_1 + \sigma_2^2/n_2}$ and, with a two-sided alternative, to reject the null hypothesis if $|z| > z_{\alpha/2}$. With a one-sided alternative, we reject the null hypothesis if $z > z_\alpha$ or if $z < -z_\alpha$, depending on the alternative. Using these results, we have the procedures for *testing the equality of the means of two normal populations with known (unequal) variances on the basis of independent random samples.*

For the two-sided alternative, the test is summarized as:

$$H_0 : \mu_1 = \mu_2$$

$$H_A : \mu_1 \neq \mu_2$$

Level of significance: α

Sample sizes: n_1, n_2

Test statistic: $z = \dfrac{\bar{x}_1 - \bar{x}_2}{\sqrt{\sigma_1^2/n_1 + \sigma_2^2/n_2}}$

Decision rule: reject H_0 if $|z| > z_{\alpha/2}$

For the one-sided alternative that $\mu_1 - \mu_2$ is greater than 0 (i.e., that μ_1 is greater than μ_2), the hypothesis test is summarized as above except that the alternative and decision rule are rewritten as

$$H_A : \mu_1 > \mu_2$$

Decision rule: reject H_0 if $z > z_{\alpha}$

For the alternative that μ_1 is less than μ_2 the changes in the test summary produce

$$H_A : \mu_1 < \mu_2$$

Decision rule: reject H_0 if $z < -z_{\alpha}$

EXAMPLE 8-1 Two ore assessment processes are such that they produce assessments that are normally distributed, each with mean equal to the actual available ore in a fixed quantity of raw material but with different standard deviations, $\sigma_1 = 4.75$ and $\sigma_2 = 5.25$. The first process is used in one area to assess the available ore potential μ_1, and the second is used in another area to assess the ore potential μ_2. We wish to determine whether there is sufficient evidence at the 5% level that the areas differ. Independent samples of 18 assessments from the first area and 12 assessments from the second area are available. To make the determination we perform the following hypothesis test:

$$H_0 : \mu_1 = \mu_2$$

$$H_A : \mu_1 \neq \mu_2$$

Level of significance: 0.05

Sample sizes: 18, 12

Test statistic: $z = \dfrac{\bar{x}_1 - \bar{x}_2}{\sqrt{4.75^2/18 + 5.25^2/12}}$

Decision rule: reject H_0 if $|z| > z_{0.025} = 1.96$

Suppose that the mean of the 18 sample assessments from the first area is 82.35 and the mean of the sample from the second area is 84.80. The test statistic then becomes $z = (82.35 - 84.80)/\sqrt{4.75^2/18 + 5.25^2/12} = -1.30$. Since $|z| = 1.30$ does not exceed $1.96 = z_{0.025}$, we cannot reject the hypothesis; thus, there is insufficient evidence at the 5% level to declare that the areas do have different ore potentials. ▲

Testing a Specific Difference for μ_1 and μ_2

We may also *test the hypothesis that the difference between μ_1 and μ_2 is a particular value, say δ* (the Greek lowercase letter delta) by changing the test statistic z to $(\bar{x}_1 - \bar{x}_2 - \delta)/\sqrt{\sigma_1^2/n_1 + \sigma_2^2/n_2}$. From our general distribution result, if $\mu_1 - \mu_2$ does equal δ, then z will follow a standard normal distribution. Accordingly, we use the following test formats.

The two-sided test is summarized as:

$$H_0 : \mu_1 - \mu_2 = \delta$$

$$H_A : \mu_1 - \mu_2 \neq \delta$$

Level of significance: α

Sample sizes: n_1, n_2

Test statistic: $z = \dfrac{\bar{x}_1 - \bar{x}_2 - \delta}{\sqrt{\sigma_1^2/n_1 + \sigma_2^2/n_2}}$

Decision rule: reject H_0 if $|z| > z_{\alpha/2}$

For the one-sided alternative hypothesis that $\mu_1 - \mu_2$ is less than δ, the test is as above with H_A and the decision rule modified to

$$H_A : \mu_1 - \mu_2 < \delta$$

Decision rule: reject H_0 if $z < -z_\alpha$

For the one-sided alternative hypothesis that $\mu_1 - \mu_2$ exceeds δ, H_A and the decision rule are modified to

$$H_A : \mu_1 - \mu_2 > \delta$$

Decision rule: reject H_0 if $z > z_\alpha$

EXAMPLE 8-2 In a study to determine whether there is sufficient evidence at the 1% level that the mean calcium concentration in milligrams per cubic decameter at a downstream station exceeds that of another station upstream by more than 5, independent sample analyses are to be made at the two stations. Personnel at the first station (downstream) are to take a sample of 20 measurements for analysis. Personnel at the second station are to take 15. Measurements taken at the first station are known to follow a normal distribution with mean equal to the true concentration and standard deviation 1.8. Measurements taken at the second station are independent of those in the first and behave similarly except that the standard deviation is 2.4. The appropriate procedure is to perform a hypothesis test on the difference of two means as follows:

$$H_0 : \mu_1 - \mu_2 = 5$$

$$H_A : \mu_1 - \mu_2 > 5$$

Level of significance: 0.01

Sample sizes: 20, 15

Test statistic: $z = \dfrac{\bar{x}_1 - \bar{x}_2 - 5}{\sqrt{1.8^2/20 + 2.4^2/15}}$

Decision rule: reject H_0 if $z > z_{0.01} = 2.326$

Suppose now that the measurements produce $\bar{x}_1 = 21.3$ and $\bar{x}_2 = 14.2$. We then have $z = (21.3 - 14.2 - 5)/\sqrt{1.8^2/20 + 2.4^2/15} = 2.84$. Since $2.84 > 2.326$, the hypothesis may be rejected. We do have sufficient evidence at the 1% level that the calcium concentration at the first station is at least 5 greater than that at the second. ▲

Confidence Interval for $\mu_1 - \mu_2$

Using the previously stated result on the distribution of $\bar{x}_1 - \bar{x}_2$ when $\bar{x}_1$ and $\bar{x}_2$ are means of independent samples of sizes n_1 and n_2 from two normal populations with means μ_1 and μ_2 and variances σ_1^2 and σ_2^2, we can produce a confidence interval for the difference $\mu_1 - \mu_2$ as follows.

Since $z = [(\bar{x}_1 - \bar{x}_2) - (\mu_1 - \mu_2)]/\sqrt{\sigma_1^2/n_1 + \sigma_2^2/n_2}$ follows a standard normal distribution, there is a presample probability $1 - \alpha$ that it will take value between $-z_{\alpha/2}$ and $z_{\alpha/2}$. Multiplying by the denominator in z, we thus find that there is a presample probability $1 - \alpha$ that $(\bar{x}_1 - \bar{x}_2)$ and $(\mu_1 - \mu_2)$ will differ by at most $z_{\alpha/2}\sqrt{\sigma_1^2/n_1 + \sigma_2^2/n_2}$. Accordingly, we have post sample confidence $100(1 - \alpha)\%$ that $(\bar{x}_1 - \bar{x}_2)$ and $(\mu_1 - \mu_2)$ do differ by at most this value. As a result, we have the following confidence interval:

$$(\bar{x}_1 - \bar{x}_2) \pm z_{\alpha/2}\sqrt{\sigma_1^2/n_1 + \sigma_2^2/n_2}$$

EXAMPLE 8-3   A 98% confidence interval for the difference between the concentrations of calcium at the two stream stations in Example 8-2 is $(21.3 - 14.2) \pm 2.326\sqrt{1.8^2/20 + 2.4^2/15}$ and we can be 98% confident that the concentration at the first station exceeds that at the second by at least 5.4 and at most 8.8. ▲

Variances Known and Equal

The results given above are modified slightly if the two known variances are equal. The denominator of the test statistic in each hypothesis test is $\sqrt{\sigma_1^2/n_1 + \sigma_2^2/n_2}$. Suppose that the two known variances have common value σ^2. We may then rewrite the square root factor as $\sqrt{\sigma_1^2/n_1 + \sigma_2^2/n_2} = \sqrt{\sigma^2/n_1 + \sigma^2/n_2}$ $= \sqrt{\sigma^2(1/n_1 + 1/n_2)} = \sigma\sqrt{1/n_1 + 1/n_2}$. This factor also appears in the confidence interval for $\mu_1 - \mu_2$ and is modified there as well.

> For independent samples from two normal distributions with known equal variances, the initial test procedures are modified by simplifying the denominator of the test statistic to $\sigma\sqrt{1/n_1 + 1/n_2}$, where σ is the known standard deviation common to both populations. Confidence intervals are also modified accordingly.

Nonnormal Populations

It should be noted that as a result of the central limit theorem, the foregoing results will be valid for nonnormal populations, provided that the samples are reasonably large. (A usual rule of thumb is that a sample of 30 or more is large.)

Variances Unknown (Unequal) — Large Samples

For independent random samples from two populations with unknown (possibly unequal) variances, the procedures given above for hypothesis tests and confidence intervals must be modified. For large samples, we need only substitute sample variances for unknown population variances. Again, because of the Central Limit Theorem, the results will be valid even for nonnormal populations.

Hypothesis Tests and Confidence Intervals

> For *large* (*independent*) *samples* from populations with *unknown* population variances, we use the foregoing procedures except that we *substitute s_1^2 and s_2^2 for σ_1^2 and σ_2^2, respectively.*

EXAMPLE 8-4 An orientation program is being studied to determine whether it has any effect on the mean anxiety level of subjects undergoing a particular experience. A control group is to undergo the experience without orientation and an experimental group is to undergo the experience after orientation. To study the effect (if any) of the orientation program, it is appropriate to study $\mu_1 - \mu_2$, the difference between the mean anxiety levels for the control and experimental groups.

Let us assume that individuals produce anxiety levels that follow a normal distribution and that the control and experimental groups are independent. We may then use the foregoing results to study $\mu_1 - \mu_2$ on the basis of $\bar{x}_1 - \bar{x}_2$.

Let us now suppose that a sample of 40 individuals in a control group produced anxiety levels with mean $\bar{x}_1 = 43.3$ and standard deviation $s_1 = 4.6$ and that a sample of 30 different individuals in an experimental group produced anxiety levels with a mean $\bar{x}_2 = 41.7$ and a standard deviation $s_2 = 5.2$. We then have a 95% confidence interval for $\mu_1 - \mu_2$ as $(\bar{x}_1 - \bar{x}_2) \pm z_{0.025}\sqrt{s_1^2/n_1 + s_2^2/n_2}$, which is $(43.3 - 41.7) \pm 1.96\sqrt{4.6^2/40 + 5.2^2/30} = 1.6 \pm 2.3$ or -0.7 to 3.9.

If we are interested in the hypothesis of equal means, we note that the 95% confidence interval for $\mu_1 - \mu_2$ includes 0. Because 0 is one of the plausible values for $\mu_1 - \mu_2$, we cannot reject the hypothesis of equal means in a two-sided 5% test. With these data, there is insufficient evidence to declare that the program has any effect on the mean anxiety level. ▲

Variances Unknown (Unequal) — Small Samples

For small independent samples from populations with unknown variances, we again substitute sample variances for population variances. Because we do not have large samples we must now have populations that are (at least approximately) normal. Furthermore, the test statistic will no longer be based on a standard normal. Instead, the distribution of the test statistic becomes an *approximate t-distribution*. The test statistic is denoted as t, and the critical values, z_α and $z_{\alpha/2}$, must be replaced with approximate *t*-values. There are a few possible approximations. The one that we choose is to replace the *z*-values with *t*-values based on approximate degrees of freedom.

Hypothesis Tests

The hypothesis test for the equality of μ_1 and μ_2 with a two-sided alternative becomes

$H_0 : \mu_1 = \mu_2$

$H_A : \mu_1 \neq \mu_2$

Level of significance: α

Sample sizes: n_1, n_2

Test statistic: $t = \dfrac{\bar{x}_1 - \bar{x}_2}{\sqrt{s_1^2/n_1 + s_2^2/n_2}}$

Decision rule: reject H if $|t| > t_{\alpha/2, \, m}$,
where m, the value for the degrees of freedom, is the

closest integer to

$$
\frac{\left(s_1^2/n_1 + s_2^2/n_2\right)^2}{\left[\left(s_1^2/n_1\right)^2 / (n_1 - 1)\right] + \left[\left(s_2^2/n_2\right)^2 / (n_2 - 1)\right]}
$$

The one-sided tests are as summarized for the two-sided test except that H_A becomes $\mu_1 > \mu_2$ with rejection region $t > t_{\alpha, m}$, or H_A becomes $\mu_1 < \mu_2$ with rejection region $t < -t_{\alpha, m}$.

EXAMPLE 8-5   A Canadian motorist has noticed an apparent improvement in mileage when driving in the United States after a number of motor trips under similar driving conditions in both Canada and the United States in a vehicle using unleaded gasoline at all times. As a result of ten refills after using U.S. gasoline, the mileage figures produced a mean of 18.3 mpg (converted to Imperial gallons) and a standard deviation of 2.5, whereas 14 refills after using Canadian gasoline produced mileage figures with a mean of 16.0 and a standard deviation of 2.1.

To determine whether the data do provide sufficient evidence that U.S. gasoline does produce better mileage, it is appropriate to test $H_0 : \mu_1 = \mu_2$ against $H_A : \mu_1 > \mu_2$, where μ_1 and μ_2 are the "true" mean miles per gallon with U.S. and Canadian gasoline, respectively. On the assumption of independent samples, it is appropriate to reject the null hypothesis if $t > t_{\alpha, m}$, where t is the test statistic $t = (\bar{x}_1 - \bar{x}_2)/\sqrt{s_1^2/n_1 + s_2^2/n_2}$. The test statistic is then found to be $t = (18.3 - 16.0)/\sqrt{2.5^2/10 + 2.1^2/14} = 2.372$. The approximate value for the degrees of freedom m is the closest integer to

$$
\frac{\left(\dfrac{2.5^2}{10} + \dfrac{2.1^2}{14}\right)^2}{\dfrac{\left(2.5^2/10\right)^2}{9} + \dfrac{\left(2.1^2/14\right)^2}{13}} = 17.3
$$

Critical values are thus obtained from the t-distribution with 17 degrees of freedom. Using a 5% level of significance, the critical value is $t_{0.05, 17} = 1.740$; using a 1% level, it is $t_{0.01, 17} = 2.567$. There is sufficient evidence at the 5% level, but not the 1% level, that, for this particular car, U.S. gasoline produces better mileage than Canadian gasoline.

Noting also the critical value $t_{0.025, 17} = 2.110$, we may determine that $t_{0.05}$ is the most extreme tabulated percentile exceeded by the test statistic. Using the p-level approach, we say that the difference is significant with $p < 0.05$.   ▲

For a test of a null hypothesis that two means differ by a specific amount δ, we modify the foregoing test result as we did for the test of $H_0 : \mu_1 - \mu_2 = \delta$, where we had known population variances.

To test a specific difference δ for $\mu_1 - \mu_2$, the tests above are modified by changing the test statistic to

$$t = \frac{\bar{x}_1 - \bar{x}_2 - \delta}{\sqrt{s_1^2/n_1 + s_2^2/n_2}}$$

Confidence Intervals

Similarly, the previous confidence interval is modified to produce a confidence interval for $\mu_1 - \mu_2$ in the case of small independent samples from two normal populations with unknown variances as

$$(\bar{x}_1 - \bar{x}_2) \pm t_{\alpha/2,m}\sqrt{s_1^2/n_2 + s_2^2/n_2}$$

where m is as defined in the hypothesis test above.

EXAMPLE 8-6 For the gasoline data of Example 8-5, a 90% confidence interval for $\mu_1 - \mu_2$ is $(18.3 - 16.0) \pm 1.740\sqrt{2.5^2/10 + 2.1^2/14} = 2.3 \pm 1.7$ or 0.6 to 4.0. The motorist can be 90% confident that, for this car, U.S. gasoline produces an improvement of between 0.6 and 4.0 mpg. ▲

Variance Unknown but Equal — Small Samples

For small independent samples from two normal distributions with equal but unknown variances, the initial statistic is modified to a quantity with an exact t-distribution. If $\sigma_1^2 = \sigma_2^2 = \sigma^2$ (say), $\bar{x}_1 - \bar{x}_2$ will have variance $\sigma^2(1/n_1 + 1/n_2)$. We used this result to produce the factor $\sigma\sqrt{1/n_1 + 1/n_2}$ for tests and confidence intervals with two known equal variances. We must now modify the result further since σ is unknown. The unknown variance σ^2 is replaced with a pooled estimate based on the two sample variances s_1^2 and s_2^2. Each sample variance is based on a sample from a population with variance σ^2; hence, each provides an estimate of σ^2. We pool the information in both estimates to provide an overall estimate for σ^2. The pooled estimate is a weighted mean of the sample variances using degrees of freedom as weights and produces the following estimate of the standard deviation σ.

$$s_p = \sqrt{\frac{(n_1 - 1)s_1^2 + (n_2 - 1)s_2^2}{n_1 + n_2 - 2}}$$

Introducing s_p for σ in the original z produces the following:

$$t = \frac{[(\bar{x}_1 - \bar{x}_2) - (\mu_1 - \mu_2)]}{s_p\sqrt{1/n_1 + 1/n_2}}$$

is a variable with $n_1 + n_2 - 2$ degrees of freedom (if the value for the degrees of freedom exceeds 30, we may use z values).

Hypothesis Tests

Tests on the equality of the means of two independent normal populations with equal variances may be summarized in the usual form:

$H_0: \mu_1 = \mu_2$

$H_A: \mu_1 \neq \mu_2$

Level of significance: α

Sample sizes: n_1, n_2

Test statistic: $t = \dfrac{\bar{x}_1 - \bar{x}_2}{s_p\sqrt{1/n_1 + 1/n_2}}$

Decision rule: reject H_0 if $|t| > t_{\alpha/2, n_1+n_2-2}$

Modifications produce one-sided tests with

$H_A: \mu_1 > \mu_2$ and rejection region $t > t_{\alpha, n_1+n_2-2}$

or $H_A: \mu_1 < \mu_2$ and rejection region $t < -t_{\alpha, n_1+n_2-2}$.

Also, as before, the tests can be modified to tests on a specific difference δ by changing the null hypothesis to $H_0: \mu_1 - \mu_2 = \delta$ and the test statistic to

$$t = \frac{\bar{x}_1 - \bar{x}_2 - \delta}{s_p\sqrt{1/n_1 + 1/n_2}}$$

Confidence Intervals

The t-distribution variable is also the basis for a confidence interval for the difference $\mu_1 - \mu_2$. The *confidence interval* for $\mu_1 - \mu_2$ is found as

$$(\bar{x}_1 - \bar{x}_2) \pm t_{\alpha/2, n_1+n_2-2} s_p\sqrt{1/n_1 + 1/n_2}$$

EXAMPLE 8-7   In a small study of two methods of teaching a freshman university course, one method was used with 16 students chosen at random and their final marks produced a mean of 70.2 and a standard deviation of 8.4. The second method was used with 14 other students chosen at random and their marks produced a mean of 62.7 and a standard deviation of 7.7. It is assumed that these are independent samples from normal populations with equal variances.

A 95% confidence interval for $\mu_1 - \mu_2$, the difference of the mean marks produced by methods one and two, may be produced as follows. The pooled estimate of the standard deviation σ is found as

$$s_p = \sqrt{\frac{(n_1 - 1)s_1^2 + (n_2 - 1)s_2^2}{n_1 + n_2 - 2}} = \sqrt{\frac{(15 \times 8.4^2) + (13 \times 7.7^2)}{16 + 14 - 2}} = 8.08$$

The appropriate t-value from tables is $t_{\alpha/2, n_1 + n_2 - 2} = t_{0.025, 28} = 2.048$. The confidence interval is then found as $(\bar{x}_1 - \bar{x}_2) \pm t_{\alpha/2, n_1 + n_2 - 2} s_p \sqrt{1/n_1 + 1/n_2}$, which is $(70.2 - 62.7) \pm 2.048 \times 8.08\sqrt{1/16 + 1/14} = 7.5 \pm 6.1$ or 1.4 to 13.6.

Note that 0 is not in this interval; hence, we may reject $H_0 : \mu_1 - \mu_2 = 0$ and adopt the alternative $H_A : \mu_1 \neq \mu_2$ in a 5% test.

Furthermore, if we had calculated the test statistic $t = (70.2 - 62.7)/8.08$ $\sqrt{1/16 + 1/14} = 2.536$, we would have found the difference significant with $p < 0.02$. (Note that although t is between $t_{0.01}$ and $t_{0.005}$, we have not stated that $p < 0.01$ because we have a two-sided test and must consider the subscript on t to be $\alpha/2$.) ▲

Paired Comparisons — Samples Dependent Through Pairing

Finally, we consider a special case of *two normal populations* for which samples are not independent but are *dependent through pairing*, such as in "before and after" studies. In such cases, we study pairs of observations in which one member of each pair comes from the first population and the other comes from the second. Such an analysis may be referred to as a *paired comparison analysis*.

Hypothesis Tests

EXAMPLE 8-8   An experiment is performed to determine at the 2% level of significance whether a drug has an effect on maze scores (time to complete a maze run) when administered to rats. The experiment consists of running ten rats through the maze with and without the drug. The rats are randomly selected such that half run the maze first with the drug then without, while the others first run without the drug then with. To avoid residual drug effect or learning effect, there is a time lapse between runs for each rat. Suppose that such an experiment produces the following scores:

Rat:	1	2	3	4	5	6	7	8	9	10
Score with drug	39	41	46	50	43	30	37	43	40	51
Score without drug	30	37	40	48	36	33	35	34	42	45

The two scores obtained by each rat form a pair of observations that are not independent—they both reflect the abilities of that particular rat. In order to determine whether the drug has an effect, we test the hypothesis that it has no effect on the average, and in order to test this hypothesis, we consider the ten individual differences, *score with drug* minus *score without*, and we test the null hypothesis that the true mean of all such differences is 0. By taking a pair of observations for each rat (approximately randomized and adjusted for residual effect) we have a design that reduces potential experimental error. Each individual difference includes variability of performance of one rat. The difference *with* minus *without* is free of the variability from rat to rat that would be present in an experiment with independent samples.

Analyzing the single set of individual differences, we may reduce the test to a hypothesis test on a single mean, as developed in Section 7-3.

The $n = 10$ individual differences are 9, 4, 6, 2, 7, -3, 2, 9, -2, and 6, and produce mean 4.0 and standard deviation 4.22. Since n is small, the test statistic may be denoted as t and written as $t = (\bar{x} - \mu_0)/(s/\sqrt{n}) = (4 - 0)/(4.22/\sqrt{10}) = 3.00$. For the 2% two-sided test, the critical t-value is $t_{0.01, 9} = 2.821$. With $3.00 > 2.821$ the hypothesis may be rejected. We may conclude that the drug does have an effect on maze scores. ▲

Although such a test is just a repeat of procedures considered in Section 7-3 after the data have been obtained as individual differences, it is useful to present a general form which can be extended to other cases or to confidence intervals.

From n pairs of observations from two normal populations, the first value in the ith pair is denoted as x_{1i} and the second as x_{2i}. Their difference is $d_i = x_{1i} - x_{2i}$. The sample of n d's produces a mean $\bar{d}$ and standard deviation s_d.

The two-sided test of equality ($\mu_1 = \mu_2$) is a two-sided test of no difference on the average for both values in a pair. This test is interpreted as a test of the null hypothesis that $\mu_d = 0$, where μ_d is the mean of the population of all individual differences and is equal to $\mu_1 - \mu_2$. The test is summarized as:

$H_0 : \mu_d = 0$

$H_A : \mu_d \neq 0$

Level of significance: α

Sample size: n pairs   (i.e., $2n$ observations)

Test statistic: $t = \dfrac{\bar{d}}{s_d/\sqrt{n}}$

Decision rule: reject H_0 if $|t| > t_{\alpha/2, n-1}$

Modifications to the alternative and decision rule produce the one-sided tests with

$H_A : \mu_d > 0$ and rejection region $t > t_{\alpha, n-1}$

or $H_A : \mu_d < 0$ and rejection region $t < -t_{\alpha, n-1}$.

To test the hypothesis that differences
on the average equal a specific value
δ (i.e., that $\mu_d = \delta$), the procedure above
is used with the test statistic modified to $t = (\bar{d} - \delta)/(s_d/\sqrt{n})$.

EXAMPLE 8-9 If we wish to determine whether data provide sufficient
evidence at the 5% level of the claim that an adjustment to a machine setting
will improve mean output by more than 10 units, we might consider applying
the foregoing procedures in a before–after study on a few machines.

Suppose that we decide to compare before adjustment output and after
adjustment output for each of 15 machines. If we let d denote the difference
after minus *before*, the claim is that $\mu_d > 10$. Since the burden of proof is on
the claim, the claim becomes the alternative hypothesis and the test becomes a
test of $H_0 : \mu_d = 10$ against $H_A : \mu_d > 10$. With an assumption of normal
distributions, we have test statistic $t = (\bar{d} - 10)/(s_d/\sqrt{15})$ and a decision rule
to reject H_0 if $t > t_{0.05, 14} = 1.761$.

Suppose that, after obtaining the data, we find that the 15 differences
produce a mean of 13.3 and a standard deviation of 4.2. The test statistic is
then $t = (13.3 - 10)/(4.2/\sqrt{15}) = 3.04$, which exceeds $1.761 = t_{0.05, 14}$. We
may reject the null hypothesis. At the 5% level of significance, the data
substantiate the claim that the mean output is increased by more than 10.   ▲

Confidence Intervals

Again, after producing the individual differences, we are essentially working with a
single sample problem and the confidence intervals developed in Section 6-4 apply.
The only difference in form is that we are using the symbol d instead of x for the n
values ultimately analyzed. The confidence interval for μ_d is, thus,

$$\bar{d} \pm \frac{t_{\alpha/2, n-1} s_d}{\sqrt{n}}$$

EXAMPLE 8-10 For the machine data in Example 8-9, we may develop a
95% confidence interval for μ_d as $13.3 \pm 2.145 \times 4.2/\sqrt{15}$ or 11.0 to 15.6
(since $t_{0.025, 14} = 2.145$). We may be 95% confident that the adjustment will
produce an average increase per machine of between 11.0 and 15.6 units.   ▲

PRACTICE PROBLEMS

P8-1 For the range of pressures being considered, two pressure gauges produce
measurements that are normally distributed with mean supposedly equal to the
actual pressure and with standard deviations of $\sigma_1 = 1.5$ kilopascals (kPa) and
$\sigma_2 = 2.0$ kPa, respectively. It is suspected, however, that the two gauges do not

have equal "zeros"; that is, that they would produce different means from repeat measurements on the same pressure. Determine whether there is sufficient evidence to substantiate the suspicion at the 5% level if independent samples of $n_1 = 10$ and $n_2 = 15$ measurements of the same pressure produce means $\bar{x}_1 = 32.6$ and $\bar{x}_2 = 33.1$.

P8-2 In a study on attitude in achieving relief from pain, 40 subjects were told to rest quietly until they felt relief. A further 30 subjects were given a placebo and told that it was a pain killer. They, too, were told to rest quietly until they felt relief. The times until reported relief for the first group produced mean and standard deviation of 123 minutes and 35 minutes. The times for the second group produced a mean and standard deviation of 77 minutes and 28 minutes. Determine a 95% confidence interval for the difference in mean times until relief.

P8-3 The numbers of pollen per flower (in thousands) were determined from samples on two varieties of a flower. For the first variety, 10 counts produced a mean of 17.22 and a standard deviation of 3.74. For the second variety, 15 counts produced a mean of 14.68 and a standard deviation of 2.44. Assuming independent samples from normal populations, determine whether there is sufficient evidence at the 5% level that pollen counts for the first variety would exceed those for the second on the average by more than 0.50:
(a) assuming equal population variances
(b) without assuming equal population variances

P8-4 A sample of 12 students with apparent numerical skills deficiencies were given a diagnostic test and produced the following scores: 17, 15, 18, 13, 12, 16, 14, 13, 14, 16, 15, and 11. Following a remedial training program these students were given an equivalent test and their scores were (in the same order) 19, 22, 25, 18, 18, 19, 20, 20, 19, 20, 21, and 19. Is there sufficient evidence at the 5% level that the program increases mean score by more than 5? What distribution assumption is made in applying this test?

EXERCISES

8-1 Under usual conditions, particular assessments on individuals within any given population follow a normal distribution with a standard deviation of 15. Suppose that independent samples of ten individuals are taken from each of two populations and that their assessments produce means of 103.2 and 118.4.
(a) Determine, at the 5% level, whether the average assessments of the two populations differ.
(b) Determine a 95% confidence interval for the difference between the mean assessments of the two populations.

8-2 It is claimed that a particular diet modification will increase weight gain in cattle by more than 100 g/day. A control group of 50 cattle on the original diet produced a mean gain of 1.056 kg/day with a standard deviation of 103 g. An experimental group of 35 cattle on the modified diet produced a mean gain of 1.197 kg/day with a standard deviation of 112 g. Do these data substantiate the claim at the 5% level?

8-3 Determine an appropriate one-sided 95% confidence interval to estimate at least how great the average weight gain is in Exercise 8-2.

8-4 In a comparison of mean ice thickness in two regions, a sample of measurements at 16 points in the first region produced a mean of 23.04 and a standard deviation of 3.13, and a sample of measurements at 12 points in the second region produced a mean of 18.21 and a standard deviation of 2.83.

(a) Assuming independent samples from normal distributions with unequal variances, calculate an interval of values such that you can be 95% confident that the difference between the "true" mean thickness of each of the two regions will be included in the interval.

(b) Using this interval, determine, at the 5% level, whether mean ice thickness is the same in the two regions.

8-5 Repeat Exercise 8-4 with the assumption of equal variances.

8-6 Suppose that 16 rats have learned two mazes in order I, then II. The number of trials to criterion for each rat in each maze is as follows:

Rat:	1	2	3	4	5	6	7	8
I	30	21	29	28	24	28	26	28
II	28	20	19	22	16	32	27	23
Rat:	9	10	11	12	13	14	15	16
I	33	31	36	29	27	27	24	35
II	29	32	30	25	24	26	18	38

Determine whether learning maze I helped in learning maze II. If there is a significant difference, indicate the p-level.

8-7 An employer claims that it is justifiable to pay male employees more than female employees on the grounds that they are more productive. Can the employer's claim be substantiated at the 5% level if the hourly outputs of a sample of 50 male employees produced a mean of 350 units and a standard deviation of 18 units and the hourly output of 30 female employees produced a mean of 345 and a standard deviation of 21?

8-8 In a study on the effects of noradrenalin, eight rats injected with noradrenalin had measured blood flows with a mean of 88.9 ml/min and a standard deviation of 9.1 ml/min. For 12 rats without the injection, the mean and standard deviation were 118.1 and 10.8. Find a 98% confidence interval for the difference in mean blood flows.

8-9 A new measurement process is rejected on the basis of a claim that it produces values that are "on the average" more than 5 below those produced on a current, more difficult, but reliable process. Each of ten units of test material is measured with the current process and the new process, with the following results:

Test unit:	1	2	3	4	5	6	7	8	9	10
Current process	79	74	95	85	81	99	87	71	75	86
New process	73	70	86	78	75	91	82	66	67	78

(a) Using a 1% level of significance, determine whether the observed differences substantiate the claim.

(b) Find a 98% confidence interval for the mean difference between the two processes.

P8-1 We test $H_0 : \mu_1 = \mu_2$ against $H_A : \mu_1 \neq \mu_2$ with test statistic $z = (\bar{x}_1 - \bar{x}_2)/\sqrt{\sigma_1^2/n_1 + \sigma_2^2/n_2} = (32.6 - 33.1)/\sqrt{1.5^2/10 + 2.0^2/15} = -0.35$. Since $|z| = 0.35 < 1.96 = z_{0.025}$, there is insufficient evidence that the gauges have different zeros.

P8-2 We use the large-sample confidence interval for the case of unknown population variances to produce $(\bar{x}_1 - \bar{x}_2) \pm z_{\alpha/2}\sqrt{s_1^2/n_1 + s_1^2/n_2} = (123 - 77) \pm 1.96\sqrt{35^2/40 + 28^2/30} = 46 \pm 15$ or 31 to 61.

P8-3 (a) We test $H_0 : \mu_1 - \mu_2 = 0.5$ against $H_A : \mu_1 - \mu_2 > 0.5$ with test statistic

$$t = \frac{(\bar{x}_1 - \bar{x}_2 - 0.50)}{s_p\sqrt{1/n_1 + 1/n_2}}$$

where

$$s_p = \sqrt{\frac{[(n_1 - 1)s_1^2 + (n_2 - 1)s_2^2]}{(n_1 + n_2 - 2)}} = \sqrt{\frac{[(9 \times 3.74^2) + (14 \times 2.44^2)]}{(10 + 15 - 2)}}$$

$$= 3.016$$

We thus have

$$t = \frac{(17.22 - 14.68 - 0.50)}{3.016\sqrt{1/10 + 1/15}} = 1.657$$

Since t does not exceed $t_{\alpha, n_1+n_2-2} = t_{0.05, 23} = 1.714$, we may not reject the null hypothesis. At the 5% level, there is not sufficient evidence that the mean pollen count for the first variety exceeds that of the second by more than 0.5.

(b) Without the assumption of equal variance, the critical t value is $t_{\alpha, m}$, where m, the value for the degrees of freedom, is the closest integer to

$$\frac{[s_1^2/n_1 + s_2^2/n_2]^2}{\dfrac{(s_1^2/n_1)^2}{n_1 - 1} + \dfrac{(s_2^2/n_2)^2}{n_2 - 1}} = \frac{[3.74^2/10 + 2.44^2/15]^2}{\dfrac{(3.74^2/10)^2}{9} + \dfrac{(2.44^2/15)^2}{14}} = 14.1$$

The critical t-value is $t_{0.05, 14} = 1.761$. The test statistic is $t = (\bar{x}_1 - \bar{x}_2 - 0.50)/\sqrt{s_1^2/n_1 + s_2^2/n_2} = 1.522$. Since 1.522 does not exceed 1.761, we may not reject the null hypothesis. There is insufficient evidence that the mean pollen count for the first variety exceeds that of the second by more than 0.5.

P8-4 We have a paired comparisons analysis since we are comparing pairs of scores achieved by individual students. Using the $n = 12$ individual differences, *second minus first*, 2, 7, 7, 5, 6, 3, 6, 7, 5, 4, 6, and 8, we obtain $\bar{d} = 5.5$ and $s_d = 1.78$. In the test of the null hypothesis that $\mu = 5.0$ against the alternative that $\mu > 5.0$, we obtain the test statistic $t = (\bar{d} - 5.0)/(s_d/\sqrt{n}) = (5.5 - 5.0)/(1.78/\sqrt{12}) = 0.971$, which does not exceed $1.833 = t_{0.05, 9} = t_{\alpha, n-1}$. At the 5% level there is insufficient evidence that the mean increase exceeds 5. In applying this test, we have assumed that, even though discrete, individual scores follow a distribution that is well approximated by a normal distribution.

8-3 COMPARING VARIANCES

Variances are compared in a ratio rather than a difference, with equality producing a ratio of 1 rather than a difference of 0. In Chapter 7 we used the fact that the sample variance s^2 has a sampling distribution related to the χ^2- distribution if the sample is drawn at random from a normal population. The ratio of two independent χ^2 variables divided by their respective degrees of freedom has a positively skewed sampling distribution referred to as an F-distribution. This result leads us to the use of the ratio of variances.

Testing the Hypothesis of Equality

If s_1^2 and s_2^2 are the variances of two independent random samples of sizes n_1 and n_2, respectively, from *two normal populations* with variances σ_1^2 and σ_2^2, respectively, then $(s_1^2/\sigma_1^2)/(s_2^2/\sigma_2^2)$ will have a sampling distribution that is the F-distribution with $n_1 - 1$ degrees of freedom for the numerator and $n_2 - 1$ degrees of freedom for the denominator.

Percentiles of the F-distribution are tabulated in Table VIIa (95th percentile) and Table VIIb (99th percentile) of Appendix B. Figure 8-1 illustrates Table VII.

The $100(1 - \alpha)$th percentile of the F-distribution with f_1 and f_2 degrees is denoted as $F_{\alpha;\, f_1,\, f_2}$ and is found in the appropriate part of Table VII in the column labeled f_1 (degrees of freedom for the numerator) and the row f_2 (degrees of freedom for the denominator).

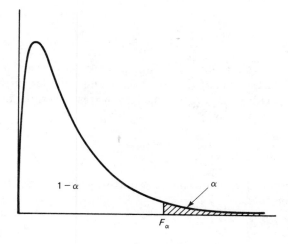

Figure 8-1 Illustration of Table VII.

Table VII provides only upper percentiles and we must find lower percentiles from the following relationship:

$$F_{1-\alpha; f_1, f_2} = \frac{1}{F_{\alpha; f_2, f_1}}$$

EXAMPLE 8-11

$$F_{0.95; 3, 12} = \frac{1}{F_{0.05: 12, 3}} = \frac{1}{8.74} = 0.114 \qquad \blacktriangle$$

If $\sigma_1^2 = \sigma_2^2$, they will cancel out in the ratio above, with the following result.

> If $\sigma_1^2 = \sigma_2^2$, the sample variance ratio s_1^2/s_2^2 has the F-distribution with $n_1 - 1$ and $n_2 - 1$ degrees of freedom.

As a result, we can develop a test for $H_0: \sigma_1^2 = \sigma_2^2$ (i.e., $\sigma_1 = \sigma_2$) on the basis of s_1^2/s_2^2 as a test statistic. In a two-sided test we reject the null hypothesis if the variance ratio is unusually smaller than 1 (i.e., if $s_1^2/s_2^2 < F_{1-\alpha/2; n_1-2, n_2-1}$) or if the ratio is unusually larger than 1 (i.e., if $s_1^2/s_2^2 > F_{\alpha/2; n_1-1, n_2-1}$).

If we wish to use only upper percentiles, we modify the test procedure to have the test statistic as s_1^2/s_2^2 or s_2^2/s_1^2, whichever is larger, and the rejection region as $s_1^2/s_2^2 > F_{\alpha/2; n_1-1, n_2-1}$ or $s_2^2/s_1^2 > F_{\alpha/2; n_2-1, n_1-1}$, accordingly.

We may thus summarize the two-sided test as follows:

$$H_0: \sigma_1^2 = \sigma_2^2$$

$$H_A: \sigma_1^2 \neq \sigma_2^2$$

Level of significance: α

Sample sizes: n_1, n_2

Test statistic: s_1^2/s_2^2 or s_2^2/s_1^2, whichever is larger

Decision rule: reject H_0 if $s_1^2/s_2^2 > F_{\alpha/2; n_1-1, n_2-1}$ or if

$$s_2^2/s_1^2 > F_{\alpha/2; n_2-1, n_1-1}.$$

For a one-sided test with alternative $\sigma_1^2 > \sigma_1^2$, we will only reject the null hypothesis if s_1^2 is unusually larger than s_2^2 (i.e., if s_1^2/s_2^2 is unusually larger than 1). Making the test statistic automatically s_1^2/s_2^2, no matter which is larger, the test is as above with new H_A and decision rule as

$$H_A: \sigma_1^2 > \sigma_2^2$$

Decision rule: reject H_0 if $s_1^2/s_2^2 > F_{\alpha; n_1-1, n_2-1}.$

Similarly, for the alternative $\sigma_1^2 < \sigma_2^2$, the test statistic is automatically s_2^2/s_1^2 and the new H_A and decision rule are

$$H_A: \sigma_1^2 < \sigma_2^2$$

Decision rule: reject H_0 if $s_2^2/s_1^2 > F_{\alpha; n_2-1, n_1-1}.$

The null hypothesis and alternative hypothesis may be expressed in terms of the standard deviations instead of the variances and the test procedures will, of course, be unchanged.

EXAMPLE 8-12 Suppose it is claimed that the smoothing out of automobile design to reduce air resistance will not only decrease average fuel consumption, but will also eliminate some of the variability. Independent samples of fuel consumptions for various conditions are to be considered to determine whether the claim can be substantiated at the 5% level.

A sample of 16 fuel consumptions in liters per hundred kilometers with the ordinary design produced a standard deviation of 3.1. A sample of 12 consumptions with the modified design produced a standard deviation of 1.8. With the assumption of independent random samples from normal distributions, we test $H_0: \sigma_1 = \sigma_2$ against $H_A: \sigma_1 > \sigma_2$ with test statistic $s_1^2/s_2^2 = 3.1^2/1.8^2 = 2.97$, which exceeds $2.72 = F_{0.05;\, 15,\, 11}$. We may reject H_0 and conclude that the sample data do substantiate the claim. ▲

Confidence Intervals

With appropriate use of the general result on a ratio of variances, we produce a confidence interval for the ratio of the variances of two normal populations on the basis of independent random samples as follows:

A confidence interval for the ratio σ_1^2/σ_2^2 is (using only upper percentiles)

$$\frac{s_1^2/s_2^2}{F_{\alpha/2;\, n_1-1,\, n_2-1}} \quad \text{to} \quad \frac{s_1^2}{s_2^2} \times F_{\alpha/2;\, n_2-1,\, n_1-1}$$

By taking square roots, we produce a confidence interval for the ratio of the standard deviations as

$$\frac{s_1/s_2}{\sqrt{F_{\alpha/2;\, n_1-1,\, n_2-1}}} \quad \text{to} \quad \frac{s_1}{s_2} \times \sqrt{F_{\alpha/2;\, n_2-1,\, n_1-1}}$$

The ratio can be inverted to a ratio of the second variance to the first by simply interchanging the subscripts, and the confidence interval would change accordingly.

EXAMPLE 8-13 Suppose that we wish to compare the precision of the measurements taken in two different laboratories by comparing the variances of measurements taken on the same material in each lab. Suppose that 15 measurements taken in the first lab produced a variance of 2.384 and 25 measurements taken in the second lab produced a variance of 3.433. We assume that the samples are independent and that measurements are normally distributed.

From these measurements, a 90% confidence interval for the ratio of the variance of measurements taken in the first lab to that for the second lab is obtained as $(2.384/3.433)/F_{0.05; \, 14, 24}$ to $(2.384/3.433) \times F_{0.05; \, 24, 14}$. From Table VIIa, we find $F_{0.05; \, 24, 14} = 2.35$. There is no value given for $F_{0.05; \, 14, 24}$, however. The Table includes $F_{0.05; \, 12, 24} = 2.18$ and $F_{0.05; \, 15, 24} = 2.11$. We must interpolate between these two values. To obtain a simple approximate value we interpolate linearly. Since 14 is two-thirds of the way from 12 to 15, we will consider the F value for 14 degrees of freedom in the numerator to be about two-thirds of the way from the F for 12 degrees of freedom to the F for 15 degrees of freedom. The value about two-thirds of the way from 2.18 to 2.11 is 2.13. We thus take 2.13 as the approximate value for $F_{0.05; \, 14, 24}$. The confidence interval then becomes $(2.384/3.433)/2.13$ to $(2.384/3.433) \times 2.35$ or 0.33 to 1.63. Note that with this result we would be unable to reject the null hypothesis $\sigma_1^2 = \sigma_2^2$ in a 10% two-sided test since the interval includes 1 as a possible value for the ratio σ_1^2/σ_2^2.   ▲

Testing Other Hypotheses

We might also be interested in testing a relationship other than equality for σ_1^2 and σ_2^2. We may, for example, wish to determine whether σ_1^2 is twice as large as σ_2^2, that is, whether $\sigma_1^2 = 2\sigma_2^2$ or $\sigma_1^2/\sigma_2^2 = 2$. One possibility is, of course, to use the confidence interval. If the (two-sided) confidence interval for σ_1^2/σ_2^2 includes 2, we may accept the hypothesis that $\sigma_1^2 = 2\sigma_2^2$; otherwise, we may reject the hypothesis in a two-sided test.

We will consider a one-sided test of $H_0 : \sigma_1^2 = \delta\sigma_2^2$ for some specific δ. By interchanging subscripts we may also use the procedure to test $H_0 : \sigma_2^2 = \delta\sigma_1^2$. We will only consider the alternative $\sigma_1^2 > \delta\sigma_2^2$, since $\sigma_1^2 < \delta\sigma_2^2$ can be considered as $\sigma_2^2 > (1/\delta)\sigma_1^2$, which is of the same form with interchanged subscripts.

We know that $(s_1^2/\sigma_1^2)/(s_2^2/\sigma_2^2)$ follows the F-distribution; therefore, if $\sigma_1^2 = \delta\sigma_2^2$, we may substitute $\delta\sigma_2^2$ for σ_1^2 in the ratio to obtain the result that $s_1^2/\delta s_2^2$ will follow the F-distribution. This adjusted ratio thus becomes an appropriate test statistic. If σ_1^2 exceeds $\delta\sigma_2^2$, we would expect s_1^2 to exceed δs_2^2; that is, we would expect $s_1^2/\delta s_2^2$ to exceed 1. The rejection region thus consists of unusually large values of $s_1^2/\delta s_2^2$. The test may be summarized as follows:

$$H_0 : \sigma_1^2 = \delta\sigma_2^2$$

$$H_A : \sigma_1^2 > \delta\sigma_2^2$$

Level of significance: α

Sample sizes: n_1, n_2

Test statistic: $s_1^2/\delta s_2^2$

Decision rule: reject H_0 if $s_1^2/\delta s_2^2 > F_{\alpha; \, n_1-1, \, n_2-1}$.

EXAMPLE 8-14 Suppose that, due to cost considerations, a variance reducing modification is not to be applied unless we can substantiate, at the 1% level, the claim that the modification will reduce the standard deviation by more than 25%. If σ_1 denotes the current standard deviation and σ_2 denotes the standard deviation with modification, the required condition is that σ_2 be smaller than σ_1 by more than 25% (i.e., that $\sigma_2 < 0.75\sigma_1$). In terms of the variances, the condition is that $\sigma_2^2 < (0.75\sigma_1)^2$, which we may rewrite as $(0.75)^2\sigma_1^2 > \sigma_2^2$ or $\sigma_1^2 > \sigma_2^2/(0.75)^2$ (i.e., $\sigma_1^2 > 1.78\sigma_2^2$).

We now have the required condition expressed in the same form as the alternative of our hypothesis-testing procedure. Since the burden of proof is on the alternative, we have the appropriate form. We consider the problem as a test of the null hypothesis $\sigma_1^2 = 1.78\sigma_2^2$ against the alternative hypothesis $\sigma_1^2 > 1.78\sigma_2^2$ with a 1% level of significance. Suppose that we plan to test the hypothesis by comparing 15 sample output values with the modified process to 25 sample values without. The test has test statistic $s_1^2/1.78s_2^2$ and decision rule to reject H_0 if $s_1^2/1.78s_2^2 > F_{0.01; 24, 14} = 3.43$.

Now let us suppose that the output values without modification produce a standard deviation of 3.62 and the values with modification produce a standard deviation of 1.64. The test statistic is $3.62^2/(1.78 \times 1.64^2) = 2.74$, which does not exceed 3.43. We may not reject H_0; therefore, we have not substantiated the claim that the modification reduces the standard deviation by more than 25% and we will not introduce the modification. ▲

PRACTICE PROBLEMS

P8-5 To determine whether the variances of two normal populations may be assumed equal in tests of hypotheses on the difference of means, available data from independent samples are to be investigated for possible variance differences. A sample of 16 observations from the first population produced a variance 12.83. An independent sample of 25 observations from the second produced a variance of 21.75. Test the hypothesis of equality at the 2% level.

P8-6 Find a 98% confidence interval for the ratio of the two population standard deviations in Practice Problem P8-5.

P8-7 A management consultant firm claims that the amount of overtime awarded to employees in a plant varies so much as to cause unrest. In fact, the firm claims that the standard deviation for the plant exceeds that for a second plant, using one of their systems, by more than 150%. Is the claim substantiated at the 5% level if the records for a sample of 21 employees in the first plant produced a standard deviation of 17.6, while a sample of 25 employees in the plant with the firm's system showed a standard deviation of 5.9?

EXERCISES

8-10 Suppose that it is claimed that instructor A is less consistent (i.e., more varied) than instructor B with regard to marks assigned on examinations. A random sample of 25 marks awarded by A produced a variance of 189.7. A random sample of 25 marks awarded by B produced a variance of 138.6. If marks can be assumed to follow a normal distribution and if the two instructors' marks are independent, determine whether the sample results substantiate the claim at the 5% level.

8-11 An archeologist has measured the depths of findings in two different areas. The depths for 16 findings in the first region produced a standard deviation of 6.49 and the depths for 21 findings in the second region produced a standard deviation of 4.61. Depths are assumed to be normally distributed.

(a) Find a 98% confidence interval for the ratio of the standard deviation of all (potential) depths of findings in the first region to the standard deviation in the second.

(b) Using the results of part (a), test the hypothesis of equal standard deviations in a 2% two-sided test.

(c) Using the results of part (a), test the hypothesis that the first region shows twice the standard deviation of the second in a 2% two-sided test.

8-12 A new packaging process is under investigation to determine whether it will produce packed weights which are more consistent (i.e., less variable) than those produced by the current process. As a trial, the two processes are run and samples are taken from the productions of each machine and weighed with the following results:

Old process:	16.07,	15.79,	15.83,	16.14,	16.22,	16.09,
	16.18					
New process:	16.10,	16.26,	16.10,	16.04,	16.17,	15.94,
	16.19,	16.12,	16.01,	15.84,	15.85	

Determine whether there is evidence at the 5% level that the new process is more consistent. Assume normality.

8-13 Two possible control procedures are being investigated to determine how well they control fluctuations in water temperature. A sample of 24 readings with the first procedure produced temperatures with a variance of 1.15. A sample of 24 readings with the second procedure produced a variance of 7.93. Assuming normal distributions, determine whether these results provide proof, at the 5% level, that the second procedure exhibits a variance of more than three times that for the first.

8-14 Is there reason not to assume equal variances at the 2% level in a test of equal mean percent sugar reductions for two preparations of insulin administered to rabbits if 10 rabbits receiving the first preparation showed reductions with a standard deviation of 8.3 while 12 rabbits receiving the second showed a standard deviation of 10.1?

SOLUTIONS TO PRACTICE PROBLEMS

P8-5 In the two-sided test of $H_0 : \sigma_1^2 = \sigma_2^2$ against $H_A : \sigma_1^2 \neq \sigma_2^2$, we note from the data that s_2^2 exceeds s_1^2. We thus use the test statistic $s_2^2/s_1^2 = 21.75/12.83 = 1.695$, which we compare to $F_{\alpha/2; \, n_2 - 1, \, n_1 - 1} = F_{0.01; \, 24, \, 15} = 3.29$. Since 1.695 does not exceed 3.29, we may accept the null hypothesis of equal variances.

P8-6 The 98% confidence interval for σ_1/σ_2 is found as $(s_1/s_2)/\sqrt{F_{0.01;\,15,\,24}}$ to $(s_1/s_2) \times \sqrt{F_{0.01;\,24,\,15}}$, which is $(12.83/21.75)/\sqrt{2.89}$ to $(12.83/21.75) \times \sqrt{3.29}$ or 0.45 to 1.39.

P8-7 To investigate the claim, we test the null hypothesis that $\sigma_1 = 2.5\sigma_2$ (i.e., that $\sigma_1^2 = 6.25\sigma_2^2$) against the alternative that $\sigma_1 > 2.5\sigma_2$ (i.e., $\sigma_1^2 > 6.25\sigma_2^2$). (Note that if σ_1 exceeds σ_2 by more than 150%, the magnitude of σ_1 is more than 250% of that for σ_2, hence the multiplier 2.5.) The test statistic is $s_1^2/6.25s_2^2 = 17.6^2/(6.25 \times 5.9^2) = 1.42$, which does not exceed $2.03 = F_{0.05;\,20,\,24}$. The claim is not substantiated.

8-4   CHAPTER REVIEW

8-15 If we wish to test $H_0: \sigma_1^2 = \sigma_2^2$ against $H_A: \sigma_1^2 > \sigma_2^2$ and we obtain $s_1^2 = 3.9$ and $s_2^2 = 1.3$, the numerical value of the test statistic is _____.

8-16 Twenty different cars are driven over a given route with each of two different fuels to compare the average mileage from the two fuels. An appropriate test on equality of the fuels would be
(a) a paired comparison t-test with 19 degrees of freedom
(b) a z-test on independent populations with equal variances
(c) a t-test on independent populations with equal variances
(d) a t-test on independent populations with unequal variances

8-17 If a pair of samples produces a 98% confidence interval for the ratio of two variances as 0.57 to 3.68, then, in a two-sided 2% test, the hypothesis that the standard deviations are equal may be rejected.
(a) True
(b) False

8-18 The difference of two independent random variables from two normal populations has a distribution with a variance equal to
(a) the sum of the population variances
(b) the difference of the population variances

8-19 For a test of the hypothesis that two means are equal based on the assumption of two unknown but equal variances, the degrees of freedom for the appropriate test statistic will be equal to _____.

8-20 Suppose that a number of subjects were given the following test on recall. Each subject was presented with a tray of objects to study for a fixed length of time. The tray was then removed and, in a fixed length of time, the subject had to recall as many of the objects as possible. The presentation was made twice. For one presentation, the objects were all the same color; for the other, each object was a different color. The same objects were offered for each presentation.

Assuming normal distributions, test the appropriate hypothesis at the 1% level to determine whether color differentiation helps in recall from the following data, in which values are the numbers of objects recalled in each case.

Subject:	1	2	3	4	5	6	7	8
One color	27	26	33	22	21	20	34	34
Different color	18	32	41	24	21	33	38	32

Subject:	9	10	11	12	13	14	15	16	17
One color	16	22	30	40	10	22	18	21	26
Different color	21	18	28	36	16	25	32	16	38

8-21 Suppose that sample groups of students from two regions were given the same test with the following results. The marks in group one were 44, 50, 56, 57, 61, 66, 68, 72, 72, 74, 76, 76, 78, 81, and 85. The marks in group two were 52, 60, 64, 65, 70, 71, 73, 75, 75, 77, 79, 79, 79, 80, 82, 83, 84, 87, 88, and 94. Suppose that the marks can be assumed to follow a normal distribution.

(a) Find a 95% confidence interval for the difference between the mean marks that would be achieved by all of the students in the two regions.

(b) Using the confidence interval in part (a), test at the 5% level the hypothesis that the overall means are equal against the alternative that they are different.

8-22 Two methods of iron preparation are to be compared in terms of the variability of the melting point of the iron produced. Method A is somewhat cheaper and will be used unless method B can be proved to be more consistent (i.e., to have less variability). Determine from the following data which method should be used. A sample of nine preparations from method A produced the following melting points: 1493, 1515, 1518, 1510, 1511, 1489, 1509, 1507, and 1496. A sample of six preparations from method B produced the following melting points: 1511, 1493, 1513, 1484, 1507, and 1489.

8-23 Suppose that in a study on the effects of alcohol, an experimental group of ten drivers attempted a test driving course after consuming a quantity of alcohol. The drivers were assigned numbers of demerit points producing a mean of 17.3 and a standard deviation of 4.1. A control group of 15 drivers drove the course without alcohol and were assigned demerit points with a mean of 11.2 and a standard deviation of 2.3. Using an appropriate hypothesis test, determine at the 5% level whether demerit points obtained by drinking drivers are more varied than those obtained by nondrinking drivers. What distribution assumption is made?

8-24 On the basis of the data in Exercise 8-23, determine a 95% confidence interval for the difference between mean numbers of demerit points for drinking and nondrinking drivers.

8-25 In a study on the relative sizes of the left and right hemispheres of the brain, the numbers of cells (in thousands) for 13 albino mice produced the following data.

Mouse:	1	2	3	4	5	6	7
Right	2.73	2.02	2.36	2.29	2.38	2.91	2.55
Left	2.17	2.27	1.18	1.95	1.76	2.55	2.09

Mouse:	8	9	10	11	12	13
Right	2.69	2.37	2.28	1.87	2.64	2.36
Left	2.33	2.96	2.65	2.05	2.19	2.09

Determine at the 5% level whether the left and right hemispheres differ with regard to mean number of cells.

8-26 Determinations on stone sizes in two river beds produced the following results. Sizes from 25 points in the first river bed produced a variance of 36.8. Sizes from

25 points in the second produced a variance of 21.2 Determine a 90% confidence interval for the ratio of the two population variances.

8-27 Independent samples of size 8 are to be taken from each of two normal populations with known variance 100. The sample results are to be used to test the hypothesis of equal population means against the alternative that the first exceeds the second. The test is to have a 5% level of significance. What is the probability of a type II error if the first population mean exceeds the second by 12.5?

8-28 Determine the general expressions for one-sided confidence intervals for the mean difference of paired observations.

8-29 Determine the general expression for the one-sided confidence interval indicating at least how large a ratio of standard deviations is.

8-5 GENERAL REVIEW

8-30 For continuous data, $P(x \leqslant 2.6) - P(x < 2.6)$ will equal _____.

8-31 If d and s are statistics and δ is a parameter such that $(d - \delta)/s$ has a t-distribution with f degrees of freedom, the expression for a $1 - \alpha$ confidence interval for δ is _____ ± _____.

8-32 The variance is
(a) a statistic
(b) a measure of location
(c) a parameter
(d) may be either (a) or (c)

8-33 If s_1^2 and s_2^2 are to be found from independent samples of sizes 25 and 20 from two normal populations with the same variance, then $P(s_1^2/s_2^2 > F_{0.05;\, 24,\, 19}) = $ _____.

8-34 If $P(A) = 1/4$, $P(B) = 1/5$, and $P(A|B) = 1/9$, then $P(B|A) = $ _____.

8-35 If a number of sample means vary from 56 to 77, the value 50 for $s_{\bar{x}}^2$ would appear to be
(a) too high
(b) reasonable
(c) too low

8-36 If the null hypothesis is true in a hypothesis test with level of significance 0.05, the probability of not committing a type I error is
(a) 0.05
(b) 0.95
(c) dependent on the alternative hypothesis

8-37 If a set of data with standard deviation equal to 7.5 produces a coefficient of variation equal to 0.75, the mean is equal to _____.

8-38 If there are 5005 ways of choosing 6 objects from a set of 15, the number of ways of choosing 9 objects from a set of 15 is
(a) $5005 \times (9!/6!)$
(b) 5005
(c) $5005 \times (6!/9!)$

8-39 If the odds in a game are 5 to 3 in favor of your opponent and if you pay your opponent $1.50 when you lose, then for the game to be fair, when you win your opponent should pay you _____.

8-40 If A and B are exhaustive events such that the odds against A are 1 to 2 and the odds against B are 2 to 3, find the odds against obtaining both events together.

8-41 In a study on the durability of a particular electronic component, a number of the components were tested in seven different machines. In the first, 20 components produced a mean lifetime of 1630.4 minutes. In the second, 15 produced a mean of 1632.2. In the third, fourth, fifth, sixth, and seventh, 5, 10, 5, 10, and 10 components, respectively, produced means of 1625.0, 1624.1, 1631.6, 1630.7, and 1628.0. Considering all the machines, what was the mean lifetime of the components?

8-42 In the investigation of pattern effects on recall, subjects were asked to recall the order of a set of letters intermixed in a list with digits. One group of 15 subjects was presented with the letters and digits arranged so that most of the letters were clustered near one end of the list. A second group of 15 subjects was presented with a list in which the letters were more spread among the digits. The percentages of correct recalls for the first group showed a mean of 70.0 and standard deviation of 8.9. The mean and standard deviation for the second group were 78.2 and 7.3.
(a) Determine whether variability of correct responses differs for the two techniques.
(b) Choosing an appropriate procedure on the basis of the conclusion in part (a), determine whether the means differ significantly. Indicate the p-level if the difference is significant.

8-43 To estimate the percentage of consumers in a given region that use a particular product, a researcher intends to use the corresponding percentage of consumers in a sample.
(a) With no prior knowledge on the population percentage, how large a sample must be taken to have a 90% chance that the estimate will be within 6 percentage points of the true value?
(b) If it can be assumed that at most 30% of all consumers use the product, how large a sample must be taken?

8-44 Repeat Exercise 8-43 with the added condition that the population in the region consists of only 4000 consumers.

8-45 A card is to be drawn at random from an ordinary deck of 52 playing cards and the suit is to be noted (spade, heart, diamond, or club). A balanced die is then to be rolled and the face noted (1, 2, 3, 4, 5, 6).
(a) Find the probability of obtaining
(1) a spade and an even number
(2) a heart or a number less than 3
(3) neither a club nor a 6
(b) Find the odds against obtaining
(1) a 5 and a red card
(2) a black card or an odd number

8-46 Suppose that a number of hospital records produced the following distribution of lengths of stay in hospital to the nearest half-day for patients staying at least 1 day.

Length of stay	Number of patients
1–2	39
2.5–4	61
4.5–6	72
6.5–8	56
8.5–10	50
10.5–15	63
15.5—20	18
20.5–30	5
over 30	3

(a) Determine the median length of stay.

(b) Convert the data to an "or more" frequency distribution and illustrate this distribution with an ogive.

8-47 Suppose that 25 carafes have been filled with wine such that 15 of the carafes contain an imported wine and the other 10 contain an identical-looking domestic wine. The carafes have become mixed up so that visibly it is impossible to determine the type of wine and ten of the carafes are to be selected at random for immediate use.

(a) How many ways can the ten carafes be selected?

(b) What is the probability that all of the ten selected will contain domestic wine?

(c) What is the probability that of the ten selected, six will contain imported wine and four will contain domestic wine?

8-48 The active ingredient in insect repellent is referred to as DEET. One brand is advertised as being 95% DEET. Test the hypothesis that the repellent is as claimed if sample readings on the percentages in 100 bottles produced a mean of 94.9 and a standard deviation of 0.7.

8-49 A stock of 20 couplings includes 2 that are defective and more likely than others to breakdown. The probability of breakdown with a defective coupling is 6/25, but the probability of breakdown with a nondefective is only 1/20.

(a) If two couplings are chosen at random (without replacement) what is the probability that
  (1) both will be defective?
  (2) only one will be defective?
  (3) neither will be defective?

(b) If two couplings are put into service, what is the probability that neither will fail
  (1) if both are defective?
  (2) if only one is defective?
  (3) if neither is defective?
  (Assume that couplings are independent of each other)

(c) If two couplings are chosen at random and put into service and neither fails, what is the probability that
  (1) both were defective?
  (2) neither was defective?

8-50 In a study on whether the application of heat would aid in the reduction of diseased cells through chemical treatment, a number of diseased tissues were separated into two halves and treated such that half received the treatment with heat and the other without. The following values represent the amount of reduction in numbers of diseased cells.

Sample tissue:	1	2	3	4	5	6	7	8
With heat	9.0	6.3	8.7	8.4	7.2	8.1	7.8	8.3
Without heat	8.4	6.0	5.7	6.6	7.3	7.5	6.9	8.5

Sample tissue:	9	10	11	12	13	14	15
With heat	9.9	9.3	6.9	7.4	7.0	9.2	8.5
Without heat	9.8	8.8	6.1	7.6	6.8	9.0	8.1

Determine whether there is sufficient evidence at the 1% level that heat does aid reduction.

8-51 A traffic analyst has determined that, in heavy traffic periods, traffic counts may be assumed to follow a normal distribution. Two main arteries into an urban area are being considered for improvement and it is of interest to determine whether their traffic rates differ enough to merit improving one artery ahead of the other. If a difference in traffic rates cannot be substantiated, a choice will have to be made on other grounds. Traffic rates from 16 sample counting times for the first artery produced a mean of 92.1 and a standard deviation of 22.84. Rates from 12 counting times for the second artery produced a mean of 73.8 and standard deviation of 17.57.

(a) Find a 95% confidence interval for the difference between mean traffic rates for the two areas.

(b) In order for a meaningful difference to be declared and a decision made on the basis of traffic counts alone, a mean difference of at least 4.0 must be substantiated at the 5% level. Using the results of part (a), determine whether there is a meaningful difference. Should a decision be made on the basis of traffic counts alone?

(c) What distribution assumptions were made in parts (a) and (b)?

8-52 Independent samples of size n are to be taken from each of two normal populations with standard deviation σ. The samples are to be used to test the hypothesis of equal population means against the alternative that the first exceeds the second. Show that in order to have a level of significance α and to have at most a probability of β of a type II error when the first mean exceeds the second by at least $k\sigma$ for some positive k, the sample sizes should each be at least $2 \times [(z_\alpha + z_\beta)/k]^2$.

8-53 The amounts of caffeine (in milligrams) in a sample of 750-ml bottles of a cola-flavored soft drink were as follows:

71.73	71.23	71.42	71.51	71.12	71.43	71.32	71.54
71.47	71.66	71.84	71.54	71.40	71.22	71.65	71.98
71.07	71.75	71.89	71.59	71.56	71.52	71.20	71.67
71.33							

(a) Subtract 71 from each value in the data and multiply these values by 100. Find the mean and standard deviation of the resulting values.

(b) Using the results of part (a), determine the mean and standard deviation for the original caffeine values.

(c) Calculate a 95% confidence interval for the mean amount of caffeine in 750-ml bottles of this soft drink.

8-54 According to the category of service contract sold, a salesperson receives a commission of $50, $25, $10, or $5. If no contract is sold, the commission is, of course, $0. Past records indicate the following percentage breakdown of contracts sold for the total number of potential clients contacted.

Contract sold	Commission	Percentage of cases
Type A	$50	2
Type B	25	7
Type C	10	21
Type D	5	13
No sale	0	57

(a) On the basis of past experience, what is the expected commission on the next sale attempted?

(b) If each sale attempted involves a cost of $0.50 that the salesperson must pay, what is the expected net commission on the next sale attempted?

(c) On the next ten sales attempted, what is
  (1) the maximum possible commission?
  (2) the maximum possible cost?
  (3) the expected net commission?

8-55 An instrument for assessing the weight of protein available in a fixed quantity of material produces measurements that follow a normal distribution with mean μ equal to the actual mean protein weight and standard deviation $\sigma = 0.050$ mg.

(a) If $\mu = 1.480$ mg and if four independent measurements on the same material are taken and produce a sample mean $\bar{x}$, what is $P(|\bar{x} - 1.480| \leqslant 0.030)$?

(b) If μ is unknown and if nine independent measurements are taken on the same material, what is $P(|\bar{x} - \mu| \leqslant 0.020)$?

8-56 In Exercise 8-55, the mean $\bar{x}$ of a sample of n independent measurements taken on the same material is to be used to estimate μ. It is a requirement of the estimation process that there be a 95% chance that the estimate will deviate from μ by at most 0.025. How large must n be?

8-57 For the estimation problem in Exercise 8-56, only $n = 10$ measurements are available and they produce $\bar{x} = 1.514$ mg. Determine a 95% confidence interval for μ.

8-58 (a) Suppose in Exercise 8-55 that σ is unknown but it is required that the process be sufficiently precise that σ is less than 0.075. Is there sufficient evidence at the 5% level that the process is sufficiently precise if 25 independent measurements on the same material produced a sample standard deviation of 0.046?

(b) Determine a 90% confidence interval for σ on the basis of the results of the sample in part (a).

8-59 Past experience indicates that a welding process occasionally produces faulty welds such that the probability that any individual weld will be faulty is 0.0001.

(a) In a random selection of 35,000 welds, what is the expected number of faulty welds?

(b) In a random selection of 35,000 welds, what is the probability that the number of faulty welds will be

(1) 0?

(2) 3?

(3) at least 4?

8-60 It is claimed that a carburetor modification will improve fuel efficiency in more than 60% of engines. The claim is to be tested on the basis of a sample of 15 engines. If there is improvement in only 11 or fewer sample cases, the null hypothesis that there is improvement in at most 60% of all cases will be accepted. If there is improvement in 12 or more sample cases, the null hypothesis will be rejected and the claim will be considered to be substantiated.

(a) What is the level of significance in this hypothesis test?

(b) What is the probability of making a type II error if the percentage of all cases that would show improvement is

(1) 70%?

(2) 80%?

(3) 90%?

8-61 (a) For the investigation in Exercise 8-60, set up a decision rule for a test of the hypothesis as given if the sample size is to be 150 and if the level of significance is to be $\alpha = 0.05$.

(b) For the test in part (a), find β, the probability of a type II error, if the true percentage is 65, 70, 75, or 80%. Draw the operating characteristic curve.

COMPARISONS BASED ON PROPORTIONS

9-1 INTRODUCTION

In this chapter we develop a number of procedures for comparisons based on proportions. We consider techniques for comparing proportions of different populations. It may be of interest, for example, to compare the proportion of residents of Newfoundland opposed to the hunting of seals with the corresponding proportion(s) in some other region(s).

Through an analysis of proportional allocations to various categories, we will develop procedures for testing independence as opposed to association of two attributes. We may, for example, test for an association between masculine self-expectation and involvement in juvenile delinquency.

As well, we analyze the proportional allocation of data to numerical classes to determine whether a particular distribution provides a reasonable fit to the data. In particular, many of our procedures in previous chapters are based on the assumption of a normal distribution, and we may wish to test the validity of that assumption.

9-2 COMPARING TWO PROPORTIONS

For the comparison of two proportions, we consider cases for which the binomial model applies and for which the sample sizes are sufficiently large to use the normal approximation to the binomial. The comparison of several proportions is considered as a special case of the procedures in the next section.

Independent Samples

We use the normal approximation to the binomial as developed in Section 5-5 and the results for combining two independent normal variables as noted in Section 8-2 to develop the appropriate procedures for comparing two proportions. For independent (large) random samples of sizes n_1 and n_2 from two populations with proportions of "successes" p_1 and p_2, respectively, the difference of the two sample success proportions $x_1/n_1 - x_2/n_2$ has an approximate normal distribution with mean $p_1 - p_2$ and variance $p_1(1 - p_1)/n_1 + p_2(1 - p_2)/n_2$, thus producing the following result:

$$z = \frac{(x_1/n_1 - x_2/n_2) - (p_1 - p_2)}{\sqrt{p_1(1 - p_1)/n_1 + p_2(1 - p_2)/n_2}}$$

has an approximate standard normal distribution.

This result forms the basis for our comparisons.

Testing Equality of p_1 and p_2

If each of two populations has a proportion of "successes" equal to a common value p, then, for sufficiently large independent samples, the two sample success proportions x_1/n_1 and x_2/n_2 will have approximate normal distributions both with mean p and with variances $p(1 - p)/n_1$ and $p(1 - p)/n_2$, respectively. Replacing p_1 and p_2 with the common value p in z above, we note that, in such a case, $(x_1/n_1 - x_2/n_2)/\sqrt{p(1 - p)(1/n_1 + 1/n_2)}$ will have a standard normal distribution. This quantity would thus be an appropriate test statistic to test $H_0 : p_1 = p_2$ except that the common value p is generally not specified.

To form a test statistic, an estimated value is substituted for p. This estimate is obtained by pooling the information in both samples as if there were only one sample to produce the estimate as $\hat{p} = (x_1 + x_2)/(n_1 + n_2)$.

Denoting the "true" success proportions in the populations as p_1 and p_2, we have the two-sided test procedure for testing the equality of two proportions on the basis of two independent samples.

$H_0 : p_1 = p_2$

$H_A : p_1 \neq p_2$

Test statistic: $z = \dfrac{x_1/n_1 - x_2/n_2}{\sqrt{\hat{p}(1 - \hat{p})(1/n_1 + 1/n_2)}}$

where $\hat{p} = (x_1 + x_2)/(n_1 + n_2)$

Decision rule: reject H_0 if $|z| > z_{\alpha/2}$

The one-sided test procedures are straightforward modifications producing a new alternative and decision rule:

$$H_A : p_1 > p_2$$

Decision rule: reject H_0 if $z > z_\alpha$

or, alternatively,

$$H_A : p_1 < p_2$$

Decision rule: reject H if $z < -z_\alpha$

EXAMPLE 9-1 A product can be made with natural rubber or a synthetic substitute and it is of interest to know whether the proportion of times that the product deteriorates under friction is different for the two materials. A two-sided test on the equality of two proportions is appropriate in this case. Suppose that a 5% level of significance is chosen so that $z_{\alpha/2} = z_{0.025} = 1.96$.

If, in a sample of $n_1 = 40$ occasions using rubber, deterioration occurred $x_1 = 27$ times, and in a sample of $n_2 = 15$ occasions using the synthetic material, deterioration occurred $x_2 = 8$ times, the pooled estimate $\hat{p}$ is $(27 + 8)/(40 + 15) = 35/55$ and the test statistic is $z = (27/40 - 8/15)/\sqrt{(35/55) \times (20/55) \times (1/40 + 1/15)} = 0.97$ which is less than 1.96.

The results of the sample do not provide sufficient evidence to prove that the two types of material are different and we may accept the null hypothesis that they are equivalent with regard to the proportion of times that deterioration occurs under friction (or we may reserve judgment). ▲

Confidence Intervals

Using the normal approximation again, we can modify the procedures described above to produce a *confidence interval* for $p_1 - p_2$ as

$$\left(\frac{x_1}{n_1} - \frac{x_2}{n_2} \right) \pm z_{\alpha/2} \sqrt{ \frac{(x_1/n_1)(1 - x_1/n_1)}{n_1} + \frac{(x_2/n_2)(1 - x_2/n_2)}{n_2} }$$

Note that in the expression for the confidence interval we have accounted for different population proportions by replacing p_1 and p_2 separately with $\hat{p}_1 = x_1/n_1$ and $\hat{p}_2 = x_2/n_2$.

EXAMPLE 9-2 For Example 9-1, a 95% confidence interval for $p_1 - p_2$ is

$$\left(\frac{27}{40} - \frac{8}{15} \right) \pm 1.96 \sqrt{ \frac{(27/40) \times (13/40)}{40} + \frac{(8/15) \times (7/15)}{15} }$$

or 0.142 ± 0.291 (i.e., -0.149 to 0.433). Note that the confidence interval includes 0. As determined in Example 9-1, we may not reject $H_0 : p_1 = p_2$. ▲

Testing Particular Differences

To test for a particular difference, we may use a confidence interval to test the null hypothesis of a particular difference δ, by determining whether the interval includes the value δ. We may also form a test based on a test statistic that incorporates δ:

We test $H_0: p_1 - p_2 = \delta$ with test statistic

$$z = \frac{(x_1/n_1 - x_2/n_2) - \delta}{\sqrt{\dfrac{(x_1/n_1)(1 - x_1/n_1)}{n_1} + \dfrac{(x_2/n_2)(1 - x_2/n_2)}{n_2}}}$$

For the alternatives $p_1 - p_2 \neq \delta$, $p_1 - p_2 > \delta$, and $p_1 - p_2 < \delta$, the rejection regions are $|z| > z_{\alpha/2}$, $z > z_\alpha$, and $z < -z_\alpha$, respectively.

EXAMPLE 9-3 Although not denying the use of drugs elsewhere, a school trustee claims that the local schools have a more serious problem and that the percentage of local students using drugs exceeds that of a neighboring region by more than five percentage points. Confidential random samples of 100 students in the local schools and 75 in neighboring schools produced responses indicating that in the sample groups, 26 and 12 students, respectively, used drugs. Using a 5% level of significance, the trustee's claim is investigated with a test of the null hypothesis $p_1 - p_2 = 0.05$ against the alternative $p_1 - p_2 > 0.05$ with test statistic

$$z = \frac{[(26/100 - 12/75) - 0.05]}{\sqrt{\dfrac{(26/100) \times (74/100)}{100} + \dfrac{(12/75) \times (63/75)}{75}}} = 0.820$$

which does not exceed $1.645 = z_{0.05}$. The trustee's claim is not substantiated with these data. ▲

Testing Equality of Two Proportions With Dependent Samples

The foregoing procedures for comparing proportions are not applicable in cases for which we do not have two independent sample groups, but have instead one group in two different conditions: say "before" and "after." In such cases, we have dependent samples and must use a different procedure.

It is not sufficient to note merely the proportion of interest for each sample. It becomes necessary to note how many changes took place between the before and after conditions and to note what kinds of changes took place. If no individuals change at all, then the before and after proportions will be equal. As well, however, if a number of individuals change from the category of interest to the complementary category while an equal number make the reverse change, there will still be no difference in the proportions. There will be a difference only if more than half of those who change make the same change.

We therefore test the null hypothesis of equal proportions by analyzing only those who change and by testing the null hypothesis that of those who change, half change from the category of interest to the complementary. We will develop the general notation by considering an example.

EXAMPLE 9-4 The campaign manager for a candidate for election wishes to test the degree of success of the candidate's speeches to large rallies by comparing the proportions of voters who support the candidate before and after a speech. The same sample of voters is used in each case.

The speech may change a voter from a nonsupporter to a supporter; it may not change the voter's position; or it might, unfortunately, change the voter from a supporter to a nonsupporter.

Suppose that the campaign manager first obtains a random sample of 150 voters prior to a speech and, of these, 54 support the candidate and 96 do not. After the speech, the campaign manager consults the same 150 sample voters. Of the 54 prior supporters, $x_{11} = 46$ are still supporters but $x_{12} = 8$ have changed to nonsupporters; of the original 96 nonsupporters, $x_{21} = 21$ have changed to supporters and $x_{22} = 75$ have remained unchanged. In the notation a subscript 1 indicates the category of interest "supporter," and a subscript 2 indicates the complementary category "nonsupporter." The first subscript indicates the category before and the second the category after.

The results may be summarized in tabular form as follows:

| | Before | After | |
		Supporters	Nonsupporters
Supporters	54	$x_{11} = 46$	$x_{12} = 8$
Nonsupporters	96	$x_{21} = 21$	$x_{22} = 75$

The sample thus includes $x_{11} + x_{22} = 121$ nonchangers and $x_{12} + x_{21} = 29$ who have changed. We analyze the 29 who changed and test the null hypothesis that of all those who change, the proportion changing from nonsupporter to supporter is $1/2$. Using the format for a test on a proportion developed in Section 7-5, n corresponds to $x_{12} + x_{21}$, x corresponds to x_{21}, and p_0 corresponds to $1/2$. The large-sample test statistic, using the normal approximation with continuity correction, is $z = (x \pm 1/2 - np_0)/\sqrt{np_0(1 - p_0)}$. In this case, z becomes $(x_{21} \pm 1/2 - (x_{12} + x_{21})/2)/\sqrt{(x_{12} + x_{21})/4}$ and is compared to an appropriate z_α or $z_{\alpha/2}$ value.

Assuming that the campaign manager uses the alternative that the speeches increase voter support, we use $-1/2$ as the continuity correction to produce $z = (21 - 1/2 - 29/2)/\sqrt{29/4} = 2.228$. Using a level of significance of 5%, z exceeds $z_\alpha = z_{0.05} = 1.645$ and we have evidence that voter support increases following the candidate's speeches. ▲

The test statistic as it appears in the example is modified with some simple algebraic manipulation to the forms appearing in the general test format that follows.

For *dependent random samples* the two-sided test of equality of proportions is:

$$H_0 : p_{\text{after}} = p_{\text{before}}$$

$$H_A : p_{\text{after}} \neq p_{\text{before}}$$

Test statistic: $z = \dfrac{|x_{21} - x_{12}| - 1}{\sqrt{x_{21} + x_{12}}}$

where x_{12} is the number changing from the category of interest to the complementary category and x_{21} is the number making the reverse change.

Decision rule: reject H_0 if $|z| > z_{\alpha/2}$

The one-sided tests have revised alternatives, test statistics, and decision rules as

$$H_A : p_{\text{after}} > p_{\text{before}}$$

Test statistic: $z = \dfrac{x_{21} - x_{12} - 1}{\sqrt{x_{21} + x_{12}}}$

Decision rule: reject H_0 if $z > z_\alpha$

or, for the reverse alternative, as

$$H_A : p_{\text{after}} < p_{\text{before}}$$

Test statistic: $z = \dfrac{x_{21} - x_{12} + 1}{\sqrt{x_{21} + x_{12}}}$

Decision rule: reject H_0 if $z < -z_\alpha$

PRACTICE PROBLEMS

P9-1 In a survey of traveler satisfaction, sample travelers were asked whether they used the mode of travel that they preferred or had to use some other mode of travel. Of 150 plane travelers surveyed at random, 46 would rather have used another mode of travel. Of 120 train travelers, 16 would have rather traveled otherwise. Do the proportions of dissatisfied travelers differ significantly at the 1% level?

P9-2 Determine a 95% confidence interval for the difference of the proportions in Practice Problem P9-1.

P9-3 It is claimed that because of improper preparation and the use of water from unsanitary sources, infants in Third World countries fed with milk made from powdered formulas experience health problems much more frequently than breast-fed infants in the same regions. Is there sufficient evidence at the 1% level that the percentage of formula-fed infants with serious health problems exceeds that for breast-fed infants by at least 25 percentage points if in a random sample of 117 formula-fed infants, 81 had serious problems, and in a random sample of 166 breast-fed babies, 43 had problems?

P9-4 A prison exposure program is supposed to reduce the proportion of problem juveniles becoming involved in serious crimes. Determine whether there is sufficient evidence at the 1% level of at least a short-term success in change of attitude if prior to exposure, 51 problem juveniles of a sample of 63 indicated little or no concern about prison and 12 were sufficiently concerned that they might be expected to not embark on serious criminal acts, but after a short-term experimental exposure, 47 of the first group of 51 became concerned while 2 of the other 12 thought the experience not too bad and lost their concern.

EXERCISES

9-1 Suppose that in a random sample of 500 Americans who expressed an opinion, 340 did not believe that police officers should have the right to strike. In a similar sample of 300 Canadians, 234 did not believe that police officers should have the right to strike.

Test the null hypothesis that the proportions of Americans and Canadians who do not believe that police officers should have the right to strike are equal at the 1% level.

9-2 Find a 99% confidence interval for the difference of the two proportions in Exercise 9-1. Use this interval to do the test in Exercise 9-1.

9-3 A researcher has found that 185 of 250 nonsmokers tested were able to succeed at an endurance test, whereas only 116 of 200 smokers tested were successful. Test the hypothesis of equal success proportions in a two-sided 5% test.

9-4 (a) On the basis of the sample results in Exercise 9-3, determine a 95% confidence interval for the difference in success proportions for nonsmokers and smokers.

(b) Using the results of part (a) test the hypothesis that the success proportion for nonsmokers exceeds that for smokers by 0.10 in a 5% two-sided test.

9-5 Suppose that for two animal strains, A and B, it is claimed that the proportion of type A who are capable of learning a specific set of tasks exceeds that for type B by more than 0.10. Of 100 animals of type A tested, 82 learned the tasks. Of 80 animals of type B tested, 52 learned the task. By setting up an appropriate null hypothesis and alternative hypothesis, determine whether the claim is substantiated at the 5% level on the basis of these results.

9-6 In a study to determine whether a drug will affect the proportion of subjects achieving a performance threshold, 180 subjects were tested without the drug and again with the drug. The testing was not a new experience for the subjects; hence, any "with drug" as opposed to "without drug" differences were not attributed to experience. Using a 5% level of significance, determine whether the drug has an effect if 122 subjects achieved the threshold without the drug and, of these, only 89 achieved the threshold with the drug, while 58 subjects failed to achieve the threshold without the drug and, of these, 13 were successful with the drug.

9-7 Develop a *one-sided* 95% confidence interval indicating *at least* how much of an improvement can be expected from using interferon as an anticancer agent if 21 patients of an experimental group of 33 receiving interferon were free from metastasis after $2\frac{1}{2}$ years, whereas 22 of a control group of 67 not receiving interferon were free from metastasis after the same time period.

SOLUTIONS TO PRACTICE PROBLEMS

P9-1 Considering plane travelers as the first group and the "success" proportion as the proportion of dissatisfied travelers, we have $n_1 = 150$, $x_1 = 46$, $n_2 = 120$, $x_2 = 16$, and $\hat{p} = 62/207$. We then test the null hypothesis of equal proportions with test statistic

$$z = \frac{46/150 - 16/120}{\sqrt{(62/270) \times (208/270) \times (1/150 + 1/120)}} = 3.365$$

which exceeds $2.576 = z_{0.005}$. We thus conclude that there is a difference.

P9-2 The 95% confidence interval for the plane proportion minus the train proportion is

$$\left(\frac{46}{150} - \frac{16}{120}\right) \pm 1.96\sqrt{\frac{(46/150) \times (104/150)}{150} + \frac{(16/120) \times (104/120)}{120}}$$

$$= 0.173 \pm 0.096$$

or 0.077 to 0.269. We may be 95% confident that the *percentage* of dissatisfied plane travelers exceeds that for train travelers by between 7.7 and 26.9 percentage points.

P9-3 We test $H_0: p_1 - p_2 = 0.25$ against $H_A: p_1 - p_2 > 0.25$ with test statistic

$$z = \frac{81/117 - 43/166 - 0.25}{\sqrt{[(81/117) \times (36/117)]/117 + [(43/166) \times (123/166)]/166}} = 2.443$$

which exceeds $2.326 = z_{0.01}$. We thus reject the null hypothesis and conclude that there is sufficient evidence.

P9-4 Letting the first category, the category of interest, be "sufficiently concerned not to embark on serious crimes," we have $x_{21} = 47$ and $x_{12} = 2$. We then test $H_0: p_{\text{after}} = p_{\text{before}}$ against $H_A: p_{\text{after}} > p_{\text{before}}$ with test statistic $z = (47 - 2 - 1)/\sqrt{49} = 6.286$, which exceeds $2.326 = z_{0.01}$. We may reject the null hypothesis and conclude that there is sufficient evidence of at least short-term success.

9-3 CONTINGENCY TABLES

In this section we develop, through an example, a procedure that we may interpret as a test of independence or association. The procedure may also be interpreted as a test of the homogeneity of proportion allocations in several populations. In a special case, the test may also be considered to be a test of the equality of two or more proportions.

Testing Independence or Association

A problem frequently encountered in statistical work is one in which each given individual (person, object, etc.) under study displays certain attributes at different "levels" and it is of interest to know whether two (or more) attributes are associated or independent (e.g., are reading habits associated with the sex of the reader? is degree of success of a product related to the marketing technique?, etc.)

This problem can be treated through a procedure in which a number of cases are studied and each is classified according to its level for each attribute. The proportional allocations to categories are then studied for significant differences across attribute levels.

EXAMPLE 9-5 Suppose that we wish to analyze the degree of success that a forthcoming product might have as related to varying degrees of consumer exposure to preliminary marketing pressures. Degree of exposure to marketing pressure was classified into four categories I, II, III, and IV according to the nature of advertising experienced and the frequency of exposure to advertising. A sample of 100 people was selected at random and each classified according to the marketing pressure experienced and the reaction to the product. Twenty people in the sample were found to have experienced marketing pressure classified as category I; 12 of the 20 said that they would definitely buy the product, 6 said that they were indifferent, and 2 would definitely not buy the product. For the 30 people who were classified in the category II, the numbers were 12, 8, and 10; for the 23 people classified in category III, the numbers were 6, 11, and 6; and for the 27 people classified in category IV, the numbers were 18, 5, and 4.

For each of the 100 individual cases (persons tested) we consider two attributes. The first is marketing pressure experienced and it has four "levels" corresponding to the four categories. The second attribute is degree (or anticipated degree) of success of the product and it has three "levels," corresponding to the three different reactions. There are, therefore, $3 \times 4 = 12$ different combinations (e.g., one combination is *second category — indifferent*, and there are eight individual cases in this combination). The data are usually summarized in a table such as the following:

	Category I	II	III	IV	Row total
Definitely buy	12	12	6	18	48
Indifferent	6	8	11	5	30
Definitely not buy	2	10	6	4	22
Column total	20	30	23	27	100

Observed Frequencies

The figures in the main part of the table (all but the last row and last column) are the *observed frequencies* from the data and are denoted as O_{ij}, where i identifies the row and j identifies the column. The column totals are denoted as C_j and the row totals are denoted as R_i The total number of individual cases is denoted as N.

EXAMPLE 9-6   For the marketing example we have, as illustrations, $O_{11} = 12$, $O_{23} = 11$, $C_1 = 20$, $R_2 = 30$, and so on, and $N = 100$.   ▲

Expected Frequencies

To determine whether the attributes are independent or associated, we compare the observed frequencies with the frequencies that we would expect on the average if the attributes were independent.

EXAMPLE 9-7   The overall proportion in the marketing sample who will definitely buy the product is 48/100, and we estimate the corresponding population proportion to be 48/100. Similarly in the sample of 100, 23 experienced the third category and we thus estimate the proportion of the full population who have experienced this category to be 23/100. If the two attributes are in fact independent then for any member of the population selected at random the joint probability P(*category III* and *definitely buy*) should equal the product P(*category III*) × P(*definitely buy*). We estimate each of the two probabilities in the product with the corresponding sample proportions to produce an estimate of the joint probability as (48/100) × (23/100). In random samples of size 100 we should find (48/100) × (23/100) × 100 = (48 × 23)/100 = 11.04 persons on the average who are in category III and will definitely buy the product. Because this "expected" frequency corresponds to row 1 and column 3 of the data table we label it E_{13}. Compared with $O_{13} = 6$ we would estimate that we should expect $E_{13} = 11.04$ on the average in such samples if we did have independence.

A similar argument for category IV and the reaction "definitely buy" would produce $E_{14} = (48/100) \times (27/100) \times 100 = (48 \times 27)/100$; that is, compared with $O_{14} = 18$, we would expect (48 × 27)/100 = 12.96 to definitely buy. Similarly, since 30 of the 100 were indifferent and 20 experienced the first category, compared to $O_{21} = 6$, we would expect (30 × 20)/100 = 6.00 to be indifferent and in category I if the category and response are not associated. Other expected values are found in the same manner.

The full table of expected values is:

	I	II	III	IV
Definitely buy	9.60	14.40	11.04	12.96
Indifferent	6.00	9.00	6.90	8.10
Definitely not buy	4.40	6.60	5.06	5.94

▲

These few calculated examples indicate a general pattern of expected frequencies corresponding to each of the observed frequencies.

In general, we designate the *expected frequencies* as E_{ij} and each is calculated as

$$E_{ij} = \frac{R_i \times C_j}{N}$$

Hypothesis of Independence

We now compare each of the observed frequencies to the corresponding expected frequency to determine whether the discrepancies could have reasonably come about due to chance or are so large as to be too unusual to be reasonable under the null hypothesis of independence: hence, indicating some association.

In general, for each "cell" in the main part of the table, we calculate a measure of discrepancy (deleting subscripts) as $(O - E)^2/E$, which is the squared difference relative to the expected value. We then calculate an aggregate measure of discrepancy for the full table as

$$\chi^2 = \frac{\Sigma (O - E)^2}{E}$$

where the sum is over all cells in the table.

If this measure exceeds a critical value, we may say that there is too great a discrepancy between what we observed and what we would expect if the attributes were independent, and we may declare that the attributes are associated.

This is, in fact, a hypothesis test. The null hypothesis is that the attributes are independent and the alternative is that they are associated (i.e., they are not independent). The overall measure of discrepancy χ^2 is the test statistic. If the hypothesis is true, χ^2 will follow an approximate χ^2-distribution.

The χ^2-distribution that we use is the one with degrees of freedom equal to (number of rows minus 1) × (number of columns minus 1). Intuitively, we may justify this value as follows. Once we have the row and column totals (which are used to determine the expected frequencies), we find that the last row and last column are determined by the entries in the other rows and columns. If we have a table with r rows and c columns, we may assign values to the first $r - 1$ rows and $c - 1$ columns. We may assign $(r - 1) \times (c - 1)$ values. The remaining values are then determined on the basis of the initial values and the row and column totals. Since we may have $(r - 1) \times (c - 1)$ "free" choices, the table has $(r - 1) \times (c - 1)$ degrees of freedom.

EXAMPLE 9-8   In Example 9-5 the first three entries in the first row could be 10, 10 and 7. To make a total of 48, the last entry would then have to be $48 - (10 + 10 + 7) = 21$. If the first three entries in the second row were then 8, 9, and 10, the last value would have to be $30 - (8 + 9 + 10) = 3$, to make a total of 30. If the values above occurred, the first column would have, as its first two entries, 10 and 8. To make the total 20, the last value would have to be $20 - (10 + 8) = 2$. Remaining values would be found in the same manner, to produce the following table, in which underlined values were assigned as "free" choices and values with an asterisk are then determine by the free choices and the totals.

					Total
	10	10	7	21*	48
	8	9	10	3*	30
	2*	11*	6*	3*	22
Total	20	30	23	27	100

If we considered all possible tables that could be produced with the $(r - 1) \times (c - 1)$ free choices and all of the corresponding possible values of χ^2, we would find that these values follow a distribution close to a χ^2 with $(r - 1) \times (c - 1)$ degrees of freedom if the null hypothesis of independence is true. If the actual value of χ^2 appears too large to have resulted from such a distribution, it provides evidence to contradict the null hypothesis. As a result, the null hypothesis of independence is rejected for significantly large values of χ^2. The test is summarized as follows:

H_0: the attributes are independent

H_A: the attributes are associated

Test statistic: $\chi^2 = \dfrac{\Sigma (O - E)^2}{E}$

Decision rule: reject H_0 if $\chi^2 > \chi^2_{\alpha, (r-1) \times (c-1)}$, where r and c are the numbers of rows and columns in the main part of the contingency table (i.e., r and c are the numbers of levels of the two attributes)

EXAMPLE 9-9 The observed frequencies of Example 9-5 and corresponding expected values found in Example 9-7 are combined in the following table (the expected values are in brackets).

	I	II	III	IV
Definitely buy	12	12	6	18
	(9.60)	(14.40)	(11.04)	(12.96)
Indifferent	6	8	11	5
	(6.00)	(9.00)	(6.90)	(8.10)
Definitely not buy	2	10	6	4
	(4.40)	(6.60)	(5.06)	(5.94)

The value of the test statistic is

$$\chi^2 = \frac{(12 - 9.60)^2}{9.60} + \cdots + \frac{(4 - 5.94)^2}{5.94} = 12.863$$

For this example, $r = 3$ and $c = 4$. The degrees-of-freedom value is $(3 - 1) \times (4 - 1) = 2 \times 3 = 6$. We thus compare $\chi^2 = 12.863$ to upper percentiles of the χ^2-distribution with 6 degrees of freedom and we find that 12.863 exceeds $\chi^2_{0.05,6} = 12.592$, but not $\chi^2_{0.01,6} = 16.812$. At the 5% level, we have sufficient evidence that degree of success is related to marketing pressure, but at the 1% level, the data do not provide sufficient evidence; that is, we may reject the hypothesis of independence at the 5% level, but not at the 1% level. ▲

Small Expected Frequencies

In analyses such as this, the χ^2 approximation may not be good if many of the expected frequencies are small. A rule of thumb is at least 80% of the expected values should exceed 5 and none should be less than 1. In Example 9-9 only one of the expected frequencies is less than 5 and it is only slightly less than 5. We will thus consider the approximation to be reasonable.

In examples such as the above, and in similar procedures developed in this and the following section, if a number of categories have extremely small expectations, similar or "adjacent" attribute levels may be combined, and their observed and expected frequencies pooled, to improve the approximation.

Contingency Tables

In general, in analyses such as the above, the table of observed frequencies (the main part of the table) is referred to as a *contingency table*. For two attributes, the number of levels r and c may be indicated by referring to the tables as $r \times c$ (read "*r* by *c*") *contingency tables*. Examples 9-5 through 9-9 involved a 3×4 table.

Contingency Coefficient

The hypothesis-testing procedure for a contingency table provides an indication of how significantly the observed results differ from what would be expected if the null hypothesis of independence were true. If we do not have independence, but rather some degree of association, it is useful to have some measure of the strength of the association. One measure that is sometimes used is the *contingency coefficient*.

The *contingency coefficient* for two attributes is equal to

$$\sqrt{\frac{\chi^2}{\chi^2 + N}}$$

where χ^2 is calculated from the contingency table for the attributes and N is the total frequency of the contingency table.

If the observed results agree perfectly with those expected when the attributes are independent, χ^2 will be 0 and the contingency coefficient will be 0. As the degree of association increases, the observed values will tend to differ from the expected values. The value of χ^2 will increase and the contingency coefficient will increase to a maximum possible value $\sqrt{(q-1)/q}$, where q is equal to r or c, whichever is smaller.

> For a $r \times c$ contingency table, the contingency coefficient takes values between 0 and $\sqrt{(q-1)/q}$, where q is the smaller of r and c. Values close to 0 indicate independence or very little association. Values close to $\sqrt{(q-1)/q}$ indicate a strong dependence or association.

EXAMPLE 9-10 For the 3×4 contingency table for the data of Example 9-5, $q = 3$ and the maximum possible value of the contingency coefficient is $\sqrt{2/3} = 0.816$. In Example 9-9, χ^2 was found to be 12.863. The total frequency is $N = 100$. The contingency coefficient is actually found to be $\sqrt{12.863/(12.863 + 100)} = 0.338$. This value differs from 0 but is not very close (relatively) to the maximum possible 0.816. We thus note that there is some association, but it is not an extremely strong association. This conclusion is in agreement with the results of the hypothesis test in Example 9-9, in which we found significance at the (moderate) 5% level, but not the (extreme) 1% level. ▲

Testing Homogeneity

In Examples 9-5 through 9-9, we had a random sample from a population of potential consumers of a product. The members of the sample were cross-classified according to the marketing pressures to which they were exposed and the reactions that they demonstrated. Analysis of a contingency table based on the cross-classification provided a test for association between the attribute level according to one classification (marketing pressure) and the attribute level according to the other (reaction to the product).

The same form of analysis can be used to compare several populations for differences or homogeneity of proportional allocations to categories within the populations. Rather than cross-classify individual members of one population, we consider a "superpopulation" of several populations and we cross-classify individuals according to population and category of interest within the population. The contingency table analysis thus provides a test of homogeneity across several populations of the proportional allocations to various categories within populations. The populations are considered to be levels of one attribute, and the various categories the levels of the other. We assume that we have independent random samples from the populations.

EXAMPLE 9-11 In an analysis of housing conditions, independent samples of housing units were obtained from five municipalities. Each housing unit was assessed in terms of a set of standards and classified according to whether it failed to meet standards, was marginal and could be relatively easily brought up to standards, met standards, or exceeded standards. The resulting numbers of units within each category for the five municipalities were as follows:

	I	II	Municipality III	IV	V
Fail standards	28	18	24	13	10
Marginal	8	15	7	15	17
Meet standards	108	120	105	126	130
Exceed standards	6	7	4	6	8

It is of interest to know whether these sample data indicate that housing conditions vary across the minicipalities with regard to proportions of units in each category. To determine whether the data do provide such an indication, we apply a contingency table analysis. We consider the five populations (the municipalities) to be the levels of one attribute and the four categories to be the levels of another. The data then form a 4×5 contingency table. The expected values and marginal totals for the table are as follows:

	I	II	III	IV	V	Total
Fail	18.0	19.2	16.8	19.2	19.8	93
Marginal	12.0	12.8	11.2	12.8	13.2	62
Meet	114.0	121.6	106.4	121.6	125.4	589
Exceed	6.0	6.4	5.6	6.4	6.6	31
Total	150	160	140	160	165	775

The test statistic is then calculated as $\chi^2 = (28 - 18.0)^2/18.0 + \cdots + (8 - 6.6)^2/6.6 = 21.846$. Using $(4 - 1) \times (5 - 1) = 3 \times 4 = 12$ degrees of freedom, we find the 95th percentile of the χ^2 distribution to be $\chi^2_{0.05, 12} = 21.026$. The $97\frac{1}{2}$th percentile is $\chi^2_{0.025, 12} = 23.337$. The χ^2 value from the data is significant at the 5% level, but not the $2\frac{1}{2}\%$ (or more extreme) level. We may say that $\chi^2 = 21.846$ is significant with $p < 0.05$. If we choose to use a 5% level of significance, we may reject the null hypothesis of homogeneity across the municipalities and we may adopt the alternative that there are differences in proportional allocations to categories across the municipalities.

As a measure of the degree to which proportional allocation to housing categories is associated with municipalities, we might calculate the contingency coefficient. For this table with $r = 4$ and $c = 5$, q the smaller of r and c is 4.

The maximum possible contingency coefficient for such tables is $\sqrt{(q-1)/q} = \sqrt{0.75} = 0.866$. The actual contingency coefficient for the data is $\sqrt{x^2/(x^2+N)} = \sqrt{21.846/(21.846+775)} = 0.166$. It is somewhat different from 0, but not close to the maximum. As was also indicated by the degree of significance of x^2, there is some association, but it is only moderately significant. ▲

Testing Equality of Proportions for Several Populations

For the special case that there are only two complementary categories within each population, we need only specify one as the category of interest. Using a contingency table analysis with either two rows or two columns, the test becomes a test of the equality of success proportions from several binomial populations. We form a test of the equality of proportions from k populations as

$H_0: p_1 = p_2 = \cdots = p_k$

$H_A: p_1, p_2, \ldots, p_k$ are not all equal

Test statistic: $x^2 = \sum (O - E)^2/E$ is found from the data summarized in a $2 \times k$ or $k \times 2$ contingency table

Decision rule: reject H_0 if $x^2 > x^2_{\alpha, k-1}$

EXAMPLE 9-12   In a study on voting behavior, people of $k = 3$ different national origins were presented with two fictitious candidates in a mock election. One of the candidates was of the same national origin as the "voter" and the other was not. Suppose that the following table represents the numbers of persons voting for the candidate of the same national origin or of another origin:

Origin of candidate	Origin of voter A	B	C
Same	28	10	10
Other	7	10	15

The data form a 2×3 contingency table and we may use the x^2 analysis to test whether voting behavior is associated with the national origin of the voter. We may also interpret the test as a test on the equality of three proportions, p_1, p_2, and p_3, where p_1 is the true proportion of people of origin A who would vote for a candidate of the same origin as themselves and p_2 and p_3 are analogous for B and C.

The contingency table with observed frequencies, expected frequencies (in brackets), and corresponding totals is:

	A	B	C	Total
Same	28	10	10	48
	(21)	(12)	(15)	
Other	7	10	15	32
	(14)	(8)	(10)	
Total	35	20	25	80

The test statistic is then calculated as $\chi^2 = (28 - 21)^2/21 + \cdots + (15 - 10)^2/10 = 10.83$. Since $\chi^2 = 10.83$ is greater than $\chi^2_{0.005, 2} = 10.597$, we have a significant result with $p < 0.005$. We may conclude that voting behavior is related to national origin or, in other words, we may conclude that the proportions of people of the three different origins who would vote for a candidate of their own origin are not all the same. ▲

2 × 2 Tables

For the case that the table has two rows and two columns, the contingency table analysis is equivalent to a two-sided test on the equality of two proportions as considered in Section 9-2. For this case, there is a simplified method of calculating χ^2. If the observed frequencies in the 2 × 2 table are denoted as a, b, c, and d and occupy positions in the table as

$$
\begin{array}{cc}
a & b \\
c & d
\end{array}
$$

and the total frequency is denoted as n, then

$$\chi^2 = \frac{n(ad - bc)^2}{(a + b)(c + d)(a + c)(b + d)}$$

EXAMPLE 9-13 For Example 9-1, the data may be presented in a contingency table as follows:

	Rubber	Synthetic material	Total
Deterioration	27	8	35
No deterioration	13	7	20
Total	40	15	55

Using the simplified method of calculation for these values, we obtain $\chi^2 = 55[(27 \times 7) - (8 \times 13)]^2/[35 \times 20 \times 40 \times 15] = 0.946$, which is less than $\chi^2_{0.05, 1} = 3.841$. As in Section 9-2, we accept the null hypothesis of equal proportions.

The equivalence of the two test procedures holds because the test statistic χ^2 in the second test is equal to the square of z, the test statistic in the first test, and the critical value for the second test, $\chi^2_{0.05, 1} = 3.84$, is the square of $z_{0.025} = 1.96$, the critical value for the first test. ▲

This relationship would hold true in any other similar example, since a χ^2 variable with 1 degree of freedom is, by definition, the square of a standard normal.

PRACTICE PROBLEMS

P9-5 The two sections of a connector are manufactured in different parts of a plant but are combined on the same day that they are made. In an inspection sampling of 1000 connectors, 814 had no flaws in either the inner or outer section, 26 had no flaws in the inner section but minor flaws in the outer section, and 2 had no flaw in the inner section and a major flaw in the outer. As well, 78 connectors had minor flaws in the inner and, of these, the numbers with no flaws, minor flaws, or major flaws in the outer section were 8, 60, and 10. Finally 80 had major flaws in the inner section and, of these, the numbers with no flaws, minor flaws or major flaws in the outer were 28, 14, and 38. Determine whether there is an association between flaws in the inner and outer sections. Use a 1% level of significance. Calculate the contingency coefficient.

P9-6 A study on home size versus income included a number of single-family households consisting of married couples with children under age 16 living at home. Is there sufficient evidence at the 5% level that the proportions of families with more than one person per room differ for income levels if the study produced the following results?

Income (dollars)	Number of families sampled	Number of families with more than one person per room
Less than 10,000	33	10
10,000–19,999	161	36
20,000–24,999	44	7
25,000 or more	12	2

P9-7 Repeat Practice Problem P9-1 with a 2×2 contingency table analysis and compare the results with those of the solution to P9-1.

EXERCISES

9-8 As a follow-up to a campaign on the need for resource conservation, a sample survey of several establishments was conducted in each of four regions. In the first region, improvement was found in 33 cases, little or no change was found in

36 cases, and deterioration was found in 16 cases. In the second region, there were 28 cases of improvement, 54 cases of little or no change, and 13 cases of deterioration. In the third and fourth regions, the corresponding figures were 37, 42, 11 and 28, 48, 14, respectively. Is there sufficient evidence at the 5% level that reaction to the campaign differed across the four regions (i.e., that reaction is associated with region)?

9-9 At the 5% level, test the null hypothesis that belief as to the ideal number of children in a family is independent of the age of the respondent if a survey of 600 individuals produced the following table of numbers of individuals in age categories and according to belief as to the ideal number of children in a family.

	Ideal number of children		
Age	2 or fewer	3	4 or more
18–30	131	71	23
31–50	106	59	35
Over 50	73	60	42

Calculate the contingency coefficient for these data.

9-10 Using the approach of a 2×2 contingency table,
(a) repeat Exercise 9-1
(b) repeat Exercise 9-3

9-11 Suppose that in a study on rate of divorce, a sociologist classifies survey respondents according to four categories. In the first category, 14 of the 445 respondents were divorced, in the second 9 of 305 respondents were divorced, in the third, 15 of 630 were divorced, and in the fourth, 7 of 420 were divorced. Test the null hypothesis that the proportions of divorced persons in the four categories are all equal using a 5% level of significance.

9-12 Determine whether there is a difference in infant mortality rates for inbred and noninbred zoo animals on the basis of the following sample data.

	Inbred animals	Noninbred animals
Number surviving 6 months	28	47
Number not surviving	22	16

9-13 A political analyst suspects a difference in the proportion of uncommitted voters according to the method of conducting polls. To investigate method differences, the analyst simultaneously conducted three surveys: one by telephone, one based on street interviews, and one based on personal in-home interviews. Of 94 people surveyed by telephone, 31 were uncommitted; of 96 surveyed in street interviews, 21 were uncommitted; and of 86 interviewed personally in their homes, 17 were uncommitted. Do these data provide sufficient evidence at the 1% level that there are differences in proportions of uncommitted voters according to method of survey?

P9-5 The data can be summarized in a contingency table as follows (with expected values in parentheses):

Outer section	Inner section		
	No flaw	Minor flaw	Major flaw
No flaw	814 (715.7)	8 (66.3)	28 (68.0)
Minor flaw	26 (84.2)	60 (7.8)	14 (8.0)
Major flaw	2 (42.1)	10 (3.9)	38 (4.0)

The test statistic is $\chi^2 = (814 - 715.7)^2/715.7 + \cdots + (38 - 4.0)^2/4.0 = 819.1$, which greatly exceeds $\chi^2_{0.01,4} = 13.277$. There is association at the 1% level. In fact, consulting the χ^2 table in Appendix B we note that we have significance with $p < 0.005$. The value is so significant that we need not be concerned about the two small expectations 3.9 and 4.0. The contingency coefficient is $\sqrt{\chi^2/(\chi^2 + N)} = \sqrt{819.1/(819.1 + 1000)} = 0.67$.

P9-6 To test the $k = 4$ proportions for equality, we analyze the data in the following 4×2 contingency table (with expected values in parentheses):

Income (dollars)	Number of families with more than one person per room	Number without	Total number of families sampled
Less than 10,000	10 (7.26)	23 (25.74)	33
10,000–19,999	36 (35.42)	125 (125.58)	161
20,000–24,999	7 (9.68)	37 (34.32)	44
25,000 or more	2 (2.64)	10 (9.36)	12
Total	55	195	250

The test statistic is then $\chi^2 = (10 - 7.26)^2/7.26 + \cdots + (10 - 9.36)^2/9.36 = 2.772$, which does not exceed $\chi^2_{0.05,3} = 7.815$. There is insufficient evidence of association between income and number of persons per room.

P9-7 Summarized in a 2×2 table, the data produce:

	Train	Plane
Satisfied	104	104
Prefer another mode of travel	46	16

The test statistic is calculated as $\chi^2 = 270[(104 \times 16) - (104 \times 46)]^2/[208 \times 62 \times 150 \times 120] = 11.323$ which exceeds $\chi^2_{0.01,1} = 6.635$. We again conclude that there is a difference. (Note that $\sqrt{\chi^2} = \sqrt{11.323} = 3.365 = z$ of Practice Problem P9-1 and $\sqrt{\chi^2_{0.01,1}} = \sqrt{6.635} = 2.576 = z_{0.005}$.)

9-4 TESTING GOODNESS OF FIT

Testing the Form of a Distribution

Besides testing hypotheses on the possible values of individual parameters of populations or distributions, we may also test the form of the distribution. We may, for example, test the hypothesis that a particular set of data comes from a distribution that is normal. We test such a hypothesis by testing how well the hypothesized distribution fits the empirical distribution of the data; accordingly, we refer to such a test as a *goodness-of-fit test*.

The χ^2 Test

There are a number of possible procedures for testing goodness of fit, but we will only develop one as an illustration of the general concept. We consider a test procedure that is based on the same principles as the contingency table analysis. We assign a set of classes for the data that we have obtained in a random sample and, in a χ^2 analysis, we compare the frequencies that we observe in the classes with those that we would expect to observe on the average if the hypothesized distribution were the true distribution.

EXAMPLE 9-14   Suppose that it is hypothesized that, through a random scatter of seeds, plants of a particular type should be scattered across a plain area according to a Poisson distribution with a mean of $\lambda = 4$ plants per unit of area. The null hypothesis of a Poisson with mean 4 is to be tested at the 1% level. Suppose that 1000 units of area were chosen at random and the numbers of plants in the individual units counted. Suppose further that, of the 1000 units, 26 contained no plants, 104 contained 1 plant, 172 contained 2 plants, and so on, with the complete results summarized in the following table of classes and observed frequencies:

Number of plants in a unit	0	1	2	3	4	5	6	7	8	9	10 or more
Frequency of occurrence	26	104	172	223	180	147	74	46	15	6	7

We now calculate expected frequencies for the classes. If the null hypothesis is true, we have a Poisson distribution with $\lambda = 4$ and the probability that the number of plants in an individual unit of area will be x is $f(x) = 4^x e^{-4}/x!$. For example, the probability that the number of plants in a unit of area will be 2 is $f(2) = 4^2 e^{-4}/2! = 0.146$. There is a 14.6% chance of obtaining 2 plants in a unit of area and we therefore expect 14.6% of the units in such samples to contain 2 plants on the average. In particular, in a sample of 1000 units of area, we would "expect" 14.6% of 1000; that is, $1000 \times 0.146 = 146$ units to contain 2 plants, and, in general, for a sample of n units we would expect $n \times f(x)$ units to contain x plants.

The expected frequencies for the first ten classes in the example are found as $1000f(x)$, with x taking values $0, 1, 2, \ldots, 9$. Since $P(x \geq 10) =$

$\sum_{x=10}^{\infty} f(x) = 1 - \sum_{x=0}^{9} f(x) = 0.008$, the expected frequency for the class "*10 or more*" is 8.

The observed frequencies, the probabilities corresponding to the hypothesized distribution, and the expected frequencies are summarized as follows:

Number of plants, x	Observed frequency of occurrence	Hypothesized probability, $f(x)$	Expected frequency, $1000\,f(x)$
0	26	0.018	18
1	104	0.074	74
2	172	0.146	146
3	223	0.195	195
4	180	0.196	196
5	147	0.156	156
6	74	0.104	104
7	46	0.060	60
8	15	0.030	30
9	6	0.013	13
10 or more	7	0.008	8

For each class, we denote the observed and expected frequencies as O and E, respectively, and we calculate a total measure of discrepancy, or test statistic, as $\chi^2 = \Sigma (O - E)^2 / E = (26 - 18)^2/18 + \cdots + (7 - 8)^2/8 = 49.508$ which must be compared to an appropriate value of $\chi^2_{0.01}$. In this example the test statistic has 10 degrees of freedom and since 49.508 exceeds $\chi^2_{0.01,\,10} = 23.209$, we reject the null hypothesis and conclude that plants are not scattered according to the Poisson distribution with mean $\lambda = 4$. ▲

General Procedure

In general, to test the null hypothesis that data followed a particular distribution, we assign the data to k categories and for each category we determine the observed frequency O and the frequency E that we would expect if the hypothesized distribution were the true distribution.

If, as in the example, the hypothesis specifies a distribution completely, then, if the hypothesis is true, χ^2 will follow an approximate χ^2 distribution with $k - 1$ degrees of freedom.

In such cases the test statistic and decision rule are

Test statistic: $\chi^2 = \dfrac{\Sigma (O - E)^2}{E}$

Decision rule: reject the hypothesized distribution if $\chi^2 > \chi^2_{\alpha,\,k-1}$

If a distribution is not fully specified in the hypothesis but has m parameters unspecified, then to calculate the expected frequencies, it is necessary to estimate the values of these parameters from the data. This reduces the degrees of freedom by m to $k - m - 1$. The decision rule then becomes

Decision rule: reject the hypothesized distribution if $\chi^2 > \chi^2_{\alpha,\,k-m-1}$.

EXAMPLE 9-15 If in Example 9-14, the null hypothesis were only that the plants should follow a Poisson distribution (without specifying λ), then λ would have to be estimated from the data as $\bar{x}$ because λ is the mean of the Poisson. The expected frequencies and the test statistic χ^2 would have to be recalculated on the basis of this new value for λ. The test would then be based on a χ^2 with 9 degrees of freedom.

Suppose now that we choose a 5% level of significance. Suppose further that the observed counts in the units with 10 or more were 10, 10, 11, 12, 12, 15, and 18. The mean number per unit is $\bar{x} = 3.6$ and we estimate λ to be 3.6. Using $\lambda = 3.6$, the expected frequencies become 27, 98, 177, 212, 191, 138, 83, 42, 19, 8, and 5, producing $\chi^2 = (26 - 27)^2/27 + (104 - 98)^2/98 + \cdots + (7 - 5)^2/5 = 5.846$ which does not exceed $16.919 = \chi^2_{0.05,\,9}$. We may thus accept the null hypothesis that the data follow a Poisson distribution. ▲

EXAMPLE 9-16 For another example, suppose that it is hypothesized that the lifetimes (in hours) of light bulbs of a particular type follow a normal distribution and that the hypothesis is to be tested at the 1% level. Suppose further that a sample of 220 such light bulbs produced lifetimes with the following frequency distribution:

Class	Frequency	Class	Frequency
775–824	6	1025–1074	42
825–874	10	1075–1124	16
875–924	21	1125–1174	10
925–974	32	1175–1224	8
975–1024	66	1225–1274	9

To calculate expected frequencies for these classes, we must calculate appropriate normal probabilities for the classes. Each class is regarded as extending from boundary to boundary except for the first and last classes. Since the normal distribution extends (theoretically) from $-\infty$ to ∞, we regard the first class as starting at $-\infty$ and the last class as ending at ∞.

The probability of the first class is $P(x \leqslant 824.5)$. The probability of the last class is $P(1224.5 \leqslant x)$. The probabilities of the other classes are similar to the probability of the second class, which is $P(824.5 \leqslant x \leqslant 874.5)$. To calculate these probabilities, we require the mean and standard deviation of the normal distribution and, since they were not specified in the hypothesis, we must estimate them with the mean and standard deviation of the frequency distribution which are 1011.5 and 97.6, respectively. The probability of the second class becomes

$$P\left(\frac{824.5 - 1011.5}{97.6} \leqslant \frac{x - 1011.5}{97.6} \leqslant \frac{874.5 - 1011.5}{97.6}\right)$$

$$= P(-1.92 \leqslant z \leqslant -1.40) = 0.0534$$

Multiplying this probability by the total frequency, we find the expected frequency for the second class as $220 \times 0.0534 = 11.7$. With similar calculations for all the classes, the expected frequencies for the ten classes (first to last) are 6.0, 11.7, 23.3, 36.4, 43.9, 41.9, 29.6, 16.6, 7.2, and 3.2. Because the expected frequency of the last class is quite small, we may combine the last two classes to produce a combined observed frequency of 17 and a combined expected frequency of 10.4.

For the adjusted set of classes, the observed frequencies, normal probabilities [using $\mu = \bar{x} = 1011.5$, $\sigma = s = 97.6$ to find P(lower boundary $\leqslant x \leqslant$ upper boundary) for each class], and expected frequencies (found as $220 \times$ probability) are as follows:

Class	Observed frequency	Normal probability	Expected frequency
824 or less	6	0.0274	6.0
825–874	10	0.0534	11.7
875–924	21	0.1059	23.3
925–974	32	0.1653	36.4
975–1024	66	0.1997	43.9
1025–1074	42	0.1905	41.9
1075–1124	16	0.1348	29.7
1125–1174	10	0.0755	16.6
1175 or more	17	0.0475	10.4

From this adjusted set of classes, we calculate the test statistic $\chi^2 = (6 - 6.0)^2/6.0 + \cdots + (17 - 10.4)^2/10.4 = 25.26$. The final set has $k = 9$ classes and we estimated $m = 2$ parameters. We therefore refer to the χ^2-distribution with $9 - 2 - 1 = 6$ degrees of freedom. Since $\chi^2_{0.01,6} = 16.812$, we may reject the hypothesis at the 1% level of significance. We may conclude that lifetimes of such bulbs do not follow a normal distribution. ▲

PRACTICE PROBLEMS

P9-8 Prizes awarded to children fishing at a fish pond are awarded on the basis of the markings on the fish caught. Fish are marked 0, 1, 2, 3, 4, or 5 and the prizes range from nothing for "0" through an inexpensive toy for "1," with increasing values, to a valuable prize for "5." The probabilities for 0, 1, 2, 3, 4, and 5, respectively, are supposed to be 0.35, 0.25, 0.20, 0.15, 0.04, and 0.01, respectively.

A games inspector has observed that the results of 500 catches produced frequencies 193, 130, 99, 69, 7, and 2. Test the null hypothesis that the probabilities are as they should be. Indicate the p-level if there is significance at a level of significance at least as extreme as 10%.

P9-9 An experimenter knows that a sampling scheme to determine small values will produce observations with mean μ equal to the quantity of interest and with standard deviation $\sigma = 0.150$. The distribution of observations is known to be a positively skewed distribution, but the experimenter believes that, through the central limit theorem, samples of size $n = 9$ will be sufficiently large that the distribution of the sample mean will be normal and that confidence intervals for μ based on the normal distribution, will be applicable. Test the null hypothesis that the distribution of sample means is normal with standard deviation $\sigma/\sqrt{n}$ $= 0.150/3 = 0.050$, as the experimenter expects, if 150 samples of size $n = 9$ on the same substance produced the following frequency distribution for the 150 sample means. Prior to the construction of the frequency distribution, the overall mean of the sample means was found to be 0.2495.

Class	Frequency	Class	Frequency
0.000–0.159	4	0.260–0.279	24
0.160–0.179	5	0.280–0.299	19
0.180–0.199	8	0.300–0.319	10
0.200–0.219	18	0.320–0.339	6
0.220–0.239	21	0.340–	6
0.240–0.259	29		

EXERCISES

9-14 A technician in a medical laboratory believes that the values in a particular set of data should have resulted from a normal distribution with mean 100. The set of data consists of 500 values in the following frequency distribution:

Class	Frequency	Class	Frequency
75–79	6	105–109	73
80–84	19	110–114	45
85–89	47	115–119	27
90–94	69	120–124	10
95–99	97	125–129	5
100–104	102		

Suppose that the technician plans to test his belief with a goodness-of-fit test.
(a) (1) What parameter is specified in the null hypothesis?
    (2) What parameter must be estimated from the data?
(b) Test the hypothesis at the 5% level.

9-15 Particle emissions are believed to follow a Poisson distribution with a mean rate of λ emissions per unit of time. The number of emissions observed in 500 units of time were as follows:

Number of emissions	Frequency	Number of emissions	Frequency
0	3	8	41
1	5	9	20
2	19	10	11
3	48	11	3
4	88	12	2
5	107	13	1
6	83	14	1
7	68		

(a) Estimate the parameter λ.

(b) Test the hypothesis that the data result from a Poisson distribution.

9-16 A game is based on the result of a spinner which, when spun, can stop pointing at any number from 1 to 10 inclusive. There is to be no bias in the spinner, so that it should be just as likely to stop at one number as any other. The results of 100 spins are as follows:

Number	1	2	3	4	5	6	7	8	9	10
Frequency	15	10	6	4	4	8	11	12	16	14

Determine whether these results provide proof at the 1% level that the spinner is biased.

SOLUTIONS TO PRACTICE PROBLEMS

P9-8 Multiplying $n = 500$ by the hypothesized probabilities produces the following expected frequencies 175, 125, 100, 75, 20, and 5. Comparing the observed frequencies to these, we find the test statistic χ^2 as $(193 - 175)^2/175 + \cdots + (2 - 5)^2/5 = 12.79$, which is between $\chi^2_{0.05, 5} = 11.070$ and $\chi^2_{0.025, 5} = 12.832$. We thus have significance with $p < 0.05$. (We have used 5 degrees of freedom because we had six classes, 0, 1, 2, 3, 4, and 5, and the distribution was fully specified.) We may reject the hypothesis of a good fit and we conclude that the probabilities are not as claimed.

P9-9 The class boundaries, $0.1595, 0.1795, \ldots, 0.3395$, are converted to z-values by subtracting the overall sample mean 0.2495 and dividing the differences by the standard deviation 0.05 of the hypothesized distribution to produce -1.8, $-1.4, \ldots, 1.4, 1.8$. From the normal table we find $P(z \leqslant -1.8) = 0.0359$, $P(-1.8 \leqslant z \leqslant -1.4) = 0.0449, \ldots, P(1.8 \leqslant z) = 0.0359$, and, multiplying these probabilities by 150 (the number of sample means), we obtain the expected values 5.4, 6.7, 11.7, 17.3, 22.0, 23.8, 22.0, 17.3, 11.7, 6.7, and 5.4 and then find the test statistic $\chi^2 = (4 - 5.4)^2/5.4 + (5 - 6.7)^2/6.7 + \cdots + (6 - 5.4)^2/5.4 = 3.910$. Since we have 11 classes and estimated one parameter (we estimated μ with the overall sample mean), we have $11 - 1 - 1 = 9$ degrees of freedom. Now

$\chi^2 = 3.910$ does not exceed $\chi^2_{0.05, 9} = 16.919$; hence, we may accept the null hypothesis that means of samples of size 9 follow a normal distribution with standard deviation 0.05.

9-5 CHAPTER REVIEW

9-17 A test of independence with a contingency table is based on
(a) a uniform scale of measurement
(b) deviations from the true mean
(c) observed frequencies
(d) sample variances of two populations

9-18 If we observed 15 successes in a sample of size 20 from one population and 21 successes in a sample of size 30 from another, the pooled estimate of the overall proportion of successes is _____ .

9-19 If we wish to determine whether the proportion of absentees has changed over time by considering one random sample of 200 individuals at different times, we use
(a) a 2×2 contingency table
(b) a test on proportions from dependent samples
(c) a test on proportions from independent samples

9-20 The number of degrees of freedom in a χ^2 goodness-of-fit test with 15 categories and with two parameters being estimated from the data is _____ .

9-21 In a contingency table analysis to test whether two attributes are independent or associated, the null hypothesis is that _____ .

9-22 The test of equality of success proportions from five populations is a χ^2 test based on _____ degrees of freedom.

9-23 Suppose that, in a study of immigrants (with relatives in Canada) settling in Toronto and Montreal, it was found that, of a sample of 63 families settling in Montreal, 21 displayed a strong familial system (i.e., weekly visits to relatives, etc.) and 42 displayed a weak familial system and that, of a sample of 89 families settling in Toronto, 40 displayed a strong familial system and 49 a weak familial system. Determine at the 5% level whether the proportion of immigrant families with strong familial systems is greater in Toronto than in Montreal.

9-24 Suppose that 102 tosses of a die produced 10 ones, 16 twos, 16 threes, 14 fours, 17 fives, and 29 sixes. If the die were balanced, how many ones, twos, and so on, would be "expected" to occur in 102 tosses? Is there sufficient evidence at the 5% level to declare that the die is not balanced?

9-25 In a study on the effects of stress, 300 random subjects were divided randomly into four groups of 75 each and then required to attempt a task under varying degrees of stress. The first group experienced no stress and of the 75 in this group, 48 were successful, 17 were partially successful, and 10 were unsuccessful. The second group experienced low stress and of the 75 in this group, 42 were successful, 18 were partially successful, and 15 were unsuccessful. The third group experienced medium stress and the numbers as above were 39, 18, and 18, respectively. The fourth group experienced a high level of stress and the numbers were 30, 16, and 29, respectively. Determine at the 1% level whether stress affects the ability to succeed at the task.

9-26 In a study on changing life-styles, 350 individuals sampled at random were asked, among other things, whether they smoked. The survey was conducted at two points in time with the same group of individuals. At the first point in time 190 of those surveyed were smokers. Of these, 42 had become nonsmokers by the time of the follow-up survey. Of the 160 nonsmokers in the first survey, 23 had become smokers by the time of the second survey. Do these sample data indicate a difference in time for the whole population with regard to the proportion of smokers?

9-27 Determine whether there is a significant difference with regard to reconvictions between an experimental group of prisoners granted amnesty and a control group released prior to the amnesty program on the basis of the following data:

	Number with no reconvictions	Number reconvicted without sentence	Number reconvicted with sentence
Amnesty group	68	35	56
Control group	64	36	35

9-28 Determine whether there is an association between the density of two species if a sample of 400 locations produced the following numbers:

Second species	First species		
	Low density	Medium	High
Low	36	28	65
Medium	47	23	32
High	107	41	21

9-29 In 120 trial strategy decisions, two trainees made, respectively, 102 and 93 successful decisions. Determine a 95% confidence interval for the difference of their success rates.

9-30 Can an instructor's marks be reasonably believed to follow a normal distribution if a sample of marks produced the following data?

Mark	Frequency	Mark	Frequency
0–24	3	70–74	23
25–39	6	75–79	16
40–49	14	80–84	11
50–54	8	85–89	6
55–59	10	90–94	3
60–64	17	95–99	1
65–69	28		

9-31 The mode of a set of data can be found easily from a frequency distribution with class intervals of 5.
(a) True
(b) False

9-32 If A, B, and C are mutually exclusive and exhaustive, then
(a) $P(A \text{ and } B \text{ and } C) = 1$
(b) $P(A \text{ and } B \text{ and } C) = 0$
(c) $0 < P(A \text{ and } B \text{ and } C) < 1$
(d) $P(A \text{ and } B \text{ and } C) = P(A) + P(B) + P(C)$

9-33 The number of degrees of freedom in a contingency table is based on
(a) the total number of observations
(b) the numbers of rows and columns
(c) the row and column totals
(d) the expected frequencies

9-34 If an experiment produces a 98% confidence interval for the ratio of two standard deviations as 0.59 to 3.78, then if σ_1, σ_2 are the population standard deviations, $H_0 : \sigma_1 = \sigma_2$ may be _____ in a two-sided 2% test.

9-35 If x is chosen from a normal population with mean 50.0 and standard deviation 15.0, then $P(x > 72.5) = P(z > \underline{\hspace{1cm}})$.

9-36 If two samples produce s_1^2 and s_2^2 such that s_1^2/s_2^2 is not significant, the conclusion that $\mu_1 = \mu_2$ is
(a) correct
(b) incorrect
(c) irrelevant

9-37 $P(A \text{ or } B)$ must always be
(a) less than $P(A) + P(B)$
(b) equal to $P(A) + P(B)$
(c) greater than $P(A) + P(B)$
(d) none of these

9-38 If each value in a set of data is tripled, the new standard deviation will be _____.

9-39 In a rose diagram in which frequency impressions are obtained from areas of wedge-shaped pieces in a circle or part circle, the radii producing the wedges should be proportioned to the
(a) squares of the frequencies
(b) the frequencies
(c) the square roots of the frequencies

9-40 The confidence interval for the difference of two proportions involves a pooled proportion estimate.
(a) True
(b) False

9-41 (a) Use the values given to determine population sizes and calculate the overall mean number of telephones per 100 population.
(b) Using a bar graph or pictograph, illustrate the number of telephones per 100 population from the following 1973 data:

Country	Number of telephones (thousands)	Number of telephones per 100 population
Australia	4,659	35.5
Canada	11,665	52.8
France	11,337	21.7
Japan	38,698	35.7
New Zealand	1,444	47.5
United Kingdom	19,095	34.0
United States	138,286	65.7
USSR	14,261	5.7
West Germany	17,803	28.7

9-42 Assessments on the amount of a contaminant in potential drinking water are useful only if the measurement process is sufficiently precise that the standard deviation of repeat measurements on a given quantity of water does not exceed 2.0 units. It has been claimed that the measurement process is not sufficiently precise. Set up an appropriate hypothesis and alternative to judge the precision on the basis of the following sample measurements on a fixed quantity of water (assume that a normal distribution applies).

8.4 3.8 7.2 2.8 5.2 3.0 7.8 6.4 7.6 3.8 6.4 8.2
6.8 7.2 1.8 3.6

Do these data provide sufficient evidence at the 5% level to substantiate the claim?

9-43 A survey is to be conducted to investigate the claim that fewer than 30% of the subscribers to a magazine would favor a proposed change in format. In the survey, 200 subscribers are to be asked whether they favor the proposed change. Set up a decision rule to determine, from the results of the sample, whether the claim can be substantiated at the 5% level of significance.

9-44 Suppose that the following numbers taken as a sample represent the change in value per share over a specific time of 100 stocks.

−1	+2	−2	−10	+6	+6	+11	−6	+6	+8
+4	−5	−13	+8	+2	−5	−2	−3	+9	−3
0	0	+5	+2	−6	−10	+12	+3	−2	+1
+3	−3	−10	−10	+5	−5	+1	−10	−3	−2
0	−2	−9	−1	+4	−15	+4	−5	−2	−6
+9	+13	+14	−14	+13	+8	+5	0	−1	+7
−2	−8	−3	+2	−14	+6	−3	−1	−6	+2
+8	+9	+8	+4	+14	+2	+1	+4	+9	+1
−3	+13	+14	−8	+12	+6	0	+3	+2	+3
+6	+5	+4	+4	+5	−13	+13	+4	+5	0

(a) Calculate the mean and median change.

(b) Calculate the standard deviation of these values.

(c) (1) If one of these stocks is chosen at random, what is the limit on the probability, according to Chebyshev's theory, that its change will be within 10 of the mean?

(2) What is the actual probability?

(d) If the stocks are expected to continue to follow the same pattern and one is chosen at random, what are the odds against a resulting loss (negative change)?

9-45 Test the null hypothesis that the data in Exercise 9-44 resulted from a normal distribution.

9-46 An experimental group of 24 steel plates with a test spraying of a protective coating resisted corrosion against a corrosive agent with maximum strengths showing a mean of 1.18 and a standard deviation of 0.18. A control group of 36 plates with the standard coating resisted corrosive strengths with mean and standard deviation 1.07 and 0.17. Find a one-sided 95% confidence interval to indicate at least how much of a gain can be expected on the average with the new coating.

9-47 A vinyl manufacturing process produces sheets of vinyl for which the thickness varies according to a normal distribution with a standard deviation $\sigma = 0.10$ micron. If the process is under control, the mean of the normal distribution is $\mu = 120.00$ microns. The null hypothesis that the process is under control is to be tested against the alternative that it is out of control (i.e., μ is not 120.00). The hypothesis is to be tested on the basis of the mean of a sample of 10 values.

(a) Set up a testing procedure with a 5% level of significance.

(b) If $\mu = 120.08$, what is the probability that the testing procedure will result in a type II error?

(c) Repeat part (b) for $\mu = 119.88, 119.90, 119.92, 119.94, 119.96, 119.98, 120.02, 120.04, 120.06, 120.10,$ and 120.12.

(d) Sketch the operating characteristic curve for this testing procedure.

9-48 Suppose that records on repairs to small vehicles indicate that, in the recent past, 35% of vehicles serviced were North American made and one-seventh of these were station wagons; 35% were European made and one-fifth of these were station wagons; and 30% were Japanese made and two-fifths were station wagons.

(a) What is the overall percentage of small vehicles serviced in the recent past that were station wagons?

(b) Suppose that the pattern in the near future is expected to follow that of the recent past. If so, what is the probability that a given car to be serviced will be

(1) North American made?

(2) a Japanese made station wagon?

(3) European made or a station wagon?

(4) neither European made nor a station wagon?

9-49 In a study on the effects of vitamin C, an experimental group of 100 subjects received large doses of vitamin C and a control group of 300 subjects received a placebo. The 400 subjects experienced similar conditions for the same length of time. Is there a significant difference at the 5% level in the incidence of colds if

47 of those in the experimental group and 161 of those in the experimental group suffered from colds?

9-50 Test the usefulness of vitamin C if the data in Exercise 9-49 were classified as follows:

	No cold	Mild cold	Severe cold
Vitamin C	53	35	12
Placebo	139	64	97

9-51 Of several thousand experimental units available for use in the analysis of new treatment procedures, only 80% are suitable for analysis.
  (a) If 10 of these units are to be selected at random, what is the probability that at most 3 will be unsuitable for analysis?
  (b) For a sample of 100 units to be selected at random:
    (1) what are the mean and variance of the possible numbers of unsuitable units in the sample?
    (2) what is the probability that the number of unsuitable units will be at most 30?

9-52 A school trustee has claimed that nonstandard tests administered at a particular school are too difficult and produce marks which are on average well below those that would be obtained on another school's accredited standard test. The school board has agreed that the first school should withdraw its nonstandard test if it can be proven at the 5% level of significance that average marks would be more than 2 below those that would be produced on the standard test. To investigate the claim, 15 students randomly selected were given the two tests, with the following resulting marks:

Student:	1	2	3	4	5	6	7	8
Standard test	47	89	66	77	55	70	72	85
Nonstandard test	44	83	64	76	53	63	67	78

Student:	9	10	11	12	13	14	15
Standard test	75	82	66	87	63	72	68
Nonstandard test	70	77	67	80	57	68	70

Should the nonstandard test be withdrawn? Explain your conclusion. What distribution assumptions are made in the analysis?

9-53 A game consists of spinning a pointer that can come to rest at any one of 50 equally likely points. Each point is indicated by a color (red, green, blue, white, or black) and a number $1, 2, \ldots,$ or 10). Each number occurs once with each color. A player places $1.00 on any point and wins $25.00 if the pointer comes to rest at that point or $2.50 if it comes to rest at any other point of the same color. What can the operator of the game expect to win on the average in an 8-week summer period if the game will be played by about 7000 players per week? (An individual playing twice counts as 2 players.)

9-54 In a survey of 250 parents in a community, 100 were sympathetic to teachers threatening strike action over classroom conditions. After the teachers' organization and the school board made public policy statements, the parents were surveyed again. Of the original 100 supporting the teachers, only 67 still supported them. Of the original 150 nonsympathizers, 24 now supported the teachers. Do these data indicate a change in the overall proportion of teacher supporters in the population?

9-55 In a study on parent care and child access to substances producing lead in the blood, children from low and high care-giving environments were assessed for the amount of lead in the bloodstream in micrograms per decaliter. Determine whether there is a significant difference and, if so, indicate the p-level if the measurements for 16 children from the low environment produced a mean of 40.3 and a standard deviation of 7.1, while the measurements for 12 children in the high environment produced mean and standard deviation 24.3 and 6.8.

9-56 Find a 95% confidence interval for the mean length of the femur in a species of terrestrial traveling primates if a sample of 12 produced a mean of 177.50 mm and a standard deviation of 17.66 mm.

9-57 A survey is to be conducted to estimate the percentage of members of a population who make regular use of an electronic calculator. It is a requirement of the survey that there be at least a 90% chance that the estimate will be within five percentage points of the true population percentage. How many members of the population should be included in the survey?

9-58 Repeat Exercise 9-57 if the population is a population of 7000 students and if it can be safely assumed that at least 5000 regularly use electronic calculators.

9-59 In a given time period a particular grocery wholesale company will handle 1.5 million cases of bananas. If there is only 1 chance in 1 million that an individual case of bananas will contain a poisonous spider, what is the probability that, in the time period, more than two of the company's cases will contain a spider?

9-60 Referring to the data in Example 9-11, determine a 95% confidence interval for the proportion of housing units in all five municipalities combined that meet or exceed the standards.

ANALYSIS
OF
VARIANCE

10-1 INTRODUCTION

In this chapter we use the results of sampling distributions for means and variances to develop a method of statistical inference referred to as *analysis of variance*. This procedure involves analyzing sample variances, but the ultimate purpose is to make inferences on the possible equality of two or more population means.

The procedure is based on general concepts that we have already considered in Chapters 2, 3, and 4: sampling distributions for statistics, measures of variability, and "usual" versus "unusual" events as defined by probability statements. The procedure also requires theory from Chapters 5, 6, and 7, but the intuitive concepts in this and the next two sections may be grasped on the basis of Chapters 1 through 4.

In the first three sections, we study the analysis of variance on an intuitive basis and consider only the simplest possible form of application. Our purpose in these sections is to develop an appreciation of the inference procedure prior to developing the full theory. The basic ideas of analysis of variance are illustrated in the following example.

EXAMPLE 10-1   Suppose that we are interested in determining whether three exercising devices, a bicycle, a treadmill, and a stepping device, may be considered equivalent. We are concerned with the time that a man requires to achieve maximum oxygen uptake when exercising with any of the devices. To be equivalent, the devices need not produce exactly the same time in every instance; rather, they should produce equal times on the average. To make the determination of equivalence, we consider three hypothetical populations, consisting of all of the required times that would result if all possible men exercised on each of the three devices. We then attempt to determine whether the means of these three populations are all equal. Obviously, we cannot determine the actual means of the three populations. Instead, we obtain, and study, an independent random sample from each.

We might, for example, study the performances of three homogeneous groups of five men each; that is, 15 men whose capabilities and performances on exercise devices are such that there should be no differences on the average from group to group if each group used the same device.

Suppose that we have three such groups selected at random and that we assign one group to the bicycle, one to the treadmill, and the other to the stepping device. We then obtain times for five men on each of the three devices. We have a sample of size 5 from each population. Suppose further that the mean times (in minutes) from the three samples are 11.4, 15.4, and 18.8. The sample means are quite different and appear to indicate that the population means that they represent are also different; however, to confirm this indication requires further analysis.

We consider the sampling distribution for means of samples of size 5 to see how much variability such sample means *should usually exhibit* if they come from populations with equal means. For the population structures that we consider, this variability would be the same as would be found if they *all come from the same population*. We then find the variability—the sample variance—that the sample means *do exhibit* and we compare it to the anticipated usual variability. If this sample variance appears to be "usual" for such cases, we can say that it is plausible that the three populations have equal means and that any apparent differences just represent usual chance variation; however, if this sample variance is unusually large, we can say that there is a low probability of obtaining this result when sampling from populations whose means are all equal and we can, therefore, conclude that the population means are not all equal.

Furthermore, since the groups of men were homogeneous, we can attribute any differences that exist among the population means to the effects of the exercise devices. ▲

We use analysis of variance for many such studies in which we wish to determine whether two or more populations are equivalent in the sense that they have equal means. Some other examples of studies using analysis of variance are: grass yield trials in Saskatchewan, a Peterborough study to determine whether orientation programs help reduce anxiety before entering a hospital, a study on flavor and aroma assessments on foods with varying preparation procedures, a study on the average percentage of single females in metropolitan areas for varying regions of the United States, an Australian study on memory skills of aboriginal children and white children, nitrogen fertilizer studies in Alberta, and a University of Wisconsin study of information retention capabilities after varying lengths of time.

10-2 BASIC CONCEPTS OF ANALYSIS OF VARIANCE

We now continue with Example 10-1 to develop the general methods of analysis of variance and the conditions that must exist for analysis of variance to be valid.

EXAMPLE 10-2 Suppose that the data for Example 10-1 are as follows:

Bicycle	11	8	14	9	15
Treadmill	17	19	14	12	15
Step	22	21	17	15	19

Each value or *observation* in the set of data is denoted as x_{ij}. The first subscript, i, refers to the exerciser, with bicycle, treadmill, and step represented by 1, 2, and 3, respectively. The second subscript, j, refers to the observation in a group. For example, x_{24} is the time for the fourth member of the treadmill group; accordingly, $x_{24} = 12$.

The sample mean for the first device is denoted as $\bar{x}_{1.}$, where the dot indicates that the subscript j is not present (since all values of j are represented in the values producing the mean). The subscript 1 is the value for i. The mean is calculated as

$$\bar{x}_{1.} = \frac{x_{11} + x_{12} + x_{13} + x_{14} + x_{15}}{5}$$

which may be written as

$$\bar{x}_{1.} = \sum_{j=1}^{5} \frac{x_{1j}}{5} = \frac{57}{5} = 11.4$$

Similarly, the sample means for the other exercisers are

$$\bar{x}_{2.} = \sum_{j=1}^{5} \frac{x_{2j}}{5} = \frac{77}{5} = 15.4$$

$$\bar{x}_{3.} = \sum_{j=1}^{5} \frac{x_{3j}}{5} = \frac{94}{5} = 18.8$$

The differences among the three sample means appear to indicate that the three exercisers are not equivalent. The bicycle, for example, appears to produce maximum uptake in a shorter time. Confirmation of the suspected differences requires further analysis. ▲

Assumptions

For this analysis to be valid, we must make some assumptions about the populations. We must assume that the populations all have the same form of probability/frequency curve, namely the normal distribution discussed in Chapter 5. The procedures will be reasonably valid provided that we have unimodel symmetric distributions with a bell shape similar to that for the normal. We also assume that the population variances are all equal. The conclusions from analysis of variance will still be valid if these assumptions are not exactly true, but they should be at least approximately true.

We make one further assumption—that the samples from the populations are independent as discussed in Chapter 4. Intuitively, we assume that the sample drawn from one of the populations does not affect the probabilities of the samples drawn from the others.

Intuitive Analysis

In the general comparison of three populations on the basis of samples of size 5, we wish to decide which of the following conditions is true: Either all three population means are equal or they are not. The second condition may occur because two means are equal and the other different or because all three means are different. If we denote the population means as μ_1, μ_2, and μ_3, the first condition is as illustrated in Figure 10-1(a), in which the solid curve represents the original populations which have a variance of σ^2, and the dashed curve represents the sampling distribution for means of samples of size 5. As we know from Chapter 4, this sampling distribution has a variance of σ^2/n, which in our case is $\sigma^2/5$. Possible examples of the two variations of the second condition are illustrated in Figure 10-1(b) and (c).

In the analysis that follows, we will not be able to distinguish between the two variations of the second condition. Our decision will involve a choice of one of two

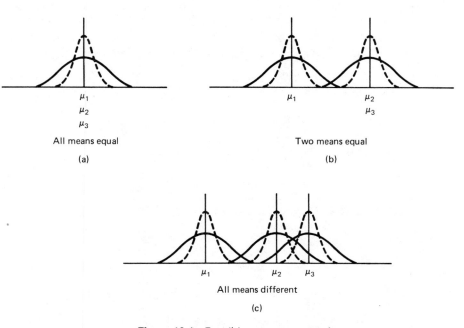

Figure 10-1 Possible arrangements of μ_1, μ_2, μ_3.

alternatives, either the population means are all equal or they are not all equal. To specify further requires procedures developed in Section 10-5.

Let us first consider the implications of the first alternative. If the population means are all equal, then, as illustrated in Figure 10-1(a), the three $\bar{x}$'s may be considered as coming from a single sampling distribution (i.e., a population) with a variance of $\sigma^2/5$. Furthermore, as illustrated in Chapter 4, the variance of a sample is, on the average, equal to the variance of the population from which the sample was drawn. Considering the $\bar{x}$'s as one sample of three values from a population (sampling distribution) with a variance of $\sigma^2/5$, we would thus expect the variance of three such $\bar{x}$'s to equal $\sigma^2/5$ on the average. In particular, we would anticipate that the variance of the three $\bar{x}$'s in the example would approximate $\sigma^2/5$.

If, on the other hand, the population means are not all equal, the three $\bar{x}$'s have come from a more varied situation such as illustrated in Figure 10-1(b) or (c). Since any such situation involves greater variation than that in Figure 10-1(a), the three $\bar{x}$'s will have come from a combined population (sampling distribution) with a variance greater than $\sigma^2/5$. We would thus expect the variance of three such $\bar{x}$'s to be greater than $\sigma^2/5$. In particular, we would, in such cases, anticipate that the variance of the $\bar{x}$'s in the sample would be greater than $\sigma^2/5$.

In summary, if the first condition is true, then subject to sampling variation, the three $\bar{x}$'s should fall within the sampling distribution illustrated in Figure 10-1(a) and exhibit a variance of approximately $\sigma^2/5$. If either variation of the second condition is true, the $\bar{x}$'s should be more varied since they come from a wider spread of sampling distributions as shown in Figure 10-1(b) and (c), and they should, therefore, exhibit a variance greater than $\sigma^2/5$. To arrive at a conclusion, we must calculate $s_{\bar{x}}^2$, the sample variance of the $\bar{x}$'s, to determine whether it is approximately $\sigma^2/5$ or greater.

We let $\bar{x}_{..}$ denote the overall mean or, equivalently, the mean of the three sample means, and the variance of the sample means becomes

$$s_{\bar{x}}^2 = \sum_{i=1}^{3} \frac{(\bar{x}_{i.} - \bar{x}_{..})^2}{2}$$

To find $s_{\bar{x}}^2$, we must first determine $\bar{x}_{..}$ which is calculated as

$$\bar{x}_{..} = \sum_{i=1}^{3} \frac{\bar{x}_{i.}}{3}$$

EXAMPLE 10-3   For the exercise data, $\bar{x}_{..} = 15.2$. We may equivalently find $\bar{x}_{..}$ as

$$\bar{x}_{..} = \frac{57/5 + 77/5 + 94/5}{3}$$

$$= \frac{57 + 77 + 94}{15} = \frac{228}{15} = 15.2$$

That is, the mean of the three $\bar{x}$'s is the sum of the 15 observations divided by 15, and is thus the overall mean of the 15 observations.   ▲

In mathematical notation, each $\bar{x}_i$. can be expressed as

$$\bar{x}_i = \sum_{j=1}^{5} \frac{x_{ij}}{5}$$

and we can write

$$\bar{x}.. = \sum_{i=1}^{3} \frac{\sum_{j=1}^{5} x_{ij}/5}{3}$$

$$= \frac{\sum_{i=1}^{3} \sum_{j=1}^{5} x_{ij}}{15}$$

The expression in the numerator is an example of *double summation notation* and is a form of grand total found by summing a number of subtotals. As with ordinary summation notation, we can modify the expression for partial sums.

EXAMPLE 10-4   If we want the overall total for treadmill and step, we find

$$\sum_{i=2}^{3} \sum_{j=1}^{5} x_{ij} = \sum_{j=1}^{5} x_{2j} + \sum_{j=1}^{5} x_{3j}$$

$$= (x_{21} + x_{22} + x_{23} + x_{24} + x_{25}) + (x_{31} + x_{32} + x_{33} + x_{34} + x_{35})$$

$$= 77 + 94 = 171$$

or, if we want the total time for the first two men in each group, we find

$$\sum_{i=1}^{3} \sum_{j=1}^{2} x_{ij} = \sum_{j=1}^{2} x_{1j} + \sum_{j=1}^{2} x_{2j} + \sum_{j=1}^{2} x_{3j}$$

$$= (x_{11} + x_{12}) + (x_{21} + x_{22}) + (x_{31} + x_{32})$$

$$= 19 + 36 + 43 = 98$$ ▲

Having found $\bar{x}..$, we may calculate $s_{\bar{x}}^2$ as defined; however, in general, we might consider the calculating formula from Section 3-3 that used the result

$$\sum (x - \bar{x})^2 = \sum x^2 - \frac{(\sum x)^2}{n},$$

The analogy for $s_{\bar{x}}^2$ for three groups is

$$\sum (\bar{x}_i - \bar{x}..)^2 = \sum \bar{x}_i^2 - \frac{(\sum \bar{x}_i)^2}{3}$$

producing the calculating formula

$$s_{\bar{x}}^2 = \frac{\sum \bar{x}_i^2 - (\sum \bar{x}_i)^2/3}{2}$$

EXAMPLE 10-5   For the exercisers, $s_{\bar{x}}^2$ becomes

$$\frac{\sum \bar{x}_{i.}^2 - (\sum \bar{x}_{i.})^2/3}{2} = \frac{11.4^2 + 15.4^2 + 18.8^2 - 45.6^2/3}{2} = \frac{27.44}{2} = 13.72$$

▲

Since our purpose in finding the value of $s_{\bar{x}}^2$ is to determine whether or not the variability of the sample means is usual, we must now compare this value with $\sigma^2/5$ or, equivalently, we must compare $5s_{\bar{x}}^2$ with σ^2, the variance of the original population. Unfortunately, we do not know σ^2; however, as noted in the previous work, if we have s^2 from a sample, it will serve as an estimate of σ^2. We, in fact, have several such estimates available in this example, one from each of the three samples.

We have assumed that each of the three populations exhibits the same variance σ^2. As a result, we know that variances from samples from any one of the populations will follow a sampling distribution with a mean of σ^2. We can thus consider any such sample variance as a reasonable estimate of the unknown σ^2.

In particular, we can estimate σ^2 with the variance of the first sample. Denoting this variance as s_1^2, we have

$$s_1^2 = \sum_{j=1}^{5} \frac{(x_{1j} - \bar{x}_{1.})^2}{4}$$

Again using the calculating formula approach of Section 3-3, we can also find this variance as

$$s_1^2 = \frac{\sum\limits_{j=1}^{5} x_{1j}^2 - \left(\sum\limits_{j=1}^{5} x_{1j}\right)^2 \Big/ 5}{4}$$

EXAMPLE 10-6   For the bicycle times, the sample variance is

$$s_1^2 = \frac{(11^2 + 8^2 + 14^2 + 9^2 + 15^2) - 57^2 \Big/ 5}{4} = 9.3$$

▲

We can also estimate σ^2 with the variance of the second sample,

$$s_2^2 = \frac{\sum\limits_{j=1}^{5} x_{2j}^2 - \left(\sum\limits_{j=1}^{5} x_{2j}\right)^2 \Big/ 5}{4} = 7.3$$

or we can estimate σ^2 with the variance of the third sample,

$$s_3^2 = \frac{\sum\limits_{j=1}^{5} x_{3j}^2 - \left(\sum\limits_{j=1}^{5} x_{3j}\right)^2 /5}{4} = 8.2$$

A better estimate is one that uses all the information available to us. To obtain the estimate of σ^2, we pool the information in the three sample values by taking their mean,

$$\overline{s^2} = \frac{s_1^2 + s_2^2 + s_3^2}{3}$$

EXAMPLE 10-7   For the exercise times, $\overline{s^2} = (9.3 + 7.3 + 8.2)/3 = 8.27$ ▲

Now, in order to determine whether $s_{\bar{x}}^2$, the variability among the sample means, is usual or larger than usual for samples from populations with equal means, we make the comparison between $5s_{\bar{x}}^2$ and $\overline{s^2}$, the estimate of σ^2.

EXAMPLE 10-8   Upon making the comparison for the exercisers, we find that $5s_{\bar{x}}^2 = 5 \times 13.72 = 68.6$ is considerably larger than $\overline{s^2} = 8.27$. In fact, it is 8.30 times larger, as determined by the variance ratio $5s_{\bar{x}}^2 / \overline{s^2} = 68.6/8.27 = 8.30$. This result appears to indicate that the variation among the sample means is larger than would be expected if the populations means were all equal and suggests that the three exercisers are not all equivalent.   ▲

Formal Decision

Before making a final decision, we note that the variance ratio, $5s_{\bar{x}}^2 / \overline{s^2}$, which is formed from sample statistics, is itself a sample statistic and is subject to chance variation. We expect this ratio to be close to 1 if the population means are all equal, but even if the population means are all equal it is still possible for the variance ratio to be quite large. The sample means may be spread out across the full range of the sampling distribution producing an unusually large $s_{\bar{x}}^2$. As well, the individual samples may be well clustered within the original populations, producing an unusually small s_i^2 for each sample and hence an unusually small $\overline{s^2}$. For either or both reasons, we may have a large variance ratio even if the population means are all equal. Although it is possible to obtain a large variance ratio in this manner, it is not very probable. We must now use probability to assist us in the final decision. We must consider one more sampling distribution.

If the population means are all equal, the variance ratio has a sampling distribution which is an F-distribution. The F-distribution, also discussed in Chapter 8, takes on different forms depending on the number of populations being compared and the size of each sample. The appropriate F-distribution for the variance ratio when five observations are taken from each of three populations with equal means is shown in Figure 10-2, in which the 95th and 99th percentiles are indicated.

EXAMPLE 10-9   The sample ratio 8.30 for the exercisers is in the upper 1%, well beyond the 99th percentile as illustrated in Figure 10-2. As such, it is an

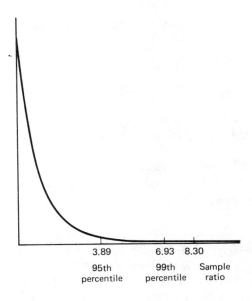

| 3.89 | 6.93 | 8.30 |
| 95th
percentile | 99th
percentile | Sample
ratio |

Figure 10-2 Sampling distribution for the variance ratio for samples of size 5 from three populations.

unusually large value to obtain from this distribution. If the population means are all equal, there is less than a 1% chance of obtaining a sample variance ratio as large as 8.30. We therefore do not expect to observe such a large value if the population means are all equal. Accordingly, we formally come to the conclusion that we originally suspected—that the population means are not all equal and that the exercisers are not equivalent. ▲

In general, whenever we wish to assess the possible equality of the means of two or more populations that can be assumed to be (approximately) normal with (approximately) equal variances, we obtain independent samples of size n from each and calculate the variance ratio $ns_{\bar{x}}^2 / \overline{s^2}$ and compare this with either the 95th or 99th percentile of the appropriate F-distribution. If the variance ratio exceeds the chosen percentile, we conclude that the population means are not all equal.

If the variance ratio does not exceed the chosen percentile, it is not classified as unusually large. The variation from group to group is not unusually large in comparison to variation within groups and does not provide sufficient evidence to declare that population means are not all equal. We therefore conclude that the population means may be equal.

The choice of percentile (95th or 99th) is a personal choice and depends on how unusually large a result must be (in the upper 5% or upper 1%) before we consider it too large to have resulted by chance from populations with equal means.

We denote the number of populations being investigated as k; then with a sample of size n from each population, we find the 95th and 99th percentiles, $F_{0.05;\, k-1,\, k(n-1)}$ and $F_{0.01;\, k-1,\, k(n-1)}$, respectively, in the row labeled with the value of $k(n-1)$ and the column labeled with the value of $k-1$ in Tables VIIa and

VIIb of Appendix B, respectively. Interpolation between tabulated values is required for values of $k(n - 1)$ and $k - 1$ not included in the tables.

EXAMPLE 10-10 For the exerciser data, $k = 3$ and $n = 5$ and we find that $F_{0.05;2,12} = 3.89$ in Table VIIa in the row labeled 12 and the column labeled 2, and we find $F_{0.01;2,12} = 6.93$ in the row labeled 12 and the column labeled 2 in Table VIIb.

▲

In general, then, the process of analysis of variance may be summarized in the following decision rule.

> The means of k normal populations with equal variances are declared to be not all equal on the basis of independent samples of size n if the variance ratio $ns_{\bar{x}}^2 / s^2$ exceeds $F_{0.05;\, k-1,\, k(n-1)}$ or $F_{0.01;\, k-1,\, k(n-1)}$, where the choice of F_α is a personal choice of the experimenter, and where $s_{\bar{x}}^2$ is the variance of the sample means and s^2 is the mean of the sample variances.

The general procedure is illustrated further in another example.

EXAMPLE 10-11 We wish to determine whether children with four different home backgrounds require different numbers of trials to learn a specific task adequately. Although the results will not come from a continuous distribution, suppose that it is not unreasonable to believe that the frequency distribution of all such numbers of trials would be well approximated by a normal curve. Suppose further that we will conclude that there are differences in the populations if the data in the samples produce a variance ratio that could occur at most 5% of the time when sampling from populations with equal means. We will, in other words, conclude that there are differences among the groups if the variance ratio exceeds $F_{0.05;\, 3,\, 4(n-1)}$ because we have $k = 4$ groups.

Let us suppose that seven children selected at random from each background were tested with the following results (numbers of trials to learn the task):

	Background		
1	2	3	4
19	24	23	26
27	25	16	21
23	18	16	20
26	23	19	18
22	19	15	23
25	26	21	25
20	23	22	19

These data produce the following means and variances:

| | Background | | | |
	1	2	3	4
$\bar{x}_{i\cdot}$	23.143	22.571	18.857	21.714
s_i^2	9.143	8.952	10.476	9.238

From the four $\bar{x}$'s, we obtain $s_{\bar{x}}^2 = 3.619$, and from the four s^2's we obtain $\overline{s^2} = 9.452$. The variance ratio is thus $n s_{\bar{x}}^2 / \overline{s^2} = 7 \times 3.619/9.452 = 2.68$.

The variation across groups is almost three times the corresponding variation within groups. With $k - 1 = 3$ and $k(n - 1) = 4 \times 6 = 24$ we find $F_{0.05;3,24} = 3.01$ in Table VIIa. In this case, as 2.68 is less than 3.01, the observed variance ratio does not exceed the corresponding 95th percentile. As illustrated in Figure 10-3, the variance ratio is not in the upper 5% of the sampling distribution that would apply if seven observations were obtained from four populations with equal means. (*Note*: This positively skewed distribution is more representative of the shape of the distribution for comparisons of four or more populations. The distribution in Figure 10-2 was particular to the comparison of three populations.) The variance ratio is not large enough to declare the population means not all equal.

Using $F_{0.05}$, we do not have sufficient evidence to declare the populations means to be not all equal. ▲

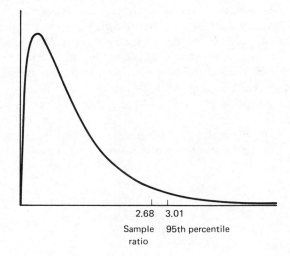

2.68   3.01

Sample   95th percentile
ratio

Figure 10-3 Sampling distribution for the variance ratio for samples of size 7 from four populations.

PRACTICE PROBLEMS

P10-1 Three extractants were each used for ten sample extractions of soil minerals. The quantities of zinc (in parts per million) produced the following results: For the first extractant, the ten sample values produce a mean of 2.40 and a standard deviation of 1.69. For the second, the mean and standard deviation were 3.19 and 2.35. For the third, the mean and standard deviation were 4.72 and 2.75. Do the three extractants differ with regard to the mean amount of zinc extracted? Use $F_{0.05}$ in the analysis.

P10-2 In a conditioning-for-stress analysis, 35 subjects were assigned to five groups of seven each. The members of each group experienced a conditioning program—a different program for each group—and then had their anxiety levels assessed with the following results:

		Group		
I	II	III	IV	V
59	58	50	44	51
54	52	46	36	45
52	46	39	50	51
64	53	43	49	56
68	43	44	51	60
60	51	42	49	48
57	49	54	43	61

Determine whether mean anxiety levels differ for the different programs using $F_{0.01}$.

10-1 Suppose that in an assessment of how densely packed snow was in three different regions, a researcher melted equal volumes of snow from ten sample points in each of the three regions and then compared the resulting volumes of water. The resulting means and variances for the three samples of ten volumes were:

$$\bar{x}_1. = 2.37, \ s_1^2 = 0.85$$

$$\bar{x}_2. = 3.41, \ s_2^2 = 0.98$$

$$\bar{x}_3. = 4.13, \ s_3^2 = 0.96$$

Using the $F_{0.01}$ value from tables, determine whether there are differences among the three regions.

10-2 In a study of four different preparations of lubricating oil, 6 liters were chosen at random from each of the four preparations and the viscosities (in centipoise) were as follows:

	Preparation		
I	II	III	IV
218	239	224	227
228	227	225	218
227	243	211	217
226	235	224	219
221	234	219	212
233	229	226	221

(a) Calculate the mean and variance for each group.
(b) Using the results of part (a), determine whether there are differences among the preparations (use $F_{0.05}$).

10-3 Suppose that independent samples of size 6 are to be taken from each of four normal populations with equal variances. If the population means are also all equal, what is the probability that the sample variance ratio will exceed
(a) $F_{0.05;3,20} = 3.10$?
(b) $F_{0.01;3,20} = 4.94$?

10-4 The following data represent the numbers of colonies of bacteria counted in different groups of soil portions—each group being treated with a different solution.

| | Group | | |
I	II	III	IV
63	62	68	67
61	67	66	71
64	64	64	68
65	65	65	67
61	63	69	69
62	66	68	70
64	63	67	67
60	64	66	70

(a) Calculate the mean number of colonies in each sample group.
(b) Calculate the variance for each sample.
(c) Determine whether the mean numbers of colonies differ for the different solutions.

10-5 A sociologist studying age at first marriage for males has information available from 15 marriages in each of three regions. In the first region the 15 ages produced a mean of 24.0 and a standard deviation of 3.2; in the second region, the 15 ages produced a mean of 25.4 and a standard deviation of 2.9; and in the third region, the 15 ages produced a mean of 23.6 and a standard deviation of 3.6. On the basis of these sample results, determine whether the mean age differs across the three regions.

10-6 What are the assumptions made in applying the analysis of variance?

SOLUTIONS TO PRACTICE PROBLEMS

P10-1 The $k = 3$ sample means produce $\Sigma \bar{x} = 10.31$ and $\Sigma \bar{x}^2 = 38.2145$, thus producing $s_{\bar{x}}^2 = (38.2145 - 10.31^2/3)/2 = 1.3912$. The three sample variances produce $\overline{s^2} = (1.69^2 + 2.35^2 + 2.75^2)/3 = 5.3137$. With $n = 10$ the variance ratio is thus $10 \times 1.3912/5.3137 = 2.62$. Consulting Table VIIa for $k - 1 = 2$ and $k(n - 1) = 27$ degrees of freedom, we find that $F_{0.05;2,27}$ falls between $F_{0.05;2,25} = 3.39$ and $F_{0.05;2,30} = 3.32$. As a result, 2.62 will not exceed $F_{0.05}$ and the extractants do not appear to differ.

P10-2 The means for the $k = 5$ groups are 59.143, 50.286, 45.429, 46.000, and 53.143. The variance of these five values is $s_{\bar{x}}^2 = 31.812$. The variances from the $n = 7$

observations in each of the groups are 30.810, 23.905, 25.952, 28.667, and 36.476. The mean of these five values is $s^2 = 29.162$. The variance ratio is thus $7 \times 31.812/29.162 = 7.64$. Consulting Table VIIb with $k - 1 = 4$ and $k(n - 1) = 30$ degrees of freedom, we find $F_{0.01;\,4,\,30} = 4.02$. Since 7.64 exceeds 4.02, we may conclude that the conditioning programs are not all the same.

10-3 CALCULATING FORMULAS AND ANALYSIS-OF-VARIANCE TABLE

The methods of calculation used in Section 10-2 served to explain the rationale for analysis of variance and, since the set of data was particularly simple, the calculations were not difficult; however, in general, data will not be as simple and calculating formulas similar to those previously used for variances are required to make the calculations more manageable.

Treatment Sum of Squares and Mean Square

In Section 10-2 we referred to the result $\Sigma (x - \bar{x})^2 = \Sigma x^2 - (\Sigma x)^2/n$, and we modified this result for the analysis of variance. For the variance of the means, $s_{\bar{x}}^2$ we apply the result to the numerator to produce

$$\Sigma (\bar{x}_{i.} - \bar{x}_{..})^2 = \Sigma \bar{x}_{i.}^2 - \frac{(\Sigma \bar{x}_{i.})^2}{k}$$

To modify further this expression, we introduce more notation for subtotals and totals.

Individual sample subtotals are expressed as

$$T_{i.} = \sum_{j=1}^{n} x_{ij}$$

and the overall total is expressed as

$$T_{..} = \sum_{i=1}^{k} T_{i.} = \sum_{i=1}^{k} \sum_{j=1}^{n} x_{ij}$$

We may thus express $\bar{x}_{i.}$ as $T_{i.}/n$ and $\bar{x}_{i.}^2$ as $T_{i.}^2/n^2$. As well, $\Sigma \bar{x}_{i.} = \Sigma T_{i.}/n = T_{..}/n$ and $(\Sigma \bar{x}_{i.})^2 = T_{..}^2/n^2$. Introducing these expressions into the numerator for $s_{\bar{x}}^2$ and multiplying by n, we obtain the numerator for $ns_{\bar{x}}^2$ as $\Sigma T_{i.}^2/n - T_{..}^2/N$, where $N = nk$ is the total number of observations.

This quantity is expressed as a difference, but as can be seen in the original expression, it is a sum of squared differences—it is a sum of squares. It is used in the measure of variation among (or across) the individual samples. These samples represent the individual populations which will, henceforth, be referred to as *treatments* and the expression is referred to as the *treatment sum of squares* and denoted as SS_{Tr}.

To complete the calculation of $ns_{\bar{x}}^2$, we must now introduce the denominator $k - 1$ to obtain $ns_{\bar{x}}^2 = SS_{Tr}/(k - 1)$. This quantity is referred to as the *treatment mean square* and is denoted as MS_{Tr}.

Error Sum of Squares and Mean Square

Using the same methods for the individual sample variances, we have the numerator for s_i^2 as $\Sigma \, (x_{ij} - \bar{x}_{i.})^2 = \Sigma \, x_{ij}^2 - (\Sigma \, x_{ij})^2/n = \Sigma \, x_{ij}^2 - T_{i.}^2/n$.

If we consider $\overline{s^2} = (s_1^2 + s_2^2 + \cdots + s_k^2)/k$, we may write

$$\overline{s^2} = \frac{\Sigma \, (x_{1j} - \bar{x}_{1.})^2/(n - 1) + \cdots + \Sigma \, (x_{kj} - \bar{x}_{k.})^2/(n - 1)}{k}$$

$$= \frac{\Sigma \, (x_{1j} - \bar{x}_{1.})^2 + \cdots + \Sigma \, (x_{kj} - x_{k.})^2}{k(n - 1)}$$

Again, we apply the previous calculating formula to obtain the numerator as

$$\left(\sum_{j=1}^{n} x_{1j}^2 - \frac{T_{1.}^2}{n} \right) + \cdots + \left(\sum_{j=1}^{n} x_{kj}^2 - \frac{T_{k.}^2}{n} \right)$$

$$= \sum_{j=1}^{n} x_{1j}^2 + \sum_{j=1}^{n} x_{2j}^2 + \cdots + \sum_{j=1}^{n} x_{kj}^2 - \frac{T_{1.}^2}{n} - \frac{T_{2.}^2}{n} - \cdots - \frac{T_{k.}^2}{n}$$

$$= \sum_{i=1}^{k} \sum_{j=1}^{n} x_{ij}^2 - \sum_{i=1}^{k} \frac{T_{i.}^2}{n}$$

We have another quantity expressed as a difference although it is actually a sum of squares. This sum of squares is used in the measurement of variation within samples. Such variation is referred to as chance or *error* variation and the sum of squares is referred to as the *error sum of squares* and denoted as SS_E.

For $\overline{s^2}$ we must use the denominator $k(n - 1)$ to obtain $\overline{s^2} = SS_E/k(n - 1)$. This quantity is referred to as the *error mean square* and is denoted as MS_E.

Total Sum of Squares

If we consider the sum $SS_{Tr} + SS_E$, we obtain

$$\Sigma \frac{T_{i.}^2}{n} - \frac{T_{..}^2}{N} + \Sigma\Sigma x_{ij}^2 - \Sigma \frac{T_{i.}^2}{n} = \Sigma\Sigma x_{ij}^2 - \frac{T_{..}^2}{N}$$

which is the sum of the squares of all the observations minus the square of the overall total divided by the number of observations. It is thus the calculating formula for $\Sigma\Sigma \, (x_{ij} - \bar{x}_{..})^2$, which would be used as the numerator of the variance if the entire set of data was considered as a single sample. As such, it is the sum of squares for the total variation in the data. It is referred to as the *total sum of squares* and is denoted as SS_T; thus,

$$SS_T = SS_{Tr} + SS_E$$

Summary of Calculating Formulas

The sums of squares are expressed in their calculating forms as

$$SS_{Tr} = \sum_{i=1}^{k} \frac{T_{i\cdot}^2}{n} - \frac{T_{\cdot\cdot}^2}{N}$$

$$SS_E = \sum_{i=1}^{k} \sum_{j=1}^{n} x_{ij}^2 - \sum_{i=1}^{k} \frac{T_{i\cdot}^2}{n}$$

$$SS_T = \sum_{i=1}^{k} \sum_{j=1}^{n} x_{ij}^2 - \frac{T_{\cdot\cdot}^2}{N}$$

Although they are expressed as differences, these quantities are sums of squares and *must always be positive*. The mean squares are expressed as

$$MS_{Tr} = \frac{SS_{Tr}}{k - 1}$$

$$MS_E = \frac{SS_E}{k(n - 1)}$$

These mean squares *must always be positive*. As noted, the numerators are sums of squares and thus positive. The denominators are also positive.

Degrees of Freedom

The denominators in the mean-squares formulas represent the numbers of independent elements involved in the sums of squares and are referred to as the *degrees of freedom*. For the *degrees of freedom for treatments* we have $k - 1$, and for the *degrees of freedom for error* we have $k(n - 1)$. If we add these together, we obtain the *total degrees of freedom* $k - 1 + k(n - 1) = N - 1$.

F-Ratio

The variance ratio is usually referred to as the *F-ratio* (recall that its sampling distribution is the *F*-distribution) and, in general, is calculated as

$$F = \frac{MS_{Tr}}{MS_E}$$

As a ratio of positive quantities, it *must always be positive*.

As stated previously, under the assumption of independent random samples from normal populations with equal variances, if all the population means are equal, the *F*-ratio follows an *F*-distribution. The particular form of the distribution depends on the number of treatments and the number of observations as reflected in the degrees of freedom.

In performing the analysis of variance, we have taken the total variation in the data and partitioned it into two components measuring variation among (or across) the individual samples and variation within the individual samples. The mean squares MS_{Tr} and MS_E are the variances $n s_{\bar{x}}^2$ and $\overline{s^2}$ used to form the variance ratio. The latter, MS_E or $\overline{s^2}$, provides an estimate of σ^2, the common variance of the original populations, and the former, $n s_{\bar{x}}^2$ or MS_{Tr}, is approximately equal to or tends to be larger than σ^2, depending on whether or not the population means are all equal. The variance ratio, $n s_{\bar{x}}^2 / \overline{s^2}$ or $F = MS_{Tr}/MS_E$, should thus be close to 1 or larger than 1, depending on whether or not the population means are all equal.

Decision Rule

The decision rule of the preceding section may be restated as follows:

> We declare the means of k normal populations with equal variances to be not all equal if, on the basis of independent random samples of size n, we find that $F = MS_{Tr}/MS_E$ exceeds either $F_{0.05;\, k-1,\, k(n-1)}$ or $F_{0.01;\, k-1,\, k(n-1)}$, where the choice of $F_{0.05}$ or $F_{0.01}$ is selected by the experimenter.

EXAMPLE 10-12 The revised notation and calculating formulas applied to the exerciser data in Example 10-2 produce

$$T_{1.} = 11 + 8 + 14 + 9 + 15 = 57$$

$$T_{2.} = 17 + \cdots + 15 = 77$$

$$T_{3.} = 22 + \cdots + 19 = 94$$

$$\sum T_{i.}^2 = 57^2 + 77^2 + 94^2 = 18{,}014$$

$$T_{..} = 57 + 77 + 94 = 228$$

$$T_{..}^2 = 51{,}984$$

$$\sum\sum x_{ij}^2 = 11^2 + 8^2 + 14^2 + \cdots + 19^2 = 3702$$

$$SS_{Tr} = \frac{18{,}014}{5} - \frac{51{,}984}{15} = 137.2$$

$$SS_E = 3702 - \frac{18{,}014}{5} = 99.2$$

$$SS_T = 3702 - \frac{51{,}984}{15} = 236.4$$

$$MS_{Tr} = \frac{137.2}{2} = 68.6$$

$$MS_E = \frac{99.2}{12} = 8.27$$

and the variance ratio is $F = 68.6/8.27 = 8.30$, as before. ▲

Analysis-of-Variance Table

After performing the various calculations for an analysis of variance, it is usual to tabulate the results in an *analysis-of-variance table* with headings and entries as follows:

Source of variation	Degrees of freedom	Sums of squares	Mean squares	F-ratio
Treatments	$k - 1$	SS_{Tr}	MS_{Tr}	MS_{Tr}/MS_E
Error	$k(n - 1)$	SS_E	MS_E	
Total	$N - 1$	SS_T		

EXAMPLE 10-13 For the exerciser data of Examples 10-2 and 10-12, the analysis-of-variance table is as follows:

Source of variation	Degrees of freedom	Sums of squares	Mean squares	F-ratio
Treatments	2	137.2	68.6	8.30
Error	12	99.2	8.27	
Total	14	236.4		

▲

Adjusting Data

Another example will further illustrate the general procedures and the effects of adjusting data.

EXAMPLE 10-14 We have six sample measurements of hardness on each of four batches of plastic material and we wish to determine whether the batches are equivalent with regard to hardness. We interpret "equivalent" as meaning that the four (hypothetical) populations of all possible hardness measurements have equal means (i.e., the batches have equal hardness on the average). This problem is suitable for analysis of variance if we can assume that the four populations have approximately normal form with equal variances and that the samples are independent. Let us suppose that we can make these assumptions and that the measurements are as follows:

	Batch		
1	2	3	4
54.0	51.0	53.1	50.7
50.8	52.2	42.0	53.8
55.0	50.0	41.5	54.9
47.1	53.9	50.8	55.8
47.6	57.5	50.0	51.2
50.0	55.2	45.8	56.7

Before performing the analysis of variance, we must decide whether to choose the critical F value as the $F_{0.05}$ or the $F_{0.01}$ point (i.e., the 95th or the 99th percentile). Making this choice now avoids possible (unconscious) bias after seeing the results of the calculations. Let us suppose that we decide that any F-ratio that is so large as to occur only 5% of the time with equal population means will cause us to declare that the means are not all equal. Such F's, in the upper 5% of the distribution, are "cut off" by the 95th percentile, $F_{0.05}$; therefore, we decide to use this as the critical value.

In analysis of variance, the calculations can often be simplified by adjusting the data as was done for calculating variances in Section 3-5. In this example, the three-digit data can be simplified to two digits by subtracting 50.0 from all the observations. The decimals can then be eliminated by multiplying all the new values by 10. The adjusted measurements are:

| | Batch | | |
1	2	3	4
40	10	31	7
8	22	-80	38
50	0	-85	49
-29	39	8	58
-24	75	0	12
0	52	-42	67

Let us consider how these adjustments affect the components of the analysis of variance table. The degrees of freedom are unchanged since they are determined only by the numbers of groups and observations. The sums of squares and the mean squares are squared measures of variation and are not affected by the relocation, but they are affected by the rescaling. They are multiplied by 100, the square of 10. The numerator and the denominator of the F-ratio are, accordingly, both multiplied by 100; thus, the F-ratio itself is unchanged. We can relocate and/or rescale the data to simplify the calculations and still produce exactly the same F-ratio.

Applying the general notation to this set of data, we have $k = 4$ treatments (the batches are the treatments) with $n = 6$ observations each for a total of $N = 24$ observations. The sample subtotals are $T_{1.} = 45$, $T_{2.} = 198$, $T_{3.} = -168$, and $T_{4.} = 231$. The overall total is $T_{..} = 306$ and the sum of the squares of all the observations is $\sum\sum x^2 = 44{,}320$. The sums of squares are:

$$SS_{Tr} = \sum \frac{T_{i.}^2}{n} - \frac{T_{..}^2}{N} = \frac{122{,}814}{6} - \frac{93{,}636}{24} = 16{,}567.5$$

$$SS_E = \sum\sum x^2 - \sum \frac{T_{i.}^2}{n} = 44{,}320 - \frac{122{,}814}{6} = 23{,}851.0$$

$$SS_T = \sum\sum x^2 - \frac{T_{..}^2}{N} = 44{,}320 - \frac{93{,}636}{24} = 40{,}418.5$$

Note that these values are all positive, as they should be, and that $SS_T = SS_{Tr} + SS_E$.

The degrees of freedom for treatments, error, and total are, respectively, $k - 1 = 3$, $k(n - 1) = 20$, and $N - 1 = 23$; thus, the necessary mean squares are

$$MS_{Tr} = \frac{SS_{Tr}}{k - 1} = \frac{16,567.5}{3} = 5522.5$$

$$MS_E = \frac{SS_E}{k(n - 1)} = \frac{23,851}{20} = 1192.55$$

and the F-ratio is

$$F = \frac{MS_{Tr}}{MS_E} = \frac{5522.5}{1192.55} = 4.63$$

Summarized in the analysis-of-variance table, we have these values as follows (note that, as may be done, we substitute "batches" for "treatments" to emphasize the fact that we are investigating possible batch differences):

Source of variation	Degrees of freedom	Sums of squares	Mean squares	F-ratio
Batches	3	16,567.5	5522.5	4.63
Error	20	23,851.0	1192.55	
Total	23	40,418.5		

If the population means are equal (i.e., if the batches are equivalent), the F-ratio should have a sampling distribution which is the F-distribution with 3 and 20 degrees of freedom.

From Table VIIa, $F_{0.05;3,20} = 3.10$, and as the sample $F = 4.63$ exceeds the chosen critical value, $F_{0.05} = 3.10$, it has (less than) a 5% chance of occurring when the means are all equal. We thus classify it as sufficiently unusual to cast doubt on the equality of the means and we declare that the batches of plastic material are not all equivalent with regard to hardness. ▲

If we had chosen $F_{0.01;3,\,20} = 4.94$ as the critical value in Example 10-14, the sample F would be considered only moderately large and would not be considered large enough to come to the conclusion that we reached; thus, we would have to conclude that the plastics may be equivalent. Note that, in this case, we cannot make as strong a statement. Because a very large sample F is not likely to arise when the means are all equal, we can, with such F's, be confident that the means are not all equal; however, as a moderately large F can occur frequently with equal or unequal means, we cannot, in these cases, make strong statements. This illustrates some of the need for deciding early what we consider extreme and what we consider moderate.

In a case such as this, with the sample F between $F_{0.05}$ and $F_{0.01}$, if we did not choose the critical value before the analysis, we might be influenced in the choice by a (subconscious) wish to arrive at a particular conclusion. This result illustrates further the importance of the prior choice of the critical value.

P10-3 Measurements on the amount of sediment on a lake bed in three different locations produced the following values:

	Location	
1	2	3
7.3	3.1	6.6
2.1	11.9	7.8
5.5	10.3	13.0
6.4	4.9	3.9
4.5	6.2	6.6
2.0	7.4	9.8
9.3	12.7	10.5
7.0	4.9	14.2

Using calculating formulas to find sums of squares, and so on, produce an analysis of variance table for these data and, using the $F_{0.05}$ value from Table VII, determine whether the mean amount of sediment varies across the three locations.

P10-4 The following data represent the percentages of recall by four groups of people after all had been presented with the same set of nonsense symbols to memorize with the methods of presentation varying from group to group.

	Group		
I	II	III	IV
62.5	71.3	65.1	57.0
68.3	77.6	70.6	51.8
58.7	69.3	73.8	58.3
61.6	67.6	63.4	63.1
55.0	71.4	59.3	52.4
61.7	63.2	66.5	61.0

Subtract 65.0 from each of the values above and then multiply these new values by 10. Perform an analysis of variance on these resulting values using the $F_{0.01}$ value and determine whether there exists any evidence of differences among the methods of presentation. Produce the analysis of variance table.

10-7 Apply calculating formulas to the task learning data in Example 10-11.

10-8 Do the analysis of variance in Exercise 10-2 using calculating formulas. Produce the analysis of variance table.

10-9 Do the analysis of variance for Exercise 10-4 using calculating formulas. Produce the analysis of variance table.

10-10 In an attempt to investigate the possible equivalence of four treatment processes with regard to the induced breaking strength, 28 sample pieces of material were obtained from a single run and grouped at random into four groups of seven. Each group was then assigned (randomly) to one of the four treatment processes and the following breaking strengths resulted.

	Treatment		
I	II	III	IV
31	26	18	38
36	29	14	45
25	19	43	39
29	11	11	37
39	21	17	30
21	14	24	41
14	24	35	34

Determine whether the treatment processes produce different mean breaking strengths. Summarize the results in an analysis of variance table.

10-11 If each of the values in a set of data to be analyzed by analysis of variance is multiplied by 100, what effect will this adjustment have on
(a) the treatment degrees of freedom?
(b) the mean square for error?
(c) the F-ratio?

10-12 In a study of different conditions involving various forms of distraction, a sample of stimulus intensities required to produce a response produced the following values.

	Condition		
I	II	III	IV
62.34	62.30	62.25	62.19
62.29	62.27	62.21	62.11
62.27	62.21	62.14	62.25
62.39	62.28	62.18	62.24
62.43	62.20	62.19	62.26
62.35	62.26	62.17	62.24

Multiply all these values by 100 and then subtract 6225 from each of the resulting values. Using calculating formulas, perform an analysis of variance on the adjusted data giving a complete analysis of variance table and determine whether there exists sufficient evidence using the $F_{0.05}$ value to declare the conditions not all equal.

10-13 In a study of four different tillage procedures, suppose that samples of a fixed quantity of soil from each of the procedures produced the following amounts of chemical fertilizer still in the soil after storm runoff (all areas received an equal application of the chemical fertilizer originally).

	Procedure		
I	II	III	IV
15.91	15.43	17.59	15.44
15.52	17.32	16.81	15.11
16.68	14.45	16.93	15.92
16.96	14.82	16.79	14.14
16.13	14.78	17.17	14.96
15.17	16.55	16.60	15.67
14.40	14.14	16.43	14.45

Multiply each of these values by 100 and then subtract 1400 from each of the resulting values. Using calculating formulas, perform an analysis of variance on the adjusted data giving a complete analysis of variance table and determine whether there exists sufficient evidence using the $F_{0.01}$ value to declare the procedures not all equal.

SOLUTIONS TO PRACTICE PROBLEMS

P10-3 For these data we have $k = 3$ groups of $n = 8$ observations each for a total of $N = 24$ observations. Summary totals are $T_1 = 44.1$, $T_2 = 61.4$, $T_3 = 72.4$, $T_{..} = 177.9$, $\sum T_{i.}^2 = 10{,}956.53$, and $\sum\sum x_{ij}^2 = 1588.57$. We thus have the sums of squares as

$$SS_{Tr} = \frac{10{,}956.53}{8} - \frac{177.9^2}{24} = 50.8825$$

$$SS_E = 1588.57 - \frac{10{,}956.53}{8} = 219.00375$$

$$SS_T = 1588.57 - \frac{177.9^2}{24} = 269.88625$$

Dividing the sums of squares by degrees of freedom to produce mean squares, we have the analysis-of-variance table:

Source of variation	Degrees of freedom	Sums of squares	Mean square	F-ratio
Locations	2	50.8825	25.44125	2.44
Error	21	219.00375	10.42875	
Total	23	269.88625		

Since 2.44 does not exceed $3.47 = F_{0.05;\, 2,\, 21}$, the mean amount of sediment does not appear to vary across locations.

P10-4 We have $k = 4$ groups of $n = 6$ observations each for a total of $N = 24$ observations. The adjusted data are:

	Group		
I	II	III	IV
-25	63	1	-80
33	126	56	-132
-63	43	88	-67
-34	26	-16	-19
-100	64	-57	-126
-33	-18	15	-40

Summary totals are: $T_{1.} = -222$, $T_{2.} = 304$, $T_{3.} = 87$, $T_{4.} = -464$, $T_{..} = -295$, $\Sigma T_i^2 = 364{,}565$, and $\Sigma\Sigma x_{ij}^2 = 105{,}479$. The treatment and error sums of squares and mean squares are:

$$SS_{Tr} = \frac{364{,}565}{6} - \frac{(-295)^2}{24} = 57{,}134.792$$

$$SS_E = 105{,}479 - \frac{364{,}565}{6} = 44{,}718.167$$

$$MS_{Tr} = \frac{57{,}134.792}{3} = 19{,}044.931$$

$$MS_E = \frac{44{,}718.167}{20} = 2235.908$$

and the variance ratio is $F = 19{,}044.931/2235.908 = 8.52$, which exceeds $F_{0.01;\,3,\,20} = 4.94$. There is evidence of differences among the methods of presentation.

10-4 GENERAL ONE-WAY ANALYSIS OF VARIANCE

In previous sections, the general ideas of analysis of variance were introduced in terms of the most basic form of problem—an analysis of the possible equality of the means of two or more populations that might differ in only one respect, the analysis being based on equal-size samples from each population. In some cases, equal-size samples may not be available. In other cases, some observations may be lost or contaminated and, therefore, not available for analysis. In this section we study the theoretical concepts of the procedure and we also modify the earlier techniques to suit cases with different sample sizes. The analysis in this section is still limited to determining whether or not the populations were all equivalent without specifying the differences if it is found that any existed.

In the next section we consider two examples of a number of techniques developed to determine where differences may exist in a set of population means. Formally, the methodology is the same as that in previous sections. The theoretical justification is based on the assumption that we have independent random samples from k normal populations with means $\mu_1, \mu_2, \ldots, \mu_k$ and common variance σ^2 and makes use of results subsequent to Chapters 1 through 4.

Formal Model

Referring to results stated in previous chapters, we recall that not only is each sample variance s_i^2 unbiased for σ^2, but, as well, each sample variance has a distribution related to a χ^2-distribution. Furthermore, since MS_E comes from a sum of the individual sample variances, it turns out that it too (as well as being unbiased for σ^2) has a distribution related to a χ^2-distribution.

In the formal development of the procedure, we introduce an "overall mean" $\mu = \sum \mu_i / k$. We then have the result that MS_{Tr} has a distribution related to a χ^2-distribution and that its expectation is $\sigma^2 + n \sum (\mu_i - \mu)^2 / (k - 1)$. As previously indicated intuitively, MS_{Tr} measures natural "error" variation within a group (as indicated by σ^2) plus variation among the groups [as indicated by $\sum (\mu_i - \mu)^2$]. If the population means are all equal, then each μ_i will equal μ, and $\sum (\mu_i - \mu)^2$ will equal 0; thus, as previously argued intuitively, MS_{Tr} will have expectation equal to σ^2. In summary, then,

> For independent samples of size n from each of k normal populations with common variance σ^2,
>
> $$E(MS_E) = \sigma^2$$
>
> $$E(MS_{Tr}) = \sigma^2 + \frac{n \sum (\mu_i - \mu)^2}{k - 1}$$
>
> If the populations all have equal means,
>
> $$E(MS_{Tr}) = \sigma^2$$

By considering the expectations, we now note formally what we previously argued intuitively; if all of the population means are equal, MS_{Tr} and MS_E will each equal σ^2 on the average and we can anticipate that their ratio will be close to 1. If the means are not all equal, MS_{Tr} will, on the average, exceed σ^2 and the ratio MS_{Tr}/MS_E will tend to exceed 1.

Furthermore, since a ratio of independent χ^2 variables modified by degrees of freedom produces a variable with an F-distribution, we have formal justification for the comparison of MS_{Tr}/MS_E with $F_{\alpha;\, k-1,\, k(n-1)}$.

Hypothesis Test

Having developed hypothesis-testing procedures in Chapter 7, we may also note now that the analysis of variance procedure is a hypothesis-testing procedure which is summarized in the usual format.

$$H_0: \mu_1 = \mu_2 = \mu_3 = \cdots = \mu_k$$

H_A: the μ_i's are not all equal

Test statistic: $F = \dfrac{MS_{Tr}}{MS_E}$

Decision rule: reject H_0 if $F > F_{\alpha;\, k-1,\, k(n-1)}$

The special case $k = 2$ is also considered in Section 8-2 as the test of H_0: $\mu_1 = \mu_2$ against H_A: $\mu_1 \ne \mu_2$ when σ_1^2 and σ_2^2 are unknown but assumed equal and the two samples are independent.

The two procedures are equivalent since the variance ratio F is equal to the square of the test statistic t in Section 8-2, which is

$$t = \frac{\bar{x}_1 - \bar{x}_2}{\sqrt{\dfrac{(n_2 - 1)s_1^2 + (n_2 - 1)s_2^2}{n_1 + n_2 - 2}\left(\dfrac{1}{n_1} + \dfrac{1}{n_2}\right)}}$$

but with $n_1 = n_2 = n$. Since $F_{\alpha;\, 1,\, 2(n-1)} = (t_{\alpha/2;\, 2(n-1)})^2$, the decision rule to reject H_0 if $|t| > t_{\alpha/2;\, n_1+n_2-2} = t_{\alpha/2;\, 2(n-1)}$ is the same as the decision rule to reject H_0 if $F > F_{\alpha;\, 1,\, 2(n-1)}$.

A variation of the model is to write $\alpha_i = \mu_i - \mu$ for the ith treatment and to interpret α_i as the "effect" of treatment i. From the definition, we note that $\sum_{i=1}^{k}\alpha_i = 0$. The ith population is then normal with mean $\mu + \alpha_i$ (because $\mu_i = \mu + \mu_i - \mu = \mu + \alpha_i$) and the hypothesis and alternative may be rewritten as

$$H_0: \alpha_i = 0 \qquad \text{for } i = 1, 2, \ldots, k$$

H_A: not all α_i's are 0

Unequal Sample Sizes

If the sample groups are not all of the same size, modifications to the procedures are required. The underlying theory is essentially as outlined above and we will not develop it further. We will, however, develop the modification to the methodology.

The previous method of analysis for equal sample sizes can be modified quite easily to accommodate the case of different sample sizes. The sample size for the ith group is denoted as n_i and the total number of observations becomes $N = \sum_{i=1}^{k} n_i$, where, as before, k is the number of populations.

The subtotal for the ith group is $T_i = \sum_{j=1}^{n_i} x_{ij}$, and the overall total is $T_{..} = \sum_{i=1}^{k} T_i = \sum_{i=1}^{k}\sum_{j=1}^{n_i} x_{ij}$. Similarly, the sum of the squares of all of the observations is $\sum_{i=1}^{k}\sum_{j=1}^{n_i} x_{ij}^2$. Note that these totals differ from the expressions in the previous sections only in the use of n_i instead of n. Means are also modified accordingly, so that the mean of the ith group is $\bar{x}_{i.} = T_{i.}/n_i$ and the overall mean becomes $\bar{x}_{..} = T_{..}/N$ or, equivalently, $\bar{x}_{..} = \sum_{i=1}^{k} n_i \bar{x}_{i.}/N$. The latter form for the overall mean indicates that it is a weighted mean of the individual sample means.

We make further use of weighting to provide the necessary modification of the methodology for analysis of variance for different sample sizes. The weights used in the treatment sum of squares are the individual samples sizes; thus,

$$SS_{Tr} = \sum_{j=1}^{k} n_i (\bar{x}_{i.} - \bar{\bar{x}}_{..})^2$$

The weights used in the error sum of squares are the individual sample degrees of freedom; thus,

$$SS_E = \sum_{i=1}^{k} (n_i - 1)s_i^2 = \sum_{i=1}^{k} \sum_{j=1}^{n_i} (x_{ij} - \bar{x}_{i.})^2$$

The total sum of squares is modified simply by using n_i in place of n; thus,

$$SS_T = \sum_{i=1}^{k} \sum_{j=1}^{n_i} (x_{ij} - \bar{x}_{..})^2$$

Calculating Formulas for Sums of Squares

Some algebraic manipulations of these expressions produce the following *calculating formulas*:

$$SS_{Tr} = \sum_{i=1}^{k} \frac{T_{i.}^2}{n_i} - \frac{T_{..}^2}{N}$$

$$SS_E = \sum_{i=1}^{k} \sum_{j=1}^{n_i} x_{ij}^2 - \sum_{i=1}^{k} \frac{T_{i.}^2}{n_i}$$

$$SS_T = \sum_{i=1}^{k} \sum_{j=1}^{n_i} x_{ij}^2 - \frac{T_{..}^2}{N}$$

Degrees of Freedom and Mean Squares

The *degrees of freedom* for the *total* and *treatments* sums of squares remain unchanged as $N - 1$ and $k - 1$, respectively, and we now express the *degrees of freedom* for *error* as $N - k$; thus, the *treatment* and *error mean squares* are:

$$MS_{Tr} = \frac{SS_{Tr}}{k - 1}$$

$$MS_E = \frac{SS_E}{N - k}$$

The analysis-of-variance table, the *F*-ratio, and the critical value, F_α, from tables are determined as in previous sections.

EXAMPLE 10-15 In a study on the effect of diet on sugar in the blood, 32 subjects were selected for their uniformity and assigned randomly to four diet groups, eight individuals being assigned to each diet.

A mishap resulted in the loss of the records for six subjects. The following data represent the results for the remaining cases:

	Diet		
I	II	III	IV
24	26	30	30
18	21	32	28
25	23	29	27
23	25	25	23
22	20	31	31
	24	33	25
	20	29	
		28	

For this set of data, we have

$$n_1 = 5, \quad n_2 = 7, \quad n_3 = 8, \quad n_4 = 6$$
$$T_{1.} = 112, \quad T_{2.} = 159, \quad T_{3.} = 237, \quad T_{4.} = 164, \quad T_{..} = 672,$$
$$\sum\sum x_{ij}^2 = 17{,}778$$

and, using the calculating formulas, we obtain

$$SS_{Tr} = \frac{112^2}{5} + \frac{159^2}{7} + \frac{237^2}{8} + \frac{164^2}{6} - \frac{672^2}{26}$$
$$= 17{,}624.163 - 17{,}368.615 = 255.548$$
$$SS_E = 17{,}778 - 17{,}624.163 = 153.837$$
$$SS_T = 17{,}778 - 17{,}368.615 = 409.385$$

and the analysis-of-variance table is

Source of variation	Degrees of freedom	Sums of squares	Mean squares	F-ratio
Treatments	3	255.548	85.183	12.18
Error	22	153.837	6.993	
Total	25	409.385		

Since 12.18 exceeds $F_{0.05;\,3,\,22} = 3.05$, we may reject the null hypothesis that the diets produce equal mean effects on sugar in the blood. ▲

PRACTICE PROBLEMS

P10-5 If independent samples of size 10 are to be taken from five normal populations with equal means μ and with equal variances σ^2 and an analysis of variance is to be performed on the results, what are the expected values of MS_{Tr} and MS_E?

P10-6 If independent samples of size 6 are to be taken from each of three normal populations with equal variances σ^2 and with means $\mu_1 = \mu + \sigma$, $\mu_2 = \mu$, and $\mu_3 = \mu - \sigma$, what are the expected values of the treatment and error mean squares that would result from an analysis of variance?

P10-7 A fabric was divided into two dozen sections and each of four dyeing processes was applied to six sections and, after some accidental losses, the following color fastness measurements resulted for the remaining sections.

	Dyeing Process		
I	II	III	IV
34	31	17	27
39	27	28	37
42	17	13	39
24	15	14	29
27	20	27	45
	13		36

Produce the analysis-of-variance table for these data and determine whether there are differences in mean color fastness.

P10-8 In a study on reaction to different prison environments, sample groups from three types of prison environments were assessed on a general scale for depression with the following results:

	Environment		
	1	2	3
n	10	28	22
$\bar{x}$	20.30	27.07	20.09
s	5.37	5.87	5.33

Do these data provide evidence of different mean depression levels at the 1% level of significance?

This problem may be solved by using the formulas for SS_{Tr} and SS_E that involve the $\bar{x}_{i\cdot}$ and s_i^2 values or the calculating formulas indicated in Exercise 10-18. Alternatively, the totals $T_{i\cdot}$, $T_{\cdot\cdot}$, $\sum_{j=1}^{n_i} x_{ij}^2$ and hence $\sum_{i=1}^{k}\sum_{j=1}^{n_i} x_{ij}^2$ may be reconstructed from the data summary. Note that the original data were whole numbers.

EXERCISES

10-14 A study was conducted to investigate the possible equality of the mean level of a contaminant in material from four sources. Varying numbers of sample readings were available from the four sources and they produced the following results:

Source			
A	B	C	D
0.032	0.036	0.026	0.038
0.036	0.038	0.039	0.032
0.035	0.033	0.034	0.045
0.039	0.043	0.033	0.036
0.029	0.046	0.032	0.042
0.041		0.038	
		0.029	

Determine at the 5% level whether there are differences across the four sources.

10-15 In an assessment of five different reading programs, a number of children judged equivalent in abilities on the basis of pretesting were assigned at random to the five programs. Assessments on the reading capabilities of the children completing the programs produced the following scores.

Program				
I	II	III	IV	V
63	81	72	59	62
67	71	77	65	71
59	74	79	70	73
60	70	83	71	67
72	73	70	67	68
58	83	82	60	61
65	79	71	62	68
	80	77		66
		73		
		78		

Determine whether the programs differ at the 5% level of significance.

10-16 Four different assessments on acidity of volcanic ash fallout produced the following summary results for the pH values.

	Assessment			
	1	2	3	4
n	10	15	5	10
$\bar{x}$	6.6	4.6	5.9	5.3
s	0.39	0.51	0.43	0.47

Are there differences at the 5% level in mean pH value?

10-17 Resort visitors were classified according to which of three activities were most important in their choice of resort and how many hours they spend in that activity, with the following results for the numbers of hours:

	Activity		
	Golf	Swimming	Tennis
n	33	40	27
$\bar{x}$	18.3	22.5	15.2
s	4.45	6.63	5.27

Do these data indicate different mean numbers of hours for the three activities?

10-18 Show algebraically that

(a) $\sum n_i(\bar{x}_{i.} - \bar{x}_{..})^2 = \sum n_i \bar{x}_{i.}^2 - \dfrac{(\sum n_i \bar{x}_{i.})^2}{N}$

$= \dfrac{\sum T_{i.}^2}{n_i} - \dfrac{T_{..}^2}{N}$

(b) $\sum(n_i - 1)s_i^2 = \sum\sum x_{ij}^2 - \dfrac{\sum T_{i.}^2}{n_i}$

SOLUTIONS TO PRACTICE PROBLEMS

P10-5 For independent samples from normal populations with equal variances σ^2 we have $E(MS_E) = \sigma^2$. If the means are also equal, we have $E(MS_{Tr}) = \sigma^2$ as well.

P10-6 As in Practice Problem P10-5, $E(MS_E) = \sigma^2$. With the population means as given, $\sum(\mu_i - \mu^2) = 2\sigma^2$, and for $k = 3$ and $n = 6$, we have $E(MS_{Tr}) = \sigma^2 + (6 \times 2\sigma^2/2) = 7\sigma^2$.

P10-7 Summary values from the data are $k = 4$, $n_1 = 5$, $n_2 = 6$, $n_3 = 5$, $n_4 = 6$, $N = 22$, $T_{1.} = 166$, $T_{2.} = 123$, $T_{3.} = 99$, $T_{4.} = 213$, $T_{..} = 601$, $\sum T_{i.}^2/n_i = 17{,}554.4$, $\sum\sum x_{ij}^2 = 18{,}467$. The sums of squares are then

$$SS_{Tr} = 17{,}554.4 - \frac{601^2}{22} = 1136.173$$

$$SS_E = 18{,}467 - 17{,}554.4 = 912.6$$

$$SS_T = 18{,}467 - \frac{601^2}{22} = 2048.773$$

and the analysis-of-variance table is

Source of variation	Degrees of freedom	Sums of squares	Mean squares	F-ratio
Processes	3	1136.173	378.724	7.47
Error	18	912.6	50.7	
Total	21	2048.773		

Since 7.47 exceeds $5.09 = F_{0.01;\,3,18}$, we have evidence of differences with $p < 0.01$.

P10-8 We may either refer to the method of calculating the (weighted) variance for a frequency distribution in Chapter 3 or use some algebra (see Exercise 10-18) to establish that $SS_{Tr} = \Sigma n_i(\bar{x}_{i.} - \bar{x}_{..})^2$ may be calculated as $\Sigma n_i\bar{x}_{i.}^2 - (\Sigma n_i\bar{x}_{i.})^2/N$. For the data, $\Sigma n_i\bar{x}_{i.}^2 = 33{,}518.2554$ and $\Sigma(n_i\bar{x}_{i.})^2/N = 32{,}804.0107$ produce $SS_{Tr} = 714.2447$. Alternatively, we may construct the group totals and overall totals by noting that $T_{i.} = n_i\bar{x}_{i.}$. If we do this we find the group totals to be 203, 757.96, and 441.98. If we knew that the assessments were whole numbers, we would realize that the sample means involved rounding error and would then adjust the totals to be 203, 758, and 442, and the overall total to be 1403. By the usual calculating formula, these values produce $SS_{Tr} = (203^2/10 + 758^2/28 + 442^2/22) - 1403^2/60 = 33{,}521.2247 - 32{,}806.8167 = 714.4080$. Since the assessments are, in fact, whole numbers, we will use this second result for SS_{Tr}. SS_E is found as $\Sigma(n_i - 1)s_i^2 = 9 \times 5.37^2 + 27 \times 5.87^2 + 21 \times 5.33^2 = 1786.4553$. Alternatively, realizing that individual observations are whole numbers we recall that since $s^2 = [\Sigma x^2 - (\Sigma x)^2/n]/(n - 1)$, we can find Σx^2 as $(n - 1)s^2 + (\Sigma x)^2/n$. Applying this procedure and rounding to whole numbers, we find the Σx^2 values for the three groups to be 4380, 21,450, and 9477, producing $\Sigma\Sigma x^2 = 35{,}307$ and $SS_E = 35{,}307 - 33{,}521.2247 = 1785.7753$. Again, we will use this alternative value. We calculate the mean squares as $MS_{Tr} = 714.4080/2 = 357.2040$ and $MS_E = 1785.7753/57 = 31.329$ and the variance ratio as $F = 357.204/31.329 = 11.40$, which exceeds $F_{0.01;\,2,\,57}$ since $F_{0.01;\,2,\,57}$ falls between $F_{0.01;\,2,\,40} = 5.18$ and $F_{0.01;\,1,\,60} = 4.98$. The data do provide evidence of different mean depression levels.

10-5 MULTIPLE COMPARISONS

In the blood sugar example in Section 10-4, we concluded that the diets were not all equivalent; however, we did not determine where the differences existed among the diets. To investigate further, we might apply a series of t-tests. There would then be $\binom{4}{2}$ different comparisons (not all independent) and from this collection of tests, the actual overall level of significance would be larger than the desired 5%.

For example, if only one mean of the four is different and we consider all the comparisons of pairs of the other three, we have $\binom{3}{2} = 3$ opportunities for making a type I error. If we do each test at the 5% level, the actual probability of making at least one type I error will be greater than 5%. In fact, if all the tests were independent (they are not), the probability of making at least one type I error would be $1 - P(all\ tests\ correct) = 1 - 0.95^3 = 0.143$; that is, the actual overall level of significance for the three tests would be 14.3%, not the chosen 5%.

Scheffé's Procedure

A number of procedures, referred to as *multiple comparison* procedures, have been developed to overcome this problem. One such procedure, Scheffé's procedure, requires only the use of F tables and is based on the same assumptions as analysis of variance.

Suppose that independent random samples are taken from each of k normal populations with the same variance σ^2. The sample of size n_i from the ith population produces mean $\bar{x}_i$. The unknown variance σ^2 is estimated with a statistic s^2 which has f degrees of freedom and is independent of the $\bar{x}_i$'s.

In simultaneous tests of equality of all the $\binom{k}{2}$ pairs of population means, an overall level of significance less than α is obtained if we declare any two population means μ_i and μ_l different whenever the two corresponding sample means $\bar{x}_i$ and $\bar{x}_l$ are such that their difference exceeds a critical value based on the F-distribution.

In this multiple comparisons procedure, we are testing a null hypothesis of equality for each pair of means. In the decision rule, we simultaneously judge all of these and declare which may be rejected. The procedure is as follows:

$$H_0 : \mu_i = \mu_l \text{ for } i \neq l;\ i, l = 1, 2, \ldots, k$$

Decision rule: declare $\mu_i \neq \mu_l$ if

$$|\bar{x}_i - \bar{x}_l| > \sqrt{(k-1)s^2 F_{\alpha;\,k-1,\,f}\left(\frac{1}{n_i} + \frac{1}{n_l}\right)}$$

In the case of one-way analysis of variance, s^2 is the error mean square, MS_E, and f is equal to the degrees of freedom for error, $N - k$.

EXAMPLE 10-16 For Example 10-15, $k = 4$, $s^2 = MS_E = 6.993$, $f = 22$, $\alpha = 0.05$, and $F_{\alpha;\,k-1,\,f} = F_{0.05;\,3,\,22} = 3.05$. We thus find the critical differences as:

$$\sqrt{3 \times 6.993 \times 3.05 \times \left(\frac{1}{n_i} + \frac{1}{n_l}\right)} = 7.999\sqrt{\frac{1}{n_i} + \frac{1}{n_l}}$$

The critical difference for comparing $\bar{x}_1$. and $\bar{x}_2$. is $7.999\sqrt{1/n_1 + 1/n_2}$ $= 7.999\sqrt{1/5 + 1/7} = 4.68$ since $n_1 = 5$ and $n_2 = 7$. Similarly, the critical differences for the other comparisons are:

$$7.999\sqrt{\frac{1}{5} + \frac{1}{8}} = 4.56$$

$$7.999\sqrt{\frac{1}{5} + \frac{1}{6}} = 4.84$$

$$7.999\sqrt{\frac{1}{7} + \frac{1}{8}} = 4.14$$

$$7.999\sqrt{\frac{1}{7} + \frac{1}{6}} = 4.45$$

$$7.999\sqrt{\frac{1}{8} + \frac{1}{6}} = 4.32$$

The four sample means are $\bar{x}_1. = 22.40$, $\bar{x}_2. = 22.71$, $\bar{x}_3. = 29.63$, and $\bar{x}_4. = 27.33$. There are six differences, $|\bar{x}_i. - \bar{x}_l.|$, which we tabulate as:

		i	
l	1	2	3
2	0.31		
3	7.23	6.92	
4	4.93	4.62	2.30

where, for example, $|\bar{x}_1. - \bar{x}_2.| = 0.31, |x_2. - x_3.| = 6.92$, and so on.

Each of the actual differences must be compared to the appropriate critical difference to determine whether it is significant (i.e., exceeds the critical difference). We can reduce the number of comparisons as follows. The differences 0.31 and 2.30 are less than 4.14, the smallest critical difference, and are, therefore, not significant; accordingly, we may accept the hypotheses $\mu_1 = \mu_2$ and $\mu_3 = \mu_4$. The differences 7.23, 4.93, and 6.92 exceed 4.84, the largest critical difference, and are, therefore, significant; accordingly, we may reject the hypotheses $\mu_1 = \mu_3$, $\mu_1 = \mu_4$, and $\mu_2 = \mu_3$. Finally, the difference $|\bar{x}_2 - \bar{x}_4| = 4.62$, which has not yet been considered, as it falls between 4.14 and 4.84, is compared to the critical difference 4.45, since $n_2 = 7$ and $n_4 = 6$. We again have a significant difference, $4.62 > 4.45$, and we may reject the hypothesis $\mu_2 = \mu_4$.

In summary, the analysis of variance provided only the information that the diets were not all equivalent; however, the multiple comparisons procedure provided the conclusion that the first and second diets were equivalent to each other and different from the third and fourth diets, which were equivalent to each other. ▲

Duncan's Multiple Range

Another multiple comparisons procedure that we will consider is based on *Duncan's multiple range*. It is again based on the same assumptions as analysis of variance and is the procedure that we will use *for cases in which the samples are all of the same size*.

In the assessment of the possible equality of a number of population means as indicated by sample means, multiple range procedures are based on the sampling distribution of the range of the sample means rather than the variance of the sample means as considered in analysis of variance. The range of a set of sample means, or any subset of the sample means, is compared to an appropriate critical value based on percentiles of the sampling distribution for the range. The appropriate critical value is the tabulated percentile for the range multiplied by $s/\sqrt{n}$, where s is, as before, the estimate of the common standard deviation (for analysis of variance models, $s = \sqrt{MS_E}$) and n is the number of observations in each sample.

In the multiple range procedure, two sample means differ significantly (leading to rejection of the null hypothesis of equality of the corresponding population means) if the range of every subset of sample means including them is significant.

The procedure based on Duncan's multiple range provides a compromise between the very general Scheffé's procedure, for which the probability of a type I error is actually less than α if all population means are equal, and the multiple set of t-tests for which the probability greatly exceeds α. Using Duncan's multiple range, the probability will exceed α, but not as much as would be the case with several t-tests. When we use a level of significance α in a multiple range test, it will be the probability of a type I error in a comparison of two sample means when the two population means are in fact equal. We will refer to α in these tests as the *base level of significance*.

The tabulated critical value of Duncan's multiple range for a base level of significance α is denoted as $r_{\alpha; p, f}$, where p is the number of means being compared and f is the number of degrees of freedom for s, the estimate of σ.

The rule for declaring a range of sample means significant is based on the critical value of $r_{\alpha; p, f}$.

For any set of p sample means, the range is declared significant with a base level of significance α if it exceeds the critical value $(s/\sqrt{n}) \times r_{\alpha; p, f}$.

For the analysis-of-variance model with k samples of n observations each, $f = k(n - 1)$. The range of any subset of p of the sample means is thus significant with base level of significance α if it exceeds $\sqrt{MS_E/n} \times r_{\alpha; p, k(n-1)}$. Values of the critical range are found in Tables VIIIa and VIIIb of Appendix B for 5% and 1% base levels of significance, respectively.

In an application of the multiple range procedure, the individual sample means are listed in order from smallest to largest and the range is tested as above. If the range is not significant, the population means may all be declared equal. If the range is significant, the population means are declared not all equal and further tests are made. The range of each of the two sets of $k - 1$ *adjacent means* is tested for significance. If either of these ranges is not significant, the corresponding population means are all declared equal and further testing among them ceases; otherwise, the testing continues until eventually (if necessary) pairs of adjacent means are tested. At each stage of testing, whenever a set of means produces a nonsignificant range, a line is drawn under the set, indicating that the hypotheses of equality may be accepted and no further comparisons *within the set* are made. At the end, any sample means not underlined by the same line are said to differ significantly and the null hypothesis that their corresponding population means are equal may be rejected.

EXAMPLE 10-17   As an illustration, suppose that in a study on the composition of the material in pottery found in five different sites, measurements on the amount of one of the components of the clay were made. Random samples of seven pieces of pottery from each site produced the following sample means: 58.39, 48.14, 44.47, 56.21, and 59.63, and corresponding sample variances: 20.382, 33.182, 29.716, 21.393, and 32.247. It is decided to test for differences among the five sites using the multiple range procedure and a 5% base level of significance. We might note that for these results, $MS_{Tr}/MS_E = ns_{\bar{x}}^2/s^2 = 313.279/27.384 = 11.44$, which exceeds $F_{0.05;4,30}$ indicating that, with an analysis of variance, we find that the compositions for the sites are not all the same on the average.

The $k = 5$ means from smallest to largest are 44.47, 48.14, 56.21, 58.39, and 59.63. The value for s^2 is the mean of the individual sample variances which is $MS_E = 27.384$, as found through analysis-of-variance procedures. Its corresponding degrees of freedom value is $f = k(n - 1) = 30$ because there are $n = 7$ observations per group.

In Table VIIIa, we find $r_{0.05;5,30} = 3.20$ and we compare the overall range $59.63 - 44.47 = 15.16$ to $(s/\sqrt{n}) \times r_{0.05;5,30} = \sqrt{MS_E/n} \times r_{0.05;5,30} = \sqrt{27.384/5} \times 3.20 = 2.34 \times 3.20 = 7.49$. Since 15.16 exceeds 7.49, we declare the population means not all equal (as also indicated by analysis of variance). Having found significance, we proceed further to test ranges of *adjacent sets* of $p = 4$ sample means in the list.

From Table VIIIa, we find $r_{0.05;4,30} = 3.13$ and calculate $2.34 \times 3.13 = 7.32$ as the critical value. The range for the first four adjacent means is $58.39 - 44.47 = 13.92$, which exceeds 7.32. We may say that the first four population means are not all equal and test them further. The range for the other four adjacent means is $59.63 - 48.14 = 11.49$, which also exceeds 7.32, leading to the same conclusions for these four populations. Having found significance in both cases, we proceed further to *adjacent sets* of $p = 3$ sample means.

From tables, $r_{0.05;3,30} = 3.03$, producing the critical value $2.34 \times 3.03 = 7.09$. The first three adjacent means have a significant range of 11.74 and may be tested further. The next three adjacent means have a significant range of 10.25 and may be tested further. The last set of three means produces a range of 3.42, which does not exceed 7.09 and hence is not significant. We draw a line under these three sample means and we may accept the null hypothesis that their corresponding populations have equal means and we perform no further tests among them. Having found significance among the first and second sets of three means, we proceed to test these means further in sets of $p = 2$.

From tables, $r_{0.05;2,30} = 2.89$, producing a critical value of $2.34 \times 2.89 = 6.76$. The range for the first two means is 3.67. Since it does not exceed 6.76, it is not significant. We draw a line under these means and we may accept the null hypothesis that their corresponding population means are equal. The next two means produce a significant range 8.07 and we declare that their corresponding population means are not equal. Since the last three means have already been declared equal, we do not test pairs among them. The following list of the sample means with corresponding pottery sites indicates the resulting grouping by underlining.

Third site, $\bar{x}_3$	Second site, $\bar{x}_2$	Fourth site, $\bar{x}_4$	First site, $\bar{x}_1$	Fifth site $\bar{x}_5$
44.47	48.14	56.21	58.39	59.63

We may accept the null hypothesis that $\mu_3 = \mu_2$; that is, the mean compositions are the same in the second and third sites. We may also accept the null hypothesis that $\mu_1 = \mu_4 = \mu_5$; that is, the mean compositions are the same for sites one, four, and five. Furthermore, we may declare that the second and third sites differ from the first, fourth, and fifth.

The steps of comparison above may be summarized in the following table, indicating which sites are compared at each stage, the range of the sample means included (i.e., largest minus smallest), the number of means p, Duncan's critical value r, the resulting critical value for the range $(s/\sqrt{n}) \times r$, and the classification (sig. for significant and n. sig. for not significant) or an indication that no test is necessary due to previous lack of significance.

Step	Sites compared	Range	p	r	$(s/\sqrt{n}) \times r$	Classification
1	1, 2, 3, 4, 5	15.16	5	3.20	7.49	sig.
2	1, 2, 3, 4	13.92	4	3.13	7.32	sig.
3	1, 2, 4, 5	11.49	4	3.13	7.32	sig.
4	2, 3, 4	11.74	3	3.03	7.09	sig.
5	1, 2, 4	10.25	3	3.03	7.09	sig.
6	1, 4, 5	3.42	3	3.03	7.09	n. sig.
7	2, 3	3.67	2	2.89	6.76	n. sig.
8	2, 4	8.07	2	2.89	6.76	sig.
9	1, 4	—		no test,	see step 6	—
10	1, 5	—		no test,	see step 6	—

▲

PRACTICE PROBLEMS

P10-9 Determine which age groups differ significantly at the 5% level with regard to mean percent body fat on the basis of the following data summary from a random sample of 356 men.

Age	n	Percent body fat	
		$\bar{x}$	s
17–22	187	16.30	5.90
23–29	96	18.85	7.25
30–39	43	22.97	6.73
40–49	30	26.00	6.19

P10-10 A study was conducted to observe the effect of altitude above sea level on the percentage area of land under woodland. Suppose that five areas were considered for each of the four altitude levels listed, with the following percentages under woodland:

0 – 249 ft	250 – 499 ft	500 – 749 ft	750 – 999 ft
41	49	47	42
45	48	47	44
43	44	51	43
44	45	48	45
41	46	49	47

(a) Calculate the mean percentage under woodland for each altitude level.
(b) Determine any significant differences among the groups with regard to mean percent under woodland. Use a 5% base level of significance.

EXERCISES

10-19 Referring to the data in Exercise 10-15, determine which programs differ using an overall 1% level of significance.

10-20 In a study, a number of subjects were classified according to their usual sleeping patterns as "very light," "light," "moderate," or "heavy" sleepers. Performance scores achieved in an assessment were as follows for the subjects in the four groups.

Very Light	Light	Moderate	Heavy
148	186	206	161
153	192	209	165
127	149	178	123
139	160	162	126
125	178	181	129
120		167	119
159		193	158
		202	
		163	

(a) Calculate the mean performance score for each sample group.
(b) Using an overall 5% level of significance, determine what differences exist among the four groups with regard to average performance.

10-21 Referring to the data in Exercise 10-1, determine which regions differ on the basis of a multiple range testing procedure with a 1% base level of significance.

10-22 Using a base significance level of 5%, determine what differences exist among the solutions referred to in Exercise 10-4.

10-23 Using an overall 5% level of significance, determine any significant differences that exist among mean numbers of days absent for participants in a community-based training program with different living arrangements:

Living arrangements	n	$\bar{x}$	s^2
With natural parents	58	5.48	13.362
With relatives	10	2.80	11.357
With foster parents	14	3.00	8.864
In community facilities	82	1.68	4.534
Living independently	15	2.80	10.110

SOLUTIONS TO PRACTICE PROBLEMS

P10-9 The table of individual differences is as follows (the marginal entries are the means themselves with sample sizes in parentheses).

	16.30 (187)	18.85 (96)	22.97 (43)
18.85 (96)	2.55		
22.97 (43)	6.67	4.12	
26.00 (30)	9.70	7.15	3.03

The estimator of σ^2 is the weighted mean of the individual sample variances, with degrees of freedom as weights.

$$s^2 = \frac{186 \times 5.90^2 + 95 \times 7.25^2 + 42 \times 6.73^2 + 29 \times 6.19^2}{352} = 41.14$$

For $F_{0.05;\, k-1,\, N-k} = F_{0.05;\, 3,\, 352}$, we use $F_{0.05;\, 3,\, \infty} = 2.60$. Critical differences are then found as

$$\sqrt{3 \times 2.60 \times 41.14 \left(\frac{1}{n_i} + \frac{1}{n_l} \right)} = 17.913 \sqrt{\frac{1}{n_i} + \frac{1}{n_l}}$$

Using the largest sample sizes $n_1 = 187$ and $n_2 = 96$, we find the smallest critical difference as $17.913\sqrt{1/187 + 1/96} = 2.25$. All actual differences exceed this value and *may* be significant. Because the first actual difference, 2.55, is based on the two largest samples, it is significant. Using the smallest sample sizes, $n_3 = 43$ and $n_4 = 30$, we find the largest critical difference as $17.913\sqrt{1/43 + 1/30} = 4.26$. This value is exceeded by 6.67, 9.70, and 7.15; hence, these three actual differences are significant. The actual difference 3.03 is not significant, as it is based on samples of size 43 and 30 and does not exceed 4.26. Finally, for the unresolved actual difference 4.12 based on sample sizes 96 and 43, we calculate the critical difference $17.913\sqrt{1/96 + 1/43} = 3.29$. Since 4.12 exceeds 3.29, it, too, is significant. We have thus found that all groups except the last two differ significantly.

P10-10 (a) The means for the four altitude levels (in order of altitude) are $\bar{x}_1. = 42.8$, $\bar{x}_2. = 46.4$, $\bar{x}_3. = 48.4$, and $\bar{x}_4. = 44.2$.

(b) With group totals $T_1. = 214$, $T_2. = 232$, $T_3. = 242$, and $T_4. = 221$ and overall sum of squares $\sum\sum x^2 = 41{,}461$, we find the error sum of square corresponding to an analysis of variance to be $41{,}461 - (214^2 + \cdots + 221^2)/5 = 56$, and the error mean square to be $s^2 = MS_E = 56/16 = 3.5$. The means listed in order of magnitude are $\bar{x}_1. = 42.8$, $\bar{x}_4. = 44.2$, $\bar{x}_2. = 46.4$, and $\bar{x}_3. = 48.4$. From Table VIII, we find $r_{0.05;\, p, 16}$ for $p = 4$, 3, and 2 to be 3.23, 3.14, and 3.00. Significant ranges are then found by multiplying these values by $s/\sqrt{n} = \sqrt{s^2/n} = \sqrt{3.5/5} = 0.84$. The step-by-step procedure is summarized as follows:

Step	Altitudes compared	Range	p	r	$(s/\sqrt{n}) \times r$	Classification
1	1, 2, 3, 4	5.6	4	3.23	2.7	sig.
2	1, 2, 4	3.6	3	3.14	2.6	sig.
3	2, 3, 4	4.2	3	3.14	2.6	sig.
4	1, 4	1.4	2	3.00	2.5	n. sig.
5	2, 4	2.2	2	3.00	2.5	n. sig.
6	3, 4	2.0	2	3.00	2.5	n. sig.

The underlined summary list is:

$\bar{x}_1.$	$\bar{x}_4.$	$\bar{x}_2.$	$\bar{x}_3.$
42.8	44.2	46.4	48.4

This result leads to some potential confusion because of the overlapping lines. Two things equal to the same thing are equal to each other in the usual rules of equality. In our listing, however, we have conclusions such as $\mu_1 = \mu_4$, $\mu_4 = \mu_2$, but $\mu_1 \ne \mu_2$. In fact, we have not violated the rules of equalities since we have not concluded definitely that $\mu_1 = \mu_4$, $\mu_4 = \mu_2$, and so on. We have merely found that they do not differ significantly; that is, we have, loosely speaking, found that they are close. Two things close to the same thing (but on opposite sides) are not necessarily close to each other. We have, in summary, determined that the first and second altitude levels differ significantly; the first and third differ significantly; and the third and fourth differ significantly.

10-6 TWO-WAY ANALYSIS OF VARIANCE

Populations may differ in more than one respect and, in investigating potential differences among means, we must take into account these other sources of variations. The mean mileage that we obtain from a vehicle may vary not only with different fuel mixtures, for example, but also with different types of vehicles.

The case that we have already considered, in which populations differ in only one respect (i.e., in only one way), is referred to as *one-way analysis of variance*. For cases in which the populations differ in two respects, the analysis is referred to as *two-way analysis of variance*.

In this section we develop the methodology for two-way analysis of variance through an example.

EXAMPLE 10-18 An experiment was conducted to investigate at the 5% level the possible equivalence of three different fuel mixtures with regard to average (i.e., mean) mileage. To study the fuels under different conditions, the experiment involved four types of vehicles, and each fuel was used once with each type of vehicle to ensure that no fuel received preferential treatment. The mileages resulting from the experiment were:

Type of vehicle, j	Fuels, i		
	1	2	3
1	19.5	16.0	23.2
2	20.7	17.4	22.5
3	26.2	15.5	22.5
4	29.4	27.6	35.6

This set of data should not be analyzed with a one-way analysis since there are two sources of variation: fuels and vehicle types. The treatment sum of squares would still be appropriate as a measure of chance variation plus variation among fuel types; however, the error mean square from one-way analysis of variance would produce a measure of chance variation plus variation among vehicle types rather than just a measure of chance variation. The error sum of squares is thus inflated by the effect of variations among vehicles, if they exist. This result is due to the fact that the four values in each fuel group represent outcomes for four different vehicle types.

To overcome this "inflation" of the error mean square that would result from the usual one-way analysis of variance, we remove the variation among the types of vehicles by finding an appropriate sum of squares for vehicles and subtracting it from the value that we would have otherwise used as the error sum of squares. Denoting the subtotal for the jth type of vehicle as $T_{\cdot j} = \sum_{i=1}^{3} x_{ij}$ and the mean for the jth type of vehicle as $\bar{x}_{\cdot j} = T_{\cdot j}/3$, the sum of squares for vehicles is analogous to that for fuels: $\sum_{j=1}^{4} 3(\bar{x}_{\cdot j} - \bar{x}_{\cdot \cdot})^2 = \sum_{j=1}^{4} T_{\cdot j}^2/3 - T_{\cdot \cdot}^2/12$. Subtracting this value from the previous error sum of squares, we find a residual sum of squares which we will use to estimate the variation due to error. Intuitively, we may argue that we have taken the total variation and removed from it variation due to fuel mixtures and variation due to vehicle types. The remaining variation or residual variation is not assigned to either source of variation but is attributed to the natural chance variation or error variation. By taking the difference of the calculating formulas, we find this residual sum of squares (i.e., the error sum of squares) in a calculating form as

$$\left(\sum_{i=1}^{3} \sum_{j=1}^{4} x_{ij}^2 - \sum_{i=1}^{3} \frac{T_{i\cdot}^2}{4} \right) - \left(\sum_{j=1}^{4} \frac{T_{\cdot j}^2}{3} - \frac{T_{\cdot\cdot}^2}{12} \right)$$

$$= \sum_{j=1}^{3} \sum_{j=1}^{4} x_{ij}^2 - \sum_{i=1}^{3} \frac{T_{i\cdot}^2}{4} - \sum_{j=1}^{4} \frac{T_{\cdot j}^2}{3} + \frac{T_{\cdot\cdot}^2}{12}$$

Since there are four types of vehicles, the value for the degrees of freedom for vehicles is 3 and the value for the degrees of freedom for error that would result from a one-way analysis is reduced by this amount to produce the new degrees of freedom for residual or error as $9 - 3 = 6$. ▲

Randomized Block Design

Just as the fuels are referred to as treatments, the vehicle types, which are the different types of experimental units, are referred to as *blocks*. With the fuels, or treatments, assigned at random to each individual vehicle, or experimental unit, in the vehicle types, or blocks, the design is referred to as a randomized block design. The sums of squares for total and treatments are denoted as before and we now use SS_{Bl} to denote the sum of squares for blocks. The residual sum of squares will continue to be denoted as SS_E because it is the value that we use to measure error variation.

EXAMPLE 10-18 (CONTINUED) With SS_T and SS_{Tr} found in the usual fashion and with SS_{Bl} and SS_E found as outlined above, we now perform a two-way analysis of variance on the data. The totals from the data are:

$$T_{1\cdot} = 95.8, \quad T_{2\cdot} = 76.5, \quad T_{3\cdot} = 103.8$$
$$T_{\cdot 1} = 58.7, \quad T_{\cdot 2} = 60.6, \quad T_{\cdot 3} = 64.2, \quad T_{\cdot 4} = 92.6$$
$$T_{\cdot\cdot} = 276.1, \quad \sum\sum x^2 = 6738.41$$

The sums of squares are:

$$SS_{Tr} = \frac{95.8^2 + 76.5^2 + 103.8^2}{4} - \frac{276.1^2}{12}$$
$$= 6451.08 - 6352.60 = 98.48$$
$$SS_{Bl} = \frac{58.7^2 + 60.6^2 + 64.2^2 + 92.6^2}{3} - 6352.60$$
$$= 6604.82 - 6352.60 = 252.22$$
$$SS_E = 6738.41 - 6451.08 - 6604.82 + 6352.60 = 35.11$$
$$SS_T = 6738.41 - 6352.60 = 385.81$$

The treatment and error mean squares are

$$MS_{Tr} = \frac{98.48}{2} = 49.24$$

$$MS_E = \frac{35.11}{6} = 5.85$$

and the F-ratio is $49.24/5.85 = 8.42$, which is significant because it exceeds $F_{0.05;2,6} = 5.14$. In this analysis we find that the fuels do differ.

The analysis-of-variance table summarizing the results is:

Source of variation	Degrees of freedom	Sums of squares	Mean squares	F-ratio
Treatments	2	98.48	49.24	8.42
Blocks	3	252.22		
Error (residual)	6	35.11	5.85	
Total	11	385.81		

Denoting the block mean square as MS_{Bl}, we may also investigate the block effect by calculating $MS_{Bl} = SS_{Bl}/3 = 252.22/3 = 84.07$ and the F-ratio $MS_{Bl}/MS_E = 84.07/5.85 = 14.37$, which exceeds $F_{0.05;3,6} = 4.76$, indicating that the types of vehicles differ with respect to average mileage.

It is interesting in this example to note the difference between the results from the appropriate two-way analysis and an inappropriate one-way analysis with an inflated error mean square.

If we analyze this set of data with a one-way analysis of variance with fuels as treatments, the sums of squares are

$$SS_{Tr} = \frac{95.8^2 + 76.5^2 + 103.8^2}{4} - \frac{276.1}{12}$$

$$= 6451.08 - 6352.60 = 98.48$$

$$SS_E = 6738.41 - 6451.08 = 287.33$$

$$SS_T = 6738.41 - 6352.60 = 385.81$$

The treatment and error mean squares are

$$MS_{Tr} = \frac{98.48}{2} = 49.24$$

$$MS_E = \frac{287.33}{9} = 31.93$$

and the F-ratio is $49.24/31.93 = 1.54$, which is not significant because it does not exceed $F_{0.05;2,9} = 4.26$.

The variation among the fuels is apparently not greater than the variation within the groups which we have, in the one-way analysis, considered to be random error. In this experiment, however, as previously noted, the variation within the groups includes not only error, but also variation among the types of vehicles and the one-way analysis uses an inappropriate value for the error mean square, one that turns out to be too large.

The appropriate two-way analysis partitions 287.33, the variation within groups, into components 252.22, the variation due to vehicle types, and 35.11, the residual variation now used for error.

▲

The procedures developed in this example can be generalized to produce the methodology for any such problem.

General Two-Way Analysis of Variance for Randomized Blocks

The general formulas and analysis of variance for the randomized block design with k treatments and m blocks for a total of $N = km$ observations are as follows:

The subtotal for the ith treatment is

$$T_{i\cdot} = \sum_{j=1}^{m} x_{ij}$$

The subtotal for the jth block is

$$T_{\cdot j} = \sum_{i=1}^{k} x_{ij}$$

The overall total is

$$T_{\cdot\cdot} = \sum_{i=1}^{k} \sum_{j=1}^{m} x_{ij}$$

The corresponding means are $\bar{x}_{i\cdot} = T_{i\cdot}/m$, $\bar{x}_{\cdot j} = T_{\cdot j}/k$, and $\bar{x}_{\cdot\cdot} = T_{\cdot\cdot}/N$. The sums of squares, together with calculating formulas, are

$$SS_{Tr} = \sum_{i=1}^{k} m(\bar{x}_{i\cdot} - \bar{x}_{\cdot\cdot})^2 = \sum_{i=1}^{k} \frac{T_{i\cdot}^2}{m} - \frac{T_{\cdot\cdot}^2}{N}$$

$$SS_{Bl} = \sum_{j=1}^{m} k(\bar{x}_{\cdot j} - \bar{x}_{\cdot\cdot})^2 = \sum_{j=1}^{m} \frac{T_{\cdot j}^2}{k} - \frac{T_{\cdot\cdot}^2}{N}$$

$$SS_E = \sum_{i=1}^{k} \sum_{j=1}^{m} (x_{ij} - \bar{x}_{i\cdot} - \bar{x}_{\cdot j} + \bar{x}_{\cdot\cdot})^2$$

$$= \sum_{i=1}^{k} \sum_{j=1}^{m} x_{ij}^2 - \sum_{i=1}^{k} \frac{T_{i\cdot}^2}{m} - \sum_{j=1}^{m} \frac{T_{\cdot j}^2}{k} + \frac{T_{\cdot\cdot}^2}{N}$$

$$SS_T = \sum_{i=1}^{k} \sum_{j=1}^{m} (x_{ij} - \bar{x}_{\cdot\cdot})^2 = \sum_{i=1}^{k} \sum_{j=1}^{m} x_{ij}^2 - \frac{T_{\cdot\cdot}^2}{N}$$

The degrees of freedom for treatments, blocks, error, and total are, respectively,

$k - 1, m - 1, (k - 1)(m - 1)$, and $N - 1$. The mean squares for treatments, blocks, and error are

$$MS_{Tr} = \frac{SS_{Tr}}{k - 1}$$

$$MS_{Bl} = \frac{SS_{Bl}}{m - 1}$$

$$MS_E = \frac{SS_E}{(k - 1)(m - 1)}$$

The analysis-of-variance table is:

Source of variation	Degrees of freedom	Sums of squares	Mean squares	F-ratio
Treatments	$k - 1$	SS_{Tr}	SS_{Tr}	MS_{Tr}/MS_E
Blocks	$m - 1$	SS_{Bl}	MS_{Bl}	MS_{Bl}/MS_E
Error (residual)	$(k - 1)(m - 1)$	SS_E	MS_E	
Total	$N - 1$	SS_T		

If in Example 10-18 we were interested in the effect of the types of vehicles as treatments with the study taken over the different fuel mixtures as differing conditions, the fuels would be the blocks. The computations and the numerical results would be the same except that the values for treatments and blocks would be interchanged.

Two-Factor Design

If we were equally interested in both fuel mixtures and vehicle types as "treatments," then we would not refer to either as treatments or blocks, but we would refer to both as *factors*, referring to fuels, say, as factor A and vehicles as factor B. The experiment then would be referred to as a *two-factor* design and SS_{Tr}, SS_{Bl}, MS_{Tr}, and MS_{Bl} would be relabeled SS_A, SS_B, MS_A, and MS_B, respectively. The number of levels for factor A (i.e., the number of treatments) would be denoted as a instead of k, and the number of levels of factor B (i.e., the number of blocks) would be denoted as b instead of m. Again the calculations and the numerical results would be unchanged.

EXAMPLE 10-19   Let us consider another example of two-way analysis of variance—this time, as a two-factor design. In this example we wish to determine at the 1% level whether average maze scores differ for $a = 3$ conditioning programs and $b = 5$ experimental rat strains. We randomly select three rats from each of the five strains and randomly assign these three rats so that there is one rat assigned to each conditioning program. After conditioning, the rats are required to run the maze.

Suppose that we have completed the experiment with the following maze scores obtained by the 15 rats:

Program	Strain A	B	C	D	E
I	49	60	48	53	51
II	50	59	56	57	51
III	63	72	66	69	61

Summarizing the data, we have

$$a = 3, \quad b = 5, \quad N = 15$$
$$T_{1.} = 261, \quad T_{2.} = 273, \quad T_{3.} = 331$$
$$T_{.1} = 162, \quad T_{.2} = 191, \quad T_{.3} = 170, \quad T_{.4} = 179, \quad T_{.5} = 163$$
$$T_{..} = 865, \quad \sum\sum x_{ij}^2 = 50{,}673$$

$$\frac{T_{..}^2}{N} = 49{,}881.67, \quad \sum \frac{T_{i.}^2}{b} = 50{,}442.20, \quad \sum \frac{T_{.j}^2}{a} = 50{,}078.33$$

The sums of squares are found to be

$$SS_A = 50{,}442.20 - 49{,}881.67 = 560.53$$
$$SS_B = 50{,}078.33 = 49{,}881.67 = 196.66$$
$$SS_E = 50{,}673 - 50{,}442.20 - 50{,}078.33 + 49{,}881.67 = 34.14$$
$$SS_T = 50{,}673 - 49{,}881.67 = 791.33$$

The analysis-of-variance table is

Source of variation	Degrees of freedom	Sums of squares	Mean squares	F-ratio
Programs	2	560.53	280.27	65.6
Strains	4	196.66	49.17	11.5
Error (residual)	8	34.14	4.27	
Total	14	791.33		

Because 65.6 exceeds $8.65 = F_{0.01; 2, 8}$, we may conclude that average maze scores do differ across the three conditioning programs. Also, as 11.5 exceeds $7.01 = F_{0.01; 4, 8}$, we may conclude that average scores differ for the five rat strains. In this set of data, we have found both factors to be significant. ▲

The basis for the formal theory of the two-way analysis can be developed from the second expression of the one-way model. In the one-way model, we rewrote $\mu_i - \mu$ as α_i and interpreted it as the effect of treatment i. We may also interpret α_i as the "effect" of the ith level of factor A.

We now introduce a term β_j to represent the effect of block j or the jth level of factor B. As with the α_i's we have a sum of 0 for the β_j's (i.e., we have $\sum \beta_j = 0$).

The population of all values produced by level i of factor A and level j of factor B (or equivalently subject to treatment i in block j) is then assumed to be normal with mean $\mu + \alpha_i + \beta_j$ and variance σ^2, where μ is the "overall mean" of all the populations.

In this model we have two possible hypothesis tests which we consider in the two-factor format rather than the randomized block format.

The hypothesis that factor A has no effect is tested as follows:

$$H_0: \alpha_i = 0 \qquad \text{for } i = 1, 2, \ldots, a$$

H_A : not all of the α_i's are 0

Decision rule: reject H_0 if $MS_A/MS_E > F_{\alpha;\, a-1,(a-1)(b-1)}$

The hypothesis that factor B has no effect is tested with

$$H_0: \beta_j = 0 \qquad \text{for } j = 1, 2, \ldots, b$$

H_A : not all of the β_j's are 0

Decision rule: reject H_0 if $MS_B/MS_E > F_{\alpha;\, a-1,(a-1)(b-1)}$

The tests may again be justified on the basis of expectation.

The expected values of MS_A, MS_B, and MS_E are, respectively, $\sigma^2 + b\sum \alpha_i^2/(a-1)$, $\sigma^2 + a\sum \beta_j^2/(b-1)$, and σ^2.

If all the α's are 0, then MS_A and MS_E have the same expectation, σ^2, and we might anticipate a ratio near 1; otherwise, $\sum \alpha^2$ will introduce an "inflation" into MS_A on the average and we might anticipate a ratio in excess of 1. The consideration of MS_B is analogous.

Interaction

In Example 10-18 we have assumed that fuel effect and vehicle effect are *additive*, that is, the fuel effect is the same for all types of vehicles and the vehicle effect is the same for all fuel mixtures. This assumption is implicit in the assumptions of the model which state that for fuel i and vehicle j, the mean mileage is $\mu + \alpha_i + \beta_j$. The effect α_i of the particular fuel is added to the overall mean in the same way regardless of which vehicle effect β_j is present, and vice-versa.

It is possible, however, that there is some *interaction* of the two effects and that the fuel effect is different for different types of vehicles and the vehicle effect is different for different fuel mixtures. Rather than rewrite the α's and β's to account for such interaction, we introduce a new term to represent the interaction effect. For the ith level of factor A and jth level of factor B, we denote the interaction effect as $(\alpha\beta)_{ij}$. We now assume that, for level i of factor A and level j of factor B, the mean response is $\mu + \alpha_i + \beta_j + (\alpha\beta)_{ij}$, where μ, α, and β are as before; $(\alpha\beta)_{ij}$ is as defined above and is such that

$$\sum_{i=1}^{a} (\alpha\beta)_{ij} = \sum_{j=1}^{b} (\alpha\beta)_{ij} = 0$$

In view of this new aspect to the model, we reexamine the residual sum of squares that we have used as the error sum of squares. Unlike the error term in one-way analysis of variance which was formed from variations within groups of observations in identical conditions, the residual error term in the two-way analysis is not as readily interpreted intuitively as a measure of natural variation. In fact, it is merely the residual part of the total sum of squares after removing components for treatments and blocks or for factors A and B. It is a valid estimate of σ^2 and does equal σ^2 on the average if there is no interaction present; however, as might be suspected from its expression involving "crossovers" of factors A and B, MS_E measures any interaction that is present as well as natural "error" variation, and its expectation is not σ^2 in this case. If interaction is present, the expectation of the residual sum of squares is $\sigma^2 + \sum_{i=1}^{a}\sum_{j=1}^{b}(\alpha\beta)_{ij}^2/(a-1)(b-1)$.

Since the value that we have used to measure error variation is also a measure of any interaction that might be present, to use it as error only, *we must assume that the interaction does not exist or is negligible.*

If we wish to consider possible interaction, we must obtain another estimate of error variation. To do so, we repeat or *replicate* the entire experiment so that we have more than one observation in each cell (treatment–block combination). We then obtain the error estimate from the within-cell variation. As with the one-way analysis of variance, the "within" sum of squares is seen intuitively, and justified theoretically, as natural "error" variation since it represents variation among responses obtained under identical conditions.

Two-Way Analysis of Variance with Replication

We develop the general case of *two-way analysis of variance with replication* as a two-factor design (appropriate changes of labels makes the experiment a randomized block design). In the general case, we have a levels of factor A and b levels of factor B and n replications; that is, we have n observations for each of the ab factor combinations, producing a total of $N = nab$ observations. We now have a triple-subscript notation so that the kth observation in the ijth cell (i.e., the combination of the ith level of factor A and the jth level of factor B) is x_{ijk}.

The subtotal for the ijth cell is

$$T_{ij\cdot} = \sum_{k=1}^{n} x_{ijk}$$

The subtotals for the ith level of factor A and the jth level of factor B and the overall total are, respectively:

$$T_{i..} = \sum_{j=1}^{b} T_{ij.} = \sum_{j=1}^{b} \sum_{k=1}^{n} x_{ijk}$$

$$T_{.j.} = \sum_{i=1}^{a} T_{ij.} = \sum_{i=1}^{a} \sum_{k=1}^{n} x_{ijk}$$

$$T_{...} = \sum_{i=1}^{a} T_{i..} \left(\text{or } \sum_{j=1}^{b} T_{.j.}\right) = \sum_{i=1}^{a} \sum_{j=1}^{b} \sum_{k=1}^{n} x_{ijk}$$

(Note in the overall total the introduction of triple summation, which is a natural extension from double summation.) The corresponding means are $\bar{x}_{ij.} = T_{ij.}/n$, $\bar{x}_{i..} = T_{i..}/nb$, $\bar{x}_{.j.} = T_{.j.}/na$, and $\bar{x}_{...} = T_{...}/N$.

The sums of squares for A, B, interaction, error, and total, respectively, are (together with calculating formulas):

$$SS_A = \sum_{i=1}^{a} nb(\bar{x}_{i..} - \bar{x}_{...})^2 = \sum_{i=1}^{a} \frac{T_{i..}^2}{nb} - \frac{T_{...}^2}{N}$$

$$SS_B = \sum_{j=1}^{b} na(\bar{x}_{.j.} - \bar{x}_{...})^2 = \sum_{j=1}^{b} \frac{T_{.j.}^2}{na} - \frac{T_{...}^2}{N}$$

$$SS_{AB} = \sum_{i=1}^{a} \sum_{j=1}^{b} n(\bar{x}_{ij.} - \bar{x}_{i..} - \bar{x}_{.j.} + \bar{x}_{...})^2$$

$$= \sum_{i=1}^{a} \sum_{j=1}^{b} \frac{T_{ij.}^2}{n} - \sum_{i=1}^{a} \frac{T_{i..}^2}{nb} - \sum_{j=1}^{b} \frac{T_{.j.}^2}{na} + \frac{T_{...}^2}{N}$$

$$SS_E = \sum_{i=1}^{a} \sum_{j=1}^{b} \sum_{k=1}^{n} (x_{ijk} - \bar{x}_{ij.})^2$$

$$= \sum_{i=1}^{a} \sum_{j=1}^{b} \sum_{k=1}^{n} x_{ijk}^2 - \sum_{i=1}^{a} \sum_{j=1}^{b} \frac{T_{ij.}^2}{n}$$

$$SS_T = \sum_{i=1}^{a} \sum_{j=1}^{b} \sum_{k=1}^{n} (x_{ijk} - \bar{x}_{...})^2$$

$$= \sum_{i=1}^{a} \sum_{j=1}^{b} \sum_{k=1}^{n} x_{ijk}^2 - \frac{T_{...}^2}{N}$$

The corresponding values for degrees of freedom are $a - 1$, $b - 1$, $(a - 1)(b - 1)$, $ab(n - 1)$, and $N - 1$. The mean squares for A, B, interaction, and error are, thus,

$$MS_A = \frac{SS_A}{a - 1}$$

$$MS_B = \frac{SS_B}{b - 1}$$

$$MS_{AB} = \frac{SS_{AB}}{(a - 1)(b - 1)}$$

$$MS_E = \frac{SS_E}{ab(n - 1)}$$

The analysis-of-variance table is

Source of variation	Degrees of freedom	Sums of squares	Mean squares	F-ratios
Factor A	$a - 1$	SS_A	MS_A	MS_A/MS_E
Factor B	$b - 1$	SS_B	MS_B	MS_B/MS_E
Interaction	$(a - 1)(b - 1)$	SS_{AB}	MS_{AB}	MS_{AB}/MS_E
Error	$ab(n - 1)$	SS_E	MS_E	
Total	$N - 1$	SS_T		

EXAMPLE 10-20   Suppose that the mileage experiment is performed once again to provide 12 more observations. Denoting fuel as factor A and vehicles as factor B, we have $a = 3$ levels of A, $b = 4$ levels of B, and $n = 2$ replications for a total of $N = 24$ observations. Suppose that the full set of data with appropriate subtotals and the overall total is as indicated in the following table:

Vehicle B, j	k	Fuel A, i 1	2	3		$\Sigma\Sigma$	
1	1	19.5	16.0	23.2			
	2	19.8	13.5	25.1			
	Σ	39.3	29.5	48.3	||	117. 1	
2	1	20.7	17.4	22.5			
	2	21.3	16.1	20.7			
	Σ	42.0	33.5	43.2	||	118. 7	
3	1	26.2	15.5	22.5			
	2	26.6	20.6	23.2			
	Σ	52.8	36.1	45.7	||	134. 6	
4	1	29.4	27.6	35.6			
	2	31.3	30.3	35.0			
	Σ	60.7	57.9	70.6	||	189. 2	
$\Sigma\Sigma$		194.8	157.0	207.8	|||	559.6	$\Sigma\Sigma\Sigma$

Using calculating formulas, the sums of squares are:

$$SS_A = \frac{194.8^2 + 157.0^2 + 207.8^2}{8} - \frac{559.6^2}{24}$$
$$= 13{,}222.11 - 13{,}048.007 = 174.103$$

$$SS_B = \frac{117.1^2 + 118.7^2 + 134.6^2 + 189.2^2}{6} - 134{,}048.007$$

$$= 13{,}619.317 - 13{,}048.007 = 571.310$$

$$SS_{AB} = \frac{(39.3^2 + 42.0^2 + \cdots + 70.6^2)}{2} - 13{,}222.11 - 13{,}619.317$$

$$+ 13{,}048.007$$

$$= 13{,}850.46 - 13{,}222.11 - 13{,}619.317 + 13{,}048.007 = 57.04$$

$$SS_E = 19.5^2 + 19.8^2 + 20.7^2 + \cdots + 35.6^2 + 35.0^2 - 13{,}850.46$$

$$= 13{,}877.04 - 13{,}850.46 = 26.58$$

$$SS_T = 13{,}877.04 - 13{,}048.007 = 829.033$$

Finally, the analysis-of-variance table is:

Source of variation	Degrees of freedom	Sums of squares	Mean squares	F-ratios
Fuels	2	174.103	87.052	39.30
Vehicles	3	571.310	190.437	85.98
Interaction	6	57.04	9.507	4.29
Error	12	26.58	2.215	
Total	23	829.033		

Because the three F-ratios exceed $F_{0.05;\, 2,\, 12} = 3.98$, $F_{0.05;\, 3,\, 12} = 3.49$, and $F_{0.05;\, 6,\, 12} = 3.00$, respectively, we may conclude that there are differences among the fuel mixtures and among the types of vehicles and that there is some interaction between fuel mixtures and vehicle types. ▲

Replicated Two-Way Analysis without Interaction

If we have a replicated two-factor (or randomized block) design in which it is reasonable (on the basis of prior knowledge) to assume that the factors do not interact, we modify the foregoing general results to eliminate the sum of squares, and so on, for interaction. We find the new residual or error degrees of freedom, sum of squares, and mean square as: $abn - a - b + 1$,

$$SS_E = \sum_{i=1}^{a} \sum_{j=1}^{b} \sum_{k=1}^{n} x_{ijk}^2 - \sum_{i=1}^{a} \frac{T_{i..}^2}{nb} - \sum_{j=1}^{b} \frac{T_{.j.}^2}{na} + \frac{T_{...}^2}{N}$$

and $MS_E = SS_E/(abn - a - b + 1)$. All other sums of squares and so on are unchanged.

PRACTICE PROBLEMS

P10-11 In a study on the penetration capabilities of shells of different shapes, four shapes were tested with each of three charges and the following penetrations resulted:

Charge	Shape A	B	C	D
I	6.2	4.9	4.6	7.1
II	6.7	4.3	3.9	6.4
III	6.3	5.1	4.2	6.8

Determine at the 1% level whether there are differences due to shapes. Are there differences due to charges?

P10-12 In two-way analysis-of-variance models assuming additivity, multiple comparison techniques can be employed to determine which treatments (or blocks) differ significantly. The residual error estimate (i.e., MS_E) is used as s^2 in such an application and n is the number of observations in each treatment (or block). Use the procedure based on Duncan's multiple range to determine which shapes differ significantly in Practice Problem P10-11. Use a 1% base level of significance.

P10-13 In a study on the effect of carbon dioxide on plant life, the diffusive resistance of wax on leaves was studied in an ordinary atmosphere and after exposure for a brief time to pure carbon dioxide. In each case leaves were studied in a dark environment, low light, and high light. The results of three assessments in each situation produced the following data:

Atmosphere	Light level Dark	Low	High
Ordinary	41	24	22
	43	27	24
	46	29	20
Carbon dioxide	58	58	63
	54	59	64
	55	63	60

(a) Determine at the 1% level whether resistance differs for the two atmospheres.
(b) Determine at the 1% level whether resistance differs across light levels.
(c) Determine whether light and atmosphere interact.

P10-14 In a study of retention over varying time intervals and the effect of interference in memory, a number of subjects were asked to memorize a set of three consonants and to repeat them after either 10, 20, or 30 seconds. In one set of trials, the subject was asked to count backward from a randomly selected number. In another, the experimenter spoke briefly to the subject on the nature of memory studies. On the assumption of no interaction, determine whether the percentages of correct responses for several trials differ with time interval and/or degree of interference if the following percentages were obtained by six subjects. (Assume that interference, timing, and letters were randomly administered with sufficient rest periods so that there was no applicable learning effect from one set of trials to another.)

Interference	Time		
	10	20	30
Counting	50.3	46.3	46.7
	49.8	48.1	43.2
	50.1	47.2	39.8
	48.1	46.6	41.3
	51.4	48.8	44.1
	47.9	50.2	40.5
Conversation	49.3	48.1	44.6
	54.2	53.3	43.0
	53.6	50.3	47.8
	55.3	49.9	41.7
	52.8	54.5	48.3
	50.9	56.7	49.6

10-24 The following eight observations represent coded chemical conversions in runs made at four temperatures for two durations:

Duration (minutes)	Temperature (°F)			
	100	125	150	175
30	−4	−2	0	3
60	−3	2	3	5

Determine whether temperature has a (statistically) significant effect on the chemical conversion.

10-25 In order to assess the average effects of four different processing techniques, four sample units were drawn at random from each of three batches of raw material and, after processing, the following results were observed:

Batch	Process			
	A	B	C	D
I	8.6	8.7	9.6	8.1
II	10.7	9.1	10.1	10.4
III	8.3	7.5	8.7	7.3

(a) Do the data provide evidence at the 5% level that the processes differ with regard to average results?

(b) Do the data provide evidence at the 1% level of differences among the batches?

(c) What modification to the sampling procedure would permit an assessment of whether the batch and process effects interact?

10-26 In an experiment in light awareness, subjects viewed a small light signal at one of three levels of brightness. At a particular time, either the light was extinguished or a shield was placed between the subject and the light. The subject was required to press a button upon becoming aware that the signal was no longer visible. The experiment was performed with 24 subjects who were assumed to have equivalent reactions. Each subject experienced only one trial under one condition. Each of the six conditions (three levels of brightness and two signal stopping procedures) was experienced by four subjects. The resulting reaction times (after coding to simplify calculations) were as follows:

	Brightness		
	1	2	3
Light	-2	-3	8
extinguished	-8	6	9
	-14	2	2
	-16	-6	-7
Shield	-10	8	10
placed	7	-6	20
	-4	7	6
	3	1	14

(a) Determine, at the 5% level, whether the levels of brightness have an effect.
(b) Determine whether the signal stopping procedures differ.
(c) Determine whether the brightness levels and stopping procedures interact.

10-27 Besides studying variation in percent body fat for various age groups, researchers also studied percent body fat for both men and women. Considering age and sex to be factors, determine which factors have significant effects and whether age and sex interact if samples of 20 men and 20 women in each age group produced the following summary figures.

	Percent body fat			
	Men		Women	
Age	$\bar{x}$	s	$\bar{x}$	s
17–22	16.2	6.13	26.1	7.56
23–29	19.3	6.89	30.7	6.98
30–39	21.8	6.92	33.8	7.46
40–49	27.2	5.87	34.4	6.32

Hint: From each mean, reconstruct the cell totals T_{ij}. From each standard deviation, reconstruct the sum of squares for the cell (i.e., $\sum_{k=1}^{20} x_{ijk}^2$) and hence determine $\sum_{i=1}^{2} \sum_{j=1}^{4} \sum_{k=1}^{20} x_{ijk}^2$.

10-28 In a study on cerebral ventricular enlargement and the effects of therapeutic drugs in cases of chronic schizophrenia, subjects with normal and enlarged ventricles were assessed on the basic psychiatric rating scale both with and without therapeutic drug treatment. Perform an analysis of variance of the following data summary assuming that the data have resulted from 40 subjects, 10 in each category.

	Ventricles			
	Normal		Enlarged	
	$\bar{x}$	s	$\bar{x}$	s
Drug	31.9	10.7	51.4	13.9
No drug	59.5	20.6	51.7	21.2

SOLUTIONS TO PRACTICE PROBLEMS

P10-11 Denoting the shapes as treatments and charges as blocks, we have $k = 4$, $m = 3$, and $N = 12$. Summary totals are $T_{1.} = 19.2$, $T_{2.} = 14.3$, $T_{3.} = 12.7$, $T_{4.} = 20.3$, $T_{.1} = 22.8$, $T_{.2} = 21.3$, $T_{.3} = 22.4$, $T_{..} = 66.5$, and $\Sigma\Sigma\, x_{ij}^2 = 383.15$. The sums of squares for the analysis of variance are:

$$SS_{Tr} = \frac{19.2^2 + 14.3^2 + 12.7^2 + 20.3^2}{4} - \frac{66.5^2}{12} = 13.649$$

$$SS_{Bl} = \frac{22.8^2 + 21.3^2 + 22.4^2}{3} - \frac{66.5^2}{12} = 0.302$$

$$SS_E = 383.15 - \frac{19.2^2 + \cdots + 20.3^2}{4} - \frac{22.8^2 + \cdots + 22.4^2}{3}$$
$$+ \frac{66.5^2}{12} = 0.678$$

$$SS_T = 383.15 - \frac{66.5^2}{12} = 14.629$$

The analysis-of-variance table is

Source of variation	Degrees of freedom	Sums of squares	Mean squares	F-ratios
Shapes	3	13.649	4.550	40.27
Charges	2	0.302	0.151	1.34
Error (residual)	6	0.678	0.113	
Total	11	14.629		

Because 40.27 exceeds $9.78 = F_{0.01;\,3,6}$, there are differences across the shapes. Because 1.34 does not exceed $5.14 = F_{0.05;\,2,6}$, there is insufficient evidence of differences among the charges.

P10-12 The four shapes produce means $\bar{x}_{1.} = 6.40$, $\bar{x}_{2.} = 4.77$, $\bar{x}_{3.} = 4.23$, and $\bar{x}_{4.} = 6.77$. We use $s^2 = MS_E = 0.113$ with 6 degrees of freedom and $n = m = 3$. For $p = 4$, 3, 2, we find $r_{0.01;\,p,6} = 5.55$, 5.44, and 5.24. We then produce $(s/\sqrt{n}) \times r$ in each case as 1.08, 1.06, and 1.02 and the final underline summary is

C	B	A	D
4.23	4.77	6.40	6.77

We thus conclude that shapes B and C are similar and differ from shapes A and D, which are similar.

P10-13 We analyze the study as a two-factor analysis with atmosphere as factor A and light level as factor B. Cell totals and factor level totals are as indicated in the following table:

	Light			
Atmosphere	Dark	Low	High	Total
Ordinary	130	80	66	276
Carbon dioxide	167	180	187	534
Total	297	260	253	810

As well, $\Sigma\Sigma\Sigma x^2 = 41{,}036$. For this analysis, we have $a = 2$, $b = 3$, $n = 3$, and $N = 18$. Sums of squares are

$$SS_A = \frac{276^2 + 534^2}{9} - \frac{810^2}{18} = 3698.00$$

$$SS_B = \frac{297^2 + 260^2 + 253^2}{6} - \frac{810^2}{18} = 186.33$$

$$SS_{AB} = \frac{130^2 + \cdots + 187^2}{3} - \frac{276^2 + 534^2}{9}$$
$$- \frac{297^2 + 260^2 + 253^2}{6} + \frac{810^2}{18} = 637.00$$

$$SS_E = 41{,}036 - \frac{130^2 + \cdots + 187^2}{3} = 64.67$$

$$SS_T = 41{,}036 - \frac{810^2}{18} = 4586.00$$

The analysis-of-variance table is

Source of variation	Degrees of freedom	Sums of squares	Mean squares	F-ratios
Atmosphere	1	3698.00	3698.00	686.09
Light	2	186.33	93.17	17.29
Interaction	2	637.00	318.50	59.09
Error	12	64.67	5.39	
Total	17	4586.00		

(a) $686.09 > 9.33 = F_{0.01; 1, 12}$. Resistance differs across atmospheres.
(b) $17.29 > 6.93 = F_{0.01; 2, 12}$. Resistance differs across light levels.
(c) $59.09 > 6.93 = F_{0.01; 2, 12}$. Light and atmosphere interact.

P10-14 To simplify calculations we adjust the data by subtracting 45 and multiplying by 10 to produce new data as follows, with cell totals and factor totals as indicated:

Interference		Time 10	Time 20	Time 30	Total
Counting		53	13	17	
		48	31	-18	
		51	22	-52	
		31	16	-37	
		64	38	-9	
		29	52	-45	
	Total	276	172	-144	304
Conversation		43	31	-4	
		92	83	-20	
		86	53	28	
		103	49	-33	
		78	95	33	
		59	117	46	
	Total	461	428	50	939
	Total	737	600	-94	1243

The overall sum of squares is $\Sigma\Sigma\Sigma\, x^2 = 105{,}573$. Letting A denote interference and B denote time, we have $a = 2$, $b = 3$, $n = 6$, and $N = 36$. Sums of squares for analysis of variance are

$$SS_A = \frac{304^2 + 939^2}{18} - \frac{1243^2}{36} = 11{,}200.69$$

$$SS_B = \frac{737^2 + 600^2 + (-94)^2}{12} - \frac{1243^2}{36} = 33{,}082.39$$

$$SS_E = 105{,}573 - \frac{304^2 + 939^2}{18} - \frac{737^2 + 600^2 + 94^2}{12}$$

$$+ \frac{1243^2}{36} = 18{,}371.89$$

$$SS_T = 105{,}573 - \frac{1243^2}{36} = 62{,}654.97$$

The analysis-of-variance table then becomes

Source of variation	Degrees of freedom	Sums of squares	Mean squares	F-ratios
Interference	1	11,200.69	11,200.69	19.51
Time	2	33,082.39	16,541.20	28.81
Error (residual)	32	18,371.89	574.12	
Total	35	62,654.97		

Because 19.51 exceeds $7.37 = F_{0.01; 1, 36}$ and 28.81 exceeds $5.26 = F_{0.01; 2, 36}$, percentages differ for time intervals and degrees of interference. (The F values are obtained by interpolation.)

10-7 CHAPTER REVIEW

10-29 The purpose of analysis of variance is to investigate the equality of the _____ of a number of populations.

10-30 When we do an analysis of variance, we assume that all of the populations have equal _____.

10-31 If the *sample* means for each group in analysis of variance are identical, the F-ratio should be
(a) 1.0
(b) 0.0
(c) between 0 and 1
(d) negative

10-32 If the means of a number of populations are all equal and we perform an analysis of variance on samples from these populations using the $F_{0.05}$ value from tables, the probability of obtaining an F-ratio beyond the critical point is _____.

10-33 If the F-ratio from analysis of variance is less than 1, then we declare the groups
(a) different
(b) equal
(c) (a) or (b), depending on how much less than 1

10-34 The F-ratio in analysis of variance
(a) is always greater than 1
(b) is always less than 1
(c) may be -3
(d) none of these

10-35 If a number of 5% hypothesis tests are performed on a collection of hypotheses, all of which are true, the probability of erroneously rejecting at least one of these hypotheses is
(a) 0.05
(b) less than 0.05
(c) greater than 0.05

10-36 Simple one-way analysis of variance is modified for different sample sizes by
(a) choosing an average sample size
(b) using weighting in the sums of squares
(c) discarding observations to make groups the same size

10-37 If the F-ratio from an analysis of variance is greater than the value in the F tables, we conclude that the population _____ are _____.

10-38 If we use one-way analysis of variance for a set of data that requires two-way analysis of variance, the resulting error mean square will be
(a) too small
(b) too large
(c) appropriate

10-39 Using multiple comparison techniques, determine at the 5% level what differences exist among the methods of presentation in Practice Problem P10-4.

10-40 The following data represent yields per unit of area from a number of plots of ground planted with a given crop with three different fertilizers being applied to three groups

	Fertilizer	
I	II	III
20.9	29.1	31.3
21.7	29.5	32.4
18.2	25.2	25.9
19.6	22.9	26.3
17.9	25.6	26.7
17.0	23.6	28.0
22.5	27.3	
	28.5	

(a) Perform an analysis of variance on these values giving a complete analysis-of-variance table.

(b) Determine, at the 5% level, which fertilizers, if any, are different.

10-41 A study of three insulating materials was conducted by using each material in four houses, one house of each of four different types of construction. The 12 houses were all in the same area and all heated to the same degree with the same fuel. The results of the study are given in the following table, where each entry represents annual heating cost after appropriate coding to simplify the data. Determine whether there are any differences among the insulating materials.

Insulating material	Type of Construction			
	I	II	III	IV
A	−4	11	4	14
B	−3	9	−7	16
C	−8	10	−2	10

10-42 On each of four days, six similar cloud formations were chosen at random: Two were left unseeded, two received type A seeding, and two received type B seeding. The following data represent precipitation in inches.

Treatment day	No seeding		Type A		Type B	
1	0.03	0.06	0.07	0.05	0.09	0.08
2	0.25	0.29	0.39	0.41	0.40	0.43
3	0.18	0.16	0.21	0.18	0.26	0.24
4	0.33	0.39	0.33	0.36	0.35	0.37

Determine at the 5% level whether there are any differences among the three treatments, where "no seeding" is considered to be a treatment.

10-43 A sampling of the prices of 3-year-old midsize cars in five cities produced the following data:

		City		
A	B	C	D	E
$3175	$3400	$3775	$4150	$4275
3475	4250	3900	3350	3950
3375	3675	3350	3900	4100
3750	3900	3750	3575	3875
3525	3875	3650	3675	4350

Adjust the data by subtracting $3700 from each price and dividing the result by 25.

(a) Find the mean adjusted price for each city and the overall mean adjusted price. Convert these adjusted means to produce the appropriate means for the original values.

(b) Find the variance of the adjusted prices for each city. Find the variance of the complete set of adjusted mean prices. Convert these variances to the corresponding values for the original prices.

(c) Making appropriate use of the variances in part (b), determine whether this set of data provides sufficient evidence of price differences among the five cities.

(d) If there are differences according to the analysis of variance, identify the differences on the basis of Duncan's multiple range with 5% base level of significance.

(e) Explain the reason for restricting the sampling of the cars to the same age and size for the analysis in part (c).

10-44 In a study on iconic memory in mentally retarded adults, subjects were shown a Chinese character to be identified later. Subsequently, each subject was shown sets of five Chinese characters and asked whether the original character was in the set. In half the sets, "yes" was the correct answer. In the other half, "no" was the correct answer. In the study, a group of mentally retarded adults (MR) was compared to a group of children matched for mental age (MA) and a group of adults matched for chronological age (CA). The percentages of correct answers in the "yes" and "no" categories were noted for each subject. Perform a complete analysis of variance on the following data. Use a 5% level of significance for testing F-ratios.

Correct response	% Correct responses of group		
	MR	MA	CA
Yes	63	48	57
	57	52	54
	67	54	52
	59	46	59
	65	44	61
	61	50	57

Correct response	% Correct responses of group		
	MR	MA	CA
No	43	74	57
	46	76	57
	39	69	52
	43	70	63
	44	69	61
	48	72	59

10-8 GENERAL REVIEW

10-45 To test $H_0 : \sigma_1^2 = \sigma_2^2$ against $H_A : \sigma_1^2 > \sigma_2^2$, the test statistic is
(a) $s_1^2 - s_2^2$
(b) the larger of s_1^2/s_2^2 or s_2^2/s_1^2
(c) $(n - 1)s^2$
(d) s_1^2/s_2^2

10-46 Any set of data has a unique mode.
(a) True
(b) False

10-47 If we look at the means of all possible samples of a given size from a population, then
(a) most of them should equal the population mean
(b) at least a few should equal the population mean
(c) it is possible that none will equal the population mean

10-48 In analysis of variance, if the F-ratio is significantly greater than 1, this is attributed to chance.
(a) True
(b) False

10-49 If A, B, and C are exhaustive, the probabilities $P(A) = 0.36$, $P(B) = 0.21$, and $P(C) = 0.38$
(a) are impossible
(b) are possible only if A, B, and C are mutually exclusive
(c) are possible

10-50 A normal distribution has
(a) no parameters
(b) three parameters
(c) an infinite number of parameters
(d) two parameters

10-51 Making use of $(n + 1)/2$, we find 26 as the median of 51 numbers ranging from 82 to 105. Is this reasonable?
(a) Yes
(b) No

10-52 When the error mean square in analysis of variance is 27.2 and the treatment mean square is only -6.8, we conclude that _____.

10-53 In two-way analysis of variance without replication, the interaction is assumed to be _____ .

10-54 A test with a 2 × 2 contingency table is the same as
(a) a test on a binomial population
(b) a test on two proportions
(c) a continuity correction
(d) a test on a variance

10-55 For a probability function $f(x)$, the expectation is calculated by the sum _____ .

10-56 If A and B are independent, such that the odds against A are 3 to 2 and the odds against B are 4 to 1, $P(AB) =$ _____ .

10-57 If a one-sided confidence interval has an upper limit indicating at most how large a variance may be, it also has a lower limit.
(a) True
(b) False

10-58 To test the equality of several proportions, we make use of a _____ table.

10-59 A sample survey is to be conducted to investigate the claim that more than three-fourths of senior high school students drink alcoholic beverages. In the survey, 100 students are to be asked whether they drink alcoholic beverages.
(a) Set up a hypothesis test with a 5% level of significance to determine whether the claim can be substantiated.
(b) (1) If the true percentage of senior high school students who drink is 75%, what is the probability of a type I error in part (a)?
(2) If the true percentage is 70%, what is the probability of a type I error in part (a)?
(3) If the true percentage is 85%, what is the probability of a type II error in part (a)?

10-60 In a traffic study a count was made of the numbers of vehicles passing a given point in each of 100 time units, with the following results:

Number of vehicles per time unit	Frequency of occurrence
0	5
1	11
2	27
3	22
4	16
5	8
6	4
7	3
8 or more	4

Find a range of values such that you can be 95% confident that this range includes the "true" proportion of times in which three or more vehicles pass the counting point.

10-61 Suppose that it is believed in Exercise 10-60 that x, the number of vehicles per unit time, should have a Poisson distribution with mean 2.5 [i.e., $f(x) =$

$2.5^x e^{-2.5}/x!$] with the following values:

x	0	1	2	3	4	5	6	7
$f(x)$	0.082	0.205	0.257	0.214	0.134	0.067	0.028	0.010

Also, $P(x \geqslant 8) = 0.003$. Test the belief at the 5% level.

10-62 A sample of term marks was taken to determine whether there was any difference between groups working in the first and second weeks in a class with fortnightly work sessions with the following results. For 11 students in the first week, term marks were

$$52 \quad 60 \quad 70 \quad 59 \quad 83 \quad 69 \quad 80 \quad 62 \quad 79 \quad 54 \quad 76$$

For 10 students in the second week, term marks were

$$68 \quad 53 \quad 88 \quad 76 \quad 70 \quad 78 \quad 60 \quad 67 \quad 66 \quad 71$$

(a) Determine whether the data indicate different variances for the two weeks.
(b) Assuming normal distributions, determine whether there is sufficient evidence at the 5% level to declare that performance differed from first to second week on the average.

10-63 Suppose that a researcher studying four different educational potential assessments with five children from each of three different aboriginal tribes obtained the following results in coded form:

Tribe	Assessment 1	2	3	4
1	4	2	9	14
	6	6	-7	11
	-4	2	11	-3
	-5	-6	0	5
2	-6	5	10	12
	-3	-2	1	6
	-1	-9	3	11
	-14	4	5	1
3	-4	5	10	4
	4	-13	7	1
	-5	1	11	6
	-10	-9	-2	-3

(a) Determine at the 5% level whether there are differences across the assessment procedures.
(b) Determine whether there are differences across the tribes.
(c) Determine whether there is any tribe–procedure interaction.
(d) Calculate the mean for each assessment procedure and, using the error mean square as the variance estimate, determine which assessment procedures differ at the 5% level.

10-64 In a typesetting study, a sample of 250 pages produced the following summary of number of characters per page.

Number of characters	Number of pages
2900–2924	6
2925–2949	11
2950–2974	19
2975–2999	32
3000–3024	39
3025–3049	36
3050–3074	24
3075–3099	29
3100–3124	18
3125–3149	12
3150–3174	13
3175–3199	8
3200–3224	3

(a) Sketch a histogram for these values.

(b) Determine the mean, median, and standard deviation for these values.

10-65 Requirements for a stream are such that the minimum average dissolved oxygen content that can be tolerated is 5.0 parts per million.

(a) If a sample of ten readings produced the following data, determine whether there is evidence at the 5% level that the oxygen content is too low:

$$4.7 \quad 5.0 \quad 5.1 \quad 4.9 \quad 4.5 \quad 4.9 \quad 5.1 \quad 4.9 \quad 4.6 \quad 4.5$$

(b) If the "true" standard deviation of such readings is known to be 0.10 and if the oxygen level is only 98% of the tolerable minimum (i.e., only 4.9 parts per million), what is the probability that a 5% test on a sample of 10 readings will lead to declaring the oxygen level to be tolerable (i.e., at least 5.0 parts per million)?

10-66 Consider the following data on numbers of farms sampled and average farm incomes.

Region	Number of farms in the sample	Average income per farm
Atlantic	73	2060
East	291	2627
Central	682	2603
West	1181	1891
Pacific	96	2941

Find the average income of all the farms in the sample. (Interpret "average" as mean.)

10-67 In a comparison of hiring practices of four organizations with approximately equal proportions of females among applicants, the numbers of males and females hired for a sampling of recently filled positions produced the following data:

Organization	Number of male applicants hired	Number of female applicants hired
I	37	19
II	35	29
III	33	11
IV	20	16

(a) Determine whether the organizations differ at the 5% level with regard to proportion of hired applicants who are females.

(b) For all organizations combined, determine a 95% confidence interval for the proportion of all positions filled by female applicants.

10-68 In a sample survey, 40 men and 40 women were asked whether they agreed with the statement: "The women's liberation movement will improve the position of women in society." Of the women, 15 agreed, 18 disagreed, and 7 were not sure; of the men, 19 agreed, 16 disagreed, and 5 were not sure. Is there sufficient evidence at the 5% level to indicate that men and women in the population have different opinions on the statement?

10-69 (a) The probability that an individual casing will crack under pressure is 0.001. If 5000 casings are in use under pressure, what is the probability that at most four will crack?

(b) If a batch of two dozen casings includes four that are cracked and if five casings are selected at random from the batch, what is the probability that at most two will be cracked?

(c) A crack-sensing device is such that there is a 90% chance that it will indicate a crack if there is one present. There is a 1% chance that it will falsely indicate a crack even if there is no crack present.

   (1) If one casing of the type in part (a) is put under pressure and the sensing device indicates a crack, what is the probability that the casing actually is cracked?

   (2) If three casings of the type in part (a) are put under pressure and tested with the sensing device, what is the probability that the sensing device will indicate a least one crack?

10-70 If individual units of material have melting points that follow a normal distribution with a mean of 150.0°C and a standard deviation of 2.5°C, and if x is the melting point of an individual unit selected at random, find

(a) $P(x \leqslant 153.0)$

(b) $P(x \leqslant 146.5)$

(c) $P(148.0 \leqslant x \leqslant 154.5)$

10-71 Prior to a public debate, 150 members of a religious group were asked whether they condoned euthanasia for terminally ill persons who were suffering. The same individuals were surveyed after the debate. Is there evidence at the 5% level of a shift in the proportion condoning euthanasia if 62 originally condoned euthanasia and of these 12 changed their views while 18 of the remaining 88 changed their views to condone euthanasia?

10-72 A study of tumors has indicated a relationship between elevation of estrone and the seriousness of the tumor. Estrone is found to elevate in 76% of cases of certain tumors. In 68% of these cases, death or active disease is present within 2 years. Death or active disease occurs in only 18% of cases without elevated estrone.

(a) What is the probability that, in the case of a given tumor, estrone will elevate and death or active disease will occur within 2 years?

(b) If death or active disease does not take place within 2 years, what is the probability that estrone did not elevate?

10-73 Assuming normality, determine whether gamma-ray determinations of snow-water equivalents generally exceed gravimetric determinations if determinations by both methods at 12 sites produced the following:

Site:	1	2	3	4	5	6	7	8	9	10	11	12
Gamma-ray	63	98	79	138	62	102	83	130	88	110	118	65
Gravimetric	39	72	81	120	47	88	80	97	95	92	79	58

10-74 (a) Determinations on available sugar are assumed to be normally distributed with mean μ equal to the "true" average available sugar content and with standard deviation $\sigma = 2.00$. How many sample determinations are required to provide a 95% chance that the sample mean will differ from μ by at most 0.50?

(b) If only 20 sample determinations are available and they produce $s = 2.85$, do these data provide sufficient evidence at the 1% level that σ is greater than 2.00?

(c) If the 20 sample determinations in part (b) also produced $\bar{x} = 17.59$ and if σ cannot be assumed to be 2.00, calculate a 95% confidence interval for μ.

10-75 Draw a 2-in-1 bar graph on a percentage basis to illustrate the following population changes for Canadian (1961–71) and American (1960–70) metropolitan regions.

	Number of regions	
Percent	Canada	United States
− 9.9 to −0.1	2	23
0.0 to 9.9	4	57
10.0 to 19.9	9	84
20.0 to 39.9	17	65
40.0 or more	2	14
	34	243

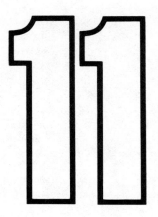

REGRESSION
AND
CORRELATION

11-1 INTRODUCTION

In the analysis of variance in Chapter 10, an attempt was made to determine whether different "treatments" produce different effects (on the average). No attempt was made to determine the form of any quantitative relationship between the treatments and the effects. It is often necessary to determine the nature of such relationships, if they exist, to be able to establish any general "trend," and to be able to predict values of one variable that would occur with specific values of another.

In many cases, we represent a general trend as a straight line about which actual outcomes are scattered. Values of one variable are predicted for values of another by use of the equation of the trend line.

One procedure for determining such relationships forms part of a method of analysis referred to as linear regression. It is a procedure for finding an equation that gives the average value of one variable as a linear function of the value(s) of some other variable(s). Regression analysis has been used in studies such as a Canadian study of the amount of biomass accumulated in a stand of timber as a function of the age of the stand; a study on the relationship between parental assistance and achievement in U.S. school children; a U.K. study on a correspondence between increasing pollutant residues and decreasing thickness of heron egg shells; a study on procedures for predicting the age of living giant sequoias in the Sierra Nevada by finding, from shallow cores, the relationship between basal area and years of growth; a study on the effect of water temperature on survival time of various species of trout; a study on the relationship between the final first-year grades of Queen's University students and other performance measures, such as their high school grades; an estimate of the relationship between savings and factors such as income and marital status; a study of the relationship between population density and distance from an urban center; and a study on the relationship between concentration of heavy metals in water and in algae blooms in order to predict the metal content in the water on the basis of the (easier to assay) content of the blooms.

Regression analysis produces an estimate of the relationship which exists on the average but does not assess the strength (and hence usefulness) of the relationship; that is, the regression analysis does not determine how well a linear pattern fits the data. Correlation analysis is a procedure for assessing the strength of such relationships, if they exist at all.

11-2   SIMPLE REGRESSION

In this section we introduce, through an example, the procedures used in the simplest form of regression analysis. The analysis will be fairly descriptive. More involved analyses requiring probability and other theory are discussed in later sections.

Scatter Diagram — Fitting a Line by Eye

EXAMPLE 11-1   Suppose that we have a sample of eight trees with one at each age from 1 through 8 years and that the heights in feet of the trees, youngest to oldest, are 3.6, 3.7, 7.8, 7.6, 10.3, 14.0, 14.6, and 18.1. From these data, we wish to estimate the average height of all such trees as a function of age.

To obtain a quick impression of the relationship, we plot the trees on a graph according to their ages and heights, with age on the horizontal scale and heights on the vertical scale, as shown in Figure 11-1. The eight trees produce a scatter of eight points and the plot is referred to as a *scatter plot* or a *scatter diagram*. The points are scattered in a pattern fairly close to a straight line, suggesting that there is a linear relationship between age and average height. We can obtain a first estimate of this relationship (or "growth line") by moving a straightedge until it appears to be as close as might reasonably be expected to all of the points and then drawing the line indicated. This procedure is referred to as *fitting a line by eye*. One such line (there are many since the choice of the position is a matter of personal judgment) is the dotted line shown.

The points are scattered around the line since *the line represents the relationship indicating the averages of the heights that we might expect for various ages* and, through natural variation, some values occur above the average and some occur below. Using this line, we can estimate various average heights. For example, for 6-year-old trees, we draw a line from 6 on the horizontal scale up to the growth line and then, from there, we draw a line to

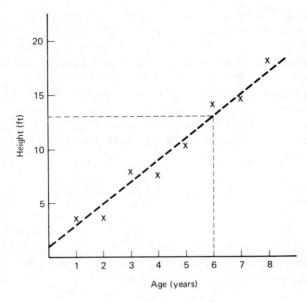

Figure 11-1 Scatter diagram for tree data.

the vertical scale at about 13 feet. We thus estimate the average height of a 6-year-old tree of this type to be about 13 feet. ▲

The method described above is quick and easy, but because of the personal judgment, it does not provide consistent results. Also, the line determined this way is not suitable for more involved statistical analyses to be considered in later sections.

Linear Regression

The analysis that we will develop is the simplest form of regression analysis—*simple linear regression*. It is most suitable in situations such as the tree example, in which two variables under study produce a set of pairs of numbers which, when plotted, forms a scatter that follows an approximate straight-line relationship.

The first step in the analysis is to express the relationship between two quantities in mathematical terms. We are trying to predict the average of one as a linear function of the other; thus, we refer to one as the *dependent* or *predicted* variable and we denote it as y, and we refer to the other as the *independent* or *predictor* variable and we denote it as x. We then produce an equation for the relationship as

$$y = b_0 + b_1 x$$

where b_1 is the *slope* of the line and b_0 is the value of y where the line crosses the vertical scale (or y-axis) and is referred to as the *y-intercept* or (since we do not use the x-intercept) simply as the *intercept*.

EXAMPLE 11-2   In Example 11-1 the dependent variable y is height and the independent variable x is age. We note, from inspection of Figure 11-1, that the line drawn by eye has an intercept of about 1 and a slope of about 2 and may thus be expressed as $y = 1 + 2x$; that is, we have $b_0 = 1$ and $b_1 = 2$ for the line drawn by eye. When we introduce the value $x = 6$, we obtain the value $y = 1 + (2 \times 6) = 13$, which is the value of the previous estimate of the average height of trees of this type at age 6.

▲

Least Squares Solution

We use the general equation to introduce a mathematical solution to the problem of fitting a line—*the least squares solution*. For each observed value x_i of the independent variable, we consider the corresponding observed value y_i of the dependent variable, which we denote as y_{obs}, and the corresponding fitted value y_{fit} which would result from fitting a line to the data; that is, for a particular x_i, y_{fit} is found as $b_0 + b_1 x_i$ for some b_0, b_1. We then attempt to find a pair of values b_0 and b_1 that produce a set of y_{fit}'s as close as possible in some collective sense to the set of y_{obs}'s. We choose as the best line the line for which the pair b_0, b_1 minimizes the sum of the squared differences between corresponding members of these sets of y's.

As the leasts squares solution, we choose the pair b_0, b_1 that minimizes

$$\sum (y_{obs} - y_{fit})^2 = \sum_{i=1}^{n} [y_i - (b_0 + b_1 x_i)]^2$$

The differences for Example 11-2 are shown as vertical lines in Figure 11-2.

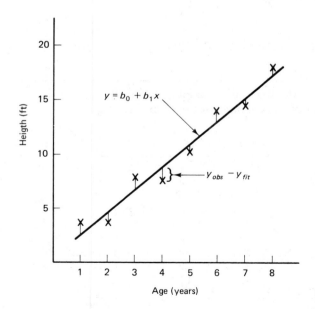

Figure 11-2 Fitted line for tree data.

The solution is referred to as the least squares solution because it minimizes the sum of squared differences between observed and fitted values of the dependent variable. The resulting line may be classified as the "best fit" because it is the line that minimizes an aggregate measure of the deviations between the actual observed points and the corresponding points on the line.

Normal Equations and General Solution

The solutions for b_0 and b_1 may be found through the use of calculus. It turns out that the required values for b_0 and b_1 are the solutions to the following *normal equations*:

$$\sum y = nb_0 + b_1 \sum x$$
$$\sum xy = b_0 \sum x + b_1 \sum x^2$$

where n is the number of x, y pairs in the data and $\sum x, \sum y$, and so on, are obtained from the observed data. The general solutions to these equations are

$$b_0 = \frac{(\sum y)(\sum x^2) - (\sum x)(\sum xy)}{n\sum x^2 - (\sum x)^2}$$

$$b_1 = \frac{n\sum xy - (\sum x)(\sum y)}{n\sum x^2 - (\sum x)^2}$$

If b_1 is calculated first, then b_0 can be found more simply as

$$b_0 = \frac{\sum y - b_1 \sum x}{n} = \bar{y} - b_1 \bar{x}$$

where $\bar{x}$ and $\bar{y}$ are the means of the observed x's and y's.

EXAMPLE 11-3 For the tree data in Example 11-1, $n = 8$ and the sums are obtained as the totals of the columns in the following table:

	x	y	x^2	xy
	1	3.6	1	3.6
	2	3.7	4	7.4
	3	7.8	9	23.4
	4	7.6	16	30.4
	5	10.3	25	51.5
	6	14.0	36	84.0
	7	14.6	49	102.2
	8	18.1	64	144.8
Σ	36	79.7	204	447.3

Using the first set of solutions, we have

$$b_0 = \frac{(79.7 \times 204) - (36 \times 447.3)}{(8 \times 204) - (36 \times 36)} = 0.4643$$

$$b_1 = \frac{(8 \times 447.3) - (36 \times 79.7)}{(8 \times 204) - (36 \times 36)} = 2.1107$$

or, if b_1 is calculated first, we can find b_0 from the second solution as

$$b_0 = \frac{79.7 - (2.1107 \times 36)}{8} = 0.4644$$

The slight difference in the two solutions for b_0 is due to the rounding in b_1 in the second solution. We thus have the least squares regression line as

$$y = 0.4643 + 2.1107x$$

Using this line, the predicted *average* height for 6-year-old trees is $y = 0.4643 + (2.1107 \times 6) = 13.1285$, which we might round to two decimals (i.e., to 13.13) because the data are only to one decimal. The value 13.13 is very close to the previous estimate of approximately 13. The regression line compares quite well with the line fitted by eye, which serves as a rough check on the calculations used in producing the regression line. ▲

Validity of the Regression Line

If we insert the value $x = 0$ in Example 11-3, we obtain $y = 0.46$ feet, an unrealistic "instant" height of over 5 inches, or if we use the age $x = 30$, we obtain $y = 63.79$ feet, which may also be unrealistic—according to the regression line, the trees have no height limit. The problem is that the regression line is generally useful only over the range of the observed x values, and extrapolation very far beyond this range may produce unreasonable results.

The true growth trend may be more like the curve shown in Figure 11-3 with the regression line superimposed on it. The relationship is reasonably linear in the range sampled and, for this range, the regression line is quite useful; however, beyond this range, the curvature of the true relationship makes extrapolation of the regression line unreasonable.

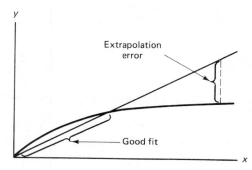

Figure 11-3 Extrapolation error with a curved trend.

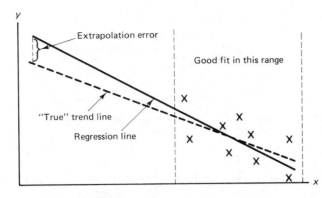

Figure 11-4 Extrapolation error with a linear trend.

Even if the average relationship between two variables is completely linear, extrapolation can produce errors. The sample result will almost always differ from the "true" result and even a slight difference in the slope can lead to large extrapolation errors, as shown in the scatter plot with corresponding lines in Figure 11-4.

Figure 11-4 illustrates a possible problem in using a line fitted by eye as a rough check on the calculations for the regression line. If the range of the x's is very far removed from the y-axis (vertical scale), differences between the two intercepts can be large merely because of the effect of the extrapolation. The primary use of the fit-by-eye line is to produce a reasonably close check on the slope—to determine whether it is positive (the line slopes up to the right), as with the tree data, or negative (the line slopes down to the right), as in Figure 11-4, and to obtain an approximate value for its magnitude. The approximation from the fit-by-eye line will not always be as close to the regression line slope as it is with the tree data, for which the scatter of points is very close to a straight line.

The effectiveness of the line, as well as the checks on the calculations, depends on how close the data points are to a straight-line pattern, as discussed in the next section.

PRACTICE PROBLEM

P11-1 To study the tensile strength (considered in terms of mass supported in suspension per unit area of cross section) of a certain type of wire, the following pairs of observations were recorded, where x is the diameter in centimeters and y is the mass supported in kg/cm.

x	0.6	0.8	1.0	1.2	1.4	1.6	1.8	2.0	2.2	2.4
y	14	26	50	56	42	98	82	88	134	124

(a) Plot a scatter diagram of these data (x versus y) and draw a line (by eye) to fit the data.

(b) Calculate the least-squares regression line $y = b_0 + b_1 x$. Draw it on the scatter diagram and compare it to the line fit by eye as a check on the calculations.

(c) Find the predicted tensile strength for a wire with diameter 1.5 cm.
(d) What does the regression line represent?
(e) What does the prediction above represent?
(f) If we extrapolate the regression line to a diameter of 0.2 cm, we obtain a negative strength. How is this explained?

11-1 In the following data, x represents the mark obtained by each of a sample of 13 students in a precourse diagnostic test and y represents the corresponding mark obtained on a final exam at the end of a course.

x	31	44	51	54	56	63	67	68	69	73	76	85	95
y	12	54	51	40	67	42	65	79	85	53	76	93	89

(a) Plot a scatter diagram of y versus x.
(b) Fit a line by eye and determine the approximate slope of this line.
(c) Find the regression line $y = b_0 + b_1 x$ and compare b_1 to the slope of the line fit by eye.
(d) Draw the regression line on the scatter diagram.
(e) Estimate the average final mark obtained by students with a mark of 65 on the diagnostic test.

11-2 In the following sample data, x represents time (actually year minus 1971) and y represents average personal expenditure on recreation (in constant 1971 dollars)

x	-4	-3	-2	-1	0	1	2	3	4
y	173	192	207	215	239	263	281	302	330

(a) Plot a scatter diagram of y versus x.
(b) Find the least-squares regression line $y = b_0 + b_1 x$.
(c) Predict average expenditure for 1977.

11-3 In the following pairs of observations, x is the intensity of a sound stimulus and y is reaction time required by a subject before responding to the stimulus.

x	y	x	y
1.3	1.70	6.2	0.61
2.1	1.62	6.9	0.70
2.4	1.48	7.4	0.44
3.0	1.00	7.4	0.31
3.8	1.19	8.1	0.33
4.5	1.23	9.6	0.30
5.1	0.80	10.0	0.18
5.7	0.59	10.9	0.15
6.0	0.35		

(a) Plot a scatter diagram of y versus x and fit a line by eye.
(b) Fit the least-squares regression line $y = b_0 + b_1 x$ and plot it on the scatter diagram.
(c) Predict average time to react to a stimulus of 12.0. How can this unrealistic result be explained?

11-4 A study of lead content in plants near a highway produced the following readings from different locations. In the data, x is the traffic flow in vehicles per hour and y is a measure of the lead content in plants.

x	207	432	589	807	939	1147	1301	1412
y	9.2	14.3	51.6	74.8	48.1	73.2	113.2	83.5

Estimate the average content for plants in a location experiencing traffic flow of 1000 vehicles per hour.

11-5 Solve the normal equations algebraically to verify the general solutions for b_0 and b_1.

SOLUTION TO PRACTICE PROBLEM

P11-1 (a) The scatter diagram with a line fit by eye (dashed line) is shown in Figure P11-1.

(b) Summary totals from the data are $n = 10$, $\sum x = 15$, $\sum x^2 = 25.8$, $\sum y = 714$, and $\sum xy = 1278$, producing

$$b_1 = \frac{(10 \times 1278) - (15 \times 714)}{(10 \times 25.8) - 15^2} = 62.727$$

$$b_0 = \frac{714 - (62.727 \times 15)}{10} = -22.691$$

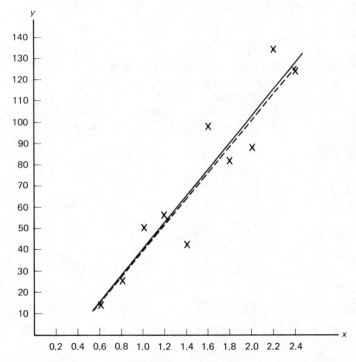

Figure P11-1 Regression line for tensile strength.

The regression line $y = -22.691 + 62.727x$ is shown as the solid line in Figure P11-1 and compares favorably to the line fit by eye, indicating no serious error in the calculations.

(c) For a diameter $x = 1.5$, the predicted tensile strength is $y = -22.691 + (62.727 \times 1.5) = 71.40$.

(d) The regression line represents the relationship between average tensile strength and diameter of wire for wires with diameters in the range sampled.

(e) The prediction represents an estimate of the average strength of wires with a diameter of 1.5.

(f) The true relationship is probably curvilinear and only approximately linear in the range sampled. The negative strength is an extrapolation error resulting from extrapolating beyond the sampling range.

11-3   INTRODUCTION TO CORRELATION

For regression to produce reasonable predictions, the "true" underlying trend or average relationship should be at least approximately linear over the range of values being considered. The scatter plot provides some general indication of the nature of the relationship, but it is most useful for fairly obvious cases of linearity or lack of linearity. With the tree data of Section 11-2, for example, the approximate linearity is quite obvious.

Variation about the Line and Variation Due to Regression

We developed probability to provide a precise measure for otherwise vague descriptions of uncertainty. Similarly, we develop a more precise measure of the strength of the linearity of the relationship between two variables. This measure is based on an analysis of the degree to which the points are scattered or varied about the "best" possible straight line (where "best" refers to the least-squares fit).

EXAMPLE 11-4   The data of Example 11-1 will serve as an example to develop the measure. The degree of scatter can be assessed by measuring the amount of variation among the points. By this we mean the total variation, which is the total sum of squared deviations about the overall mean

$$SS_T = \Sigma (y - \bar{y})^2$$

$$= \Sigma y^2 - \frac{(\Sigma y)^2}{n}$$

$$= 988.11 - \frac{79.7^2}{8}$$

$$= 194.099$$

We then attempt to analyze or "explain" this variation. In the analysis of variance in Chapter 10, we partitioned the total variation into components: One component was natural chance variation, the other was variation across different groups or populations. Similarly, we partition the total variation in the tree heights.

Natural chance variation or "error" variation is represented by the variation of the actual heights around their corresponding mean heights. For the mean heights corresponding to various ages we use the fitted values produced by the regression line in Example 11-3. The error variation is then the sum of the squared deviations about the line

$$SS_E = \Sigma(y_{obs} - y_{fit})^2$$

This value is found as the sum of the last column in the following table:

Age, x	Height, y_{obs}	$y_{fit} =$ $0.4643 + 2.1107x$	$y_{obs} - y_{fit}$	$(y_{obs} - y_{fit})^2$
1	3.6	2.575	1.025	1.051
2	3.7	4.686	−0.986	0.972
3	7.8	6.796	1.004	1.008
4	7.6	8.907	−1.307	1.708
5	10.3	11.018	−0.718	0.516
6	14.0	13.129	0.871	0.759
7	14.6	15.239	−0.639	0.408
8	18.1	17.350	0.750	0.563
				6.985 $= SS_E$

The balance of the total variation is the variation across the different age groups.

We can "explain" the total variation in the height by saying that some is just natural variation and that some is due to the fact that we have observed trees at different ages. The effect of the different ages is represented by the regression line; therefore, we refer to the variation due to different ages (x values) as the variation due to the regression. As a sum-of-squares component, this value is the remainder of the total variation after removing the component due to "error" and is

$$SS_R = SS_T - SS_E$$
$$= 194.099 - 6.985$$
$$= 187.114$$

The proportion of the total variation in the heights that is explained by, or due to, the regression is

$$\frac{SS_R}{SS_T} = \frac{187.114}{194.099}$$
$$= 0.964$$

or converted to a percent is 96.4%.

The obvious approximate linearity in the tree data is confirmed with this measure. Through this linear relationship between height and age, variation in age explains 96.4% of the variation in heights.

▲

Coefficient of Determination

The least-squares regression line is chosen to produce the minimum $\sum (y_{obs} - y_{fit})^2$ $= SS_E$ and accordingly produces the maximum SS_R of any straight line. Since this fit is, therefore, the best linear fit that we can obtain between these two variables, we may use its measure, SS_R / SS_T, as the measure of the strength of the linearity between x and y in sample data. This measure is referred to as the *coefficient of determination* and we will denote it as r^2.

In summary, then, we find total variation in the y's, variation of the y's about the line, and variation of the y's due to or explained by the regression as

$$SS_T = \sum (y_i - \bar{y})^2$$

$$SS_E = \sum (y_{obs} - y_{fit})^2$$

$$SS_R = SS_T - SS_E$$

We will shortly introduce a calculating formula for r (and hence for r^2), but it may also be noted that we can avoid the calculation of $\sum (y_{obs} - y_{fit})^2$ by using the previously calculated values of b_1 and $\sum (x - \bar{x})^2 = n \times$ (denominator of b_1) in the calculating formula

$$SS_R = b_1^2 \sum (x - \bar{x})^2$$

We then have the coefficient of determination in percentage form as

$$r^2 = 100 \left(\frac{SS_R}{SS_T} \right) \%$$

$$= 100 \left(\frac{b_1^2 \sum (x - \bar{x})^2}{\sum (y - \bar{y})^2} \right) \%$$

Note that r^2 may be expressed as a decimal or fraction rather than a percent value; however, it is easier to interpret in percentage terms.

Interpreting r^2

Since it is a proportion, r^2 takes values between 0 and 1, or between 0% and 100%. If there is no linearity present as in Figure 11-5(a) and (b), the total variation is natural error variation. The remaining variation is 0 and $r^2 = 0$, indicating no linear relationship. This situation could result from two different circumstances. In Figure 11-5(a) there appears to be no relationship between x and y. In Figure 11-5(b), on the other hand, there appears to be a very strong relationship between x and y. This

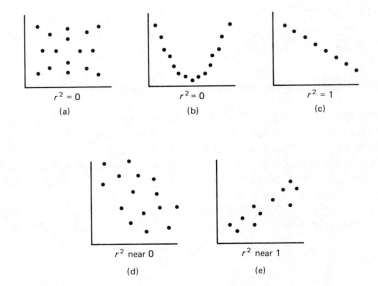

Figure 11-5 Various values for r^2.

relationship, however, is not at all linear over the range of the x's and, as a result, we still obtain $r^2 = 0$.

If all the points in a scatter plot lie exactly on a straight line as in Figure 11-5(c), there is no error variation around the line and all of the variation is explained by the line, producing $r^2 = 1$ or 100%.

Other values of r^2 occur for varying degrees of linearity. The set of data producing Figure 11-5(d) would produce r^2 close to 0 and the set of data producing Figure 11-5(e) would produce r^2 close to 1.

Correlation Coefficient

A more commonly used measure of linearity is the square root of the coefficient of determination, which is referred to as the *correlation coefficient* or simply *correlation* and denoted as r. The positive square root is used if the slope of the corresponding regression line is positive and the negative square root is used if the slope is negative.

EXAMPLE 11-5 For Example 11-4, the correlation between age and height is $r = \sqrt{0.964} = 0.982$. ▲

Interpreting Correlation

Since r^2 takes values between 0 and 1, r, its square root, takes values between -1 and $+1$. A perfect straight line produces $r = 1$ or $r = -1$, depending on whether the slope is positive or negative. Sets of data exhibiting no linearity produce $r = 0$ (since $r^2 = 0$). Values of r close to 1 indicate a strong linear relationship with positive slope. Positive values of r close to 0 indicate a weak linear relationship with positive slope. Values of r close to -1 indicate a strong linear relationship with negative slope, and negative values of r close to 0 indicate a weak linear relationship with negative slope. The sets of data for Figure 11-5(a) and (b) would produce $r = 0$; the data for Figure 11-5(c) would produce $r = -1$; for Figure 11-5(d), r would be close to 0 and negative; and for Figure 11-5(e), r would be close to 1.

Correlation is often interpreted mistakenly as a measure of any form of relationship between two quantities, but the correlation coefficient r is strictly a measure of the linearity of the relationship. To say that two quantities are highly correlated means that they exhibit a strong linear relationship. To say that they are uncorrelated or only slightly correlated indicates that they are not related linearly—there may, however, still be a strong relationship, as illustrated in Figure 11-5(b).

A strong correlation is also often interpreted as an indication of a strong cause–effect relationship between two quantities. It is true that some relationships are cause–effect—a high correlation between weight gain and the number of calories in a diet may, for example, reflect a cause–effect relationship; however, some strong correlations are not a result of such a relationship. If yellow beans and green beans are planted side by side in pairs of rows in a number of different areas, the yields of the yellow beans from area to area will be highly correlated with the yields of the green beans. This does not mean that, by growing well, the yellow beans cause the green beans to grow well also. Both quantities are simultaneously and similarly affected by other causes, such as the changes in the soil from area to area.

Calculating Formula

The direct calculation of r^2 as previously illustrated provides an explanation of the ideas involved and of how r^2 or r may be interpreted, but it is not the easiest method of calculation. It is preferable to use the following *calculating formula for the correlation coefficient*, which also automatically determines the sign:

$$r = \frac{n\sum xy - \sum x \sum y}{\sqrt{n\sum x^2 - (\sum x)^2} \times \sqrt{n\sum y^2 - (\sum y)^2}}$$

The numerator is the same as the numerator of the calculating formula for the slope b_1 of the regression line and thus has the same sign. The square roots in the denominator are positive square roots; thus, the sign of the correlation is the same as that of the slope.

Deleting the "100%" and taking a square root, we may also change the calculating formula for r^2 to another calculating formula for r, which would be useful for calculators that produce regression slopes and standard deviations but not correlations.

$$r = b_1 \frac{\sqrt{\sum (x - \bar{x})^2}}{\sqrt{\sum (y - \bar{y})^2}}$$

$$= b_1 \frac{s_x}{s_y}$$

where s_x and s_y are the standard deviations for the x's and the y's and result from dividing each value under the square root signs by $n - 1$.

EXAMPLE 11-6   Using the first calculating formula for the tree data in Example 11-1 produces

$$r = \frac{(8 \times 447.3) - (36 \times 79.7)}{\sqrt{(8 \times 204) - (36 \times 36)} \sqrt{(8 \times 988.11) - (79.7 \times 79.7)}}$$

$$= \frac{709.2}{\sqrt{336 \times 1552.79}}$$

$$= 0.982$$ ▲

Adjusting Data

Some cases may involve necessary adjustment of the data such as changes from degrees Fahrenheit to degrees Celsius. In other cases it may be desirable to adjust data, if possible, to make computations easier just as with previous variance calculations. We will only consider changes of location and scale as before. Such changes are referred to mathematically as linear transformations. If two quantities are related linearly and a linear transformation is applied to either or both of them, then, although they may produce a different slope and intercept, they will still be related linearly.

This result is as we should expect—if the amount of liquid produced by some process is correlated with the temperature involved, the degree of correlation should not depend on whether we are considering the number of gallons as related to the temperature in degrees Fahrenheit or the number of liters as related to the temperature in degrees Celsius.

We will not prove the general results but will demonstrate, intuitively, through diagrams, the effects of such changes on the results of regression and correlation calculations.

We first consider relocations of the data. Suppose that a set of pairs x, y produces a scatter diagram and resulting regression line, $y = b_0 + b_1 x$, as shown in Figure 11-6. If we relocate the x's by subtracting l from each x to produce $x_i' = x_i - l$, we move each point l units closer to the vertical scale (or y-axis, i.e., to the left), which is equivalent to moving the vertical scale l units to the right, as

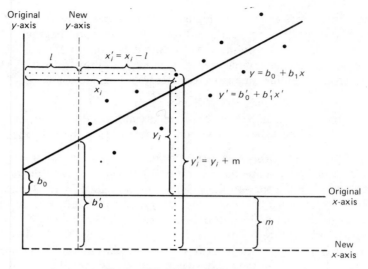

Figure 11-6 Relocating x and y.

indicated in the diagram. Similarly, relocating the y's by, say, adding a constant m to each y to produce $y_i' = y_i + m$ is equivalent to lowering the horizontal scale (or x-axis) by m units, as shown. In terms of the new data, the regression line is $y' = b_0' + b_1' x'$.

We have achieved the changes on the diagram without moving the position of the points or the regression line—what we have changed are the positions of the reference axes. We can see by inspection that the new intercept is different from the original intercept; however, since the points and the line are unchanged, the new slope must be equal to the original slope and, further, since the variation of the points around the line is unchanged, the correlation coefficient is unchanged. We thus have the following result.

> Relocation of the x and/or y values does not affect the slope or the correlation; such changes only affect the intercept.

We now consider changes of scale. Figure 11-7(a) shows original data and Figure 11-7(b) shows the same data with the x values multiplied by a constant, $c(c > 1)$. The slope is not as steep in Figure 11-7(b)—to increase y by 1 unit, x must

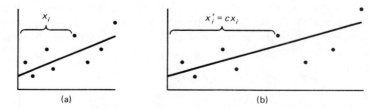

Figure 11-7 Rescaling x.

increase c times as much in Figure 11-7(b) as in Figure 11-7(a); thus, if we multiply all of the x's by c, the slope is divided by c. In both diagrams, the y values are the same. Although the points have different horizontal spread, the vertical spread is unchanged; thus, the intercept is unchanged, and further, with y variation (both total and around the line) unchanged, the correlation is unchanged.

> If we multiply (divide) all of the x values by a constant,
> the slope will be divided (multiplied) by that constant
> and the intercept and correlation will be unchanged.

Figure 11-8(a) shows a set of data with corresponding regression line; Figure 11-8(b) shows the same data with the y's all divided by a constant $k(k > 1)$. As can be seen, the intercept and the slope are both divided by the same constant as the y's. Moving from Figure 11-8(b) to 11-8(a), we see the effect of multiplying all of the y values by k—the slope and intercept are also multiplied by k. Since the variation in the y's has changed with this type of adjustment (multiplying or dividing by a constant), it might appear, at first, that the correlation changes also; however, even though variations in the y's have changed, they have all changed by the same factor, and variation around the line is still the same proportion of the total variation as it was originally. As a result, the correlation is once again unchanged.

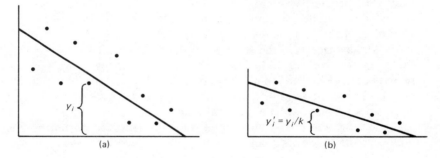

Figure 11-8   Rescaling y.

> If we multiply (divide) all of the y values by a constant,
> the slope and the intercept are also multiplied (divided)
> by that constant and the correlation is unchanged.

As we have demonstrated, relocations and/or rescalings of the x and/or y values have no effect on the correlation. The effects on the slope and intercept are readily apparent for single adjustments; however, for a combination of adjustments, converting the results from adjusted data to the appropriate results from original data (or vice versa) is more easily achieved by the methods illustrated in the following example.

EXAMPLE 11-7   In the following set of data, X represents the dosage of a drug and Y represents the corresponding time until a particular response is observed. We wish to estimate average response times for various dosages on the basis of this sample.

X	Y	X	Y	X	Y
1.5	1.70	5.0	0.80	8.0	0.33
2.0	1.32	5.5	0.59	8.5	0.50
2.5	1.48	6.0	0.35	9.0	0.27
3.0	1.00	6.5	0.61	9.5	0.30
3.5	1.41	7.0	0.70	10.0	0.18
4.0	1.19	7.5	0.44	10.5	0.15
4.5	1.23				

In order to obtain a general impression of the relationship between dosage and average response time, we first draw a scatter diagram as shown in Figure 11-9. The relationship might be expected to follow a form of curve as illustrated (dashed curve), but over the range sampled, it appears reasonably linear and we fit a straight line as shown. This line appears to cross the Y-axis at about $Y = 1.75$, indicating an intercept of about 1.75. It crosses the X-axis at about $X = 11.0$; thus, we appear to have a decrease in Y of about 1.75, together with an increase in X of about 11. The slope, therefore, is approximately $-1.75/11 = -0.16$. From the line fit by eye the relationship appears to be approximately

$$Y = 1.75 - 0.16X$$

Before proceeding further with regression or correlation calculations, we note that the data can be simplified considerably with some adjustments of location and scale. If we subtract 6.0 from each dosage value and then multiply by 2, we can produce single-digit numbers with no decimals. If we subtract

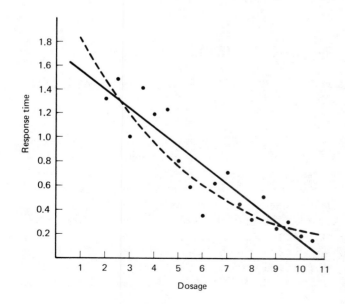

Figure 11-9 Response time versus dose.

1.00 from each of the times and multiply by 100, we can produce two-digit numbers involving no decimals. Accordingly, we produce new values $x = 2(X - 6.0)$ and $y = 100(Y - 1.00)$, listed in the following table together with other values necessary for the calculations. Column totals are also given.

X	x	Y	y	x^2	xy	y^2
1.5	−9	1.70	70	81	−630	4,900
2.0	−8	1.32	32	64	−256	1,024
2.5	−7	1.48	48	49	−336	2,304
3.0	−6	1.00	0	36	0	0
3.5	−5	1.41	41	25	−205	1,681
4.0	−4	1.19	19	16	−76	361
4.5	−3	1.23	23	9	−69	529
5.0	−2	0.80	−20	4	40	400
5.5	−1	0.59	−41	1	41	1,681
6.0	0	0.35	−65	0	0	4,225
6.5	1	0.61	−39	1	−39	1,521
7.0	2	0.70	−30	4	−60	900
7.5	3	0.44	−56	9	−168	3,136
8.0	4	0.33	−67	16	−268	4,489
8.5	5	0.50	−50	25	−250	2,500
9.0	6	0.27	−73	36	−438	5,329
9.5	7	0.30	−70	49	−490	4,900
10.0	8	0.18	−82	64	−656	6,724
10.5	9	0.15	−85	81	−765	7,225
	0		−445	570	−4625	53,829

Using the appropriate column totals, we calculate the correlation between the x and y values as

$$r = \frac{n\sum xy - \sum x \sum y}{\sqrt{n\sum x^2 - (\sum x)^2} \times \sqrt{n\sum y^2 - (\sum y)^2}}$$

$$= \frac{19 \times (-4625)}{\sqrt{19 \times 570} \times \sqrt{(19 \times 53829) - (445 \times 445)}} \quad \left(\text{since } \sum x = 0\right)$$

$$= -0.930$$

Since location and scale changes do not affect the correlation, the correlation between x and y is also the correlation between X and Y. This correlation, close to -1, indicates a strong linear relationship with a negative slope and agrees with the impression from the scatter diagram. Using the first form of assessment, the coefficient of determination, we find that $r^2 = 0.865$; that is, through the linear relationship, variation in dosage explains 86.5% of the variation in response time *for this set of data*.

For the present we will consider that there is, in general, a good linear relationship between dosage and response time for the range sampled and proceed with the estimation of the values of the slope and intercept of that relationship. In later sections, we discuss more formal techniques for determining whether this high sample correlation is due to a strong linear relationship in the general population or merely due to chance.

To estimate the relationship, $y = b_0 + b_1 x$, for the adjusted values, we calculate

$$b_1 = \frac{n\Sigma\,xy - \Sigma\,x\Sigma\,y}{n\Sigma\,x^2 - (\Sigma\,x)^2}$$

$$= \frac{n\Sigma\,xy}{n\Sigma\,x^2} = \frac{\Sigma\,xy}{\Sigma\,x^2} \qquad (\text{since } \Sigma\,x = 0)$$

$$= \frac{-4625}{570} = -8.11$$

$$b_0 = \frac{\Sigma\,y - b_1\Sigma\,x}{n} = \frac{\Sigma\,y}{n} \qquad (\text{since } \Sigma\,x = 0)$$

$$= \frac{-445}{19} = -23.42$$

We thus have the estimated relationship

$$y = -23.42 - 8.11x$$

What we want, however, is a relationship, $Y = B_0 + B_1 X$ for the original data. Rather than try to convert b_0 and b_1 to B_0 and B_1 directly, it is easier to substitute the expressions for x and y in terms of X and Y into the relationship that we have calculated to obtain

$$100(Y - 1.00) = -23.42 - [8.11 \times 2(X - 6.0)]$$

Multiplying through the brackets produces

$$100Y - 100 = -23.42 - 16.22X + 97.32$$

which may be rearranged to

$$100Y = 173.9 - 16.22X$$

Finally, dividing through by 100 produces

$$Y = 1.74 - .162X$$

This is very close to the line fit by eye and we can feel confident that the calculations are correct.

Using this line as an estimate of the general relationship, we can predict average response times for given dosages. For example, for a dosage of 8.25, we predict an average response time of $1.74 - (0.162 \times 8.25) = 0.40$.

If we extrapolate beyond the sample range to a dosage such as 13.0, we obtain an unrealistic value, -0.37, for the response time. We have an unrealistic value because a linear relationship is only reasonable in the range sampled. Over a more extended range, it is more reasonable to believe that the relationship is curved as originally shown in Figure 11.9. ▲

P11-2 (a) Calculate the correlation coefficient for the data in Practice Problem P11-1.

(b) In Practice Problem P11-1, what percentage of variation in strength is explained by variation in diameter through a linear relationship?

P11-3 The following table represents approximate values in 1971 dollars.

X, Year	Y, Real output per capita
1966	3770
1967	3820
1968	3980
1969	4130
1970	4170
1971	4360
1972	4560
1973	4820
1974	4880
1975	4780

(a) Adjust the data to produce new values $x = 2(X - 1970.5)$ and $y = (Y - 4200)/10$.

(b) What percentage of variation in output is explained by change in time?

(c) Predict real output per capita (in 1971 dollars) for 1976 on the basis of a continuing trend.

11-6 A study was conducted on the relationship between parental assistance and children's verbal achievement capabilities for various socioeconomic groupings and children's age categories. For one group and age category, a number of assessments on ten children and their parents produced the following values for a mean parental assistance index x and a mean child's verbal achievement score y:

x	53.9	68.1	52.6	51.7	64.3	63.6	67.8	54.8	62.7	59.7
y	112.0	197.5	117.5	107.5	157.0	186.5	182.0	137.0	175.0	132.5

(a) (1) Sketch a scatter diagram of x and y.

(2) Does the relationship appear to be linear?

(b) Subtract 50 from each x and multiply the resulting values by 10. Subtract 100 from each y and double the resulting values.

(1) Using the adjusted values, find the correlation between x and y.

(2) Using a linear relationship, what proportion of variation in verbal achievement can be explained by variation in parental assistance?

(c) (1) Determine the regression line for the adjusted data and hence for x and y.

(2) Predict the mean verbal achievement score for children whose parents produce an assistance index of 60.

11-7 Determine the strength of the linearity of the trend in Exercise 11-2 by determining the proportion of variation in average expenditure that can be explained by variation in time using a linear relationship.

11-8 Referring to the data in Exercise 11-4, calculate the correlation between lead content in plants near a highway and the number of vehicles per hour traveling that part of the highway.

11-9 In the following data, x is a measure of time and y is a measure of efficiency observed at the corresponding time.

x	1	2	3	4	5	6	7	8	9	10
y	5	17	20	22	26	25	24	22	19	16

(a) Plot a scatter diagram of efficiency versus time. Does there appear to be a strong relationship? Does it appear to be linear?
(b) Calculate the correlation between efficiency and time.

11-10 The following data represent the results of a study for which x is the quantity of a food supplement in the diet of chickens and y is a measurement of the hardness of shells of eggs laid by the chickens.

x	0.12	0.21	0.28	0.34	0.15	0.48	0.61	0.42	0.13	0.10	0.17
y	0.70	0.98	1.04	1.16	0.78	1.40	1.75	1.31	0.76	0.65	0.82
x	0.19	0.21	0.26	0.34	0.51	0.71	0.68	0.62			
y	0.90	0.95	1.10	1.24	1.51	1.95	1.83	1.75			

(a) Plot a scatter diagram of x and y and fit a line by eye.
(b) Subtract 1.20 from each y-value, then multiply each of these values and each of the x-values by 100. Calculate the correlation coefficient for x and y.
(c) The regression line is only to be used if it is such that at least 90% of the variation in hardness can be attributed to variation in the supplement to the data.
  (1) Should the line be used?
  (2) If so, estimate the average shell hardness that would occur with 0.50 unit of the food supplement.

11-11 If length expansion with heat produces a regression line with slope 0.001 when Y is expansion in inches and X is temperature in degrees Fahrenheit, what is the corresponding slope for x, expansion in centimeters, and y, temperature in degrees Celsius? (1 in. = 2.54 cm, 1 Fahrenheit degree = $\frac{5}{9}$ Celsius degrees.)

SOLUTIONS TO PRACTICE PROBLEMS

P11-2 (a) Using the data summary in the solution to Practice Problem P11-1 plus $\sum y^2 = 65676$, we find that

$$r = \frac{(10 \times 1278) - (15 \times 714)}{\sqrt{(10 \times 25.8) - 15^2} \times \sqrt{(10 \times 714) - 65{,}676}} = 0.940$$

(b) $100(r^2)\% = 88.4\%$ of variation in strength is explained by variation in diameter through a linear relationship.

P11-3 (a) The adjusted values with squares, cross products, and totals are as follows:

X	x	Y	y	x^2	y^2	xy
1966	−9	3,770	−43	81	1,849	387
1967	−7	3,820	−38	49	1,444	266
1968	−5	3,980	−22	25	484	110
1969	−3	4,130	−7	9	49	21
1970	−1	4,170	−3	1	9	3
1971	1	4,360	16	1	256	16
1972	3	4,560	36	9	1,296	108
1973	5	4,820	62	25	3,844	310
1974	7	4,880	68	49	4,624	476
1975	9	4,780	58	81	3,364	522
	$\overline{0}$		127	330	17,219	2,219

(b) With $\Sigma x = 0$ we find the correlation for x and y as

$$r = \frac{10 \times 2219}{\sqrt{10 \times 330}\sqrt{(10 \times 17{,}219) - 127^2}} = 0.9778$$

Since this is also the correlation for X and Y, $100(r^2)\% = 95.6\%$ of variation in output is explained by variation (i.e., change) in time.

(c) Finding the regression line for x and y, we have $b_1 = 2219/330 = 6.72424$ and $b_0 = 127/10 = 12.7$. The calculations are quite simple because $\Sigma x = 0$. We thus have $y = 12.7 + 6.72424x$. Changing back to X and Y, we have $(Y - 4200)/10 = 12.7 + [6.72424 \times 2(X - 1970.5)]$ or $Y - 4200 = 127 + 134.4848X - 265{,}002.30$ or $Y = -260{,}675.30 + 134.4848X$. For $X = 1976$, we find that $Y = -260{,}675.30 + (134.4848 \times 1976) = 5067$.

11-4 REGRESSION ANALYSIS

In Section 11-2 we found a best fit line to represent the linear relationship between two variables for a particular set of data. We considered using this line as a basis for predicting values of y corresponding to particular values of x. In so doing, we made an implicit assumption that there is, in general (not just in the particular data), a linear relationship between x values and corresponding average y-values and that in the full (hypothetical) population of all x's and corresponding y's, the points are scattered at random around the general linear relationship. We have thus used the results from our particular sample to produce a sample linear relationship, the least-squares solution, which is used to estimate the unknown population linear relationship.

We now proceed further, as in our other estimating procedures, to determine how well the sample line represents the population line—perhaps with appropriate confidence intervals, or to determine whether the sample results tend to support or refute a particular population line through the use of an appropriate hypothesis test.

Regression Model

To proceed further with the analysis of the regression line, we require a probability model. We assume that the y's are independent and that the relationship between x and y is such that for each fixed value of x (the independent or predicted variable), the variable y (the dependent or predicted variable) takes values at random according to a normal distribution with a mean of $\beta_0 + \beta_1 x$ and a standard deviation σ, where β_0 and β_1 are the intercept and slope, respectively, of the "true" underlying relationship between x and average y, and σ, the standard deviation, measures the chance or "error" fluctuation of the y's about the line.

> In regression analysis, we assume that the y's are independent and the distribution for y given a particular value of x is a normal distribution with mean $\mu_{y|x} = \beta_0 + \beta_1 x$ and standard deviation σ, where $\mu_{y|x}$ is the mean for y conditional on a given value x.

In this model we have an infinite number of normal distributions, one for each value of x. The distributions all have the same standard deviation σ but they have different means. These means lie on the line $y = \beta_0 + \beta_1 x$, as illustrated in Figure 11-10. The collection of populations is characterized by the three parameters β_0, β_1, and σ.

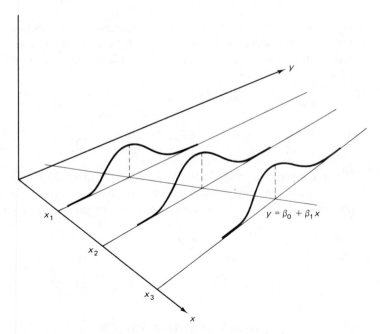

Figure 11-10 Normal distributions for y given x.

Estimating the Parameters

These parameters are estimated by three sample statistics, the two previously calculated statistics b_0 and b_1 and a new statistic s. Specifically, in the estimation of the "true" relationship, we estimate the intercept β_0 and the slope β_1 with the least-squares solutions b_0 and b_1 developed in Section 11-2, and we estimate the standard deviation σ with the new statistic

$$s = \sqrt{\frac{\Sigma\,(y_{\text{obs}} - y_{\text{fit}})^2}{n-2}}$$

which we may calculate more readily from the *calculating formula*

$$s = \sqrt{\frac{\Sigma\,y^2 - b_0\Sigma\,y - b_1\Sigma\,xy}{n-2}}$$

EXAMPLE 11-8 In a study on the elimination of idocyanine green as an indicator of hepatic blood flow after the administration of ether, the following measurements of idocyanine green (μg/ml) were noted for the times indicated (in minutes).

Time	1	2	3	4	5	6	7	8
Idocyanine	68	50	35	28	20	18	11	9

We assume a model in which the mean level of idocyanine green decreases linearly with time such that $\mu_{y|x} = \beta_0 + \beta_1 x$, where x is time and y is idocyanine green. The parameters β_0 and β_1 are estimated with b_0 and b_1, where

$$b_1 = \frac{n\Sigma\,xy - \Sigma\,x\Sigma\,y}{n\Sigma\,x^2 - (\Sigma\,x)^2} = \frac{(8 \times 742) - (36 \times 239)}{(8 \times 204) - 36^2} = -7.94048$$

$$b_0 = \frac{\Sigma\,y - b_1\Sigma\,x}{n} = \frac{239 + (7.94048 \times 36)}{8} = 65.60716$$

We assume further that individual values of idocyanine green are distributed about the mean corresponding to the appropriate time according to a normal distribution with standard deviation σ. We estimate σ with

$$s = \sqrt{\frac{\Sigma\,y^2 - b_0\Sigma\,y - b_1\Sigma\,xy}{n-2}}$$

$$= \sqrt{\frac{10{,}059 - (65.60716 \times 239) + (7.904048 \times 742)}{6}} = 6.7172 \quad \blacktriangle$$

We may proceed to confidence intervals and hypothesis tests by considering the sampling distributions for the statistics b_0, b_1, and s. Their sampling distributions (with our assumptions about normality, etc.) are such that appropriate modifications will produce statistics with t-distributions, just as $\bar{x}$ was modified in Section 6-4 to a statistic $(\bar{x} - \mu)/(s/\sqrt{n})$, which had a t-distribution.

Inferences about the Slope

If the normality assumptions are (at least approximately) correct, the statistic

$$t = \frac{b_1 - \beta_1}{s\sqrt{\dfrac{n}{n\Sigma x^2 - (\Sigma x)^2}}}$$

will have a t-distribution with $n - 2$ degrees of freedom.

Hypothesis Tests for the Slope

The t-statistic described above will be appropriate as a test statistic for testing the null hypothesis that β_1 is a specific value, say β_1^*, if we substitute β_1^* for β_1. The two-sided test procedure then becomes

$$H_0 : \beta_1 = \beta_1^*$$
$$H_A : \beta_1 \neq \beta_1^*$$

Test statistic: $t = \dfrac{b_1 - \beta_1^*}{s\sqrt{\dfrac{n}{n\Sigma x^2 - (\Sigma x)^2}}}$

Decision rule: reject H_0 if $|t| > t_{\alpha/2,\, n-2}$

The one-sided tests have alternative hypothesis and decision rule as

$$H_A : \beta_1 > \beta_1^*$$
Decision rule: reject H_0 if $t > t_{\alpha,\, n-2}$

or

$$H_A : \beta_1 < \beta_1^*$$
Decision rule: reject H_0 if $t < -t_{\alpha,\, n-2}$

EXAMPLE 11-9 Suppose that in Example 11-8 we wish to determine whether the sample results provide sufficient evidence at the 5% level that between 1 and 8 minutes, idocyanine green decreases at an average rate in excess of 5.00 μg/min. In other words, we wish to determine whether β_1, the slope of the population relationship, is less than -5.00. Since the burden of proof is on the alternative, we test $H_0 : \beta_1 = -5.00$ against $H_A : \beta_1 < -5.00$. The test statistic is calculated as

$$t = \frac{-7.94 - (-5.00)}{6.7172\sqrt{\dfrac{8}{(8 \times 204) - 36^2}}} = -2.837$$

which is less than $-t_{0.05,6} = -1.943$. We may reject H_0 and adopt H_A; that is, we have evidence that, between 1 and 8 minutes, idocyanine green decreases at an average rate in excess of 5.00 μg/min. ▲

Confidence Intervals for the Slope

Using the t-distribution and the usual procedures, we may also produce a *confidence interval for the "true" slope* β_1 as

$$b_1 \pm t_{\alpha/2,\,n-2}s\sqrt{\frac{n}{n\Sum x^2 - (\Sum x)^2}}$$

EXAMPLE 11-10 Suppose that for Example 11-8 we want a 95% confidence interval for β_1. Since $t_{0.025,6} = 2.447$, a 95% confidence interval for β_1 is $-7.940 \pm \left[2.447 \times 6.717\sqrt{8/((8 \times 204) - 36^2)} \right] = -7.940 \pm 2.536$. Thus, assuming that there is a true linear relationship between time and average amount of idocyanine green between 1 and 8 minutes and assuming a normal distribution of points around the line, we may be 95% confident that the slope of the relationship has a value between -10.476 and -5.404. ▲

Inferences about the Intercept

If the normality assumptions are (at least approximately) correct, we also have a t-distribution with $n - 2$ degrees of freedom for

$$t = \frac{b_0 - \beta_0}{s\sqrt{\dfrac{\Sum x^2}{n\Sum x^2 - (\Sum x)^2}}}$$

Hypothesis Tests for the Intercept

We may thus test the hypothesis that the value of the true intercept is equal to a particular value β_0^* by comparing the t above with appropriate values of the t-distribution with $n - 2$ degrees of freedom after substituting β_0^* for β_0. The two-sided test takes the form

$$H_0 : \beta_0 = \beta_0^*$$
$$H_A : \beta_0 \neq \beta_0^*$$

$$\text{Test statistic: } t = \frac{b_0 - \beta_0^*}{s\sqrt{\dfrac{\sum x^2}{n\sum x^2 - (\sum x)^2}}}$$

Decision rule: reject H_0 if $|t| > t_{\alpha/2,\, n-2}$

The one-sided tests have the alternative hypothesis and decision rule modified to

$$H_A : \beta_0 < \beta_0^*$$
Decision rule: reject H_0 if $t < -t_{\alpha,\, n-2}$

or

$$H_A : \beta_0 > \beta_0^*$$
Decision rule: reject H_0 if $t > t_{\alpha,\, n-2}$

EXAMPLE 11-11   Suppose that in Example 11-8 mean idocyanine green concentration at time 0 should be 70 μg/ml and that we wish to test the null hypothesis that it is so, the test to be a two-sided 5% test. We then calculate the test statistic $t = (65.607 - 70)/\left[6.7172\sqrt{204/((8 \times 204) - 36^2)}\right] = -0.839$. Thus, $|t| = 0.839$ does not exceed $t_{0.025,6} = 2.447$ and we may accept the null hypothesis that the true intercept is 70. In other words, we may accept the null hypothesis that the mean concentration at time 0 is 70 μg/ml, as it should be.   ▲

Confidence Intervals for the Intercept

Again, with appropriate use of the t-distribution, we may calculate a *confidence interval for the true intercept β_0* as

$$b_0 \pm t_{\alpha/2,\, n-2} s\sqrt{\frac{\sum x^2}{n\sum x^2 - (\sum x)^2}}$$

EXAMPLE 11-12   For the data in Example 11-8, a 95% confidence interval for β_0 is

$$65.607 \pm \left(2.447 \times 6.717\sqrt{\frac{204}{(8 \times 204) - 36^2}}\right) = 65.607 \pm 12.807$$

or 52.800 to 78.414. Note that the interval includes 70. This result is (as it should be) consistent with the acceptance of $H_0 : \beta_0 = 70$ in Example 11-11.   ▲

Confidence Intervals for the Mean y Corresponding to a Particular x

Besides finding confidence intervals for the individual parameters (i.e., the slope and intercept), we may wish to find a confidence interval for the "true average content" of idocyanine green at a particular time. Such a confidence interval is available and, in general form, *a confidence interval for the true mean y corresponding to a particular x value x_0 is*

$$(b_0 + b_1 x_0) \pm t_{\alpha/2,\, n-2} s \sqrt{\frac{1}{n} + \frac{n(x_0 - \bar{x})^2}{n\Sigma x^2 - (\Sigma x)^2}}$$

Note that the only part of the interval width that depends on the particular x-value is $(x_0 - \bar{x})^2$, the square of the difference between the particular value and the mean of the sample of x's on which the estimate is based. The width of the confidence interval increases as x_0 differs more from $\bar{x}$, indicating that (as was noted in Section 11-2) estimates in the central part of the sampling range are more reliable than those in the extremes. This emphasizes the unreliability of results that are obtained by extrapolating very far beyond the sampling range.

Figure 11-11 illustrates this effect with a sample regression line and a plot (dashed curves) of the end points of the confidence intervals for the mean y's corresponding to a range of particular x's about the mean of the sample x's.

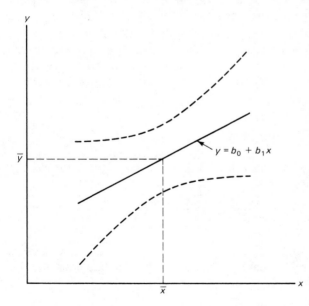

Figure 11-11   Confidence interval end points.

EXAMPLE 11-13   Suppose that, on the basis of the data in Example 11-8, we want a 95% confidence interval for the "true mean content" of idocyanine green at time 5.5 minutes. Since $\bar{x} = 4.5$ for the sample, the interval is

$$[65.607 - (7.940 \times 5.5)] \pm \left[2.447 \times 6.717\sqrt{\frac{\frac{1}{8} + 8[(5.5 - 4.5)]^2}{[(8 \times 204)] - 36^2}}\right]$$

or 15.60 to 28.28

Prediction Intervals for an Individual y Corresponding to a Particular x

We may also wish to predict the idocyanine green content of an individual at time 5.5 minutes with an interval estimate. Since individual values differ from the mean, we must account for two sources of variability: variability present in the sample producing the estimate of the mean and variability of a (future) individual value about the mean. This added source of variability increases the width of the interval estimate. We refer to such an interval as a *prediction interval*, and in the general form, a *100(1 − α)% prediction interval for an individual y corresponding to a particular x value x_0 is*

$$(b_0 + b_1x_0) \pm t_{\alpha/2,\,n-2}s\sqrt{1 + \frac{1}{n} + \frac{n(x_0 - \bar{x})^2}{n\sum x^2 - (\sum x)^2}}$$

EXAMPLE 11-14 For the example above, the 95% prediction interval for the content of a particular individual at time 5.5 minutes is $[65.607 - (7.940 \times 5.5)] \pm 2.447 \times 6.717\sqrt{1 + 1/8 + 8(5.5 - 4.5)^2/[(8 \times 204) - 36^2]}$ or 7.32 to 39.55. ▲

The extra width in this interval results from the additional "1" under the square root. This "1" represents the added variance σ^2 (s is already outside the square root) for the individual value's variability about the mean.

PRACTICE PROBLEM

P11-4 The following data represent the results of a study on ecological effects of trampling on vegetation. In the data, x is the distance in centimeters from a path center and y is the maximum height in centimeters of vegetation.

x	0	5	10	15	20	25	30	35	40	45	50
y	2	5	6	10	12	15	14	18	17	21	23

(a) Assuming that, for any distance x, maximum vegetation heights y are normally distributed about a mean $\mu_{y|x} = \beta_0 + \beta_1x$ with a standard deviation σ, estimate β_0, β_1, and σ.

(b) Find a 95% confidence interval for the slope of the relationship in part (a).

(c) Test the hypothesis that the line of means passes through the origin. (The origin is the point $x = 0$, $y = 0$; hence, for the line to pass through the origin, $\mu_{y|x}$ must equal 0 when x is 0; that is, β_0 must be 0.) Use a 5% two-sided test.

(d) Calculate a 95% confidence interval for the mean maximum height at a distance of 20 cm.

(e) Calculate a 95% prediction interval for an individual maximum height at a distance of 35 cm.

EXERCISES

11-12 In the following data, x represents blood flow and y represents pulmonary blood volume.

x	4.3	4.5	2.3	2.6	0.8	2.7	4.1	3.4	3.8	2.2
y	326	353	227	253	139	172	263	233	280	181
x	4.7	2.9	3.6	5.4	3.0	5.5	2.1	4.9	3.1	2.8
y	315	220	324	392	290	340	248	379	190	320

Assuming a linear relationship between x and mean y in the population:
(a) Find a 95% confidence interval for the slope of the relationship.
(b) Test the hypothesis that the intercept is 100 in a two-sided 5% test.

11-13 Using the data in Exercise 11-1, find a 95% confidence interval for the mean test mark corresponding to a diagnostic test score of 65.

11-14 Using the data in Exercise 11-2, find a 90% prediction interval for expenditure for 1977.

11-15 Referring to the data in Exercise 11-4, determine whether there is sufficient evidence at the 1% level that lead content increases, on the average, by more than 5 per 100 vehicles per hour increase in traffic flow.

11-16 Referring to the data in Exercise 11-6, calculate a 90% confidence interval for mean verbal achievement score for children whose parents produce an assistance index of 60.

11-17 The following values represent measurements on the amount of sodium absorbed by roots in a salt solution. In these values, x is the distance of the point measured in centimeters from the apex of the root and y is the sodium content (percent).

x	1	2	3	4	5	6	7	8	9	10
y	8.4	8.1	7.0	7.4	6.1	5.6	5.8	5.4	4.2	4.3

(a) Assuming a linear relationship between mean sodium content and distance, find a 95% confidence interval for the intercept of the relationship.
(b) Using a 5% level of significance, determine whether there is sufficient evidence that sodium content decreases at a rate of less than 0.5 per centimeter.
(c) Calculate a 90% prediction interval for an individual sodium value at a distance of 7.5 cm.

P11-4 (a) The estimates of β_1 and β_0 are the least-squares solutions b_1 and b_0 found by the usual procedures as

$$b_1 = \frac{(11 \times 4675) - (275 \times 143)}{(11 \times 9625) - 275^2} = 0.400$$

$$b_0 = \frac{143 - (0.40 \times 275)}{11} = 3.000$$

The estimate of σ is

$$s = \sqrt{\frac{\Sigma y^2 - b_0 \Sigma y - b_1 \Sigma xy}{n - 2}}$$

$$= \sqrt{\frac{2313 - (3.000 \times 143) - (0.400 \times 4675)}{9}}$$

$$= 1.247$$

(b) With $t_{0.025, 9} = 2.262$, the confidence interval is

$$b_1 \pm t_{0.025, 9} s \sqrt{\frac{n}{n \Sigma x^2 - (\Sigma x)^2}} = 0.40 \pm \left[2.262 \times 1.247 \sqrt{\frac{11}{(11 \times 9625) - 275^2}} \right]$$

$$= 0.400 \pm 0.054 \text{ or } 0.346 \text{ to } 0.454.$$

(c) To test $H_0: \beta_0 = 0$ against $H_A: \beta_0 \neq 0$, we calculate the test statistic

$$t = \frac{b_0 - 0}{s \sqrt{\dfrac{\Sigma x^2}{n \Sigma x^2 - (\Sigma x)^2}}} = \frac{3.000}{1.247 \sqrt{\dfrac{9625}{(11 \times 9625) - 275^2}}}$$

$$= 7.561 \text{ which greatly exceeds } 2.262 = t_{0.025, 9}.$$

We may reject the null hypothesis and may conclude that the line does not pass through the origin.

(d) With $x_0 = 20$ and $\bar{x} = 25$, the confidence interval is found as

$$(b_0 + b_1 x_0) \pm t_{0.025, 9} s \sqrt{\frac{1}{n} + \frac{n(x_0 - \bar{x})^2}{n \Sigma x^2 - (\Sigma x)^2}}$$

$$= [3.000 + (0.400 \times 20)] \pm \left[2.262 \times 1.247 \sqrt{\frac{1}{11} + \frac{11 \times 5^2}{(11 \times 9625) - 275^2}} \right]$$

$$= 11.000 \pm 0.892 \text{ or } 10.108 \text{ to } 11.892.$$

(e) With $x_0 = 35$, the prediction interval is found as

$$(b_0 + b_1 x_0) \pm t_{0.025, 9} s \sqrt{1 + \frac{1}{n} + \frac{n(x_0 - \bar{x})^2}{n \Sum x^2 - (\Sum x)^2}}$$

$$= [3.000 + (0.400 \times 35)] \pm \left[2.262 \times 1.247 \sqrt{1 + \frac{1}{11} + \frac{11 \times 10^2}{(11 \times 9625) - 275^2}} \right]$$

$$= 17.000 \pm 2.995 \text{ or } 14.005 \text{ to } 19.995.$$

11-5   CORRELATION ANALYSIS

The results of Section 11-4 are based on the assumption that there is some degree of linearity in the population of (x, y) pairs from which the sample was drawn. We may examine this assumption through the use of correlation. Just as we calculated, a correlation coefficient as a measure of strength of linearity in a sample, we may also consider a corresponding correlation coefficient as a measure of the linearity between x and y in the population. We attempt to make inferences on this coefficient through correlation analysis.

Although it is appropriate to consider the independent variable x to be fixed for purposes of predicting the random variable y in a regression analysis, it is necessary to consider x as well as y to be a random variable with a normal distribution in order to apply the following methods of correlation analysis.

With the assumption that x and y are both random variables with normal distributions, we introduce a new parameter, ρ (the Greek lowercase letter rho), which is the correlation between x and y in the full population of (x, y) pairs. The correlation coefficient r that we calculate from a sample of pairs is an estimate of ρ.

To assess the assumption of an underlying linear trend between x and y in the population, we might attempt to determine whether the data provide empirical "proof" that there is some linearity (i.e., that $\rho \neq 0$). We might, in other words, determine whether we can reject the hypothesis that $\rho = 0$.

Testing the Hypothesis of No Correlation

For the special case that x and y are not correlated (i.e., $\rho = 0$), the quantity

$$t = \frac{r\sqrt{n - 2}}{\sqrt{1 - r^2}}$$

will follow a t-distribution with $n - 2$ degrees of freedom.

This t serves as a test statistic to test the hypothesis that $\rho = 0$. The hypothesis test is as follows:

$$H_0 : \rho = 0$$

$$H_A : \rho \neq 0$$

Test statistic: $t = \dfrac{r\sqrt{n-2}}{\sqrt{1-r^2}}$

Decision rule: reject H_0 if $|t| > t_{\alpha/2, n-2}$

With one-sided alternative $H_A : \rho > 0$, the rejection region is $t > t_{\alpha, n-2}$, and for the alternative $H_A : \rho < 0$, the rejection region is $t < -t_{\alpha, n-2}$.

EXAMPLE 11-15 In a study on the relationship between pH levels in soil and the degree to which the soil was compacted, 16 measurements on pH and bulk of oven-dried material produced a sample correlation coefficient of $r = 0.706$. In order to determine whether there is, in general, a linear relationship, we might test $H_0 : \rho = 0$ against $H_A : \rho \neq 0$ with test statistic $t = 0.706\sqrt{14} \, / \sqrt{1 - 0.706^2} = 3.730$, which exceeds $t_{0.005, 14} = 2.977$. The correlation is significantly different from 0 with $p < 0.01$. (We use 0.01, not 0.005, because we have a two-sided test.) ▲

Equivalence of Testing $\rho = 0$ and $\beta_1 = 0$

If $\rho = 0$, there is no linear relationship at all between x and y and hence no monotonic trend of change in average y with x. We might then consider that a line representing the data would be a line parallel to the x scale, that is, a line with (population) slope $\beta_1 = 0$. It would thus seem equivalent to test the hypothesis that $\rho = 0$ or that $\beta_1 = 0$. In fact, the two tests are the same.

The test of $H_0 : \rho = 0$ is equivalent to a test of $H_0 : \beta_1 = 0$.

EXAMPLE 11-16 The following data represent the results of eight assessments on percent of stunting of cell growth in the application of X-rays for the treatment of cancer. In the data, x is the X-ray dose and y is the percent stunting.

x	5	10	15	20	25	30	35	40
y	2	11	23	29	32	46	55	56

These data produce $n = 8$, $\sum x = 180$, $\sum x^2 = 5100$, $\sum y = 254$, $\sum y^2 = 10{,}796$, $\sum xy = 7390$, $b_1 = 1.59524$, $b_0 = -4.14286$, $s = 3.14809$, and $r = 0.98905$.

The test statistic for $H_0 : \rho = 0$ is calculated as

$$t = 0.98905\sqrt{6} \,/\, \sqrt{1 - 0.98905^2} = 16.42$$

and the test statistic for $H_0 : \beta_1 = 0$ is

$$t = 1.59524/3.14809\sqrt{8/\big[(8 \times 5100) - 180^2\big]} = 16.42$$

as well.

In either case we obtain the same value, which is very significant when compared to $t_{\alpha, 6}$ for any α. ▲

Significant Versus Strong Correlation

In the test of $\rho = 0$, a "significant" sample correlation may not always be a "strong" correlation.

EXAMPLE 11-17 In Example 11-16, 0.989 is a "significant" correlation in the sense that it is significantly different from 0. It is also "strong" in the sense that 97.8% ($r^2 = 0.989^2 = 0.978$) of the variation in y (percent stunting) is explained by variation in x (dose) through the linear relationship.

In Example 11-15, we found $r = 0.706$ to be significantly different from 0, but it is only a moderate correlation since $r^2 = 0.498$ and only 49.8%, or slightly less than half, of the variation in y (pH) is explained through a linear relationship by variation in x (bulk).

Finally, as another illustration, suppose that we wish to test $\rho = 0$ in a two-sided 5% test with a sample of 30 (x, y) pairs. If the sample of $n = 30$ pairs produces a correlation coefficient $r = -0.38$, we have $t = -0.38\sqrt{28} \,/\, \sqrt{1 - 0.38^2} = -2.174$, which is more extreme than $-t_{0.025, 28} = -2.048$ and the hypothesis of no correlation may be rejected. The value -0.38 is a significant correlation; however, because only 14.4% of the variation in the y's is explained by variation in the x's, the value -0.38 is not a strong correlation at all. ▲

Testing Other Values of the Correlation Coefficient

To investigate strong correlations further, we may wish to consider values of ρ other than 0. To consider other values, we use a transformed quantity:

For inferences on ρ, we use a transformed quantity $Z(r) = \frac{1}{2} \ln[(1 + r)/(1 - r)]$ where ln is the logarithm to the base e (called the natural logarithm).

Values of $Z(r)$ for positive values of r are found in Table IX of Appendix B. For negative values of r, we use the fact that $Z(-r) = -Z(r)$.

The inferences are based on an approximate normal distribution for $Z(r)$.

If the population correlation is ρ, then $Z(r)$ will have an approximate normal distribution with mean $Z(\rho)$ and standard deviation $1/\sqrt{n-3}$ and therefore

$$z = [Z(r) - Z(\rho)] \times \sqrt{n-3}$$

will have an approximate standard normal distribution.

As a result, in order to test the hypothesis that $\rho = \rho_0$, the appropriate test statistic is z with ρ replaced by ρ_0.

The one-sided tests are as follows:

$H_0 : \rho = \rho_0$

$H_A : \rho > \rho_0$

Test statistic: $z = [Z(r) - Z(\rho_0)] \times \sqrt{n-3}$

Decision rule: reject H_0 if $z > z_\alpha$

If the alternative is

$H_A : \rho < \rho_0$

the rejection region is $z < -z_\alpha$.

We might consider the two-sided alternative

$$H_A : \rho \neq \rho_0$$

with rejection region $|z| > z_{\alpha/2}$; however, a more appropriate two-sided test for ρ is based on a one-sided test for ρ^2 (the population coefficient of determination). If we wish to determine whether ρ^2 exceeds some specific value ρ_0^2, then, in terms of the correlation, we wish to determine whether the magnitude of ρ exceeds the positive square root of ρ_0^2, that is, whether $|\rho| > \rho_0$ (where ρ_0 is positive). This is a two-sided condition since it involves two possibilities: $\rho > \rho_0$ and $\rho < -\rho_0$.

For the hypothesis test with $H_A : \rho^2 \neq \rho_0^2$, we have

$H_0 : |\rho| = \rho_0$ (we actually have $H_0 : |\rho| \leq \rho_0$)

$H_A : |\rho| > \rho_0$

Test statistic: $z = [Z(|r|) - Z(\rho_0)] \times \sqrt{n-3}$

Decision rule: reject H_0 if $|z| > z_{\alpha/2}$

EXAMPLE 11-18 Suppose that we wish to determine whether the data in Example 11-16 provide sufficient evidence at the 1% level that (in the population) variation in dose explains more than 50% of variation in percent stunting. The hypothesis is that $\rho^2 = 0.50$ (i.e., 50%) and the alternative is that $\rho^2 > 0.50$.

Since this value for ρ^2 can include positive or negative values for ρ, we use absolute values and write the hypothesis as $|\rho| = \sqrt{0.50} = 0.707$. From Table IX we find that the $Z(|r|) = Z(0.989) = 2.599$, $Z(\rho_0) = Z(0.707) = 0.8812$, and since $n = 8$, the test statistic is $z = (2.599 - 0.8812) \times \sqrt{5} = 3.841$, which exceeds $z_{0.005} = 2.576$. Since it is not reasonable to expect a negative correlation in this example, it is appropriate to change the test to a one-sided test with hypothesis $\rho = 0.707$ and alternative $\rho > 0.707$ and then compare z with $z_{0.01} = 2.326$. In either case, we may reject the hypothesis and conclude that, through a linear relationship, variation in dose explains more than 50% of variation in percent stunting. ▲

Confidence Intervals for the Correlation Coefficient

We may extend the forgoing results on $Z(r)$ to produce a confidence interval for ρ by first finding a confidence interval for $Z(\rho)$ and then using Table IX in reverse. A confidence interval for $Z(\rho)$ is

$$Z(r) \pm \frac{z_{\alpha/2}}{\sqrt{n-3}}$$

EXAMPLE 11-19 For the data in Example 11-15, a 95% confidence interval for $Z(\rho)$ is $Z(0.706) \pm 1.96/\sqrt{13}$, that is, 0.8792 ± 0.5436, or 0.3356 to 1.4228. Inspecting Table IX, we find that $1.422 = Z(0.890)$ is very close to 1.4228. We also find that $0.3361 = Z(0.324)$ is the closest value to 0.3356. The confidence limits 0.3356 and 1.4228 are very close to $Z(0.324)$ and $Z(0.890)$. The 95% confidence interval for ρ is thus from 0.324 to 0.890. ▲

EXAMPLE 11-20 As an example with negative correlation, consider the sample of 30 pairs producing $r = -0.38$ in the latter part of Example 11-17. Using the fact that $Z(-r) = -Z(r)$, we find $(-0.38) = -0.4001$ and a 90% confidence interval for $Z(\rho)$ is $-0.4001 \pm 1.645/\sqrt{27}$ or -0.7167 to -0.0835. In Table IX we find that $0.7169 = Z(0.615)$ is the closest value to 0.7167 and $Z(0.083) = 0.0832$ is the closest value to 0.0835. Adjusting for the negative values, we estimate that the limits -0.7167 and -0.0835 correspond to $Z(-0.615)$ and $Z(-0.083)$. The confidence interval for ρ is thus -0.615 to -0.083. ▲

PRACTICE PROBLEMS

P11-5 Referring to the data in Practice Problem P11-4, test
(a) $H_0 : \rho = 0$
(b) $H_0 : \beta_1 = 0$
in two-sided 1% tests. Compare the results.

P11-6 In an attempt to assess the linearity of the trend in number of indictable offenses per 100,000 population for women, the following data are to be analyzed for the degree of correlation.

Year	1935	1940	1945	1950	1955
Number of offenses	47	68	86	76	69

Year	1960	1965	1970	1975
Number of offenses	93	149	201	278

To simplify calculations, subtract 1955 from each year and divide the resulting values by 5. Calculate a 95% confidence interval for ρ on the basis of the adjusted data, and hence determine the corresponding interval for the original data.

P11-7 In a study on loss of memory over time, assessments on the amount of retained knowledge for various lengths of time after first learning material were recorded. For 20 subjects, the sample correlation between time and retained material was $r = -0.83$.

(a) What percentage of variation in retention is explained by variation in time in the sample?

(b) Do the data provide sufficient evidence at the 5% level that, in general, variation in time explains more than half of the variation in retention?

EXERCISES

11-18 In the following data, x represents the change in stream flow and y represents the change in number of insect larvae in appropriate units:

x	1.0	-3.3	-1.3	-0.7	2.1	3.8	2.5	0.1	-2.4
y	-290	-22	96	27	188	15	-112	-110	87

(a) Test the hypothesis of no correlation at the 5% level in a two-sided test.

(b) Test the hypothesis that the slope is 0 in a 5% two-sided test.

11-19 Referring to Exercise 11-6, determine whether the data provide sufficient evidence at the 5% level that (in the population) variation in parental assistance explains more than half of the variation in children's success.

11-20 (a) Is a sample correlation of $r = 0.47$ very strong?

(b) Is $r = 0.47$ significant (i.e., significantly different from 0) at the 1% level if it is based on a sample of size $n =$

(1) 5?

(2) 10?

(3) 20?

(4) 30?

11-21 Different tasks performed in different cultures produce different patterns of toothwear. If toothwear is strongly correlated with age, then, through regression procedures, toothwear can be used to predict age. In a study on such a relationship for eskimos, a toothwear index was devised. The index was a normalized measure of molar height and was expected to be negatively correlated with age. Determine whether there is a significant negative correlation in the following data. Indicate the *p*-level if the correlation is significant.

Age	Toothwear		Age	Toothwear
36	1.6		50	0.9
29	1.0		32	1.1
48	0.2		18	1.3
18	1.0		26	1.4
25	1.4		42	0.1
19	2.0		37	1.2
43	1.3		31	1.5
17	1.7		23	1.6

11-22 Calculate a 95% confidence interval for the population correlation on the basis of the data in Exercise 11-21.

11-23 The confidence interval on $Z(\rho)$ can be reexpressed as

$$\frac{1}{2}\ln\frac{1+r}{1-r} - \frac{z_{\alpha/2}}{\sqrt{n-3}} \leqslant \frac{1}{2}\ln\frac{1+\rho}{1-\rho} \leqslant \frac{1}{2}\ln\frac{1+r}{1-r} + \frac{z_{\alpha/2}}{\sqrt{n-3}}$$

Noting that $e^{\ln(x)} = x$ and $e^{a+b} = e^a \times e^b$, show that the foregoing interval can be rearranged to produce a confidence interval for ρ as

$$\frac{u-1}{u+1} \leqslant \rho \leqslant \frac{v-1}{v+1}$$

where

$$u = \frac{1+r}{1-r} \times e^{-2z_{\alpha/2}/\sqrt{n-3}}$$

$$v = \frac{1+r}{1-r} \times e^{2z_{\alpha/2}/\sqrt{n-3}}$$

11-24 Apply the results of Exercise 11-23 to Practice Problem P11-6 and compare the results of the two solutions.

SOLUTIONS TO PRACTICE PROBLEMS

P11-5 (a) The correlation coefficient is

$$r = \frac{(11 \times 4675) - (275 \times 143)}{\sqrt{(11 \times 9625) - 275^2}\sqrt{(11 \times 2313) - 143^2}} = 0.9845$$

producing

$$t = \frac{r\sqrt{n-2}}{\sqrt{1-r^2}} = \frac{0.9845 \times \sqrt{9}}{\sqrt{1 - 0.9845^2}} = 16.84$$

which exceeds $2.821 = t_{0.01,9}$. We may reject the null hypothesis.

(b) The results of Practice Problem P11-4 produce

$$t = \frac{b_0}{s\sqrt{\dfrac{n}{n\sum x^2 - (\sum x)^2}}} = \frac{0.400}{1.247\sqrt{\dfrac{11}{(11 \times 9625) - 275^2}}} = 16.82$$

which exceeds $2.821 = t_{0.01,9}$. We may again reject the null hypothesis. Except for a slight difference in the value of t due to rounding, the results are the same.

P11-6 Letting x denote coded year values and y denote number of offences, the new data are

x	-4	-3	-2	-1	0	1	2	3	4
y	47	68	86	76	69	93	149	201	278

Summary totals are $n = 9$, $\sum x = 0$, $\sum x^2 = 60$, $\sum y = 1067$, $\sum y^2 = 173{,}301$, $\sum xy = 1466$. Since $\sum x = 0$, the correlation for the coded data is found fairly easily as

$$r = \frac{9 \times 1466}{\sqrt{9 \times 60}\ \sqrt{(9 \times 173{,}301) - 1067^2}} = 0.875$$

From Table IX we find $Z(0.875) = 1.354$. The 95% confidence interval for $Z(\rho)$ is

$$Z(r) \pm \frac{z_{0.025}}{\sqrt{n - 3}} = 1.354 \pm \frac{1.96}{\sqrt{6}}$$

$$= 1.354 \pm 0.800 = 0.554 \text{ to } 2.154$$

Since 0.554 is about half way from $0.5533 = Z(0.503)$ to $0.5547 = Z(0.504)$, we estimate 0.554 to be about $Z(0.5035)$. The lower limit for ρ is thus 0.5035. Similarly, since 2.154 is about half way from $Z(0.973) = 2.146$ to $Z(0.974) = 2.165$, we estimate the upper limit to be 0.9735. The confidence interval for ρ is then from 0.5035 to 0.9735. Since the relocation and rescaling of x has no effect on the correlation, this is also the confidence interval for the original data.

P11-7 (a) The percentage of explained variation is $100(r^2)\% = 100 \times (-0.83)^2\% = 69\%$.

(b) Anticipating that any change would be a memory loss and that any correlation must be negative, we consider that proportion of variation explained will be greater than $1/2$ if $\rho < -\sqrt{1/2} = -0.707$. We then test $H_0 : \rho = -0.707$ against $H_A : \rho < -0.707$ with test statistic

$$z = [Z(r) - Z(\rho_0)] \times \sqrt{n - 3}$$

$$= [Z(-0.83) - Z(-0.707)] \times \sqrt{17}$$

$$= [-1.188 - (-0.8812)] \times \sqrt{17}$$

$$= -1.265$$

which is not less than $-1.645 = -z_{0.05}$. We may not reject the null hypothesis, and we do not have sufficient evidence at the 5% level that variation in time explains more than half of the variation in retention.

11-6 MODIFICATIONS OF THE SIMPLE LINEAR MODEL

Nonlinear Model

In Section 11-5 we considered correlation as an indicator of whether a linear relationship existed between two variables. In some cases, relationships that are not linear may be adjusted to a linear form by a suitable transformation and then analyzed by the techniques of simple linear regression.

EXAMPLE 11-21 A study on the effect of temperature on fish survival produced the following data in which x represents water temperature and t represents time until half of the fish in the aquarium died.

x	26.0	26.2	26.5	26.7	27.0	27.5	28.0	28.5	29.0	29.5
t	436	295	166	182	117	72	42	52	29	21

As indicated by the scatter diagram in Figure 11-12, there is a nonlinear relationship. An appropriate relationship to represent the mean time as a function of temperature is the exponential relationship: $t = e^{\beta_0 + \beta_1 x}$.

If we take the natural logarithm of both sides of the relationship, we produce the new relationship $\ln(t) = \beta_0 + \beta_1 x$, which we may write as the simple linear relationship $y = \beta_0 + \beta_1 x$, where y represents the natural logarithm of the time. The data with t replaced by $y = \ln(t)$ as follows are in a

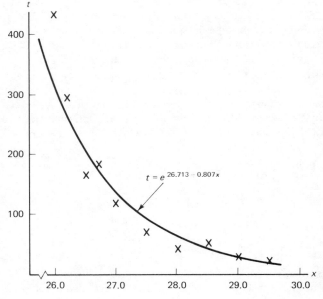

$$t = e^{26.713 - 0.807x}$$

Figure 11-12 Time t and temperature x.

form suitable for the application of the methods of simple linear regression to the modified relationship.

x	26.0	26.2	26.5	26.7	27.0	27.5	28.0	28.5	29.0	29.5
y	6.08	5.69	5.11	5.20	4.76	4.28	3.74	3.95	3.37	3.04

As indicated in Figure 11-13, the data in the (x, y) form display much more of a linear relationship than the data in the (x, t) form. The transformation produces an improvement from a moderate correlation of $r = -0.838$ for the (x, t) data to a stronger correlation of $r = -0.973$ for the (x, y) data.

For the adjusted data, the least-squares solution produces $b_0 = 26.713$ and $b_1 = -0.807$, producing the regression line, $y = 26.713 - 0.807x$ shown in Figure 11-13.

In terms of the original data, we have the relationship as $\ln(t) = 26.713 - 0.807x$; that is, $t = e^{26.713 - 0.807x}$, shown in Figure 11-12. Estimated mean times are then found from this relationship. For a temperature of 28.4°C, for example, we estimate the mean time to be $t = e^{26.713 - (0.807 \times 28.4)} = e^{3.794} = 44.4$

▲

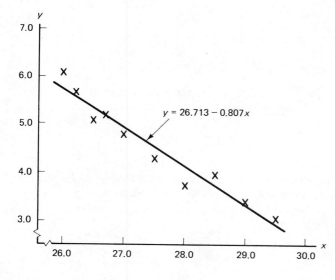

Figure 11-13 Natural logarithm of temperature y and time x.

This is only one example of the many nonlinear relationships that can be modified by an appropriate transformation to produce a linear relationship suitable for analysis with simple linear regression techniques.

Multiple Regression

The simple linear regression model may also be extended to a *multiple regression* model in which the dependent variable is predicted on the basis of a number of independent variables. A model of this type generally requires computer analysis and we do not consider a very complicated example; however, we can note the general concepts by considering the general form of the first extension to a model involving two independent variables x_1 and x_2.

The expression for the true mean y as a function of two independent variables x_1 and x_2 is

$$\mu_{y|x_1, x_2} = \beta_0 + \beta_1 x_1 + \beta_2 x_2$$

which is estimated with the relationship

$$y = b_0 + b_1 x_1 + b_2 x_2$$

The leasts-squares solution produces b_0, b_1, and b_2 by finding once again the values that minimize the sum of squared deviations $\Sigma(y_{obs} - y_{fit})^2$, which, in this case, becomes the sum $\Sigma[y_i - (b_0 + b_1 x_{1i} + b_2 x_{2i})]^2$.

The least-squares solution produces b_0, b_1, and b_2 as the simultaneous solutions of the normal equations

$$\Sigma y = nb_0 + b_1 \Sigma x_1 + b_2 \Sigma x_2$$

$$\Sigma x_1 y = b_0 \Sigma x_1 + b_1 \Sigma x_1^2 + b_2 \Sigma x_1 x_2$$

$$\Sigma x_2 y = b_0 \Sigma x_2 + b_1 \Sigma x_1 x_2 + b_2 \Sigma x_2^2$$

where Σx_1, $\Sigma x_1 y$, $\Sigma x_1 x_2$, and so on, are derived from a set of n triples (x_{1i}, x_{2i}, y).

Assumptions

As in the simple linear model, the individual y values corresponding to fixed x_1, x_2 values are assumed to follow a normal distribution with a mean of $\beta_0 + \beta_1 x_1 + \beta_2 x_2$ and a variance σ^2. As well, individual y's are assumed to be independent.

The variance σ^2 is estimated with

$$s^2 = \frac{\Sigma(y_{obs} - y_{fit})^2}{(n - 3)}$$

where the sum $\Sigma(y_{obs} - y_{fit})^2$ may be found more easily from the calculating formula,

$$\Sigma(y_{obs} - y_{fit})^2 = \Sigma y^2 - b_0 \Sigma y - b_1 \Sigma x_1 y - b_2 \Sigma x_2 y$$

EXAMPLE 11-22 A researcher is interested in predicting the average value of y, the length of time that an individual can continue a physical exercise, on the basis of two predictor variables, x_1, the number (average) of cigarettes smoked per day, and x_2, the ratio of weight in kilograms to height in meters. Data for 20 individuals are as follows:

x_1	x_2	y
24	53	11
0	47	22
25	50	7
0	52	26
5	40	22
18	44	15
20	46	9
0	45	23
15	56	15
6	40	24
0	45	27
15	47	14
18	41	13
5	38	21
10	51	20
0	43	24
12	38	15
0	36	24
15	43	12
12	45	16

The required summary totals are:

$$\Sigma y = 360 \qquad \Sigma x_1 = 200 \qquad \Sigma x_2 = 900$$
$$\Sigma y^2 = 7162 \qquad \Sigma x_1^2 = 3398 \qquad \Sigma x_2^2 = 41{,}058$$
$$\Sigma x_1 y = 2669 \qquad \Sigma x_1 x_2 = 9298 \qquad \Sigma x_2 y = 16{,}034$$

and the normal equations are:

$$360 = 20b_0 + 200b_1 + 900b_2$$
$$2669 = 200b_0 + 3398b_1 + 9298b_2$$
$$16{,}034 = 900b_0 + 9298b_1 + 41{,}058b_2$$

The simultaneous solutions of these equations are found to be $b_0 = 21.8460$, $b_1 = -0.6799$, and $b_2 = 0.0656$. Using these values and the calculating formula, we calculate s^2 as $s^2 = [7162 - (21.8460 \times 360) + (0.6799 \times 2669) - (0.0656 \times 16034)]/17 = 60.2627/17 = 3.54$.

For any given pair of values for x_1 and x_2, we estimate that y will, on the average, be equal to $21.8460 - 0.6799x_1 + 0.0656x_2$ and that individual values of y will vary about this average with a variance of 3.54. ▲

Multiple Coefficient of Determination

We may assess how well the multiple linear model fits the data in the same way that we assessed the simple linear model—through a coefficient of determination. As before, we determine the total variation of the observed y's about their mean as

$$SS_T = \Sigma\,(y - \bar{y})^2 = \Sigma\,y^2 - \frac{(\Sigma\,y)^2}{n}$$

We also determine the variation of the observed values about the fitted values as

$$SS_E = \Sigma\,(y_{\text{obs}} - y_{\text{fit}})^2$$

with calculating formula for the two-predictor model as given above and illustrated in the example. The variation that is explained, through a linear model, by variation in the predictors is, again, the difference

$$SS_R = SS_T - SS_E$$

The *multiple coefficient of determination*, denoted as R^2, is then found as

$$R^2 = \frac{SS_R}{SS_T}$$

EXAMPLE 11-23   For Example 11-22 we find that $SS_T = 7162 - 360^2/20 = 682$ and we recall from the calculation of s^2 [which is equal to $SS_E/(n - 3)$ in the two-predictor model] that $SS_E = 60.2627$. We thus find $SS_R = 682 - 60.2627 = 621.7373$ and the multiple coefficient of determination is $R^2 = 621.7373/682 = 0.912$. In the sample data, variation in the daily consumption of cigarettes and in the weight to height ratio accounts for 91.2% of the variation in endurance time through a multiple linear relationship.   ▲

Analysis of Variance Applied to Regression

In the simple linear model, we determined whether there was any linearity in the population by testing the null hypothesis that $\beta_1 = 0$ or, equivalently, that $\rho = 0$. The test on ρ was based on r, which was in turn based on the analysis of the total variation into its two components, variation about the regression and variation explained by the regression. This same analysis can be applied to a multiple regression model and an analysis of variance resulting in an F-ratio can be developed.

In this analysis, SS_T and SS_E have the same interpretation as in basic analysis of variance. The regression sum of squares SS_R is analogous to the treatment sum of squares, SS_{Tr}. If the model involves p predictors, the value of the degrees of freedom for the regression sum of squares is p and the regression mean square is SS_R/p. The value of the degrees of freedom for the error sum of squares is $n - p - 1$ and the error mean square is $SS_E/(n - p - 1)$ and is also s^2, the estimate of σ^2.

The null hypothesis that all of the predictors' regression coefficients are 0 (i.e., that $\beta_1 = \beta_2 = \cdots = \beta_p = 0$) is tested in an F-tested based on the following analysis-of-variance table.

Source of variation	Degrees of freedom	Sums of squares	Mean squares	F-ratio
Regression	p	SS_R	SS_R/p	$\dfrac{SS_R/p}{SS_E/(n-p-1)}$
Error	$n-p-1$	SS_E	$SS_E/(n-p-1)$	
Total	$n-1$	SS_T		

The hypothesis test is then written as

$$H_0: \beta_1 = \beta_2 = \cdots = \beta_p = 0$$

$$H_A: \text{the } \beta\text{'s are not all } 0$$

$$\text{Test statistic: } F = \frac{SS_R/P}{SS_E/(n-p-1)}$$

$$\text{Decision rule: reject } H_0 \text{ if } F > F_{\alpha;\, p,\, n-p-1}$$

Since $SS_R/SS_E = SS_R/(SS_T - SS_R) = (SS_R/SS_T)/(1 - SS_R/SS_T) = R^2(1 - R^2)$, the F-ratio may also be expressed in terms of the multiple coefficient of determination R^2.

$$F = \frac{R^2/p}{(1 - R^2)/(n-p-1)}$$

EXAMPLE 11-24 For the data in Example 11-22, we have $p = 2$ predictors x_1 and x_2 with regression coefficients β_1 and β_2. (The intercept is not considered to be a regression coefficient.) The joint hypothesis that $\beta_1 = 0$ and $\beta_2 = 0$ is tested on the basis of the following analysis-of-variance table.

Source of variation	Degrees of freedom	Sums of squares	Mean squares	F-ratio
Regression	2	621.7373	310.87	87.8
Error	17	60.2627	3.54	
Total	19	682.		

Since 87.8 exceeds $6.11 = F_{0.01;\, 2,\, 17}$, we may reject the null hypothesis at the 1% level of significance and we may conclude that β_1 and β_2 are not both 0. ▲

The analysis of variance can also be applied to the simple linear model in which we have $p = 1$ predictor.

EXAMPLE 11-25 For the percent stunting versus X-ray dose data in Example 11-16, we have $p = 1$ predictor with $SS_R = 2672.0372$ and $SS_E = 59.4628$. Since $n = 8$ for this example, the analysis of variance produces $F = (2672.0372/1)/(59.4628/6) = 269.62$, which exceeds $13.7 = F_{0.01; 1, 6}$. We thus may reject the hypothesis that $\beta = 0$.

As we should expect, this test is equivalent to the t-test of the null hypothesis that $\beta_1 = 0$ (i.e., that $\rho = 0$), illustrated in Section 11-5. The equivalence holds because the square of a t with f degrees of freedom is an F with 1 and f degrees of freedom. In Section 11-5 we found that $t = 16.42$, producing $t^2 = 269.62$, which is the same as F. The critical t value from Table IV was 3.707, producing $t^2 = 13.7$, which is the critical F value. ▲

In general, in the simple linear model, it is equivalent to reject the null hypothesis that $\beta_1 = 0$ (i.e., that $\rho = 0$) if $F > F_{\alpha; 1, n-2}$ or if $|t| > t_{\alpha/2, n-2}$. In fact, if we consider the general form of t based on r, it is $r\sqrt{n-2}/\sqrt{1-r^2}$. We thus have the square of t as $t = r^2(n-2)/(1-r^2) = r^2/[(1-r^2)/(n-2)] = F$ since, in the alternate form $F = R^2/[(1-R^2)/(n-p-1)]$, and in the simple linear model, $p = 1$ and $R^2 = r^2$.

PRACTICE PROBLEMS

P11-8 In bioassay, to determine drug potency, response to a drug is often considered to be a linear function of a logarithmic function of the dose. In the following data, d represents the dilution or concentration of the dose of an unknown drug being studied ($d = 1/2$ corresponds to dilution to half strength, $d = 4$ indicates concentration to quadruple strength, etc.); x represents the logarithm to base 2 of the dose (i.e., $x = \log_2 d$); and y represents the response.

d	1/8	1/4	1/2	1	2	4	8
x	-3	-2	-1	0	1	2	3
y	7	14	25	29	33	40	47

(a) Plot a scatter diagram of d and y, and a scatter diagram of x and y, and compare the apparent linearity.

(b) Find the least-squares solution:

$$y = b_0 + b_1 x$$

that is,

$$y = b_0 + b_1 \times \log_2 d$$

(c) Using the analysis-of-variance approach, test $H_0: \beta_1 = 0$ in the model $\mu_{y|x} = \beta_0 + \beta_1 x$.

P11-9 In the following data, x_1 represents number of years of formal education, x_2 represents number of years of job-related experience, and y represents salary in thousands.

(a) Find the least-squares fit $y = b_0 + b_1 x_1 + b_2 x_2$ for estimating the mean salary for given education and experience values.

(b) For the model assuming that y varies about mean $\mu_{y|x_1, x_2} = \beta_0 + \beta_1 x_1 + \beta_2 x_2$ according to a normal distribution with variance σ^2, find s^2 to estimate σ^2.

(c) For the model in b), test $H_0 : \beta_1 = \beta_2 = 0$.

(d) Estimate mean salary for 12 years of education and 7 years of experience

x_1	x_2	y
15	3	15.3
10	16	16.6
13	12	18.4
15	6	16.5
12	9	15.7
16	2	15.0
9	20	17.9

11-25 In the following data, x represents quantity of a chemical additive and y represents time required to achieve the desired effect from the additive.

x	0.005	0.010	0.015	0.020	0.025	0.030	0.035	0.040	0.045
y	225	143	84	87	56	39	52	57	34
x	0.050	0.055	0.060	0.065	0.070	0.075	0.080	0.085	0.090
y	32	45	40	28	42	32	23	36	24
x	0.095	0.100							
y	27	36							

(a) (1) Plot a scatter diagram of x and y. Does the relationship look linear?

(2) Calculate the correlation between x and y. What proportion of variation in time is explained, through linearity, by variation in chemical quantity.

(b) Transform the quantity of chemical to a new x, which is the reciprocal of the quantity (i.e., $x_{new} = 1/x_{old}$).

(1) Plot a scatter diagram of y and the new x. Does the relationship look linear?

(2) Calculate the correlation between y and the new x. What proportion of variation in time is explained, through linearity, by variation in reciprocal of quantity?

(3) Find the least-squares regression line between y and the new x and convert it to a relationship between y and the original x.

11-26 In the following data, x_1 represents age, x_2 represents number of hours watching television, and y represents score in a test.

x_1	x_2	y		x_1	x_2	y
6	14	60		10	16	63
9	14	80		9	13	72
12	20	72		12	19	81
10	14	70		6	13	50
7	7	63		9	19	47
8	10	60		11	13	83
11	18	77		11	14	72
12	12	91		8	14	70
6	12	64		7	10	60
8	7	56		10	14	71
7	21	45				

(a) Find the least-squares solution to fit

$$y = \beta_0 + \beta_1 x_1 + \beta_2 x_2$$

(b) Perform an analysis of variance to test the null hypothesis that β_1 and β_2 are both 0.

SOLUTIONS TO PRACTICE PROBLEMS

P11-8 (a) The scatter diagrams are shown in Figure P11-2(a) and (b). There is some curvature in Figure P11-2(a) and much more linearity in Figure P11-2(b).

(b) Since $\sum x = 0$, b_0 and b_1 are found quite simply as

$$b_0 = \frac{\sum y}{n} = \frac{195}{7} = 25.857$$

$$b_1 = \frac{\sum xy}{\sum x^2} = \frac{180}{28} = 6.429$$

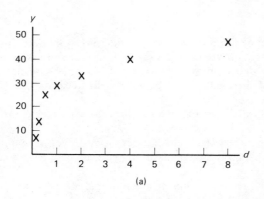

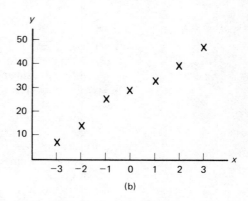

Figure P11-2 Response y against dose d and log dose x.

(c) $SS_T = \Sigma y^2 - (\Sigma y)^2/n = 6609 - (195)^2/7 = 1176.857$

$SS_E = \Sigma y^2 - b_0 \Sigma y - b_1 \Sigma xy$

$\qquad = 6609 - (195/7) \times 195 - (180/28) \times 180$

$\qquad = 19.714$

$SS_R = SS_T - SS_E = 1176.857 - 19.714 = 1157.143$

The degrees of freedom for error and regression are $n - p - 1 = 7 - 1 - 1 = 5$ and $p = 1$. The mean squares are thus $MS_E = 19.7145/3 = 3.943$ and $MS_{Tr} = 1157.143/1 = 1157.143$. The F-ratio is then $1157.143/3.943 = 293.5$, which is highly significant and we reject H_0.

P11-9 (a) After finding appropriate summary totals, we produce the normal equations as

$$115.4 = 7b_0 + 90b_1 + 68b_2$$

$$1471.7 = 90b_0 + 1200b_1 + 771b_2$$

$$1160.6 = 68b_0 + 771b_1 + 930b_2$$

The solutions to these equations are $b_0 = -0.9784$, $b_1 = 0.9672$, and $b_2 = 0.5176$.

(b) We find the variance estimate as

$$s^2 = \frac{SS_E}{n - 3} = \frac{\Sigma y^2 - b_0 \Sigma y - b_1 \Sigma x_1 y - b_2 \Sigma x_2 y}{n - 3}$$

$$= \frac{1912.36 - [(-0.9784) \times 115.4] - (0.9672 \times 1471.7) - (0.5176 \times 1160.6)}{4}$$

$$= \frac{1.1126}{4} = 0.2782$$

(c) $SS_T = \Sigma y^2 - (\Sigma y)^2/n = 1912.36 - [(195)^2/7] = 9.9086$

$SS_E = 1.1126$ \quad (from part b)

$SS_R = SS_T - SS_E = 9.9086 - 1.1126 = 8.7960$

The analysis-of-variance table then becomes

Source of variation	Degrees of freedom	Sums of squares	Mean squares	F-ratio
Regression	2	8.7960	4.3980	15.81
Error	4	1.1126	0.2782	
Total	6	9.9086		

The F-ratio exceeds $6.94 = F_{0.05; 2, 4}$ but not $18.0 = F_{0.01; 2, 4}$. The regression is significant with $p < 0.05$ and we may reject H_0 for $\alpha = 0.05$ but not $\alpha = 0.01$.

(d) For $x_1 = 12$ and $x_2 = 7$, the estimated mean is $-0.9784 + (0.9672 \times 12) + (0.5176 \times 7) = 14.25$.

11-27 If a regression line $y = b_0 + b_1 x$ is fitted to data with a scatter diagram that shows a general increase to the right, then the value of b_0 should be positive.
(a) True
(b) False
(c) Not necessarily

11-28 A regression line may be defined as a straight line about which the squared deviations are minimal.
(a) True
(b) False

11-29 If the x-values in a set of pairs (x, y) range from 0 to 100 and the y-values range from 20 to 80 with a scatter diagram showing an increase to the right, the slope of the regression line should be approximately
(a) -1.25 (b) -0.60 (c) 6.0
(d) 0.60   (e) 1.25

11-30 If a set of pairs (x, y) all lie exactly on a curve such that every increase in x occurs in conjunction with an increase in y, the sample correlation coefficient will be
(a) $+1$
(b) -1
(c) between 0 and 1 inclusive
(d) between -1 and 0 inclusive

11-31 If a sample of pairs (x, y) produces a correlation coefficient of 0.60, the variation in x explains ____ % of the variation in y.

11-32 Doubling all the x values in a set of pairs (x, y) will cause the slope to be _____.

11-33 If data set 1 of pairs (x, y) produces a regression line with a slope of 239.8 and data set 2 produces a regression line with a slope of -0.67, there is a stronger correlation in data set 1.
(a) True
(b) False
(c) Impossible to tell without more information

11-34 In a regression analysis, the prediction interval for a future y corresponding to a particular x is wider than the confidence interval for the mean y corresponding to the same x.
(a) True
(b) False

11-35 If a 95% confidence interval for the slope β_1 of a linear relationship is 3.61 to 9.74, we may reject $H_0 : \beta_1 = 0$ in a two-sided 5% test.
(a) True
(b) False

11-36 If a sample correlation coefficient $r = -0.30$ is "significantly" different from 0 in a t-test, the variables producing this r have been proven to be highly correlated.
(a) True
(b) False

11-37 In order to have sufficient evidence that the variation in x explains more than two-thirds of the variation in y through a linear relationship, we must be able to reject $H_0 : |\rho| =$ ____ in favor of $H_A : |\rho| >$ _____.

11-38 In a study of population density as related to distance measured radially from the center of a city, it is decided to fit a curve

$$Y = B_0 e^{-B_1 x}$$

where Y = number of persons per unit area
  x = distance from city center
  B_0 and B_1 are parameters to be estimated
To accomplish the goal a linear, relation $y = b_0 + b_1 x$ is used where

$$y = \ln(Y) \quad (\text{i.e., } Y = e^y)$$
$$b_0 = \ln(B_0) \quad (\text{i.e, } B_0 = e^{b_0})$$
$$b_1 = -B_1$$
$$x = x$$
$$\ln = \text{natural logarithm}$$

Suppose that population counts are taken at 11 different distances from the center of the city and the corresponding numbers of persons per unit area are recorded with the following results:

x	0	2	4	6	8	10
Y	148.3	54.2	20.7	40.1	16.7	6.3
$y = \ln(Y)$	5.00	3.99	3.03	3.69	2.82	1.84
x	12	14	16	18	20	
Y	2.7	6.5	0.9	1.6	0.5	
$y = \ln(Y)$	0.99	1.87	−0.11	0.47	−0.69	

(a) Plot a scatter diagram of x and Y.
(b) Find, by the least-squares method, the line $y = b_0 + b_1 x$ and convert this to the curve $Y = B_0 e^{-B_1 x}$ (i.e., $Y = e^{b_0} e^{b_1 x}$). Using the equation for the curve, calculate and plot the approximate values of Y corresponding to $x = 0, 4, 8, 12, 16, 20$ and sketch the curve on the same graph as the scatter diagram.

11-39 Using the data in Exercise 11-10, determine whether there is sufficient evidence at the 1% level to prove that, in general (i.e., in the population, not just the sample), over 90% of variation in hardness can be explained by variation in supplement through a linear relationship.

11-40 Using the data in Practice Problem P11-3 as a sample of a continuing long-term trend, find a 90% prediction interval for 1976 output per capita.

11-41 Referring to the data in Exercise 11-4, find a 95% confidence interval for the population correlation.

11-42 A study on energy consumption in a building over 14 heating months produced the following results in which x_1 is the number of degree-days (difference in degrees between outside temperature and 70°F times number of days), x_2 is the average windspeed in miles per hour, and y is energy consumption in megawatt hours.

x_1	x_2	y		x_1	x_2	y
1603	6.8	267		1408	7.0	257
1520	6.7	277		1695	4.3	239
1152	7.1	233		1384	5.6	214
855	6.9	224		1184	6.6	233
398	4.9	188		397	4.8	173
563	5.8	193		1527	5.4	217
1212	6.2	247		1099	5.4	217

(a) Find the least-squares regression $y = b_0 + b_1 x_1 + b_2 x_2$ to estimate the "average" relationship $\mu_{y|x_1, x_2} = \beta_0 + \beta_1 x_1 + \beta_2 x_2$.

(b) What proportion of variation in energy consumption is explained by variation in degree-days and wind speed through a linear model?

(c) Perform an analysis of variance to test the hypothesis that β_1 and β_2 are both 0.

11-43 The following eight values represent selected average prices for prescription drugs sold in each of eight consecutive years: 3.66, 3.70, 3.90, 4.06, 4.21, 4.38, 4.54, and 4.81.

(a) Determine the equation for a straight line to represent the trend and predict average price for the next year.

(b) Interpreting these data as representing a sample from a linear trend, calculate a 95% confidence interval for the slope of the trend.

11-8 GENERAL REVIEW

11-44 The proportion of a set of data that falls between the first and third quartiles is
(a) 1/3
(b) 1/2
(c) 1/4
(d) 3/4

11-45 If we test $H_0 : \mu = \mu_0$ against $H_A : \mu < \mu_0$ with test statistic t, and we obtain $t < -t_\alpha$, we have
(a) proved H_0 true
(b) proved H_0 false
(c) failed to prove H_0 false

11-46 In regression analysis, the wider interval is the
(a) confidence interval for the mean y at $x = x_0$
(b) prediction interval for an individual y at $x = x_0$

11-47 A type II error in a statistical test is the rejection of a true hypothesis.
(a) True
(b) False

11-48 If a homogeneous group is divided at random into two subgroups of 12 members each and each group follows a new different program, the appropriate procedure

for testing the equality of average performance under each program is to use
(a) a paired comparison t-test with 12 degrees of freedom
(b) a z-test on independent populations with equal variances
(c) a t-test on independent populations with equal variances
(d) a t-test on independent populations with unequal variances

11-49 If A and B are mutually exclusive, $P(A|B)$ will equal _____.

11-50 If two unbiased estimators for a parameter have sampling distributions with different variances, the preferred estimator is the one with the _____ variance.

11-51 The point $(\bar{x}, \bar{y})$ is on the least-squares regression line
(a) always
(b) sometimes
(c) never

11-52 In analysis of variance, a mean square is formed by dividing a sum of squares by the corresponding _____.

11-53 When n is very large and p is very small, binomial probabilities may be approximated with _____ probabilities.

11-54 To make a reasonable estimate of completion time for a large project, the variance of the working times for various assembly jobs must be at most 20. Suppose that a sample of 25 times is such that $\Sigma x = 1000$, $\Sigma x^2 = 40,960$.
(a) Test at the 5% level whether the variation is acceptable.
(b) Find a 95% confidence interval for the true variance of the working times.

11-55 Two balls are drawn without replacement from an urn containing two white, three black, and four green balls.
(a) What is the probability that the first is white and the second is black?
(b) What is the probability if the first ball is replaced before the second drawing?

11-56 Consider the following data from a sample of 100,000 residents of Canada:

Place of birth	Number in sample	Number with convictions of indictable offenses
Canada	74,000	160
United Kingdom and Commonwealth	6,000	70
United States	3,000	7
Other	17,000	18

Test whether there is any association between place of birth and number of convictions.

11-57 Suppose that a dental cement is such that it will hold a single crown against a force of up to 39.9 foot-pounds on the average. Suppose that an additive to the cement is considered worthwhile if it will increase the average required force to break the bond by more than 5%. If a sample of 50 observations of required forces to break the bond for the cement with the additive is such that $\Sigma x = 2210$ and $\Sigma x^2 = 99,566$, determine at the 1% level whether the additive is worthwhile.

11-58 In a test on three tire tread patterns, samples of stopping distances for a fixed set of road conditions and speed at the time of braking produced the following values in meters:

Tread A	Tread B	Tread C
31.6	41.9	30.2
36.1	35.3	31.3
33.9	39.0	34.6
30.9	39.2	37.3
39.5	42.0	37.8
33.8	36.2	35.5
30.5	34.7	37.5
31.9	36.5	35.1
31.5	37.7	32.8
35.6	40.4	28.4

Determine whether the three tread designs produce different mean stopping distances.

11-59 In the following pairs of observations, x measures the strength of a protective barrier and y measures loss from leakage through the barrier.

x	y
41	48.0
65	45.6
74	41.4
92	27.0
116	32.7
137	33.9
155	21.0
173	14.7
182	7.5

x	y
188	15.3
209	18.0
224	10.2
224	6.3
245	6.9
288	6.0
302	2.4
329	1.5

Predict average loss through barriers with a strength of 200.

11-60 Before a cold remedy is placed on the market, it is tested to determine whether it has a significant adverse effect on steadiness. Steadiness (or unsteadiness) is measured in the following manner. A stylus is held in a small hole and a count is made of how many times the stylus touches the side in a given length of time. Suppose that, in an experiment, 11 subjects had their steadiness scores recorded with and without the cold remedy with the following results:

Subject:	1	2	3	4	5	6	7	8	9	10	11
Score with remedy	57	68	38	52	42	46	45	39	32	44	36
Score without remedy	45	42	44	41	43	24	30	45	32	25	28

Is there enough evidence to prove that the drug has an adverse effect on the average?

11-61 Suppose that in a random sample of 180 links, 13 failed to hold a critical load.
(a) Estimate the probability that an individual link chosen at random would hold the load. Estimate the probability with an interval determined in such a way that, in an average of 19 out of 20 similar samples, the interval would contain the true value.
(b) If the true probability is equal to the sample success proportion in the data above, what is the probability that, in a further sample of 100 links, at most 12 will fail?

11-62 Using a base level of significance of 5%, determine which stack emission reduction schemes differ significantly with regard to the mean percent reduction in particle emissions on the basis of the following data summary from five sample reductions with each scheme under similar conditions:

Reduction scheme:	A	B	C	D	E
$\bar{x}$	52.1	46.4	42.5	41.0	46.2
s^2	8.1	6.7	7.7	6.3	7.3

11-63 A *population* of ten solutions produced the following pH values: 7.45, 7.10, 7.28, 6.86, 7.30, 6.99, 6.75, 7.06, 7.25, and 7.16. For these values, determine
(a) the mean μ
(b) the range
(c) the variance σ^2
(d) the coefficient of variation
(e) the standard score corresponding to the value 6.86

11-64 In a small pilot study on television viewing for two regions, sample families in each region were asked to keep a log of the number of hours that the television set was on. The results were as follows:

Region	Number of families sampled, n	Mean number of hours, $\bar{x}$	Standard deviation, s
I	12	23.75	5.72
II	18	21.50	4.17

(a) Assuming normality, determine whether there is evidence at the 10% level that the population variances are not equal.
(b) Assuming equal variances, calculate a 95% confidence interval for the difference between the two population means and using this interval, test the hypothesis of equal means in a two-sided 5% test.

11-65 A study on aspirin produced the following results where x is the dosage (in grains) of aspirin and y is the number of hours until relief from pain.

x	1.00	1.25	1.50	1.75	2.00	2.25	2.50	2.75	3.00	3.25
y	1.5	2.3	2.2	1.7	1.3	1.6	1.7	1.8	1.8	1.4
x	3.50	3.75	4.00	4.25	4.50	4.75	5.00	5.25	5.50	5.75
y	1.4	1.2	1.0	0.8	1.1	1.3	0.9	1.2	0.6	0.9

(a) (1) Test the null hypothesis of no correlation between x and y with three different testing procedures.

(2) Find a 95% confidence interval for the population correlation ρ.

(b) Determine a 95% confidence interval for the mean relief time corresponding to a dose of 4.50 grains.

11-66 In order to estimate the percentage of families in a given region that use a particular appliance, a researcher intends to use the corresponding percentage of families in a sample.

(a) With no prior knowledge on the population percentage, how large a sample must be taken to have a 95% chance that the estimate will be within four percentage points of the true value?

(b) If it can be assumed that at most 35% of all families use the appliance, how large a sample must be taken?

(c) Suppose that, in the study, only 250 families were sampled and, of these, 75 used the appliance. Estimate the percentage of all families who use the appliance. Estimate the percentage with an interval estimate such that you can be 95% confident that the interval includes the true population percentage.

11-67 (a), (b) Repeat Exercise 11-66 (a) and (b) if the region is, in fact, a community of 9000 families.

(c) Repeat Exercise 11-66 (c) for the community of 9000 families if a sample of 500 families included 150 using the appliance.

11-68 Times to the nearest minute for a sample of 350 runs of a particular bus route (round trip—terminal to route to terminal) produced the following cumulative distribution:

Time (minutes)	Number of trips
70 or more	350
75 or more	346
80 or more	330
85 or more	298
90 or more	240
95 or more	146
100 or more	75
105 or more	33
110 or more	9
115 or more	0

(a) Convert the data to a frequency distribution.

(b) Sketch a histogram of the distribution in part (a).

(c) Determine the mean, median, and standard deviation of the distribution. (Coding the data might be useful.)

11-69 In a comparison of two advertised coatings with standard copper plating, movement in degrees of rotation was noted between a pin and box under a torque equal to 50% of the yield strength of steel. Joints were tested both with and without a zinc compound. Perform a complete analysis of variance on the following data and determine whether the three coatings differ at the 5% level, whether there is a difference between zinc and no zinc at the 5% level, and whether there is interaction.

	Copper Plating	Coating A	Coating B
Zinc	12	22	23
	14	24	26
	13	21	24
	11	21	26
No zinc	14	23	27
	13	25	25
	15	21	25
	13	23	24

11-70 In a decision on which of two assistant managers would be promoted to general manager, 150 employees were surveyed for their preferences and 83 preferred Williams to McGregor. Each of the assistant managers served as acting general manager for a month while the other was on vacation and, following the 2-month period, the same 150 employees were questioned again. Of the original 83 preferring Williams, 18 changed their preferences to McGregor. Of the original 67 preferring McGregor, 5 changed their preference to Williams. Do these results indicate that overall opinion in the plant has changed? Use a 5% level of significance.

11-71 In the following data, x_1 represents the logarithm of objective distance in miles (actual distance traveled to a supermarket), x_2 represents the logarithm of travel time, and y represents the logarithm of cognitive distance (perceived distance to a supermarket).
(a) Use the least-squares solution to fit a multiple regression equation $y = b_0 + b_1 x_1 + b_2 x_2$.
(b) Calculate s^2 to estimate the variance σ^2 in the model.
(c) Test the null hypothesis that β_1 and β_2 are both 0.
(d) Estimate the mean value of y for $x_1 = 0.07$ and $x_2 = 1.0$.

x_1	x_2	y
-1.03	0.02	-0.05
-0.42	0.14	0.02
-0.36	0.10	0.18
-0.11	0.85	0.73
0.02	0.21	0.31
0.38	0.38	0.81
0.72	0.42	0.85
0.79	1.26	1.32
0.96	1.04	1.01

11-72 In an egg inspection, eggs with double yolks are set aside and placed in separate cartons from single-yolk eggs. The cartons are all the same in appearance. Five cartons with double-yolk eggs were inadvertently mixed in with 20 cartons of single-yolk eggs. If six of the full batch of 25 cartons are selected at random, what is the probability that
(a) four will contain double-yolk eggs?
(b) two will contain double-yolk eggs?
(c) none will contain double-yolk eggs?

11-73 It is suspected that working with hazardous materials can affect male genetic structure so that the proportion of boys born to men working with such materials will be reduced below the usual proportion, which is slightly in excess of 0.5. Is there sufficient evidence at the 5% level that the proportion of boys born to such men is less than 0.5 if the proportion of boys in a sample of babies is 0.40 and if the sample size is
(a) 10?
(b) 100?
(c) 200?
If there is significance at the 5% level, indicate the p-level.

11-74 In studies on recall capabilities and size of memory set, latency (response time per element in the memory set) has been found to be related linearly with the logarithm of the size of the memory set. In the following data, w is the number of letters in the memory set, x is the logarithm of w to the base 2, and y is the latency (milliseconds per letter).

w	2	4	6	8	10	12	14	16
x	1.0	2.0	2.6	3.0	3.3	3.6	3.8	4.0
y	488	563	630	707	762	742	758	743

(a) Plot a scatter diagram of w and y and also of x and y.
(b) Fit the least-squares solution to x and y to produce $y = b_0 + b_1 x$ and hence $y = b_0 + b_1 \log_2 w$.
(c) Estimate the mean latency for a memory set of size 8.

11-75 Experimental units deteriorate over time such that a unit selected at random from fresh stock has a 90% chance of performing adequately but a unit selected from old stock has only a 75% chance of performing adequately. A batch of experimental units has been selected at random from 48 indistinguishable batches, of which 12 were old stock and 36 fresh. Three units selected at random from this batch have performed adequately.
(a) What is the probability that the batch is fresh stock?
(b) If three more units are selected at random from this batch, what is the probability that they will all perform adequately?

NONPARAMETRIC OR DISTRIBUTION-FREE TESTS

12-1 INTRODUCTION

Up to this point, tests on population means and variances have been based on the assumption of normal (or at least approximately normal) distributions; however, in some cases, distributions may not be normal. For large samples, results based on the central limit theorem permit the use of normal distribution theory and the previous testing procedures remain valid. For small samples, however, the normal approximations may not be reasonable and other procedures must be used.

In this chapter we consider a few examples of these other procedures for testing hypotheses. These procedures are free of any assumptions on the form of the distribution; that is, they do not require any specific knowledge about the parametric form of the distribution although they do require the assumption of random sampling. Such testing procedures are referred to as *nonparametric* or *distribution-free* procedures. The contingency table analyses in Chapter 9 were examples of distribution-free procedures.

12-2 TESTS BASED ON SIGNS

The nature of nonparametric testing procedures can be illustrated quite simply with tests based on whether differences between observations and a fixed value or between pairs of observations have a positive sign or a negative sign. We will study three simple procedures based on signs, and, on the basis of one illustration, we will compare nonparametric and parametric tests.

Testing a Single Median

Interpreting average as median, we can use signs to develop a simple, nonparametric test on the value of the average of a population.

EXAMPLE 12-1 In an achievement test, average score is interpreted as median score. According to past records, groups of candidates have demonstrated a median score of 65. It is hypothesized that the average score that would be obtained by members of an experimental group is also 65. The hypothesis is to be tested on the basis of the following sample scores obtained at random from 18 members of the experimental group: 44, 32, 65, 74, 82, 60, 35, 81, 85, 37, 52, 53, 73, 58, 51, 45, 70, 93. The test is to have a two-sided alternative and a 5% level of significance.

Without making any assumptions about the form of the distribution, it is known that, if the null hypothesis is true, the numbers of scores above and below the hypothesized median, 65, will be equal. Scores equal to the hypothesized median are ignored and the remaining scores are converted to a difference, *score* minus *hypothesized median*, in this case, *score* minus 65. Each of these differences is then identified by the sign $+$ or $-$, depending on whether it is positive or negative. If the median is 65, the proportion of values that exceed 65 and produce a $+$ will be 0.5.

The true proportion of $+$'s (or $-$'s) in the population is denoted as p and the null hypothesis that the median is 65 is interpreted as the null hypothesis that $p = 0.5$. This hypothesis is then tested by the methods of Section 7-5.

The original alternative hypothesis for the example is that the average (median) is not 65. In terms of the proportion of $+$'s (or $-$'s) the alternative hypothesis is $p \neq 0.5$. One of the 18 scores in the sample is equal to 65 and is deleted from the sample. Of the remaining $n = 17$ scores, $x_0 = 7$ are identified with a $+$. If H_0 is true, the expected number of $+$'s is $np_0 = 17 \times 0.5 = 8.5$.

Referring to the methods of Section 7-5, we note that, since $x_0 = 7$ is below $np_0 = 8.5$ and since we have a two-sided test, we test for significance by comparing $P(x \leqslant x_0 | H_0 \text{ true})$ with $\alpha/2 = 0.025$. For the binomial with $n = 17$ and $p = 0.5$, $P(x \leqslant 7) = 0.240$; that is, $P(x \leqslant x_0 | H_0 \text{ true}) = 0.240$, which is not less than $0.025 = \alpha/2$. As a result, $x_0 = 7$ is not in the extreme lower $2\frac{1}{2}\%$ of the distribution and is not significant. The null hypothesis that the proportion of $+$'s is 0.5, and hence the null hypothesis that the median is 65 may be accepted. ▲

The test may be generalized in the following format.

$$H_0 : \tilde{\mu} = \tilde{\mu}_0$$
$$H_A : \bar{\mu} \neq \tilde{\mu}_0$$

Procedure: convert each difference $x - \tilde{\mu}_0$ to its sign $+$ or $-$ ignoring 0's.

Let $p = P(+) = $ the probability of a $+$ for all remaining differences in the population. Denote the number of nonzero differences as n and the number of $+$'s as x and then test $H_0: p = 0.5$ against $H_A: p \neq 0.5$ with the appropriate decision rule as given in Section 7-5. For $H_A: \tilde{\mu} < \tilde{\mu}_0$ or $H_A: \tilde{\mu} > \tilde{\mu}_0$, use $H_A: p < 0.5$ or $H_A: p > 0.5$.

This test, based on the analysis of positive and negative signs, is referred to as the *sign test*.

It should be noted that the sign test is actually based on the assumption that the population is continuous. In a discrete population with several values equal to the median, it is possible that the numbers of values above and below the median may differ. In such cases, H_0 and H_A may be restated as

$$H_0: P(x < \tilde{\mu}_0) = P(x > \tilde{\mu}_0) \quad \text{and} \quad H_A: P(x < \tilde{\mu}_0) \neq P(x > \tilde{\mu}_0)$$

Testing a Single Mean

Besides being used to test a median as above, the sign test may be used to test a value for a population mean if we make the assumption that the population is symmetric so that the mean and the median are equal.

Paired Comparisons

The sign test may also be used as a nonparametric substitute for the paired comparison test on two averages. In this case, all differences equal to 0 are deleted and the remaining differences are identified as $+$ or $-$, depending on whether they are positive or negative. If p is the population proportion of $+$'s (or $-$'s), the hypothesis of equal averages becomes the null hypothesis $p = 1/2$. The test is completed in the same manner as the test on a median.

EXAMPLE 12-2 Consider the paired comparisons analysis of the effect of a drug on rats in Example 8-8. Changing the differences to sign $+$ or $-$, we find that the ten individual differences are identified $+$, $+$, $+$, $+$, $+$, $-$, $+$, $+$, $-$, and $+$. Using the binomial probabilities with $n = 10$ and $p = 1/2$, we determine the probability of obtaining this many $+$'s and $-$'s if the null hypothesis $p = 1/2$ is true. Calculating the binomial probabilities or referring to Table I, we find that for $n = 10$ and $p = 1/2$, $P(x \geq 8) = 0.055$. Since we have a 2% two-sided test (as specified in Example 8-8), the observed value must be in the upper or lower 1% to produce a rejection. As indicated by the probability 0.055, 8 is not in the upper 1%. $P(x \geq x_0 | H_0 \text{ true}) = 0.055$ is not less than $0.01 = \alpha/2$ and the hypothesis may not be rejected.

We might also have performed the test by finding the full rejection region rather than finding the probability for the actual result. Using the binomial probabilities again with $n = 10$ and $p = 1/2$, we find that $P(x < 2) = P(x > 8) = 0.011$; thus, $P(x < 2 \ or \ x > 8 | H_0 \ \text{true}) = 0.022$, and, accordingly, an approximate 2% (actual 2.2%) level of significance may be obtained if the rejection region (for the two-sided test) is $x < 2$ or $x > 8$. The sample produces $x = 8$ +'s, which is not significant and we cannot reject the hypothesis—we do not have sufficient evidence to prove that the drug has an effect on maze scores. In view of the fact that 8 is almost significant, we might choose to reserve judgment. ▲

The general test format is as follows:

H_0: individual differences are zero on the average
H_A: individual differences differ from zero on the average.
Procedure: convert individual differences to their corresponding signs, ignoring 0's.
Let $p = P(+)$ for all remaining differences. Denote the number of nonzero differences as n and the number of +'s as x and test $H_0: p = 0.5$ against $H_A: p \neq 0.5$ with the methods of Section 7-5. For the alternative that individual differences exceed (or are less than) zero on the average, use $H_A: p > 0.5$ (or $H_A: p < 0.5$).

The null hypothesis may be interpreted as $H_0: \tilde{\mu}_d = 0$, where $\tilde{\mu}_d$ is the median of the population of all possible individual differences d. It may also be interpreted as $H_0: P(d < 0) = P(d > 0)$ to account for populations of discrete differences for which several might equal 0.

The interpretation of the null hypothesis as $H_0: \tilde{\mu}_d = 0$ emphasizes the fact that, after the data values are converted to individual differences, the paired comparisons sign test is exactly the same as the single sample sign test on a median. As such, the test is easily adapted to test for a specific value δ as the average (median) of individual differences.

To determine whether individual differences are a specified value δ on the average, we analyze the signs of the individual differences minus δ. We then apply the foregoing procedures to test $H_0: \tilde{\mu}_d = \delta$.

EXAMPLE 12-3 After being on duty for 1 hour, each of a sample group of 12 material inspectors inspected a piece of material with a given number of flaws and noted how many flaws he or she detected. This number is denoted as x_1. After being on duty for 4 hours, each inspector was required to inspect the same piece of material (unknown to the inspectors) and to again note the number of flaws detected. This number is x_2.

It is of interest to know whether the results of this experiment provide sufficient evidence at the 1% level that, for such a piece of material, x_1 would tend to exceed x_2 by more than 5. A normal distribution for the number of detected flaws cannot be assumed.

To determine whether the data do provide such evidence, we let d equal $x_1 - x_2$ for each inspector, and we use a paired comparisons sign test to test $H_0 : \tilde{\mu}_d = 5$ against $H_A : \tilde{\mu}_d > 5$. The test is performed by assigning a $+$ or $-$ to each difference minus 5 (i.e., for each inspector, we determine whether $d - 5 = x_1 - x_2 - 5$ is positive or negative) and by then testing $H_0 : p = 0.5$ against $H_A : p > 0.5$, where $p = P(+)$. In this latter test, n is the number of pairs for which $d - \delta = d - 5$ is not zero and x is the number of pairs for which $d - \delta = d - 5$ is positive.

The results of the experiment are as follows:

Inspector	x_1	x_2	$d = x_1 - x_2$	$d - \delta = d - 5$	Sign
1	43	36	7	2	$+$
2	39	30	9	4	$+$
3	47	43	4	-1	$-$
4	42	36	6	1	$+$
5	39	31	8	3	$+$
6	46	41	5	0	Delete
7	40	34	6	1	$+$
8	43	36	7	2	$+$
9	45	38	7	2	$+$
10	45	42	3	-2	$-$
11	44	36	8	3	$+$
12	41	32	9	4	$+$

After deleting the results for inspector 6, we have $n = 11$ pairs. From these 11 pairs, we have observed $x_0 = 9$ $+$'s. For the binomial distribution with $p = p_0 = 0.5$ and $n = 11$, we find that $P(x \geqslant 9) = 0.032$. We thus have the result that $P(x \geqslant x_0 | H_0 \text{ true}) = 0.032$. This value is not less than $0.01 = \alpha$. H_0 may not be rejected. These data do not provide sufficient evidence at the 1% level that the number of flaws detected in the first inspection tends to exceed that in the second by more than 5. ▲

Parametric and Nonparametric Tests

Recall that, using the t-test in Example 8-8, we could reject the null hypothesis. In Example 12-2, however, we cannot. It is difficult to determine which conclusion is correct, but there are a number of aspects of nonparametric, as opposed to parametric tests, that can be determined. For one, provided that randomness and assumptions such as independence are met as required, the assumed level of significance will be correct in a nonparametric test, whereas for a parametric test (i.e., a test based on assumptions of a particular parametric form such as a normal distribution), the level of significance will be correct only if the distribution assumptions are also satisfied. As a result, to obtain the correct level of significance (particularly in small samples), it is necessary to use nonparametric tests unless there are reasonable grounds for making parametric assumptions.

In order to be "universal" in their application, nonparametric procedures often fail to use all the information in a sample. For example, the sign test made use only of the signs of the differences and did not make use of the magnitudes of these differences: accordingly, if assumptions such as normality are reasonable, parametric tests should be used since, in these cases, both tests will have the correct level of significance, but the nonparametric test, which does not use all the information in the data, may have a higher probability of a type II error. The parametric test in such cases will have better power than the nonparametric test.

On occasion, it may be appropriate to determine a "quick and easy" solution to a hypothesis-testing problem. In such cases, if we are not overly concerned about the power of the test, then, even though parametric assumptions may be justified, we might use a simpler nonparametric test.

Testing Equality of Several Medians

If we interpret average as median we can use a contingency table analysis based on signs as a nonparametric substitute for the one-way analysis of variance discussed in Chapter 10 to test the equality of the averages of several populations.

EXAMPLE 12-4   Suppose that, from a number of small samples, we wish to determine whether average salaries for a given occupation are equal in three different regions. Suppose that, for the first region, a sample of salaries (in dollars) produced: 20,200, 16,125, 14,425, 9375, 13,825, 15,675, 14,225, 14,075, 10,650, 12,700, 16,000, and 12,250; for the second region, a sample produced: 19,600, 25,850, 16,150, 13,500, 11,900, 14,225, 14,000, 17,875, 11,750, 14,325, 14,200, 14,850, 12,500, and 14,525; and for the third region, a sample produced: 10,200, 17,600, 18,700, 12,200, 16,250, 18,700, 13,750, 10,800, 9,100, 12,800, 14,000, 13,750, and 14,600.

Since salary figures often form a skewed distribution, it may not be reasonable to assume normal distributions. It may be more appropriate to consider another test that does not require the normality assumption. Instead of testing the equality of the mean salaries, we consider the "average" to be median (as may be more appropriate for salary data) and we test the null hypothesis that the salaries in the three regions have equal medians.

To perform the test, we find the overall median for the entire set of data. There are 39 salaries in the full sample and, after the salaries have been listed from smallest to largest, the median is the 20th, which is 14,200. Next, we count the number of values above and below this median value for each of the samples (we do not include any values that are equal to the median). These numbers are the numbers of + and − signs, respectively. We summarize the results in a contingency table and then perform the usual analysis to determine whether the proportion above (below) the overall median depends on the particular group, indicating that the groups have different medians, or the proportion is independent of the group, indicating that the groups have the same median.

For the first region, there are six salaries above the overall median and six salaries below. One salary is 14,200 and is deleted in the second region. Of the remaining 13, there are eight salaries above the overall median and five below. For the third region, there are five above and eight below. The appropriate contingency table (with expected values in parentheses) is:

	First region	Second region	Third region	Total
Above	6 (6)	8 (6.5)	5 (6.5)	19
Below	6 (6)	5 (6.5)	8 (6.5)	19
Total	12	13	13	38

The test statistic is

$$\chi^2 = \frac{\Sigma (0 - E)^2}{E} = \frac{(6 - 6)^2}{6} + \cdots + \frac{(8 - 6.5)^2}{6.5} = 1.385$$

which is less than $\chi^2_{0.05, 2} = 5.991$. There is thus insufficient evidence at the 5% level to declare that the average (median) salaries are different in the three regions and we may accept the hypothesis that they are the same or reserve judgment.   ▲

This test serves as a nonparametric substitute for analysis of variance, if we interpret average as median and if we have independent random samples from the populations. It is referred to as the *median test* and is summarized as follows:

> $H_0: \tilde{\mu}_1 = \tilde{\mu}_2 = \cdots = \tilde{\mu}_k$
>
> $H_A:$ the $\tilde{\mu}_i$'s are not all equal
>
> Test statistic: $\chi^2 = \Sigma (O - E)^2/E$, where the O's are the observed numbers of values in the individual samples above and below the overall sample median (values equal to the overall median are ignored) and the E's are the corresponding expected values for a contingency table analysis.
>
> Decision rule: reject H_0 if $\chi^2 > \chi^2_{\alpha, k-1}$

PRACTICE PROBLEMS

P12-1 In a study on comfort control, individual subjects were given common standards with regard to clothing and activity and were observed in a room with average humidity and airflow. The temperature of the room was lowered for each subject until the subject found the room uncomfortably cool. Is there sufficient evidence on the basis of the following temperatures (to the nearest half-degree Celsius) for

15 subjects that the median discomfort temperature is greater than 17.0°C? Use a 5% level of significance.

18.0 17.5 19.5 19.0 18.5 19.5 19.0 15.0 17.5
17.5 18.0 16.5 17.5 15.5 19.5

P12-2 Repeat Practice Problem P8-4 without assuming normality for test scores.

P12-3 Suppose that in a study of insulin, 27 similar rabbits were divided into three groups of nine and each group received a different preparation of insulin. The following data represent the percent reductions in blood sugar after injection of insulin.

Preparation		
I	II	III
36	14	19
40	18	41
43	40	12
19	31	25
30	23	28
38	33	19
42	32	31
31	27	29
22	16	24

Using a 5% level of significance, determine whether the median percent reductions differ for the three processes.

EXERCISES

12-1 Rework Exercise 8-6 without assuming normality.

12-2 In an attempt to investigate the equivalence of four possible textbooks, an instructor divided a class of 28 students at random into four groups of seven and randomly assigned one of the texts to each groups. At the end of the year, the students obtained the following aggregate marks from a series of tests and assignments.

Text			
I	II	III	IV
68.1	41.4	54.4	44.0
69.3	47.8	44.5	51.7
60.0	54.3	51.1	69.6
64.3	65.5	56.7	59.1
71.7	73.2	59.2	55.2
67.9	44.5	49.6	66.8
75.9	48.2	41.4	61.3

Determine whether median aggregate marks differ for the four texts.

12-3 About 50 years prior to a current study, the median sperm count for healthy men was indicated to be 90 million per milliliter. Is there sufficient evidence at the 1% level of a decrease in the median if a current survey of 120 men indicated 75 with counts below 90 and 45 with counts above 90?

12-4 In a study on method of payment, 15 workers were paid hourly and produced the following numbers of units: 541, 566, 513, 547, 553, 546, 536, 542, 556, 550, 548, 536, 588, 591, and 583. Another group of 15 workers paid on a piecework basis produced the following numbers of units: 577, 595, 569, 543, 558, 604, 544, 570, 563, 534, 569, 526, 608, 550, and 527. A third group of 15 workers paid a basic hourly rate plus a piecework bonus produced the following numbers: 585, 536, 556, 563, 545, 548, 544, 573, 610, 559, 587, 571, 607, 586, and 599. Do the median outputs differ for the three payment schemes?

12-5 Two types of eyedrops were compared to determine whether they differ with regard to the length of time to sufficiently dilate the pupils for optical examination. For each patient in the study one eyedrop was used in one eye, the second in the other. The assignment of eyedrop type to each eye was random with the restriction that each eyedrop was used an equal number of times in left and right eyes. Is there sufficient evidence of a difference at the 5% level if tests on 15 patients produced the following times (nearest half-minute) (Do not assume normality.)

Patient:	1	2	3	4	5	6	7	8
Eyedrop A	5.0	4.5	6.0	5.0	7.0	6.5	4.0	4.0
Eyedrop B	5.5	5.5	5.5	6.0	7.5	6.5	4.5	5.5
Patient:	9	10	11	12	13	14	15	16
Eyedrop A	5.5	5.5	4.0	5.5	7.0	6.5	4.5	4.0
Eyedrop B	6.5	6.0	6.0	5.0	8.0	7.5	6.0	6.5

12-6 If a diagnostic test functions properly in a discrimination role, it should produce a symmetric bimodal distribution with a mean of 60. Assuming that symmetry does hold, do the following sample data support the null hypothesis that the mean is 60? 58, 37, 36, 64, 48, 74, 69, 50, 72, 66, 80, 41, 62, 50, 73, 76, 53, 68, 71, 51, 53, 75, 77, 46, 35.

12-7 Suppose that the numbers of campsites in use in two parks with equal capacity were compared over several sample weekends with the following results.

Weekend:	1	2	3	4	5	6	7	8	9	10	11	12
Camp I	23	54	78	103	120	102	96	83	62	73	67	39
Camp II	30	62	89	117	125	112	98	89	71	85	78	52

Do these data provide sufficient evidence that the number of campsites in use in the second camp generally exceeds that in the first by more than 5? If so, indicate the p-level.

12-8 The null hypothesis $\mu = \mu_0$ is to be tested against the alternative $\mu > \mu_0$ on the basis of a sample of size 10 from a normal population with known standard deviation σ. A 5% level of significance is to be used. The same hypothesis is also to be tested on the assumption only of a symmetric population—again with a sample of size 10 and a level of significance as close as can be achieved to 5%. Compare the power of the two test procedures when $\mu = \mu_0 + \sigma$ if the population actually is normal.

P12-1 Considering the differences $x - 17.0$, we find 12 positive differences. We test the null hypothesis the proportion of $+$'s is 0.5 against the alternative that it exceeds 0.5 by finding $P(x \geq 12)$ for the binomial with $n = 15$ and $p = 0.5$. This probability is $P(x \geq x_0 | H_0$ true) and is equal to 0.017, which is less than $0.05 = \alpha$. The observed result 12 is in the upper 5% and we reject the null hypothesis, thus concluding that the median temperature does exceed 17.0°C.

P12-2 Considering the sign of the individual differences minus 5, we obtain two 0's, which we delete. Of the remaining $n = 10$ differences, seven produce $+$'s and three produce $-$'s. For the one-sided test we compare $P(x \geq 7 | H_0$ true) with $\alpha = 0.05$. For the binomial with $n = 10$ and $p = 0.5$, $P(x \geq 7) = 0.172$, which exceeds 0.05. The observed result, 7, is not in the upper 5% and we cannot reject the null hypothesis. There is insufficient evidence that differences generally exceed 5.

P12-3 We determine that the overall median is 29 and deleting the single 29 from the data, we classify remaining values as above or below 29 and tabulate the results in the following contingency table (with expected values in parentheses):

	I	II	III
Above	7 (4.5)	4 (4.5)	2 (4)
Below	2 (4.5)	5 (4.5)	6 (4)

We then calculate $\chi^2 = (7 - 4.5)^2/4.5 + \cdots + (6 - 4)^2/4 = 4.889$, which does not exceed $5.991 = \chi^2_{0.05, 2}$. At the 5% level, we do not have sufficient evidence that the median percent reductions differ.

12-3 TESTS BASED ON RANKS

The tests in Section 12-2 are appealing for their simplicity, however, as a result of using very little information in the data (we have used signs and ignored magnitudes), we pay for simplicity through the loss of power.

There are other testing procedures which, although still distribution free, make use of more of the information in the data through analysis of ranking structures.

Rank Sum Test for Independent Samples

The *Mann–Whitney rank sum test* is based on the ranking structure of independent random samples from two continuous populations and may be used to test for differences in averages (means or medians—i.e., measures of location) of two populations or may be used to test for different degrees of dispersion or variation if equal means are assumed.

Comparing Locations

The Mann–Whitney test for comparing locations is based on a statistic U found by determining how many values in one sample precede each of the values in another. This statistic can be found by modifying a rank sum as follows.

Independent samples are taken from two populations and the values from both samples are combined in a single list from smallest to largest. If the two populations are identical, the values from the two samples will be randomly mixed in the combined list. If the populations are not identical but have different averages (i.e., measures of location), the members of one sample will tend to cluster at one end of the list and the members of the other sample will tend to cluster at the other end.

To test the null hypothesis of identical populations against the alternative of different averages, each value in the combined list is assigned a rank corresponding to its position in the list—the first value in the list is assigned rank 1, the next rank 2, and so on. If the values in a number of positions are equal, they are assigned a rank equal to the mean of the ranks corresponding to these positions.

The ranks assigned to the values in each sample are summed to form what are called rank sums. If the members of the first group tend to be smaller on the average than those of the second group, their rank sum will tend to be small and vice versa; accordingly, the test is based on whether a rank sum is too small, or too large, to have come from a random allocation of ranks.

The test is formulated in terms of the statistic U, which is formed from the rank sums as follows:

$$U = U_1 = n_1 n_2 + \frac{n_1(n_1 + 1)}{2} - R_1$$

or

$$U = U_2 = n_1 n_2 + \frac{n_2(n_2 + 1)}{2} - R_2$$

whichever is smaller,
where n_1 and n_2 are the sample sizes for the first and second groups, respectively, and R_1 and R_2 are the sums of the ranks assigned to the first and second samples, respectively.

Table X in Appendix B includes lower critical values of U. For samples of sizes 1 through 20 and for various values of α, Table X provides the value $U_{1-\alpha}$ such that if the two random samples are obtained from identical populations, $P(U < U_{1-\alpha}) = \alpha$. Upper critical values are found from lower critical bounds. In summary:

Table X provides $U_{1-\alpha}$.

$U_\alpha = n_1 n_2 - U_{1-\alpha}$

$P(U < U_{1-\alpha}|$ the two populations are identical$) = \alpha$

$P(U > U_\alpha|$ the two populations are identical$) = \alpha$

The test of the null hypothesis of equal populations against an alternative of different locations is formulated as follows:

H_0: the populations are identical

H_A: the populations differ with regard to location

Test statistic: $U = U_1$ or U_2, whichever is smaller

Decision rule: reject H_0 if $U < U_{1-\alpha/2}$

EXAMPLE 12-5 Two different filament structures are being considered for lights used in an extreme temperature environment and it is of interest to determine whether average lifetimes for the two structures differ at the 5% level. Because of the expense, only a few sample bulbs were available for study. Lifetimes cannot reasonably be expected to follow a normal distribution; accordingly, we consider the nonparametric Mann–Whitney test.

Lifetimes for $n_1 = 6$ bulbs with filaments of type A were found to be 1137, 1163, 1128, 1143, 1135, and 1139. Lifetimes for $n_2 = 5$ bulbs with filaments of type B were 1147, 1175, 1139, 1152, and 1142.

On the basis of these data we wish to test H_0: the populations are identical against H_A: the populations have different locations. The level of significance is $\alpha = 0.05$.

The $n_1 + n_2 = 11$ bulbs are combined in a single list and then ranked from smallest to largest with appropriate adjustment for ties. A useful approach is to present the combined list in column form with members of the two samples on opposites of a vertical line and rank allocations beside the numbers (ranking top to bottom) as follows.

Rank	Type A	Type B	Rank
1	1128		
2	1135		
3	1137		
4.5(4̸)	1139		
		1139	(5̸) 4.5
		1142	6
7	1143		
		1147	8
		1152	9
10	1163		
		1175	11
$r_1 = $ sum $= 27.5$			$38.5 = R_2 = $ sum

In this presentation, ties are at first ignored. After the full rank allocation is completed, we investigate for ties—there is one tie at the value 1139. We then delete the ranks for these ties (4 and 5 in this case) and replace them with their mean (4.5 in this case). We then add the final ranks for each sample to produce the rank sums $R_1 = 27.5$ and $R_2 = 38.5$.

The U statistics are found as

$$U_1 = n_1 n_2 + \frac{n_1(n_1 + 1)}{2} - R_1 = (6 \times 5) + \frac{6 \times 7}{2} - 27.5 = 23.5$$

$$U_2 = n_1 n_2 + \frac{n_2(n_2 + 1)}{2} - R_2 = (6 \times 5) + \frac{5 \times 6}{2} - 38.5 = 6.5$$

The smaller of these is $U = U_2 = 6.5$.

From Table X we find $U_{1-\alpha/2} = U_{1-0.025} = 4$ in the row labeled 0.025 for the case $n_1 = 6$ and $n_2 = 5$. Since $U = 6.5$ is not less than $U_{1-\alpha/2} = 4$, we cannot reject H_0. At the 5% level, these data do not provide sufficient evidence of difference in average lifetimes for the two filament structures. ▲

One-sided tests may be based on U_1 or U_2. The decision rule depends on whether U is formed from U_1 or U_2. If, for example, we have H_A: the first population has measure of location larger than that for the second, then, when H_A is true, we would expect members of the first sample to occupy upper ranks and would thus expect R_1 to be relatively large and R_2 to be relatively small. Since we form U by *subtracting* R_1 or R_2, we would expect U_1 to be relatively small and U_2 to be relatively large. The test format may then be written as:

H_0: the populations are identical

H_A: the location of the first population is larger than that for the second

Decision rule: reject H_0 if $U_1 < U_{1-\alpha}$ or if $U_2 > U_\alpha$

For the opposite one-sided alternative we just reverse the roles of U_1 and U_2.

EXAMPLE 12-6   We wish to test the null hypothesis that two populations of measurements (which cannot be assumed to be normal) are equivalent against the alternative that the measurements in the first group are smaller on the average, with a 5% level of significance.

Suppose that a random sample of $n_1 = 10$ measurements from the first group produced: 22, 25, 21, 17, 38, 25, 20, 16, 14, and 26; and an independent random sample of $n_2 = 13$ measurements from the second group produced: 39, 15, 28, 25, 23, 27, 33, 30, 43, 18, 26, 21, and 29. The combined list, from smallest to largest, with ranks and rank sums is:

Rank	First group	Second group	Rank	Rank	First group	Second group	Rank
1	14					25	(13)12
		15	2	14.5 (14)	26	26	(15)14.5
3	16					27	16
4	17					28	17
		18	5			29	18
6	20					30	19
7.5(7)	21					33	20
		21	(8)7.5	21	38		
9	22					39	22
		23	10			43	23
12(11)	25						
12(12)	25			$\sum = R_1 = 90$		$186 = R_2 = \sum$	

Since it is not necessary to calculate both forms of U, we will calculate only U_2. We now have H_0: the populations are identical; we have H_A: the location of the first is less than that for the second; and we have the decision rule to reject H_0 if $U_2 < U_{1-\alpha}$, where $\alpha = 0.05$.

In the row labeled 0.05 for the case $n_1 = 10$ and $n_2 = 13$, we find $U_{1-\alpha} = U_{1-0.05} = 38$. From the data we have

$$U_2 = n_1 n_2 + \frac{n_2(n_2 + 1)}{2} - R_2 = (10 \times 13) + \frac{13 \times 14}{2} - 186 = 35$$

Since $U_2 = 35$ is less than $U_{1-\alpha} = 38$, we may reject H_0. ▲

Normal Approximation for U

If we have larger sample sizes such that n_1 or n_2 exceeds 20, we make use of a normal approximation to find the critical values for U.

> If the null hypothesis of identical populations is true, and if the sample sizes are not small, $U(U_1 \text{ or } U_2)$ will have an approximate normal distribution with mean and standard deviation respectively of
>
> $$\mu_U = \frac{n_1 n_2}{2}$$
>
> $$\sigma_U = \sqrt{\frac{n_1 n_2(n_1 + n_2 + 1)}{12}}$$

The result above then leads to approximate critical values for U based on percentiles of the standard normal distribution.

If n_1 and n_2 are not small, critical values for U are found as

$$U_{1-\alpha} = \frac{n_1 n_2}{2} - z_\alpha \sqrt{\frac{n_1 n_2 (n_1 + n_2 + 1)}{12}}$$

$$U_\alpha = \frac{n_1 n_2}{2} + z_\alpha \sqrt{\frac{n_1 n_2 (n_1 + n_2 + 1)}{12}}$$

EXAMPLE 12-7   In a two-sided test of H_0 : two populations are identical with level of significance $\alpha = 0.01$, a researcher obtained $n_1 = 25$ measurements from the first population and $n_2 = 30$ from the second. After listing the two sets of data together and assigning ranks in the usual way, the researcher found $R_1 = 864$ and $R_2 = 676$, producing

$$U_1 = (25 \times 30) + \frac{25 \times 26}{2} - 864 = 211$$

$$U_2 = (25 \times 30) + \frac{30 \times 31}{2} - 676 = 539$$

Since U_1 is smaller than U_2, we have $U = U_1 = 211$. Using the normal approximation we find that

$$U_{1-\alpha/2} = U_{1-0.005} = \frac{n_1 n_2}{2} - z_{0.005} \sqrt{\frac{n_1 n_2 (n_1 + n_2 + 1)}{12}}$$

$$= \frac{25 \times 30}{2} - 2.576 \sqrt{\frac{25 \times 30 \times 56}{12}} = 223 \quad (\text{where } 2.576 = z_{0.005})$$

Since $U = 211$ is less than $U_{1-\alpha/2} = 223$, the researcher may reject H_0.   ▲

Comparing Measures of Variation

The rank sum test may also be used to test the null hypothesis that two populations are identical against the alternative of different measures of variation. If the two populations have different measures of variation, the members of the less varied population will tend to cluster in the central part of the list and the members of the more varied population will tend to spread to the extremes of the list. (We are assuming that the populations do not differ with regard to location.)

To investigate measures of variation, the ranks are assigned in a different way. After the two samples have been combined in a single list, the ranking proceeds from the extremes (smallest and largest) to center (i.e., "from the outside in"). The first value receives rank 1. The last receives 2. The second last receives rank 3. The second smallest and third smallest receive ranks 4 and 5. The third and fourth last receive 6 and 7, and so on, alternating from one end of the list to the other. This ranking is first done ignoring ties; then, where there are ties, the ranks are replaced with the mean of all the ranks assigned to tied values.

In this assignment, small ranks are assigned to values that vary considerably from the center of the distribution. Large ranks are assigned to values that do not vary much from the center. As a result, if the members of the first group are less varied than those of the second, then R_1 will tend to be large, and vice versa. The test is again based on whether R_1 (or R_2) is too large or too small to have resulted from a random allocation of ranks. With U defined as above, the test is as follows:

H_0 : the two populations are identical

H_A : the populations differ with regard
to measures of variation

Decision rule: reject H_0 if $U < U_{1-\alpha/2}$

A one-sided test with alternative that the first population is less varied than the second is as follows:

H_0 : the two populations are identical

H_A : the first population is less varied
than the second

Decision rule: reject H_0 if $U_1 < U_{1-\alpha}$ or if $U_2 > U_\alpha$

For the alternative that the first population is more varied than the second, we reverse the roles of U_1 and U_2.

EXAMPLE 12-8   Consider a two-sided 5% test of the null hypothesis of equal variances based on the following independent random samples of sizes $n_1 = 10$ and $n_2 = 8$ from two continuous populations: 173, 177, 184, 188, 203, 205, 209, 211, 220, 224; and 185, 191, 193, 202, 207, 208, 211, and 217. The two lists combined and ranked from the outside in are as follows:

Rank	First sample	Second sample	Rank	Rank	First sample	Second sample	Rank
1	173			18	205		
4	177					207	15
5	184					208	14
		185	8	11	209		
9	188			8.5(~~10~~)	211		
		191	12			211	(~~7~~)8.5
		193	13			217	6
		202	16	3	220		
17	203			2	224		
					$\overline{R_1 = 78.5}$		$92.5 = R_2$

Note the tied values 211, which occupy ranks 7 and 10. These two values are each assigned the mean rank $(7 + 10)/2 = 8.5$.

The two forms of U produce $U_1 = (10 \times 8) + (10 \times 11/2) - 78.5 = 56.5$ and $U_2 = (10 \times 8) + (8 \times 9/2) - 92.5 = 23.5$. Since U_2 is smaller than U_1, we have $U = U_2 = 23.5$.

In the row labeled 0.025 in Table X for the case $n_1 = 10$ ad $n_2 = 8$, we find $U_{1-\alpha/2} = U_{1-0.025} = 18$. Since $U = 23.5$ is not less than $U_{1-\alpha/2} = 18$, we cannot reject the variances. ▲

Paired Comparisons

A test that makes use of magnitudes of differences in a paired comparisons analysis through a ranking structure is the *Wilcoxon signed ranks test*. This test is a more powerful test than the paired comparisons sign test considered in Section 12-3.

In this test, the individual differences *first value* minus *second value* in a pair are ranked in the usual fashion according to their magnitudes, ignoring signs. Differences of zero are deleted from the data. Two rank sums or rank totals are then obtained.

> T_1 is the sum of the ranks assigned to the positive differences and T_2 is the sum of the ranks for negative differences.

The null hypothesis of no difference on the average in individual pairs is then tested by comparing either T_1 or T_2 with critical values from Table XI in Appendix B.

The test format for a two-sided test is as follows:

> H_0 : individual differences are 0 on the average
>
> H_A : individual differences differ from 0 on the average
>
> Test statistic: $T =$ the smaller of T_1 and T_2 (where T_1 and T_2 are as defined above)
>
> Decision rule: reject H_0 if $T \leqslant T^*$, where T^* is the value in Table XI corresponding to the number of nonzero differences and level of significance

The one-sided tests are as above with alternatives and test-statistics stated as follows.

> H_A : individual differences (first minus second) exceed 0 on the average
>
> Test statistic: $T = T_2$
>
> or
>
> H_A : individual differences (first minus second) are less than 0 on the average
>
> Test statistic: $T = T_1$

EXAMPLE 12-9 An alkaline battery is proposed as a substitute for a currently used "extra life" dry cell on the grounds that it will generally last longer under similar conditions as a power source for flashlights. To test the claim at the 5% level, one battery of each type was placed in each of 16 lights. The time until the first battery was classified as "dead" was noted. This battery was replaced with another and the light operated until the second original battery was classified as dead. The results were as follows:

Light:	1	2	3	4	5	6	7	8
"Extra Life"	828	803	796	815	863	811	796	804
Alkaline	836	821	797	810	854	842	789	810
Light:	9	10	11	12	13	14	15	16
"Extra Life"	798	835	842	807	818	800	845	791
Alkaline	813	855	840	807	831	796	856	817

The 16 differences *first* minus *second* are then -8, -18, -1, 5, 9, -31, 7, -6, -15, -20, 2, 0, -13, 4, -11, and -26. Ignoring the 0 difference for the 12th light, we assign ranks to the remaining $n = 15$ differences *according to their magnitudes*.

The differences (without signs) are listed from smallest to largest and classified according to whether they are positive or negative in the following format similar to that used in the Mann–Whitney test. Ranks are assigned in the same manner.

Rank	Positive	Negative	Rank	Rank	Positive	Negative	Rank
		1	1			11	9
2	2					13	10
3	4					15	11
4	5					18	12
		6	5			20	13
6	7					26	14
		8	7			31	15
8	9	$T_1 = \overline{23}$					$\overline{97} = T_2$

We test the claim by testing the null hypothesis of no difference against the alternative that the differences *first* minus *second* (i.e., *extra life* minus *alkaline*) are less than 0 on the average. For this alternative, the test statistic is T_1. For $n = 15$ and a one-sided test with $\alpha = 0.05$, Table XI produces the critical value $T^* = 30$. Since $T = 23$ is less than $T^* = 30$, we reject the null hypothesis and conclude that the alkaline cells are superior.

It is interesting to note that in this example a sign test would fail to detect a difference, but the *t*-test, assuming normality, would detect the difference. ▲

Comparing Several Populations

By comparing the assignment of overall ranks to members of a number of independent samples, we can test the equality of several populations using a test known as the *Kruskal–Wallis test*. For the special case of two samples, this test produces the same results as a two-sided Mann–Whitney test. This test, like the median test, is a distribution-free substitute for the one-way analysis of variance discussed in Chapter 10.

In this test, the ranking procedures used in the Mann–Whitney test are extended to k independent random samples of size $n_1, n_2, \ldots, n_k$ to produce rank sums $R_1, R_2, \ldots, R_k$. The null hypothesis of identical populations is tested by comparing the rank sums through the use of a test statistic that will follow an approximate χ^2-distribution with $k - 1$ degrees of freedom if the null hypothesis is true and if the individual sample sizes are each at least 5. The test format is as follows:

H_0 : the k populations are identical

H_A : at least one population differs from the others with regard to location (i.e., means or medians are not all equal)

Test statistic: $\chi^2 = \dfrac{12}{N(N + 1)} \displaystyle\sum_{i=1}^{k} \left(\dfrac{R_i^2}{n_i} \right) - 3(N + 1)$

where $N = \sum n_i$

Decision rule: reject H_0 if $\chi^2 > \chi^2_{\alpha, k-1}$

EXAMPLE 12-10 The recorded depths in meters of completed gas wells in four drilling areas produced the following data in an analysis of depth variability over the regions:

	Region		
I	II	III	IV
463	618	615	566
482	670	623	643
530	518	518	571
481	509	566	585
467	495	542	528
	485	557	500
		544	

After the data were combined and ranked in a single list, the ranks corresponding to each of the individual depths above were as follows:

| | Region | | |
I	II	III	IV
1	21	20	16.5
4	24	22	23
12	9.5	9.5	18
3	8	16.5	19
2	6	13	11
	5	15	7
		14	
R_i 22	73.5	110	94.5
n_i 5	6	7	6

With $k = 4$ and $N = 24$, we may test the null hypothesis of identical depth distributions for the four regions with test statistic $\chi^2 = [12/(24 \times 25)] \times (22^2/5 + 73.5^2/6 + 110^2/7 + 94.5^2/6) - (3 \times 25) = 9.282$ which falls between $\chi^2_{0.05,3} = 7.815$ and $\chi^2_{0.025,3} = 9.348$. We may thus declare that average well depths differ significantly with $p < 0.05$. ▲

Comparing Sign and Rank Tests

As indicated at the start of this section, tests based on ranks might be preferred to tests based on signs because they make use of more information in the data. The tests based on signs are somewhat simpler, however, and in some cases may be preferred for that reason.

The particular nature of the data may also produce a preference for tests based on signs. The tests based on ranks are, strictly speaking, based on the assumption that there will be no ties in the data. The approximations are good for modest numbers of ties but not for large numbers of ties. For data that are discrete with a limited number of possible values, such as subjective ratings on a scale from 1 to 5, there will tend to be many tied ranks, thus invalidating the rank sum tests. In cases such as these, the simpler tests based on signs are preferred.

PRACTICE PROBLEMS

P12-4 The following data represent the numbers of sessions missed by samples of men and women in a continuing day-care therapy program.

Men: 11   5   10   9   0   2   15   8   17   12   21
Women: 6   23   1   14   8   7   2   14   13   6   4   16   25

Is there evidence of a difference in average numbers of days missed for men and women? Use a 5% level of significance and do not assume normality.

P12-5 On the basis of the data in Practice Problem P12-4, determine whether the number of sessions missed by men vary less than those missed by women. Use a 5% level of significance.

P12-6 In order to accommodate vacation plans, management intends to reduce customer services and reassign employees to other duties either during afternoons or during mornings. Employees prefer afternoon reassignment but management proposes morning reassignment, basing the proposal on the argument that there are fewer customers in the morning. A record of customer counts for 20 sample days produced the following;

Day:	1	2	3	4	5	6	7	8	9	10
Morning	76	46	54	38	34	8	99	72	89	27
Afternoon	47	51	80	81	39	29	96	63	95	41
Day:	11	12	13	14	15	16	17	18	19	20
Morning	45	109	40	49	39	85	53	50	36	42
Afternoon	33	112	44	60	54	92	78	26	64	40

Do these data support the management argument at the 5% level of significance?

P12-7 Suppose that in order to determine whether there are differences among three child persuasion techniques, 30 children judged to be of fairly uniform attitude on the basis of pretesting and assigned randomly to three groups of ten were asked to clean up individual equivalent messes. Children in the first group faced the prospect of punishment if they did not clean up. Children in the second group were offered a reward if they cleaned up. Children in the third group had neither the prospect of punishment nor reward, but received a brief pep talk designed to convince them that cleaning up would be a nice thing to do. Do the three methods differ in their effectiveness at the 5% level if times (in seconds) to achieve satisfactory cleanup were as follows?

Punishment	373	404	391	248	415	386	263	388	341	392
Reward	197	263	313	252	361	203	317	362	293	301
Pep talk	265	327	352	206	332	401	373	298	376	259

EXERCISES

12-9 Determine whether there is a difference at the 5% level in the mosquito larvae densities for two swamps if sample densities at a number of points in the two swamps produced the following results:

First swamp:	14.2,	19.6,	19.1,	16.0,	10.7,	10.9,	13.9,
	17.3,	16.0,	18.4,	10.5,	13.3,	16.1,	10.6,
	16.6,	17.3					
Second swamp:	18.6,	16.2,	20.8,	16.5,	21.6,	18.5,	15.5,
	16.9,	20.2,	12.2,	16.1,	14.8,	17.1,	13.0,
	18.3,	20.3,	15.1,	13.3,	19.5,	17.8	

12-10 Repeat Exercise 8-25 without the assumption of normality.

12-11 Repeat Exercise 12-2 using the Kruskal–Wallis test.

12-12 Suppose that in a study on verbal reinforcement, individual applicants in a mock job interview were asked to present themselves to two interviewers, one of whom would provide verbal reinforcement such as muttering "Mmm–hmm" from time to time, and one of whom would not. Both interviewers would ask the same questions. Applicants rated their assessments of their chances of getting jobs with each of the two interviewers and the individual differences *reinforcement* minus *no reinforcement* were as follows: 7, 6, 3, −2, 9, −1, 0, 8, 5, 10, 16, −4, 10, and 12. Do these data indicate that reinforcement increases an applicant's assessment of job prospects? If so, indicate the *p*-level.

12-13 In a comparison of local management in three branch plants, a random sample of employees was selected from each plant and assessed in various ways for stress. Their aggregate stress scores were as follows:

Plant A:	186,	162,	145,	120,	149,	141,	167,	115		
Plant B:	192,	156,	196,	130,	184,	139,	199,	160,	173,	177
Plant C:	191,	188,	131,	196,	168,	157,	175,	142,	187	

Are there differences across the plants at the 5% level?

12-14 Is there a significant difference in the variation of the full span of the forepaws ("hands") of two arboreal primates if measurements on the spans in centimeters produced the following results?

Species A:	12.3,	13.1,	10.7,	12.3,	11.4,	12.9,	12.7,	11.6,
	12.9,	12.0,	9.4,	13.6,	11.5			
Species B:	10.3,	12.2,	10.4,	14.2,	6.4,	8.7,	10.2,	10.1,
	11.6,	9.4,	7.7,	9.8,	12.1,	8.5,	10.9,	13.1

12-15 (a) An urn contains $n_1 = 2$ black beads and $n_2 = 3$ red beads. The five beads are to be drawn out unseen one at a time without replacement.
   (1) Noting only color differences, in how many orders can the beads be drawn out?
   (2) What is the probability that the first two beads will be black?
   (3) What is the probability that the second and fourth beads will be black?
   (4) Let R_1 be the sum of the positions occupied by the black beads. [In (2), $R_1 = 1 + 2 = 3$; in (3) $R_1 = 2 + 4 = 6$.] Find the mean and variance for R_1 and hence find the mean and variance for $n_1 n_2 + n_1(n_1 + 1)/2 - R_1$.
(b) How are the ideas in part (a) used in the Mann–Whitney test?

SOLUTIONS TO PRACTICE PROBLEMS

P12-4 After the values are combined in a single list and ranked from smallest to largest, the values for the $n_1 = 11$ men are assigned ranks 1, 3.5, 6, 10.5, 12, 13, 14, 15, 19, 21, and 22, producing $R_1 = 137$ and $U_1 = (11 \times 13) + (11 \times 12/2) - 137 = 72$. The values for the $n_2 = 13$ women are assigned ranks 2, 3.5, 5, 7.5, 7.5, 9, 10.5, 16, 17.5, 17.5, 20, 23, and 24, producing $R_2 = 163$ and $U_2 = (11 \times 13) + (13 \times 14/2) - 163 = 71$. Since U_2 is smaller than U_1, we have $U = U_2 = 71$. For the test of H_0: the populations of missed sessions for men and women are identical against H_A: the populations have different measures of location, we refer to Table X to obtain $U_{1-\alpha/2} = U_{1-0.025} = 38$ for the case $n_1 = 11$ and $n_2 = 13$. Since $U = 71$ is not less than $U_{1-\alpha/2} = 38$, we may not reject H_0. We may accept the null hypothesis that men and women miss the same number of sessions on the average.

P12-5 To compare dispersions, we assign the ranks from the outside in, ignoring ties. We then inspect for ties and where there are ties we assign the tied values the mean of the ranks that they have been assigned. The ranks for the men are then 1, 6.5, 12, 20.5, 24, 23, 22, 19, 11, 7, and 6, producing $R_1 = 152$. [*Note*: The ranking ignoring ties assigned a rank of 5 to one "2" and a rank of 8 to the other. Each "2" was thus assigned rank $(5 + 8)/2 = 6.5$.] As in Problem P12-4, we find that $U_1 = 56$. To test H_0: the populations are identical against H_A: the first population (men) is less varied, we refer to Table X to obtain $U_{1-\alpha} = U_{1-0.05} = 43$. Since $U_1 = 56$ is not less than $U_{1-\alpha} = 43$, we may not reject H_0. The variation for men is not less than that for women.

P12-6 The 20 differences *first* minus *second* (i.e., *morning* minus *afternoon*) are: 29, -5, -26, -43, -5, -21, 3, 9, -6, -14, 8, -3, -4, -11, -15, -7, -25, 24, -28, and 2. To test the null hypothesis that the differences are 0 on the average against the alternative that they are generally less than 0 (i.e., that the afternoon count generally exceeds the morning count), we rank the differences according to magnitude (ignoring signs). We then find $T = T_1$, the sum of the ranks assigned to the positive differences. The ranks are 1, 2.5, 9, 10, 15, and 19, producing $T = 56.5$. For a one-sided test with $\alpha = 0.05$ and $n = 20$, Table XI produces $T^* = 60$. Since $56.5 < 60$, we may reject the null hypothesis and conclude that the data substantiate the management argument at the 5% level.

P12-7 The three sets of times are combined in a single list and ranked. The resulting ranks for the times in each group are as follows:

Punishment	21.5	29	26	4	30	24	7.5	25	17	27
Reward	1	7.5	13	5	19	2	14	20	10	12
Pep talk	9	15	18	3	16	28	21.5	11	23	6

For each of the $k = 3$ groups, $n_i = 10$ and the rank sums are $R_1 = 211$, $R_2 = 103.5$, and $R_3 = 150.5$, producing $N = 30$ and test statistic $\chi^2 = [12/(30 \times 31)] \times (211^2/10 + 103.5^2/10 + 150.5^2/10) - (3 \times 31) = 7.495$ which exceeds $\chi^2_{0.05,2} = 5.991$. At the 5% level we may reject the null hypothesis of no difference and we conclude that the methods do differ in their effectiveness.

12-4 RANK CORRELATION

In some cases we may wish to assess correlation without being restricted by the normality assumption, or we may be interested in a more "general correlation" that is not directly affected by the linearity of a relationship. In this section we consider a measure of correlation that can be used to determine the strength of any monotonic relationship between two quantities whether it is linear or not. [A monotonic relationship between two quantities x and y is such that y generally tends to increase (decrease) as x increases over the full range of x values being considered. If y tends to increase with increases in x over part of the range and tends to decrease in other parts, the relationship is not monotonic.] The correlation measure that we use may be regarded as a measure of association for two quantities provided that the quantities can be ordered and ranked. It is also a nonparametric measure—it does not require normal distributions.

This measure is called *rank correlation* and is obtained as follows. The x's and y's in a set of (x, y) pairs are assigned ranks in a manner similar to that used with the rank sum test. Each x_i in a set of pairs is assigned a rank r_{x_i} corresponding to its relative magnitude in the full sample of x's, and each y_i is assigned a rank r_{y_i} corresponding to its relative magnitude in the full sample of y's. If two or more x's are equal, each is assigned a rank equal to the mean of the ranks that they would otherwise receive. Equal y's are treated similarly. Each pair (x_i, y_i) is then converted to a new pair (r_{x_i}, r_{y_i}) and the rank correlation coefficient is found by calculating the correlation for these new pairs.

If every increase in x occurs in conjunction with an increase in y, the set of n pairs (r_{x_i}, r_{y_i}) consists of $(1, 1), (2, 2), \ldots, (n, n)$ (not necessarily in this order). The ranks are perfectly (positively) correlated and the rank correlation is 1. If y decreases whenever x increases, the set of pairs consists of $(1, n), (2, n - 1), \ldots, (n, 1)$. The ranks are perfectly (negatively) correlated and the rank correlation is -1. As with the usual correlation, patterns of rank pairs between these two extremes produce rank correlation values between -1 and 1.

Calculating Rank Correlation

The rank correlation coefficient, which is denoted by r', may be calculated in a simplified manner as follows:

> For each of n pairs (x_i, y_i), we determine the rank difference $d_i = r_{x_i} - r_{y_i}$ and then calculate the rank correlation coefficient as
>
> $$r' = 1 - \frac{6 \sum d^2}{n(n^2 - 1)}$$

We may test the null hypothesis that there is no general correlation or monotonic association between x and y in the population by using the fact that, if the null hypothesis is true, r' will have a sampling distribution whose percentiles are tabulated in Table XII in Appendix B.

> Table XII includes upper critical values $r'_{\alpha, n}$ such that if r' is obtained from n pairs (x, y),
>
> $$P(r' > r'_{\alpha, n} | x, y \text{ are not correlated}) = \alpha$$

Lower critical values are negatives of upper values.

The hypothesis testing procedure is as follows.

> H_0: the two variables are not correlated (i.e., are not monotonically associated)
>
> H_A: the two variables are correlated (i.e., monotonically associated)
>
> Test statistic: r'
>
> Decision rule: reject H_0 if $|r'| > r'_{\alpha/2, n}$

The foregoing testing procedure is modified to produce one-sided tests by modifying the alternative and decision rule.

$$H_A : \text{the variables are positively correlated}$$

Decision rule: reject H_0 if $r' > r'_{\alpha, n}$

or

$$H_A : \text{the variables are negatively correlated}$$

Decision rule: reject H_0 if $r' < -r'_{\alpha, n}$

EXAMPLE 12-11 We wish to determine whether there is a positive correlation between scores awarded by two judges in a type of competition in which scores are awarded on a subjective judgment basis. We wish to make the determination without requiring the assumption of normal distributions. The determination is to be made at the 1% level on the basis of the following random sample of scores received by $n = 10$ competitors, in which x is the score awarded by the first judge and y is the score awarded by the second:

Competitor:	1	2	3	4	5	6	7	8	9	10
x	85	65	55	70	75	80	60	60	90	75
y	80	70	50	60	65	75	60	55	90	85

Converted to r_x, r_y, and d, the scores produce:

r_x	9	4	1	5	6.5	8	2.5	2.5	10	6.5
r_y	8	6	1	3.5	5	7	3.5	2	10	9
$d = r_x - r_y$	1	-2	0	1.5	1.5	1	-1	0.5	0	-2.5

and we find that $r' = 1 - (6 \times 18)/(10 \times 99) = 0.89$, which exceeds $r'_{0.01, 10} = 0.7333$. We may, therefore, reject the null hypothesis of no correlation and conclude that there is some positive correlation (or monotone association) between scores awarded by the two judges at the 1% level. ▲

For large numbers of pairs ($n > 30$), r' will follow an approximate normal distribution with mean 0 and variance $1/(n - 1)$ if x and y are independent. For such cases approximate critical values may be determined as follows:

For $n > 30$,

$$r'_{\alpha, n} \doteq \frac{z_\alpha}{\sqrt{n - 1}}$$

P12-8 Researchers performed a preliminary investigation of numbers of waterfowl in marshes around a bay and on fairly close offshore islands. Simultaneous random counts in the two regions (adjusted to numbers of fowl per unit area) were as follows:

Bay	29	23	34	38	43	16	28	26	48	22	20	18
Islands	21	18	27	17	21	44	22	38	14	25	33	37

Determine whether there is a significant negative monotone association at the 1% level.

12-16 Two assessors assigned scores to each of 11 cities according to their suitabilities for the special needs of a particular conference with the following results:

City	Assessor number 1	Assessor number 2	City	Assessor number 1	Assessor number 2
Boston	83	60	New York	91	73
Los Angeles	92	96	San Francisco	88	83
Miami	61	66	San Juan	50	61
Mexico City	97	99	Toronto	66	94
Montreal	78	89	Vancouver	52	67
Nassau	53	72			

Determine, from the data, whether the assessors are in general agreement with regard to ordering by determining whether there is a significant positive correlation between their scores without assuming normality.

12-17 Suppose that each of eight subjects is given a test to measure democratic tendencies and another to determine prejudice tendencies, with the following resulting scores:

Subject:	1	2	3	4	5	6	7	8
Score on democratic tendencies test	52	37	25	38	44	51	32	33
Score on prejudice tendencies test	16	15	34	22	25	10	28	33

Without making any assumptions about distributions, determine at the 5% level whether there is a negative monotone association between the test scores.

12-18 Various indicators have been developed as measures of national quality of life. In the following data, the economic indicator measures the general economic well-being of a nation's citizens and the environmental index measures the quality of the environment. For these data determine at the 5% level whether economic well-being and environmental quality are correlated (without making such assumptions as normality).

Nation	Indicator	
	Economic	Environmental
Argentina	-0.95	0.61
Australia	0.68	2.90
Canada	1.32	1.63
Chile	-1.22	0.56
Finland	0.11	0.28
France	0.34	-0.27
East Germany	0.31	-0.51
West Germany	0.72	-0.86
Italy	0.14	-0.45
Japan	0.41	-0.87
Spain	0.20	0.17
United Kingdom	0.07	-0.89
United States	1.13	0.53
USSR	-1.07	0.58

12-19 Suppose that n red beads numbers 1 through n are to be paired at random with n blue beads numbered 1 through n.

(a) How many ways can the sets of beads be paired?

(b) Suppose that, for each pair, the difference between the numbers on the beads is squared and that x is equal to the sum of these squares. *For the special case* $n = 3$, show that the mean and variance for x are $E(x) = n(n^2 - 1)/6$ [i.e., show directly that $E(x) = 4$] and $\text{Var}(x) = n^2(n + 1)(n^2 - 1)/36$ [i.e., show directly that $\text{Var}(x) = 8$].

(c) What are the mean and variance for $6x/n(n^2 - 1)$?

(d) How are these results used in analyses such as found in Exercise 12-18?

SOLUTION TO PRACTICE PROBLEM

P12-8 Letting x denote bay count and y denote island count, the x's and y's are ranked within their own sets to produce the following r_x's and r_y's, which produce the corresponding $r_x - r_y = d$'s:

r_x	8	5	9	10	11	1	7	6	12	4	3	2
r_y	4.5	3	8	2	4.5	12	6	11	1	7	9	10
d	3.5	2	1	8	6.5	-11	1	-5	11	-3	-6	-8

We find the rank correlation coefficient

$$r' = 1 - \frac{6 \times \Sigma d^2}{n \times (n^2 - 1)} = 1 - \frac{6 \times 500.5}{12 \times 143} = -0.75$$

which is less than $-0.6713 = -r'_{0.01, 12}$. We may thus reject the null hypothesis of no correlation (i.e., no monotone association) in a one-sided test, and we conclude that there is a negative correlation (i.e., a negative monotone association).

12-20 In a rank sum test, if the value in position 16 in an ordered list is a "32" and the values in the next three positions are all "34," each "34" is assigned rank _____.

12-21 To use rank correlation, you must
(a) find the order in magnitude of two sets of data
(b) have two independent normal distributions
(c) have a straight line relationship between x and y
(d) find the percentage variation

12-22 If in a test $H_0 : \mu_1 = \mu_2$ against $H_A : \mu_1 > \mu_2$, a paired comparisons test on differences $x_1 - x_2$ yields a significantly large number of "$+$" signs and the experimenter rejects H_0, then he or she
(a) has acted contrary to the sample data
(b) has committed no error
(c) may have committed a type I error
(d) may have committed a type II error

12-23 Because they make use of more information in the data, rank sum tests are superior, even though more complicated than, tests based on signs.
(a) Always
(b) Sometimes
(c) Never

12-24 To test $H_0 : \mu = 3.00$ with a sign test, we must assume that the population is ___.

12-25 The sign test is one of a group of tests known as _____ tests.

12-26 If the averages of several groups are compared in a median test, then "average" is interpreted as _____ .

12-27 The number of degrees of freedom in a Kruskal–Wallis test based on four samples of size 7 is
(a) 4
(b) 6
(c) 3
(d) 7

12-28 If all of its underlying assumptions are satisfied, a parametric test has less chance than a nonparametric test of rejecting a true hypothesis (for the same assumed level of significance).
(a) True
(b) False
(c) Not necessarily

12-29 If members of the first sample generally exceed those of the second, then in a Mann–Whitney test comparing locations, the U statistic based on R_1 should be relatively
(a) large
(b) small

12-30 A paired comparisons test that uses more information from the data than the sign test but assumes less than the t-test is the _____ test.

12-31 It is hypothesized that the median salary in a given category is \$18,700 and of 24 salaries in a sample, 16 were above \$18,700 and 8 were below. Using a 5% level of significance, test the null hypothesis against the alternative that the median is not \$18,700.

12-32 Durability of a product is claimed to be such that over half of the individual products produced will function for more than 500 hours (i.e., such that the median functioning time exceeds 500 hours). If 178 of a sample of 300 did function for more than 500 hours, can the claim be considered proven at the 1% level?

12-33 By considering the signs of the actual differences minus 5, repeat Exercise 8-9(a) without assuming normality.

12-34 Suppose that, in an investigation of a current training program, the performance scores obtained by a sample of trainees are to be compared with scores obtained from an independent sample of trainees from the previous program, and that scores obtained are as follows:

Previous program:	9.5,	8.9,	8.7,	8.4,	8.2,	8.0,	7.9,	7.8,	7.6,	
	7.4,	7.3,	7.0,	6.9,	6.9,	6.3,	5.8,	5.0		
Current program:	8.7,	8.6,	8.3,	7.4,	7.2,	7.2,	6.9,	6.6,	6.5,	
	6.3,	5.9,	5.7,	5.2,	4.9,	4.3				

Without making any distribution assumptions, such as normality, determine, at the 5% level, whether the previous program produced higher average performance.

12-35 Weights in grams of a sample of 15 from population A and an independent sample of 20 from population B produced the following:

A: 105, 154, 193, 215, 231, 234, 251, 256, 257, 301, 306, 309, 313, 323, 343

B: 90, 100, 111, 143, 143, 150, 153, 165, 172, 179, 181, 187, 208, 209, 222, 229, 243, 253, 268, 269

Without assuming normal distributions, test the hypothesis that the populations have equal variances in a two-sided 5% test.

12-36 Four different preparation procedures are to be tested to determine whether they are equivalent with regard to average induced breaking strength. Observed breaking strengths from independent samples for the four preparation procedures were as follows:

	Preparation		
I	II	III	IV
2.34	2.30	2.17	2.19
2.29	2.27	2.21	2.11
2.35	2.21	2.14	2.25
2.39	2.28	2.18	2.24
2.43	2.20	2.19	
	2.26		

It is believed that the populations may be too skewed to assume normal distributions. Use the median test with $\alpha = 0.05$ to test the null hypothesis of equal average induced breaking strengths. What average is considered in this test?

12-37 Repeat Exercise 12-36 using the Kruskal–Wallis test.

12-38 Noting that $R_1 + R_2 = (n_1 + n_2)(n_1 + n_2 + 1)/2$ in an application of the Mann–Whitney test, show that $U_1 = n_1 n_2 - U_2$.

12-39 The following values are test scores for each twin in 14 pairs of twins. Without making any assumptions about the distributions of the scores, test at the 5% level whether there is any difference between the scores obtained by both twins in a pair.

Pair:	1	2	3	4	5	6	7
First born	66	51	57	48	71	57	71
Second born	68	57	56	44	76	45	70

Pair:	8	9	10	11	12	13	14
First born	50	51	68	67	52	59	64
Second born	45	60	61	52	50	64	57

12-40 Using the data in Exercise 12-39, determine whether there is any positive monotone association between scores obtained by both twins in a pair.

12-41 In a study to determine the importance of weight gained by a mother during pregnancy, suppose that the following data were obtained. For ten babies of mothers with no dietary restrictions, the birth weights (in pounds) were 8.1, 8.7, 9.2, 6.5, 6.8, 7.3, 7.7, 8.1, 6.8, and 5.9. For eight babies of mothers whose total weight gain was restricted to 15 lb, the corresponding figures were 6.9, 6.8, 7.7, 8.4, 9.2, 9.1, 6.3, and 6.1. At the 5% level, test whether the dietary restriction had a significant effect on average birth weight without assuming a normal distribution.

12-42 Suppose that three different growing conditions are to be compared for their effect on the yield of corn and that sample plantings for the three conditions produced the following yields:

Condition		
I	II	III
60	63	64
56	61	55
63	68	66
58	63	57
59	58	60
53	61	58
54	64	65
50	60	59
62	57	69
67	59	53
54	67	52
57	64	66

(a) Find the median yield for each condition and find the overall median.
(b) Without making the usual assumptions of normality, determine whether the true average yields differ across the three conditions.

12-43 Repeat Exercise 10-62 b) without assuming normal distributions.

12-44 Repeat Exercise 10-73 without assuming normality, but using the information in the magnitude of individual differences.

12-45 A test that serves as a possible nonparametric substitute for one-way analysis of variance is _____ .

12-46 A 95% confidence interval for the "true" correlation of 0.68 to 0.88 indicates a strong cause–effect relationship.
   (a) True
   (b) False
   (c) Not necessarily

12-47 Rank correlation is a nonparametric measure.
   (a) True
   (b) False

12-48 If, in a contingency table analysis, an experimenter finds that $\chi^2 < \chi^2_\alpha$ and declares the attributes independent, he or she
   (a) has acted contrary to the data
   (b) may have made a type I error
   (c) may have made a type II error
   (d) has committed no error

12-49 The estimate, from regression analysis, of the mean y corresponding to a value of x equal to $\bar{x}$ (the mean of the sample x's) is _____ .

12-50 The area under a normal curve to the right of $-z_\alpha$ is _____ .

12-51 If we test $H_0 : \mu = \mu_0$ against $H_A : \mu < \mu_0$ with a test statistic t and we find that $t > t_\alpha$, we should reject the hypothesis.
   (a) True
   (b) False

12-52 When we do analysis of variance we assume that all of the population _____ are _____ .

12-53 A high correlation between x and y indicates a strong _____ relationship between x and y.

12-54 To investigate the possible equality of several means, we use a technique called _____ .

12-55 To compare the equality of two means for data related in pairs without assuming underlying normal distributions, an appropriate test is the _____ _____ test.

12-56 For a probability function $f(x)$ such that $\sum xf(x) = \mu$, the value $\sum (x - \mu)^2 f(x)$ is called the _____ .

12-57 If a distribution is positively skewed, the mean is
   (a) equal to the median
   (b) less than the median
   (c) greater than the median
   (d) impossible to classify relative to the median

12-58 Suppose that, in an attempt to investigate the equivalence of three possible textbooks, an instructor divided a class of 45 students at random into three groups of 15 each and then randomly assigned one text to each group. At the end of term, the marks obtained by the students were:

Group		
I	II	III
55	41	54
68	47	55
69	54	51
60	65	56
64	73	59
71	44	49
67	48	41
75	64	61
44	51	73
59	55	69
72	68	66
65	51	56
66	47	47
83	71	55
42	55	56

Determine whether the data provide evidence of different median performance resulting from the use of different texts.

12-59 Using the data in Exercise 12-58, determine whether the location (average) of the distributions of marks differ by analyzing the rank structure.

12-60 In a competition based on personal judgment, is there evidence at the 5% level that Rogers generally awards more generous scores than Baker on the basis of the following scores awarded to 15 competitors? Use only the signs of individual differences.

Competitor:	1	2	3	4	5	6	7	8
Rogers	5.9	5.1	5.3	5.2	5.7	5.6	5.3	5.8
Baker	5.7	5.2	5.1	5.2	5.6	5.6	5.4	5.3
Competitor	9	10	11	12	13	14	15	
Rogers	5.1	5.2	5.6	4.9	5.7	5.3	5.1	
Baker	4.9	5.0	5.5	5.0	5.6	5.0	4.8	

12-61 Suppose that the following measurements of cliff-foot heights were taken on the south side of an estuary: 20.0, 14.7, 18.8, 15.6, 18.2, 13.0, 16.6, 17.2, and 18.9; and the following measurements were taken on the north side: 16.3, 17.1, 15.5, 21.9, 17.9, 20.0, 18.9, 19.8, 20.7, 18.6, and 22.3. Assuming normal distributions with equal population variances, test at the 1% level whether there is any difference between the mean heights of the two sides of an estuary.

12-62 Repeat Exercise 12-61 without assuming normality.

12-63 A sample of ten students from a large class produced the following term and examination marks:

Student:	1	2	3	4	5	6	7	8	9	10
Term	88	64	73	58	37	77	71	75	80	67
Exam	95	53	67	65	12	47	71	87	76	87

Determine whether there is a significant difference between term and exam marks at the 5% level:
(a) assuming normality
(b) without assuming normality

12-64 Suppose that the median salary in a given profession was found to be $15,000 at a given point in time. At a later point in time it is suspected that the median has increased. To investigate, a random sample of 16 salaries is to be taken.

(a) Without making any assumptions (e.g., normality), set up a test with appropriate rejection region to determine at the 5% level whether there is evidence of an increase.

(b) If the median has increased and 70% of the profession has a salary in excess of $15,000, what is the probability that the test in part (a) will fail to detect the increase?

(c) If the sample is taken and 11 of the 16 salaries are greater than $15,000, what conclusion would you draw?

12-65 An artificial sweetener is proposed as a substitute for sugar. There is opposition to the proposal on the grounds that the artificial sweetener produces a slight peculiarity to the taste and would not be well received by consumers. To test the taste effect, 15 individuals were selected at random and asked to rate taste on a scale from 1 to 10, with the following results.

Taster:	1	2	3	4	5	6	7	8
Sugar	2	7	6	6	8	10	8	3
Sweetener	3	9	6	1	5	9	4	5
Taster:	9	10	11	12	13	14	15	
Sugar	10	7	6	4	7	6	8	
Sweetener	6	6	3	5	6	5	7	

Without assuming normality, and without the use of magnitudes of differences, determine at the 1% level whether the sweetener tends to receive a lower rating.

12-66 Suppose that seeds are scattered around a given seed source and that an experiment is performed to study the relationship between the number of plants per unit area and the distance from the center of the seed source. In the experiment, the numbers of plants per unit area are counted at various distances with the following results (x is distance, y is number of plants per unit area).

x	3.0	3.5	4.0	4.5	5.0	5.5	6.0	6.5	7.0	7.5
y	31.2	32.0	23.6	19.4	24.2	20.0	18.0	19.6	12.6	11.4
x	8.0	8.5	9.0	9.5	10.0	10.5	11.0	11.5	12.0	12.5
y	15.0	10.2	6.8	8.6	6.2	4.8	2.0	3.6	6.4	1.8

(a) Determine at the 5% level whether it can be proven that variation in x explains more than 50% of the variation in y.

(b) Find the regression line $y = b_0 + b_1 x$ to estimate the average relation $\mu_{y1x} = \beta_0 + \beta_1 x$.

(c) Find a 95% confidence interval for β_1 in part (b).

(d) Find a 95% prediction interval for the number of plants to be observed at a distance of 9.0.

12-67 Suppose that our opponent will give us $4.50 each time we roll a 1 with a die and $2.50 each time we roll a 6. How much should we pay our opponent when we roll a 2, 3, 4, or 5 (the same amount for each) to make this game fair?

12-68 Measurements on a particular substance taken with an old machine produced the following results: 0.1233, 0.1240, 0.1175, 0.1211, 0.1183, 0.1178, 0.1206, 0.1185, 0.1208, and 0.1204, and measurements on the same substance taken with a new machine produced the following results: 0.1194, 0.1190, 0.1207, 0.1207, 0.1216, 0.1209, 0.1188, and 0.1201.

(a) Is there sufficient evidence to declare that the new machine produces more consistent (less varied) measurements than the old machine?

(b) Calculate a 90% confidence interval for the ratio of the two variances.

12-69 If a sample of ten items is picked from several thousand with an overall defective rate of 10%, find the probability that at most 20% of the sample will be defective.

12-70 Suppose that seven ships are to be unloaded at a given docking facility. The ships consist of one large oil tanker, two medium-size grain carriers, and four small freight carriers. The docking facilities are such that the oil tanker requires the full facilities for a full day for unloading; however, two grain carriers or three freight carriers or one of each can be unloaded in one day.

(a) By considering all the possibilities available at the start of each day based on what was unloaded on previous days, show that there are 20 different unloading schedules—all requiring 4 days. (In considering the facility limitations, assume that each ship requires a full day for unloading.)

Following are three examples of possible schedules:

(1) Two grain carriers, then the oil tanker, then three freight carriers, then the remaining freight carrier.

(2) Two grain carriers, then three freight carriers, then the last freight carrier, then the oil tanker.

(3) The oil tanker, then a grain carrier and a freight carrier, then the other grain carrier and a freight carrier, then the last two freight carriers.

(b) In how many schedules does it require exactly 4 days to unload both grain carriers? In how many does it require exactly 3? exactly 2? exactly 1?

(c) Draw a histogram for the results in part (b).

(d) Find the median and the mean number of days required to unload both grain carriers.

(e) If a schedule is selected at random, what is the probability that the grain carriers will both be unloaded in 2 days or less?

(f) If a schedule is drawn at random and it requires less than 3 days to unload both grain carriers, what is the probability of their both being unloaded in the same day?

12-71 An anthropologist studying two different races measured the diameter in centimeters of 12 skulls of members of one race with the following results: 51, 54, 56, 58, 58, 59, 62, 63, 63, 65, 68, and 71. The results for 11 skulls from the other race are 57, 59, 61, 62, 64, 66, 69, 69, 70, 73, and 75. Without making any assumptions about normality, determine whether there is sufficient evidence that the two races have different skull sizes.

12-72 One urn contains three white and three black balls; a second urn contains two white and four black balls.

(a) If one ball is chosen from each urn, what is the probability that they will be the same colour?

(b) If an urn is selected by chance and a ball drawn at random from it proves to be white, what is the probability that the urn chosen was the first one?

12-73 Three vaccines are under study such that various dilutions of the vaccine are given to mice until the highest dilution is found that produces neutrality. Each vaccine is tried with four amounts of mineral additive to try to increase antibody production. Suppose that an experiment produced the following influenza antibody responses:

Vaccine	Amount of additive			
	I	II	III	IV
A	5	7	6	5
B	6	7	7	6
C	4	3	4	5

(a) At the 5% level, determine whether the vaccines are equivalent.

(b) Determine whether the amount of additive affected the response.

12-74 Ten trout were placed in an aquarium at 29°C and the following observations resulted:

Fish	Length (mm)	Survival time (minutes)
1	59	98
2	54	85
3	53	49
4	45	39
5	51	72
6	48	15
7	60	57
8	47	52
9	49	66
10	57	28

Without making any assumptions on the distribution of lengths or survival times, determine at the 5% level whether there is any correlation between length of fish and survival time.

12-75 Suppose that, under certain conditions, stopping distances are known to follow a normal distribution with a mean of 100.00 and a standard deviation of 2.00. It is believed that a modification of the conditions might decrease the mean, although it would not affect the standard deviation. A sample of 25 distances under the modified conditions is to be sampled to investigate the possible decrease.

(a) Using a 5% level of significance, set up a decision rule to be used in determining whether the data provide sufficient evidence that the mean has decreased.

(b) Calculate the probability of accepting the null hypothesis in part (a) if the true mean is (1) 100.50 (2) 99.50 (3) 99.00 (4) 98.50

(c) Using the information in part (b), sketch the power function for the test in part (a).

12-76 A building industry considering the construction of 2100 dwelling units is interested in the number required for households of more than four persons. A sample of 210 households produced 63 with more than four persons.

(a) From this sample, estimate the proportions of households in the population with more than four persons.

(b) From this sample, find an interval estimate for the population proportion such that you can have 99% confidence in your estimate.

(c) If the 2100 households who will eventually live in the new units can be assumed to be a sample from a population with the proportion as estimated in part (a), find the probability that fewer than 600 units will be required for households with more than four persons.

12-77 (a) Suppose that, in a card game, a player has a 6, a 7, an 8, and a 9 after discarding a card. What is the player's probability of ending up with five cards in a row after receiving a replacement card?

(b) Suppose that this player has only received the 6, 7, 8, and 9 but three other players have also received four cards (all showing) such that their cards consist of one 2, one 3, one 4, two 6's, one 7, one 10, two Q's, one K, and two A's. What is the first player's probability of having five in a row upon receiving the next card?

(c) What is the probability in part (b) of having only four in a row and one of the four repeated?

12-78 A cutting machine produces items whose dimensions have a standard deviation of 0.005 and it is desired to reduce the standard deviation to less than half of this value. Suppose that after a stabilizer has been installed, 15 items had the following dimensions after cutting:

0.482	0.478	0.480	0.481	0.483	0.482	0.479	0.480
0.481	0.478	0.479	0.480	0.480	0.484	0.483	

(a) Determine whether there is sufficient evidence at the 5% level to declare that the standard deviation is less than half the original value.

(b) Calculate a range of values which you can be 90% sure will include the new standard deviation.

12-79 A study of marriage customs in two cultures produced the following results from two small samples of available information. For the first culture, a sample of 15 ages at marriage of women produced a mean of 20.8 and a variance of 8.5. For the second, a sample of 12 ages at marriage of women produced a mean of 18.4 and a variance of 11.0.

(a) Assuming independent samples from normal distributions with equal variances, find a range of values to estimate the difference in average age at marriage of women for the two cultures. Construct this range so that you can be 95% confident that it will include the true difference.

(b) Using the results of part (a), determine whether there is sufficient evidence at the 5% level that average ages at marriage of women are different for the two cultures.

12-80 In a study on whether preadmission orientation would have a beneficial effect on reducing manifest anxiety in children admitted to hospitals, three children in each of three age groups were evaluated with a sweat measurement test after admission to a hospital, and another three children in each age group were given a preadmission orientation and then were evaluated after admission. The anxiety scores (coded) for the children were as follows:

No Orientation			Orientation		
4 – 5	6 – 7	8 – 9	4 – 5	6 – 7	8 – 9
− 3	0	4	− 4	0	6
0	4	1	− 2	3	2
3	2	1	− 3	0	4

(a) Determine whether there is any evidence at the 5% level that anxiety differs between the orientation and nonorientation groups on the average.
(b) Determine whether there is evidence at the 5% level that anxiety levels differ among the age groups.
(c) Does the orientation program interact with age?

12-81 In a test to determine the proportions of a fixed quantity of a particular chemical compound that will dissolve in a fixed quantity of a given solution, quantities of each of the compound and solutions were combined in seven laboratories. In the first lab, 5 trials produced a mean of 62.50% of the compound dissolved. In the second laboratory, 15 trials produced a mean of 63.22% of the compound dissolved. In the third, fourth, fifth, sixth, and seventh labs, 10, 10, 5, 10, and 20 trials, respectively, produced means of 62.8, 63.07, 63.16, 62.41, and 63.04%, respectively. Considering all the labs, what was the mean percentage of the compound dissolved in the solution?

12-82 The following is a sample of 51 educational test scores:

114	113	134	84	108	111	80	117	88	103	139
103	126	106	118	119	136	123	115	112	92	130
108	112	115	88	145	93	123	102	95	140	100
128	106	131	128	148	95	127	121	105	133	120
114	112	106	111	110	115	100				

(a) Find the mean and median scores.
(b) Find the range and variance.
(c) Form a frequency distribution with equal class widths. For each class, give the class mark, class limits, and class boundaries.
(d) Sketch a histogram for this sample.
(e) Form a "less than" cumulative distribution for the sample.
(f) According to Chebyshev's theorem, how many of the individual values should be within 25 of the mean? How many of the actual scores are within 25 of the mean?

12-83 In a study of nine animals, blood counts (in hundreds of thousands of cells per cubic millimeter) were taken before and after administration of a drug with the following results:

| Before | 74 | 85 | 72 | 82 | 96 | 83 | 92 | 78 | 76 |
| After | 88 | 88 | 81 | 94 | 99 | 87 | 96 | 85 | 83 |

Determine whether there is sufficient evidence at the 5% level that the drug increases average blood count by more than 500,000 cells per cubic millimeter (assume normal distributions).

12-84 Repeat Exercise 12-83 without assuming normality.

12-85 To estimate the percentage of voters in a constituency who may agree with a given policy, a politician intends to use the percentage of voters in a sample who say that they agree.

(a) With no prior knowledge on the percentage, how large a sample must be taken to have a 95% chance that the estimate will be within four percentage points of the true value?

(b) If it can be assumed that at least 70% agree, how large a sample must be taken?

12-86 Repeat Exercise 12-85 if the constituency consists of 4500 voters.

12-87 Samples of pebble sizes were taken from four different beaches. All of the values were coded for simplicity by subtracting a constant and dividing by a constant, with the following results:

	Beach		
I	II	III	IV
2	-8	-14	5
14	-11	0	-4
18	0	-1	1
10	-4	1	-5
4	-7	-1	3
9		-6	
		2	

(a) Calculate the mean adjusted size for each beach.

(b) Calculate the error mean square derived from analysis of variance of the data.

(c) Determine which beaches are significantly different at the 5% level.

12-88 Yields of a manufacturing process for an alcoholic beverage are known to follow a normal distribution. Find a range of values that you can be 95% confident will include the true mean yield if a sample of 15 yields produced the following results:

70.5　76.1　72.3　73.0　75.3　73.9　73.7　71.2　74.3　71.8
73.5　74.9　77.2　73.2　72.6

12-89 Two samples of 100 people taken 10 years apart in the same region produced the following results. For the earlier sample, 65 reported English as their first language, 22 reported Spanish, and 13 reported another language. For the second sample, 55 reported English, 28 Spanish, and 17 another language.

(a) Determine, at the 5% level, whether there has been any change in the population over the 10 years with respect to the first language.

(b) Find an interval estimate for the proportion of the current population that has neither Spanish nor English as their first language. Construct this interval so that you may be 95% confident that it contains the true proportion.

12-90 Suppose that, in a study of different stimulants, 15 subjects were asked to perform a task (which they had previously practiced) after administration of each of the stimulants and the following time scores resulted:

Subject:	1	2	3	4	5	6	7	8
Stimulant I	30	29	29	24	26	31	33	28
Stimulant II	28	25	19	16	27	32	29	32
Subject:	9	10	11	12	13	14	15	
Stimulant I	36	28	27	21	25	27	24	
Stimulant II	30	22	24	20	18	30	21	

Assuming normal distributions, determine at the 5% level whether there is any difference in the effect of the two stimulants.

12-91 Repeat Exercise 12-90 without assuming normality.

12-92 Suppose that in a comparison of mean plant density in two regions a sample of measurements at 16 points in the first region produced a mean of 46.07 and a standard deviation of 10.13, and a sample of measurements at 12 points in the second region produced a mean of 37.43 and a standard deviation of 9.78.

(a) Calculate an interval of values such that you can be 95% confident that the difference between the true mean densities of the two regions will be included in this interval.

(b) Using this interval, determine at the 5% level whether densities in the two regions are equal on the average.

12-93 (a) Referring to the data in Exercise 11-9, define a new x equal to the square of the old x minus 5 [i.e., $x_{new} = (x_{old} - 5)^2$].

(1) Find the least-squares regression for y and the new x.

(2) What proportion of variation in y is explained by variation in the new x?

(b) Again referring to the data in Exercise 11-9, let x_1 be the original x and define a new variable $x_2 = x^2$.

(1) Find the least-squares regression for y with x_1 and x_2.

(2) What proportion of variation in y is explained by variation in x_1 and x_2; that is, what proportion of variation in y is explained by variations in x through a *quadratic* relationship?

APPENDIX: FURTHER READING

This book has been prepared as a noncalculus-based introduction to a broad selection of topics in statistical analysis. The examples cover many areas of application in the natural and the social sciences. The following titles represent a very brief sampling of extensions of the material that might interest the reader.

Illustrations of the General Applicability of Statistics:

TANUR, JUDITH, and others, *Statistics*: *A Guide to the Unknown* (2nd ed.). San Francisco: Holden-Day, Inc., 1978.

General Investigation of What Data Seem to Say:

TUKEY, JOHN W., *Exploratory Data Analysis*. Reading, Mass.: Addison-Wesley Publishing Co., Inc., 1977.

More Theoretical Mathematical Developments of Statistics:

FRASER, D. A. S., *Probability and Statistics*: *Theory and Applications*. North Scituate, Mass.: Duxbury Press, 1976.

FREUND, JOHN E., and RONALD E. WALPOLE, *Mathematical Statistics* (3rd ed.). Englewood Cliffs, N.J.: Prentice-Hall, Inc., 1980.

ROUSSAS, G. G., *A First Course in Mathematical Statistics*. Reading, Mass.: Addison-Wesley Publishing Co., Inc., 1973.

Statistical Inference Based on Bayesian Conditional Probabilities:

BOX, GEORGE, E. P., and GEORGE C. TIAO, *Bayesian Inference in Statistical Analysis*. Reading, Mass.: Addison-Wesley Publishing Co., Inc., 1973.

Statistics in Business and Engineering:

FREUND, JOHN E., and IRWIN MILLER, *Probability and Statistics for Engineers* (2nd ed.). Englewood Cliffs, N.J.: Prentice-Hall, Inc., 1977.

FREUND, JOHN E., and F. J. WILLIAMS, *Elementary Business Statistics*: *The Modern Approach* (3rd ed.). Englewood Cliffs, N.J.: Prentice-Hall, Inc., 1977.

GRANT, EUGENE L., and RICHARD S. LEAVENWORTH, *Statistical Quality Control* (5th ed.). New York: McGraw-Hill Book Company, 1980.

HAHN, GERALD J., and SAMUEL SHAPIRO, *Statistical Models in Engineering*. New York: John Wiley & Sons, Inc., 1967.

Analysis of Variance:

BOX, GEORGE, E. P., WILLIAM G. HUNTER, and J. STUART HUNTER, *Statistics for Experimenters*. New York: John Wiley & Sons, Inc., 1978.

COCHRAN, WILLIAM G., and GERTRUDE M. COX, *Experimental Design* (2nd ed.). New York: John Wiley & Sons, Inc., 1957.

GUENTHER, WILLIAM C., *Analysis of Variance*. Englewood Cliffs, N.J.: Prentice-Hall, Inc., 1964.

Analysis of Variance and Regression:

NETER, JOHN, and WILLIAM WASSERMAN, *Applied Linear Statistical Models*. Homewood, Ill.: Richard D. Irwin, Inc., 1974.

Regression Analysis:

DRAPER, NORMAN R., and HARRY SMITH, *Applied Regression Analysis* (2nd ed.). New York: John Wiley & Sons, Inc., 1981.

Nonparametric Methods:

CONOVER, W. J., *Practical Nonparametric Statistics* (2nd ed.). New York: John Wiley & Sons, Inc., 1980.

Sampling

COCHRAN, W. G., *Sampling Techniques* (3rd ed.). New York: John Wiley & Sons, Inc., 1977.

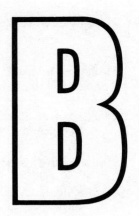

APPENDIX:
TABLES

TABLE I BINOMIAL PROBABILITIES

		p										
n	x	0.05	0.1	0.2	0.3	0.4	0.5	0.6	0.7	0.8	0.9	0.95
2	0	0.902	0.810	0.640	0.490	0.360	0.250	0.160	0.090	0.040	0.010	0.002
	1	0.095	0.180	0.320	0.420	0.480	0.500	0.480	0.420	0.320	0.180	0.095
	2	0.002	0.010	0.040	0.090	0.160	0.250	0.360	0.490	0.640	0.810	0.902
3	0	0.857	0.729	0.512	0.343	0.216	0.125	0.064	0.027	0.008	0.001	
	1	0.135	0.243	0.384	0.441	0.432	0.375	0.288	0.189	0.096	0.027	0.007
	2	0.007	0.027	0.096	0.189	0.288	0.375	0.432	0.441	0.384	0.243	0.135
	3		0.001	0.008	0.027	0.064	0.125	0.216	0.343	0.512	0.729	0.857
4	0	0.815	0.656	0.410	0.240	0.130	0.062	0.026	0.008	0.002		
	1	0.171	0.292	0.410	0.412	0.346	0.250	0.154	0.076	0.026	0.004	
	2	0.014	0.049	0.154	0.265	0.346	0.375	0.346	0.265	0.154	0.049	0.014
	3		0.004	0.026	0.076	0.154	0.250	0.346	0.412	0.410	0.292	0.171
	4			0.002	0.008	0.026	0.062	0.130	0.240	0.410	0.656	0.815
5	0	0.774	0.590	0.328	0.168	0.078	0.031	0.010	0.002			
	1	0.204	0.328	0.410	0.360	0.259	0.156	0.077	0.028	0.006		
	2	0.021	0.073	0.205	0.309	0.346	0.312	0.230	0.132	0.051	0.008	0.001
	3	0.001	0.008	0.051	0.132	0.230	0.312	0.346	0.309	0.205	0.073	0.021
	4			0.006	0.028	0.077	0.156	0.259	0.360	0.410	0.328	0.204
	5				0.002	0.010	0.031	0.078	0.168	0.328	0.590	0.774
6	0	0.735	0.531	0.262	0.118	0.047	0.016	0.004	0.001			
	1	0.232	0.354	0.393	0.303	0.187	0.094	0.037	0.010	0.002		
	2	0.031	0.098	0.246	0.324	0.311	0.234	0.138	0.060	0.015	0.001	
	3	0.002	0.015	0.082	0.185	0.276	0.312	0.276	0.185	0.082	0.015	0.002
	4		0.001	0.015	0.060	0.138	0.234	0.311	0.324	0.246	0.098	0.031
	5			0.002	0.010	0.037	0.094	0.187	0.303	0.393	0.354	0.232
	6				0.001	0.004	0.016	0.047	0.118	0.262	0.531	0.735
7	0	0.698	0.478	0.210	0.082	0.028	0.008	0.002				
	1	0.257	0.372	0.367	0.247	0.131	0.055	0.017	0.004			
	2	0.041	0.124	0.275	0.318	0.261	0.164	0.077	0.025	0.004		
	3	0.004	0.023	0.115	0.227	0.290	0.273	0.194	0.097	0.029	0.003	
	4		0.003	0.029	0.097	0.194	0.273	0.290	0.227	0.115	0.023	0.004
	5			0.004	0.025	0.077	0.164	0.261	0.318	0.275	0.124	0.041
	6				0.004	0.017	0.055	0.131	0.247	0.367	0.372	0.257
	7					0.002	0.008	0.028	0.082	0.210	0.478	0.698
8	0	0.663	0.430	0.168	0.058	0.017	0.004	0.001				
	1	0.279	0.383	0.336	0.198	0.090	0.031	0.008	0.001			
	2	0.051	0.149	0.294	0.296	0.209	0.109	0.041	0.010	0.001		
	3	0.005	0.033	0.147	0.254	0.279	0.219	0.124	0.047	0.009		
	4		0.005	0.046	0.136	0.232	0.273	0.232	0.136	0.046	0.005	
	5			0.009	0.047	0.124	0.219	0.279	0.254	0.147	0.033	0.005
	6			0.001	0.010	0.041	0.109	0.209	0.296	0.294	0.149	0.051
	7				0.001	0.008	0.031	0.090	0.198	0.336	0.383	0.279
	8					0.001	0.004	0.017	0.058	0.168	0.430	0.663

TABLE I (*Cont.*)

n	x	0.05	0.1	0.2	0.3	0.4	0.5	0.6	0.7	0.8	0.9	0.95
9	0	0.630	0.387	0.134	0.040	0.010	0.002					
	1	0.299	0.387	0.302	0.156	0.060	0.018	0.004				
	2	0.063	0.172	0.302	0.267	0.161	0.070	0.021	0.004			
	3	0.008	0.045	0.176	0.267	0.251	0.164	0.074	0.021	0.003		
	4	0.001	0.007	0.066	0.172	0.251	0.246	0.167	0.074	0.017	0.001	
	5		0.001	0.017	0.074	0.167	0.246	0.251	0.172	0.066	0.007	0.001
	6			0.003	0.021	0.074	0.164	0.251	0.267	0.176	0.045	0.008
	7				0.004	0.021	0.070	0.161	0.267	0.302	0.172	0.063
	8					0.004	0.018	0.060	0.156	0.302	0.387	0.299
	9						0.002	0.010	0.040	0.134	0.387	0.630
10	0	0.599	0.349	0.107	0.028	0.006	0.001					
	1	0.315	0.387	0.268	0.121	0.040	0.010	0.002				
	2	0.075	0.194	0.302	0.233	0.121	0.044	0.011	0.001			
	3	0.010	0.057	0.201	0.267	0.215	0.117	0.042	0.009	0.001		
	4	0.001	0.011	0.088	0.200	0.251	0.205	0.111	0.037	0.006		
	5		0.001	0.026	0.103	0.201	0.246	0.201	0.103	0.026	0.001	
	6			0.006	0.037	0.111	0.205	0.251	0.200	0.088	0.011	0.001
	7			0.001	0.009	0.042	0.117	0.215	0.267	0.201	0.057	0.010
	8				0.001	0.011	0.044	0.121	0.233	0.302	0.194	0.075
	9					0.002	0.010	0.040	0.121	0.268	0.387	0.315
	10						0.001	0.006	0.028	0.107	0.349	0.599
11	0	0.569	0.314	0.086	0.020	0.004						
	1	0.329	0.384	0.236	0.093	0.027	0.005	0.001				
	2	0.087	0.213	0.295	0.200	0.089	0.027	0.005	0.001			
	3	0.014	0.071	0.221	0.257	0.177	0.081	0.023	0.004			
	4	0.001	0.016	0.111	0.220	0.236	0.161	0.070	0.017	0.002		
	5		0.002	0.039	0.132	0.221	0.226	0.147	0.057	0.010		
	6			0.010	0.057	0.147	0.226	0.221	0.132	0.039	0.002	
	7			0.002	0.017	0.070	0.161	0.236	0.220	0.111	0.016	0.001
	8				0.004	0.023	0.081	0.177	0.257	0.221	0.071	0.014
	9				0.001	0.005	0.027	0.089	0.200	0.295	0.213	0.087
	10					0.001	0.005	0.027	0.093	0.236	0.384	0.329
	11						0.004	0.020	0.086	0.314	0.569
12	0	0.540	0.282	0.069	0.014	0.002						
	1	0.341	0.377	0.206	0.071	0.017	0.003					
	2	0.099	0.230	0.283	0.168	0.064	0.016	0.002				
	3	0.017	0.085	0.236	0.240	0.142	0.054	0.012	0.001			
	4	0.002	0.021	0.133	0.231	0.213	0.121	0.042	0.008	0.001		
	5		0.004	0.053	0.158	0.227	0.193	0.101	0.029	0.003		
	6			0.016	0.079	0.177	0.226	0.177	0.079	0.016		
	7			0.003	0.029	0.101	0.193	0.227	0.158	0.053	0.004	
	8			0.001	0.008	0.042	0.121	0.213	0.231	0.133	0.021	0.002
	9				0.001	0.012	0.054	0.142	0.240	0.236	0.085	0.017
	10					0.002	0.016	0.064	0.168	0.283	0.230	0.099
	11						0.003	0.017	0.071	0.206	0.377	0.341
	12							0.002	0.014	0.069	0.282	0.540

TABLE I (*Cont.*)

TABLE I (*Cont.*)

n	x	0.05	0.1	0.2	0.3	0.4	0.5	0.6	0.7	0.8	0.9	0.95
13	0	0.513	0.254	0.055	0.010	0.001						
	1	0.351	0.367	0.179	0.054	0.011	0.002					
	2	0.111	0.245	0.268	0.139	0.045	0.010	0.001				
	3	0.021	0.100	0.246	0.218	0.111	0.035	0.006	0.001			
	4	0.003	0.028	0.154	0.234	0.184	0.087	0.024	0.003			
	5		0.006	0.069	0.180	0.221	0.157	0.066	0.014	0.001		
	6		0.001	0.023	0.103	0.197	0.209	0.131	0.044	0.006		
	7			0.006	0.044	0.131	0.209	0.197	0.103	0.023	0.001	
	8			0.001	0.014	0.066	0.157	0.221	0.180	0.069	0.006	
	9				0.003	0.024	0.087	0.184	0.234	0.154	0.028	0.003
	10				0.001	0.006	0.035	0.111	0.218	0.246	0.100	0.021
	11					0.001	0.010	0.045	0.139	0.268	0.245	0.111
	12						0.002	0.011	0.054	0.179	0.367	0.351
	13							0.001	0.010	0.055	0.254	0.513
14	0	0.488	0.229	0.044	0.007	0.001						
	1	0.359	0.356	0.154	0.041	0.007	0.001					
	2	0.123	0.257	0.250	0.113	0.032	0.006	0.001				
	3	0.026	0.114	0.250	0.194	0.085	0.022	0.003				
	4	0.004	0.035	0.172	0.229	0.155	0.061	0.014	0.001			
	5		0.008	0.086	0.196	0.207	0.122	0.041	0.007			
	6		0.001	0.032	0.126	0.207	0.183	0.092	0.023	0.002		
	7			0.009	0.062	0.157	0.209	0.157	0.062	0.009		
	8			0.002	0.023	0.092	0.183	0.207	0.126	0.032	0.001	
	9				0.007	0.041	0.122	0.207	0.196	0.086	0.008	
	10				0.001	0.014	0.061	0.155	0.229	0.172	0.035	0.004
	11					0.003	0.022	0.085	0.194	0.250	0.114	0.026
	12					0.001	0.006	0.032	0.113	0.250	0.257	0.123
	13						0.001	0.007	0.041	0.154	0.356	0.359
	14							0.001	0.007	0.044	0.229	0.488
15	0	0.463	0.206	0.035	0.005							
	1	0.366	0.343	0.132	0.031	0.005						
	2	0.135	0.267	0.231	0.092	0.022	0.003					
	3	0.031	0.129	0.250	0.170	0.063	0.014	0.002				
	4	0.005	0.043	0.188	0.219	0.127	0.042	0.007	0.001			
	5	0.001	0.010	0.103	0.206	0.186	0.092	0.024	0.003			
	6		0.002	0.043	0.147	0.207	0.153	0.061	0.012	0.001		
	7			0.014	0.081	0.177	0.196	0.118	0.035	0.003		
	8			0.003	0.035	0.118	0.196	0.177	0.081	0.014		
	9			0.001	0.012	0.061	0.153	0.207	0.147	0.043	0.002	
	10				0.003	0.024	0.092	0.186	0.206	0.103	0.010	0.001
	11				0.001	0.007	0.042	0.127	0.219	0.188	0.043	0.005
	12					0.002	0.014	0.063	0.170	0.250	0.129	0.031
	13						0.003	0.022	0.092	0.231	0.267	0.135
	14							0.005	0.031	0.132	0.343	0.366
	15								0.005	0.035	0.206	0.463

TABLE II   VALUES OF e^{-x}

x	e^{-x}	x	e^{-x}	x	e^{-x}	x	e^{-x}
0.0	1.000	2.5	0.082	5.0	0.0067	7.5	0.00055
0.1	0.905	2.6	0.074	5.1	0.0061	7.6	0.00050
0.2	0.819	2.7	0.067	5.2	0.0055	7.7	0.00045
0.3	0.741	2.8	0.061	5.3	0.0050	7.8	0.00041
0.4	0.670	2.9	0.055	5.4	0.0045	7.9	0.00037
0.5	0.607	3.0	0.050	5.5	0.0041	8.0	0.00034
0.6	0.549	3.1	0.045	5.6	0.0037	8.1	0.00030
0.7	0.497	3.2	0.041	5.7	0.0033	8.2	0.00028
0.8	0.449	3.3	0.037	5.8	0.0030	8.3	0.00025
0.9	0.407	3.4	0.033	5.9	0.0027	8.4	0.00023
1.0	0.368	3.5	0.030	6.0	0.0025	8.5	0.00020
1.1	0.333	3.6	0.027	6.1	0.0022	8.6	0.00018
1.2	0.301	3.7	0.025	6.2	0.0020	8.7	0.00017
1.3	0.273	3.8	0.022	6.3	0.0018	8.8	0.00015
1.4	0.247	3.9	0.020	6.4	0.0017	8.9	0.00014
1.5	0.223	4.0	0.018	6.5	0.0015	9.0	0.00012
1.6	0.202	4.1	0.017	6.6	0.0014	9.1	0.00011
1.7	0.183	4.2	0.015	6.7	0.0012	9.2	0.00010
1.8	0.165	4.3	0.014	6.8	0.0011	9.3	0.00009
1.9	0.150	4.4	0.012	6.9	0.0010	9.4	0.00008
2.0	0.135	4.5	0.011	7.0	0.0009	9.5	0.00008
2.1	0.122	4.6	0.010	7.1	0.0008	9.6	0.00007
2.2	0.111	4.7	0.009	7.2	0.0007	9.7	0.00006
2.3	0.100	4.8	0.008	7.3	0.0007	9.8	0.00006
2.4	0.091	4.9	0.007	7.4	0.0006	9.9	0.00005

TABLE IIIa   NORMAL PROBABILITIES

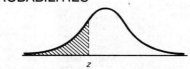

z	0.00	0.01	0.02	0.03	0.04	0.05	0.06	0.07	0.08	0.09
−1.0	0.0000									
−3.9	0.0000									
−3.8	0.0001									
−3.7	0.0001									
−3.6	0.0002									
−3.5	0.0002									
−3.4	0.0003									
−3.3	0.0005	0.0005	0.0005	0.0004	0.0004	0.0004	0.0004	0.0004	0.0004	0.0003
−3.2	0.0007	0.0007	0.0006	0.0006	0.0006	0.0006	0.0006	0.0005	0.0005	0.0005
−3.1	0.0010	0.0009	0.0009	0.0009	0.0008	0.0008	0.0008	0.0008	0.0007	0.0007
−3.0	0.0013	0.0013	0.0013	0.0012	0.0012	0.0011	0.0011	0.0011	0.0010	0.0010

TABLE IIIa (*Cont.*)

z	0.00	0.01	0.02	0.03	0.04	0.05	0.06	0.07	0.08	0.09
−2.9	0.0019	0.0018	0.0017	0.0017	0.0016	0.0016	0.0015	0.0015	0.0014	0.0014
−2.8	0.0026	0.0025	0.0024	0.0023	0.0023	0.0022	0.0021	0.0021	0.0020	0.0019
−2.7	0.0035	0.0034	0.0033	0.0032	0.0031	0.0030	0.0029	0.0028	0.0027	0.0026
−2.6	0.0047	0.0045	0.0044	0.0043	0.0041	0.0040	0.0039	0.0038	0.0037	0.0036
−2.5	0.0062	0.0060	0.0059	0.0057	0.0055	0.0054	0.0052	0.0051	0.0049	0.0048
−2.4	0.0082	0.0080	0.0078	0.0075	0.0073	0.0071	0.0069	0.0068	0.0066	0.0064
−2.3	0.0107	0.0104	0.0102	0.0099	0.0096	0.0094	0.0091	0.0089	0.0087	0.0084
−2.2	0.0139	0.0136	0.0132	0.0129	0.0125	0.0122	0.0119	0.0116	0.0113	0.0110
−2.1	0.0179	0.0174	0.0170	0.0166	0.0162	0.0158	0.0154	0.0150	0.0146	0.0143
−2.0	0.0228	0.0222	0.0217	0.0212	0.0207	0.0202	0.0197	0.0192	0.0188	0.0183
−1.9	0.0287	0.0281	0.0274	0.0268	0.0262	0.0256	0.0250	0.0244	0.0238	0.0233
−1.8	0.0359	0.0352	0.0344	0.0336	0.0329	0.0322	0.0314	0.0307	0.0300	0.0294
−1.7	0.0446	0.0436	0.0427	0.0418	0.0409	0.0401	0.0392	0.0384	0.0375	0.0367
−1.6	0.0548	0.0537	0.0526	0.0516	0.0505	0.0495	0.0485	0.0475	0.0465	0.0455
−1.5	0.0668	0.0655	0.0643	0.0630	0.0618	0.0606	0.0594	0.0582	0.0570	0.0559
−1.4	0.0808	0.0793	0.0778	0.0764	0.0749	0.0735	0.0722	0.0708	0.0694	0.0681
−1.3	0.0968	0.0951	0.0934	0.0918	0.0901	0.0885	0.0869	0.0853	0.0838	0.0823
−1.2	0.1151	0.1131	0.1112	0.1093	0.1075	0.1056	0.1038	0.1020	0.1003	0.0985
−1.1	0.1357	0.1335	0.1314	0.1292	0.1271	0.1251	0.1230	0.1210	0.1190	0.1170
−1.0	0.1587	0.1562	0.1539	0.1515	0.1492	0.1469	0.1446	0.1423	0.1401	0.1379
−0.9	0.1841	0.1814	0.1788	0.1762	0.1736	0.1711	0.1685	0.1660	0.1635	0.1611
−0.8	0.2119	0.2090	0.2061	0.2033	0.2005	0.1977	0.1949	0.1922	0.1894	0.1867
−0.7	0.2420	0.2389	0.2358	0.2327	0.2297	0.2266	0.2236	0.2206	0.2177	0.2148
−0.6	0.2743	0.2709	0.2676	0.2643	0.2611	0.2578	0.2546	0.2514	0.2483	0.2451
−0.5	0.3085	0.3050	0.3015	0.2981	0.2946	0.2912	0.2877	0.2843	0.2810	0.2776
−0.4	0.3446	0.3409	0.3372	0.3336	0.3300	0.3264	0.3228	0.3192	0.3156	0.3121
−0.3	0.3821	0.3783	0.3745	0.3707	0.3669	0.3632	0.3594	0.3557	0.3520	0.3483
−0.2	0.4207	0.4168	0.4129	0.4090	0.4052	0.4013	0.3974	0.3936	0.3897	0.3859
−0.1	0.4602	0.4562	0.4522	0.4483	0.4443	0.4404	0.4364	0.4325	0.4286	0.4247
−0.0	0.5000	0.4960	0.4920	0.4880	0.4840	0.4801	0.4761	0.4721	0.4681	0.4641
0.0	0.5000	0.5040	0.5080	0.5120	0.5160	0.5199	0.5239	0.5279	0.5319	0.5359
0.1	0.5398	0.5438	0.5478	0.5517	0.5557	0.5596	0.5636	0.5675	0.5714	0.5753
0.2	0.5793	0.5832	0.5871	0.5910	0.5948	0.5987	0.6026	0.6064	0.6103	0.6141
0.3	0.6179	0.6217	0.6255	0.6293	0.6331	0.6368	0.6406	0.6443	0.6480	0.6517
0.4	0.6554	0.6591	0.6628	0.6664	0.6700	0.6736	0.6772	0.6808	0.6844	0.6879
0.5	0.6915	0.6950	0.6985	0.7019	0.7054	0.7088	0.7123	0.7157	0.7190	0.7224
0.6	0.7257	0.7291	0.7324	0.7357	0.7389	0.7422	0.7454	0.7486	0.7517	0.7549
0.7	0.7580	0.7611	0.7642	0.7673	0.7704	0.7734	0.7764	0.7794	0.7823	0.7852
0.8	0.7881	0.7910	0.7939	0.7967	0.7995	0.8023	0.8051	0.8078	0.8106	0.8133
0.9	0.8159	0.8186	0.8212	0.8238	0.8264	0.8289	0.8315	0.8340	0.8365	0.8389
1.0	0.8413	0.8438	0.8461	0.8485	0.8508	0.8531	0.8554	0.8577	0.8599	0.8621
1.1	0.8643	0.8665	0.8686	0.8708	0.8729	0.8749	0.8770	0.8790	0.8810	0.8830

TABLE IIIa   (*Cont.*)

z	0.00	0.01	0.02	0.03	0.04	0.05	0.06	0.07	0.08	0.09
1.2	0.8849	0.8869	0.8888	0.8907	0.8925	0.8944	0.8962	0.8980	0.8997	0.9015
1.3	0.9032	0.9049	0.9066	0.9082	0.9099	0.9115	0.9131	0.9147	0.9162	0.9177
1.4	0.9192	0.9207	0.9222	0.9236	0.9251	0.9265	0.9279	0.9292	0.9306	0.9319
1.5	0.9332	0.9345	0.9357	0.9370	0.9382	0.9394	0.9406	0.9418	0.9429	0.9441
1.6	0.9452	0.9463	0.9474	0.9484	0.9495	0.9505	0.9515	0.9525	0.9535	0.9545
1.7	0.9554	0.9564	0.9573	0.9582	0.9591	0.9599	0.9608	0.9616	0.9625	0.9633
1.8	0.9641	0.9649	0.9656	0.9664	0.9671	0.9678	0.9686	0.9693	0.9699	0.9706
1.9	0.9713	0.9719	0.9726	0.9732	0.9738	0.9744	0.9750	0.9756	0.9761	0.9767
2.0	0.9772	0.9778	0.9783	0.9788	0.9793	0.9798	0.9803	0.9808	0.9812	0.9817
2.1	0.9821	0.9826	0.9830	0.9834	0.9838	0.9842	0.9846	0.9850	0.9854	0.9857
2.2	0.9861	0.9864	0.9868	0.9871	0.9875	0.9878	0.9881	0.9884	0.9887	0.9890
2.3	0.9893	0.9896	0.9898	0.9901	0.9904	0.9906	0.9909	0.9911	0.9913	0.9916
2.4	0.9918	0.9920	0.9922	0.9925	0.9927	0.9929	0.9931	0.9932	0.9934	0.9936
2.5	0.9938	0.9940	0.9941	0.9943	0.9945	0.9946	0.9948	0.9949	0.9951	0.9952
2.6	0.9953	0.9955	0.9956	0.9957	0.9959	0.9960	0.9961	0.9962	0.9963	0.9964
2.7	0.9965	0.9966	0.9967	0.9968	0.9969	0.9970	0.9971	0.9972	0.9973	0.9974
2.8	0.9974	0.9975	0.9976	0.9977	0.9977	0.9978	0.9979	0.9979	0.9980	0.9981
2.9	0.9981	0.9982	0.9982	0.9983	0.9984	0.9984	0.9985	0.9985	0.9986	0.9986
3.0	0.9987	0.9987	0.9987	0.9988	0.9988	0.9989	0.9989	0.9989	0.9990	0.9990
3.1	0.9990	0.9991	0.9991	0.9991	0.9992	0.9992	0.9992	0.9992	0.9993	0.9993
3.2	0.9993	0.9993	0.9994	0.9994	0.9994	0.9994	0.9994	0.9995	0.9995	0.9995
3.3	0.9995	0.9995	0.9995	0.9996	0.9996	0.9996	0.9996	0.9996	0.9996	0.9997
3.4	0.9997	0.9997	0.9997	0.9997	0.9997	0.9997	0.9997	0.9997	0.9997	0.9998
3.5	0.9998									
3.6	0.9998									
3.7	0.9999									
3.8	0.9999									
3.9	1.0000									
4.0	1.0000									

TABLE IIIb   Percentiles of the Standard Normal (z_α)

α	z_α
0.20	0.842
0.15	1.036
0.10	1.282
0.05	1.645
0.025	1.960
0.01	2.326
0.005	2.576
0.001	3.090

TABLE IV PERCENTILES OF THE *t*-DISTRIBUTION (VALUES OF $t_{\alpha, f}$)[a]

f	$t_{.100}$	$t_{.050}$	$t_{.025}$	$t_{.010}$	$t_{.005}$	f
1	3.078	6.314	12.706	31.821	63.657	1
2	1.886	2.920	4.303	6.965	9.925	2
3	1.638	2.353	3.182	4.541	5.841	3
4	1.533	2.132	2.776	3.747	4.604	4
5	1.476	2.015	2.571	3.365	4.032	5
6	1.440	1.943	2.447	3.143	3.707	6
7	1.415	1.895	2.365	2.998	3.499	7
8	1.397	1.860	2.306	2.896	3.355	8
9	1.383	1.833	2.262	2.821	3.250	9
10	1.372	1.812	2.228	2.764	3.169	10
11	1.363	1.796	2.201	2.718	3.106	11
12	1.356	1.782	2.179	2.681	3.055	12
13	1.350	1.771	2.160	2.650	3.012	13
14	1.345	1.761	2.145	2.624	2.977	14
15	1.341	1.753	2.131	2.602	2.947	15
16	1.337	1.746	2.120	2.583	2.921	16
17	1.333	1.740	2.110	2.567	2.898	17
18	1.330	1.734	2.101	2.552	2.878	18
19	1.328	1.729	2.093	2.539	2.861	19
20	1.325	1.725	2.086	2.528	2.845	20
21	1.323	1.721	2.080	2.518	2.831	21
22	1.321	1.717	2.074	2.508	2.819	22
23	1.319	1.714	2.069	2.500	2.807	23
24	1.318	1.711	2.064	2.492	2.797	24
25	1.316	1.708	2.060	2.485	2.787	25
26	1.315	1.706	2.056	2.479	2.779	26
27	1.314	1.703	2.052	2.473	2.771	27
28	1.313	1.701	2.048	2.467	2.763	28
29	1.311	1.699	2.045	2.462	2.756	29
inf.	1.282	1.645	1.960	2.326	2.576	inf.

[a] This table is abridged from Table 12 of the *Biometrika Tables for Statisticians*, Vol. I. (New York: Cambridge University Press, 1954) by permission of the *Biometrika* trustees.

TABLE V PERCENTILES OF THE CHI-SQUARE DISTRIBUTION (VALUES OF $\chi^2_{\alpha,\,f}$)[a,b]

f	$\chi^2_{.995}$	$\chi^2_{.99}$	$\chi^2_{.975}$	$\chi^2_{.95}$	$\chi^2_{.05}$	$\chi^2_{.025}$	$\chi^2_{.01}$	$\chi^2_{.005}$	f
1	.0000393	.000157	.000982	.00393	3.841	5.024	6.635	7.879	1
2	.0100	.0201	.0506	.103	5.991	7.378	9.210	10.597	2
3	.0717	.115	.216	.352	7.815	9.348	11.345	12.838	3
4	.207	.297	.484	.711	9.488	11.143	13.277	14.860	4
5	.412	.554	.831	1.145	11.070	12.832	15.086	16.750	5
6	.676	.872	1.237	1.635	12.592	14.449	16.812	18.548	6
7	.989	1.239	1.690	2.167	14.067	16.013	18.475	20.278	7
8	1.344	1.646	2.180	2.733	15.507	17.535	20.090	21.955	8
9	1.735	2.088	2.700	3.325	16.919	19.023	21.666	23.589	9
10	2.156	2.558	3.247	3.940	18.307	20.483	23.209	25.188	10
11	2.603	3.053	3.816	4.575	19.675	21.920	24.725	26.757	11
12	3.074	3.571	4.404	5.226	21.026	23.337	26.217	28.300	12
13	3.565	4.107	5.009	5.892	22.362	24.736	27.688	29.819	13
14	4.075	4.660	5.629	6.571	23.685	26.119	29.141	31.319	14
15	4.601	5.229	6.262	7.261	24.996	27.488	30.578	32.801	15
16	5.142	5.812	6.908	7.962	26.296	28.845	32.000	34.267	16
17	5.697	6.408	7.564	8.672	27.587	30.191	33.409	35.718	17
18	6.265	7.015	8.231	9.390	28.869	31.526	34.805	37.156	18
19	6.844	7.633	8.907	10.117	30.144	32.852	36.191	38.582	19
20	7.434	8.260	9.591	10.851	31.410	34.170	37.566	39.997	20
21	8.034	8.897	10.283	11.591	32.671	35.479	38.932	41.401	21
22	8.643	9.542	10.982	12.338	33.924	36.781	40.289	42.796	22
23	9.260	10.196	11.689	13.091	35.172	38.076	41.638	44.181	23
24	9.886	10.856	12.401	13.848	36.415	39.364	42.980	45.558	24
25	10.520	11.524	13.120	14.611	37.652	40.646	44.314	46.928	25
26	11.160	12.198	13.844	15.379	38.885	41.923	45.642	48.290	26
27	11.808	12.879	14.573	16.151	40.113	43.194	46.963	49.645	27
28	12.461	13.565	15.308	16.928	41.337	44.461	48.278	50.993	28
29	13.121	14.256	16.047	17.708	42.557	45.722	49.588	52.336	29
30	13.787	14.953	16.791	18.493	43.773	46.979	50.892	53.672	30

[a]This table is based on Table 8 of *Biometrika Tables for Statisticians*, Vol. I (New York: Cambridge University Press, 1954) by permission of the *Biometrika* trustees.

[b]For $f > 30$, use $\chi^2_{\alpha,\,f} \doteq f(1 - 2/9f + z_\alpha\sqrt{2/9f})^3$.

TABLE VIa   95% CONFIDENCE INTERVALS FOR PROPORTIONS[a]

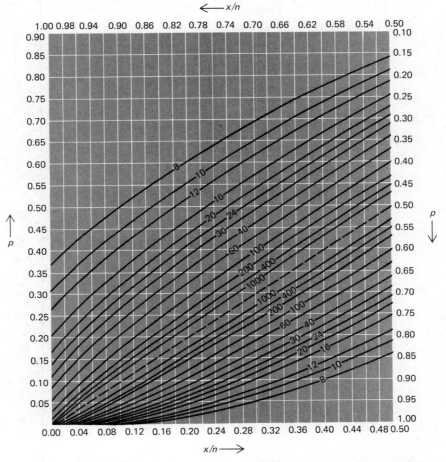

TABLE VIb 99% CONFIDENCE INTERVALS FOR PROPORTIONS[a]

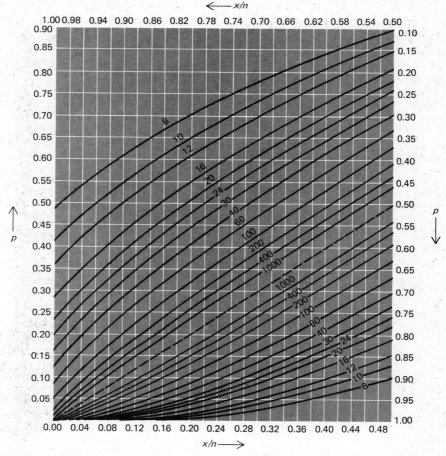

TABLE VIIa 95th PERCENTILES OF THE F DISTRIBUTION (VALUES OF $F_{0.05;\,f_1,\,f_2}$)[a]

Degrees of freedom for numerator, f_1

f_2	1	2	3	4	5	6	7	8	9	10	12	15	20	24	30	40	60	120	∞
1	161	200	216	225	230	234	237	239	241	242	244	246	248	249	250	251	252	253	254
2	18.5	19.0	19.2	19.2	19.3	19.3	19.4	19.4	19.4	19.4	19.4	19.4	19.4	19.5	19.5	19.5	19.5	19.5	19.5
3	10.1	9.55	9.28	9.12	9.01	8.94	8.89	8.85	8.81	8.79	8.74	8.70	8.66	8.64	8.62	8.59	8.57	8.55	8.53
4	7.71	6.94	6.59	6.39	6.26	6.16	6.09	6.04	6.00	5.96	5.91	5.86	5.80	5.77	5.75	5.72	5.69	5.66	5.63
5	6.61	5.79	5.41	5.19	5.05	4.95	4.88	4.82	4.77	4.74	4.68	4.62	4.56	4.53	4.50	4.46	4.43	4.40	4.37
6	5.99	5.14	4.76	4.53	4.39	4.28	4.21	4.15	4.10	4.06	4.00	3.94	3.87	3.84	3.81	3.77	3.74	3.70	3.67
7	5.59	4.74	4.35	4.12	3.97	3.87	3.79	3.73	3.68	3.64	3.57	3.51	3.44	3.41	3.38	3.34	3.30	3.27	3.23
8	5.32	4.46	4.07	3.84	3.69	3.58	3.50	3.44	3.39	3.35	3.28	3.22	3.15	3.12	3.08	3.04	3.01	2.97	2.93
9	5.12	4.26	3.86	3.63	3.48	3.37	3.29	3.23	3.18	3.14	3.07	3.01	2.94	2.90	2.86	2.83	2.79	2.75	2.71
10	4.96	4.10	3.71	3.48	3.33	3.22	3.14	3.07	3.02	2.98	2.91	2.85	2.77	2.74	2.70	2.66	2.62	2.58	2.54
11	4.84	3.98	3.59	3.36	3.20	3.09	3.01	2.95	2.90	2.85	2.79	2.72	2.65	2.61	2.57	2.53	2.49	2.45	2.40
12	4.75	3.89	3.49	3.26	3.11	3.00	2.91	2.85	2.80	2.75	2.69	2.62	2.54	2.51	2.47	2.43	2.38	2.34	2.30
13	4.67	3.81	3.41	3.18	3.03	2.92	2.83	2.77	2.71	2.67	2.60	2.53	2.46	2.42	2.38	2.34	2.30	2.25	2.21
14	4.60	3.74	3.34	3.11	2.96	2.85	2.76	2.70	2.65	2.60	2.53	2.46	2.39	2.35	2.31	2.27	2.22	2.18	2.13
15	4.54	3.68	3.29	3.06	2.90	2.79	2.71	2.64	2.59	2.54	2.48	2.40	2.33	2.29	2.25	2.20	2.16	2.11	2.07
16	4.49	3.63	3.24	3.01	2.85	2.74	2.66	2.59	2.54	2.49	2.42	2.35	2.28	2.24	2.19	2.15	2.11	2.06	2.01
17	4.45	3.59	3.20	2.96	2.81	2.70	2.61	2.55	2.49	2.45	2.38	2.31	2.23	2.19	2.15	2.10	2.06	2.01	1.96
18	4.41	3.55	3.16	2.93	2.77	2.66	2.58	2.51	2.46	2.41	2.34	2.27	2.19	2.15	2.11	2.06	2.02	1.97	1.92
19	4.38	3.52	3.13	2.90	2.74	2.63	2.54	2.48	2.42	2.38	2.31	2.23	2.16	2.11	2.07	2.03	1.98	1.93	1.88
20	4.35	3.49	3.10	2.87	2.71	2.60	2.51	2.45	2.39	2.35	2.28	2.20	2.12	2.08	2.04	1.99	1.95	1.90	1.84
21	4.32	3.47	3.07	2.84	2.68	2.57	2.49	2.42	2.37	2.32	2.25	2.18	2.10	2.05	2.01	1.96	1.92	1.87	1.81
22	4.30	3.44	3.05	2.82	2.66	2.55	2.46	2.40	2.34	2.30	2.23	2.15	2.07	2.03	1.98	1.94	1.89	1.84	1.78
23	4.28	3.42	3.03	2.80	2.64	2.53	2.44	2.37	2.32	2.27	2.20	2.13	2.05	2.01	1.96	1.91	1.86	1.81	1.76
24	4.26	3.40	3.01	2.78	2.62	2.51	2.42	2.36	2.30	2.25	2.18	2.11	2.03	1.98	1.94	1.89	1.84	1.79	1.73
25	4.24	3.39	2.99	2.76	2.60	2.49	2.40	2.34	2.28	2.24	2.16	2.09	2.01	1.96	1.92	1.87	1.82	1.77	1.71
30	4.17	3.32	2.92	2.69	2.53	2.42	2.33	2.27	2.21	2.16	2.09	2.01	1.93	1.89	1.84	1.79	1.74	1.68	1.62
40	4.08	3.23	2.84	2.61	2.45	2.34	2.25	2.18	2.12	2.08	2.00	1.92	1.84	1.79	1.74	1.69	1.64	1.58	1.51
60	4.00	3.15	2.76	2.53	2.37	2.25	2.17	2.10	2.04	1.99	1.92	1.84	1.75	1.70	1.65	1.59	1.53	1.47	1.39
120	3.92	3.07	2.68	2.45	2.29	2.18	2.09	2.02	1.96	1.91	1.83	1.75	1.66	1.61	1.55	1.50	1.43	1.35	1.25
∞	3.84	3.00	2.60	2.37	2.21	2.10	2.01	1.94	1.88	1.83	1.75	1.67	1.57	1.52	1.46	1.39	1.32	1.22	1.00

Degrees of freedom for denominator, f_2

[a] This table is reproduced from M. Merrington and C. M. Thompson, "Tables of percentage points of the inverted beta (F) distribution," *Biometrika*, Vol. 33 (1943), by permission of the *Biometrika* trustees.

TABLE VIIb 99th PERCENTILES OF THE F DISTRIBUTION (VALUES OF $F_{0.01; f_1, f_2}$)[a]

Degrees of freedom for numerator, f_1

f_2	1	2	3	4	5	6	7	8	9	10	12	15	20	24	30	40	60	120	∞
1	4,052	5,000	5,403	5,625	5,764	5,859	5,928	5,982	6,023	6,056	6,106	6,157	6,209	6,235	6,261	6,287	6,313	6,339	6,366
2	98.5	99.0	99.2	99.2	99.3	99.3	99.4	99.4	99.4	99.4	99.4	99.4	99.4	99.5	99.5	99.5	99.5	99.5	99.5
3	34.1	30.8	29.5	28.7	28.2	27.9	27.7	27.5	27.3	27.2	27.1	26.9	26.7	26.6	26.5	26.4	26.3	26.2	26.1
4	21.2	18.0	16.7	16.0	15.5	15.2	15.0	14.8	14.7	14.5	14.4	14.2	14.0	13.9	13.8	13.7	13.7	13.6	13.5
5	16.3	13.3	12.1	11.4	11.0	10.7	10.5	10.3	10.2	10.1	9.89	9.72	9.55	9.47	9.38	9.29	9.20	9.11	9.02
6	13.7	10.9	9.78	9.15	8.75	8.47	8.26	8.10	7.98	7.87	7.72	7.56	7.40	7.31	7.23	7.14	7.06	6.97	6.88
7	12.2	9.55	8.45	7.85	7.46	7.19	6.99	6.84	6.72	6.62	6.47	6.31	6.16	6.07	5.99	5.91	5.82	5.74	5.65
8	11.3	8.65	7.59	7.01	6.63	6.37	6.18	6.03	5.91	5.81	5.67	5.52	5.36	5.28	5.20	5.12	5.03	4.95	4.86
9	10.6	8.02	6.99	6.42	6.06	5.80	5.61	5.47	5.35	5.26	5.11	4.96	4.81	4.73	4.65	4.57	4.48	4.40	4.31
10	10.0	7.56	6.55	5.99	5.64	5.39	5.20	5.06	4.94	4.85	4.71	4.56	4.41	4.33	4.25	4.17	4.08	4.00	3.91
11	9.65	7.21	6.22	5.67	5.32	5.07	4.89	4.74	4.63	4.54	4.40	4.25	4.10	4.02	3.94	3.86	3.78	3.69	3.60
12	9.33	6.93	5.95	5.41	5.06	4.82	4.64	4.50	4.39	4.30	4.16	4.01	3.86	3.78	3.70	3.62	3.54	3.45	3.36
13	9.07	6.70	5.74	5.21	4.86	4.62	4.44	4.30	4.19	4.10	3.96	3.82	3.66	3.59	3.51	3.43	3.34	3.25	3.17
14	8.86	6.51	5.56	5.04	4.70	4.46	4.28	4.14	4.03	3.94	3.80	3.66	3.51	3.43	3.35	3.27	3.18	3.09	3.00
15	8.68	6.36	5.42	4.89	4.56	4.32	4.14	4.00	3.89	3.80	3.67	3.52	3.37	3.29	3.21	3.13	3.05	2.96	2.87
16	8.53	6.23	5.29	4.77	4.44	4.20	4.03	3.89	3.78	3.69	3.55	3.41	3.26	3.18	3.10	3.02	2.93	2.84	2.75
17	8.40	6.11	5.19	4.67	4.34	4.10	3.93	3.79	3.68	3.59	3.46	3.31	3.16	3.08	3.00	2.92	2.83	2.75	2.65
18	8.29	6.01	5.09	4.58	4.25	4.01	3.84	3.71	3.60	3.51	3.37	3.23	3.08	3.00	2.92	2.84	2.75	2.66	2.57
19	8.19	5.93	5.01	4.50	4.17	3.94	3.77	3.63	3.52	3.43	3.30	3.15	3.00	2.92	2.84	2.76	2.67	2.58	2.49
20	8.10	5.85	4.94	4.43	4.10	3.87	3.70	3.56	3.46	3.37	3.23	3.09	2.94	2.86	2.78	2.69	2.61	2.52	2.42
21	8.02	5.78	4.87	4.37	4.04	3.81	3.64	3.51	3.40	3.31	3.17	3.03	2.88	2.80	2.72	2.64	2.55	2.46	2.36
22	7.95	5.72	4.82	4.31	3.99	3.76	3.59	3.45	3.35	3.26	3.12	2.98	2.83	2.75	2.67	2.58	2.50	2.40	2.31
23	7.88	5.66	4.76	4.26	3.94	3.71	3.54	3.41	3.30	3.21	3.07	2.93	2.78	2.70	2.62	2.54	2.45	2.35	2.26
24	7.82	5.61	4.72	4.22	3.90	3.67	3.50	3.36	3.26	3.17	3.03	2.89	2.74	2.66	2.58	2.49	2.40	2.31	2.21
25	7.77	5.57	4.68	4.18	3.86	3.63	3.46	3.32	3.22	3.13	2.99	2.85	2.70	2.62	2.53	2.45	2.36	2.27	2.17
30	7.56	5.39	4.51	4.02	3.70	3.47	3.30	3.17	3.07	2.98	2.84	2.70	2.55	2.47	2.39	2.30	2.21	2.11	2.01
40	7.31	5.18	4.31	3.83	3.51	3.29	3.12	2.99	2.89	2.80	2.66	2.52	2.37	2.29	2.20	2.11	2.02	1.92	1.80
60	7.08	4.98	4.13	3.65	3.34	3.12	2.95	2.82	2.72	2.63	2.50	2.35	2.20	2.12	2.03	1.94	1.84	1.73	1.60
120	6.85	4.79	3.95	3.48	3.17	2.96	2.79	2.66	2.56	2.47	2.34	2.19	2.03	1.95	1.86	1.76	1.66	1.53	1.38
∞	6.63	4.61	3.78	3.32	3.02	2.80	2.64	2.51	2.41	2.32	2.18	2.04	1.88	1.79	1.70	1.59	1.47	1.32	1.00

Degrees of freedom for denominator, f_2

[a] This table is reproduced from M. Merrington and C. M. Thompson, "Tables of percentage points of the inverted beta (F) distribution," *Biometrika*, Vol. 33 (1943), by permission of the *Biometrika* trustees.

TABLE VIIIa CRITICAL VALUES FOR DUNCAN'S MULTIPLE RANGE
(VALUES OF $r_{\alpha;\,p,\,f}$ FOR $\alpha = 0.05^a$)

f \ p	2	3	4	5	6	7	8	9	10
1	17.97								
2	6.09	6.09							
3	4.50	4.52	4.52						
4	3.93	4.01	4.03	4.03					
5	3.64	3.75	3.80	3.81	3.81				
6	3.46	3.59	3.65	3.68	3.69	3.70			
7	3.34	3.48	3.55	3.59	3.61	3.62	3.63		
8	3.26	3.40	3.48	3.52	3.55	3.57	3.57	3.58	
9	3.20	3.34	3.42	3.47	3.50	3.52	3.54	3.54	3.55
10	3.15	3.29	3.38	3.43	3.47	3.49	3.51	3.52	3.52
11	3.11	3.26	3.34	3.40	3.44	3.46	3.48	3.49	3.50
12	3.08	3.23	3.31	3.37	3.41	3.44	3.46	3.47	3.48
13	3.06	3.20	3.29	3.35	3.39	3.42	3.46	3.46	3.47
14	3.03	3.18	3.27	3.33	3.37	3.40	3.43	3.44	3.46
15	3.01	3.16	3.25	3.31	3.36	3.39	3.41	3.43	3.45
16	3.00	3.14	3.23	3.30	3.34	3.38	3.40	3.42	3.44
17	2.98	3.13	3.22	3.28	3.33	3.37	3.39	3.41	3.43
18	2.97	3.12	3.21	3.27	3.32	3.36	3.38	3.40	3.42
19	2.96	3.11	3.20	3.26	3.31	3.35	3.38	3.40	3.41
20	2.95	3.10	3.19	3.25	3.30	3.34	3.37	3.39	3.41
24	2.92	3.07	3.16	3.23	3.28	3.31	3.35	3.37	3.39
30	2.89	3.03	3.13	3.20	3.25	3.29	3.32	3.35	3.37
40	2.86	3.01	3.10	3.17	3.22	3.27	3.30	3.33	3.35
60	2.83	2.98	3.07	3.14	3.20	3.24	3.28	3.31	3.33
120	2.80	2.95	3.04	3.12	3.17	3.22	3.25	3.29	3.31
∞	2.77	2.92	3.02	3.09	3.15	3.19	3.23	3.27	3.29

[a] This table is reproduced from H. L. Harter, "Critical values for Duncan's new multiple range test." It contains some corrected values to replace those given by D. B. Duncan in "Multiple Range and Multiple F Tests," *Biometrics*, Vol. 11 (1955). The above table is reproduced with the permission of the Biometric Society.

TABLE VIIIb CRITICAL VALUES FOR DUNCAN'S MULTIPLE RANGE (VALUES OF $r_{\alpha;\,p,\,f}$ FOR $\alpha = 0.01$)[a]

f \ p	2	3	4	5	6	7	8	9	10
1	90.02								
2	14.04	14.04							
3	8.26	8.32	8.32						
4	6.51	6.68	6.74	6.76					
5	5.70	5.90	5.99	6.04	6.07				
6	5.24	5.44	5.55	5.62	5.66	5.68			
7	4.95	5.15	5.26	5.33	5.38	5.42	5.44		
8	4.74	4.94	5.06	5.13	5.19	5.23	5.26	5.28	
9	4.60	4.79	4.91	4.99	5.04	5.09	5.12	5.14	5.16
10	4.48	4.67	4.79	4.88	4.93	4.98	5.01	5.04	5.06
11	4.39	4.58	4.70	4.78	4.84	4.89	4.92	4.95	4.97
12	4.32	4.50	4.62	4.71	4.77	4.81	4.85	4.88	4.91
13	4.26	4.44	4.56	4.64	4.71	4.75	4.79	4.82	4.85
14	4.21	4.39	4.51	4.59	4.66	4.70	4.74	4.77	4.80
15	4.17	4.34	4.46	4.55	4.61	4.66	4.70	4.73	4.76
16	4.13	4.31	4.43	4.51	4.57	4.62	4.66	4.70	4.72
17	4.10	4.27	4.39	4.47	4.54	4.59	4.63	4.66	4.69
18	4.07	4.25	4.36	4.45	4.51	4.56	4.60	4.64	4.66
19	4.05	4.22	4.33	4.42	4.48	4.53	4.57	4.61	4.64
20	4.02	4.20	4.31	4.40	4.46	4.51	4.55	4.59	4.62
24	3.96	4.13	4.24	4.32	4.39	4.44	4.48	4.52	4.55
30	3.89	4.06	4.17	4.25	4.31	4.36	4.41	4.45	4.48
40	3.82	3.99	4.10	4.18	4.24	4.29	4.33	4.38	4.41
60	3.76	3.92	4.03	4.11	4.18	4.23	4.37	4.31	4.34
120	3.70	3.86	3.97	4.04	4.11	4.16	4.20	4.24	4.27
∞	3.64	3.80	3.90	3.98	4.04	4.09	4.13	4.17	4.21

[a]This table is reproduced from H. L. Harter, "Critical values for Duncan's new multiple range test." It contains some corrected values to replace those given by D. B. Duncan in "Multiple Range and Multiple *F* Tests," *Biometrics*, Vol. 11 (1955). The above table is reproduced with the permission of the Biometric Society.

TABLE IX $\quad Z = Z(r) = \frac{1}{2}\ln\left(\frac{1+r}{1-r}\right)$

| | | | | | r (third decimal) | | | | | |
r	.000	.001	.002	.003	.004	.005	.006	.007	.008	.009
.00	.0000	.0010	.0020	.0030	.0040	.0050	.0060	.0070	.0080	.0090
.01	.0100	.0110	.0120	.0130	.0140	.0150	.0160	.0170	.0180	.0190
.02	.0200	.0210	.0220	.0230	.0240	.0250	.0260	.0270	.0280	.0290
.03	.0300	.0310	.0320	.0330	.0340	.0350	.0360	.0370	.0380	.0390
.04	.0400	.0410	.0420	.0430	.0440	.0450	.0460	.0470	.0480	.0490
.05	.0500	.0510	.0520	.0530	.0541	.0551	.0561	.0571	.0581	.0591
.06	.0601	.0611	.0621	.0631	.0641	.0651	.0661	.0671	.0681	.0691
.07	.0701	.0711	.0721	.0731	.0741	.0751	.0761	.0771	.0782	.0792
.08	.0802	.0812	.0822	.0832	.0842	.0852	.0862	.0872	.0882	.0892
.09	.0902	.0913	.0923	.0933	.0943	.0953	.0963	.0973	.0983	.0993
.10	.1003	.1013	.1024	.1034	.1044	.1054	.1064	.1074	.1084	.1094
.11	.1104	.1115	.1125	.1135	.1145	.1155	.1165	.1175	.1186	.1196
.12	.1206	.1216	.1226	.1236	.1246	.1257	.1267	.1277	.1287	.1297
.13	.1307	.1318	.1328	.1338	.1348	.1358	.1368	.1379	.1389	.1399
.14	.1409	.1419	.1430	.1440	.1450	.1460	.1471	.1481	.1491	.1501
.15	.1511	.1522	.1532	.1542	.1552	.1563	.1573	.1583	.1593	.1604
.16	.1614	.1624	.1634	.1645	.1655	.1665	.1676	.1686	.1696	.1706
.17	.1717	.1727	.1737	.1748	.1758	.1768	.1779	.1789	.1799	.1809
.18	.1820	.1830	.1841	.1851	.1861	.1872	.1882	.1892	.1903	.1913
.19	.1923	.1934	.1944	.1955	.1965	.1976	.1986	.1996	.2007	.2017
.20	.2027	.2038	.2048	.2059	.2069	.2079	.2090	.2100	.2111	.2121
.21	.2132	.2142	.2153	.2163	.2174	.2184	.2195	.2205	.2216	.2226
.22	.2237	.2247	.2258	.2268	.2279	.2289	.2300	.2310	.2321	.2331
.23	.2342	.2352	.2363	.2374	.2384	.2395	.2405	.2416	.2427	.2437
.24	.2448	.2458	.2469	.2480	.2490	.2501	.2512	.2522	.2533	.2543
.25	.2554	.2565	.2575	.2586	.2597	.2608	.2618	.2629	.2640	.2650
.26	.2661	.2672	.2683	.2693	.2704	.2715	.2726	.2736	.2747	.2758
.27	.2769	.2780	.2790	.2801	.2812	.2823	.2833	.2844	.2855	.2866
.28	.2877	.2888	.2899	.2909	.2920	.2931	.2942	.2953	.2964	.2975
.29	.2986	.2997	.3008	.3018	.3029	.3040	.3051	.3062	.3073	.3084
.30	.3095	.3106	.3117	.3128	.3139	.3150	.3161	.3172	.3183	.3194
.31	.3205	.3217	.3228	.3239	.3250	.3261	.3272	.3283	.3294	.3305
.32	.3316	.3328	.3339	.3350	.3361	.3372	.3383	.3395	.3406	.3417
.33	.3428	.3440	.3451	.3462	.3473	.3484	.3496	.3507	.3518	.3530
.34	.3541	.3552	.3564	.3575	.3586	.3598	.3609	.3620	.3632	.3643
.35	.3654	.3666	.3677	.3689	.3700	.3712	.3723	.3734	.3746	.3757
.36	.3769	.3780	.3792	.3803	.3815	.3826	.3838	.3850	.3861	.3873
.37	.3884	.3896	.3907	.3919	.3931	.3942	.3954	.3966	.3977	.3989
.38	.4001	.4012	.4024	.4036	.4047	.4059	.4071	.4083	.4094	.4106
.39	.4118	.4130	.4142	.4153	.4165	.4177	.4189	.4201	.4213	.4225
.40	.4236	.4248	.4260	.4272	.4284	.4296	.4308	.4320	.4332	.4344
.41	.4356	.4368	.4380	.4392	.4404	.4416	.4428	.4441	.4453	.4465
.42	.4477	.4489	.4501	.4513	.4526	.4538	.4550	.4562	.4574	.4587
.43	.4599	.4611	.4624	.4636	.4648	.4660	.4673	.4685	.4698	.4710
.44	.4722	.4735	.4747	.4760	.4772	.4784	.4797	.4809	.4822	.4834
.45	.4847	.4860	.4872	.4885	.4897	.4910	.4922	.4935	.4948	.4960
.46	.4973	.4986	.4999	.5011	.5024	.5037	.5049	.5062	.5075	.5088
.47	.5101	.5114	.5126	.5139	.5152	.5165	.5178	.5191	.5204	.5217
.48	.5230	.5243	.5256	.5269	.5282	.5295	.5308	.5321	.5334	.5347
.49	.5361	.5374	.5387	.5400	.5413	.5427	.5440	.5453	.5466	.5480

TABLE IX   (*Cont.*)

					r (third decimal)					
r	*.000*	*.001*	*.002*	*.003*	*.004*	*.005*	*.006*	*.007*	*.008*	*.009*
.50	.5493	.5506	.5520	.5533	.5547	.5560	.5573	.5587	.5600	.5614
.51	.5627	.5641	.5654	.5668	.5682	.5695	.5709	.5722	.5736	.5750
.52	.5763	.5777	.5791	.5805	.5818	.5832	.5846	.5860	.5874	.5888
.53	.5901	.5915	.5929	.5943	.5957	.5971	.5985	.5999	.6013	.6027
.54	.6042	.6056	.6070	.6084	.6098	.6112	.6127	.6141	.6155	.6169
.55	.6184	.6198	.6213	.6227	.6241	.6256	.6270	.6285	.6299	.6314
.56	.6328	.6343	.6358	.6372	.6387	.6401	.6416	.6431	.6446	.6460
.57	.6475	.6490	.6505	.6520	.6535	.6550	.6565	.6580	.6595	.6610
.58	.6625	.6640	.6655	.6670	.6685	.6700	.6716	.6731	.6746	.6761
.59	.6777	.6792	.6807	.6823	.6838	.6854	.6869	.6885	.6900	.6916
.60	.6931	.6947	.6963	.6978	.6994	.7010	.7026	.7042	.7057	.7073
.61	.7089	.7105	.7121	.7137	.7153	.7169	.7185	.7201	.7218	.7234
.62	.7250	.7266	.7283	.7299	.7315	.7332	.7348	.7365	.7381	.7398
.63	.7414	.7431	.7447	.7464	.7481	.7498	.7514	.7531	.7548	.7565
.64	.7582	.7599	.7616	.7633	.7650	.7667	.7684	.7701	.7718	.7736
.65	.7753	.7770	.7788	.7805	.7823	.7840	.7858	.7875	.7893	.7910
.66	.7928	.7946	.7964	.7981	.7999	.8017	.8035	.8053	.8071	.8089
.67	.8107	.8126	.8144	.8162	.8180	.8199	.8217	.8236	.8254	.8272
.68	.8291	.8310	.8328	.8347	.8366	.8385	.8404	.8423	.8441	.8460
.69	.8480	.8499	.8518	.8537	.8556	.8576	.8595	.8614	.8634	.8653
.70	.8673	.8693	.8712	.8732	.8752	.8772	.8792	.8812	.8832	.8852
.71	.8872	.8892	.8912	.8933	.8953	.8973	.8994	.9014	.9035	.9056
.72	.9076	.9097	.9118	.9139	.9160	.9181	.9202	.9223	.9245	.9266
.73	.9287	.9309	.9330	.9352	.9373	.9395	.9417	.9439	.9461	.9483
.74	.9505	.9527	.9549	.9571	.9594	.9616	.9639	.9661	.9684	.9707
.75	.9730	.9752	.9775	.9798	.9822	.9845	.9868	.9892	.9915	.9939
.76	.9962	.9986	1.001	1.003	1.006	1.008	1.011	1.013	1.015	1.018
.77	1.020	1.023	1.025	1.028	1.030	1.033	1.035	1.038	1.040	1.043
.78	1.045	1.048	1.050	1.053	1.056	1.058	1.061	1.064	1.066	1.069
.79	1.071	1.074	1.077	1.079	1.082	1.085	1.088	1.090	1.093	1.096
.80	1.099	1.101	1.104	1.107	1.110	1.113	1.116	1.118	1.121	1.124
.81	1.127	1.130	1.133	1.136	1.139	1.142	1.145	1.148	1.151	1.154
.82	1.157	1.160	1.163	1.166	1.169	1.172	1.175	1.179	1.182	1.185
.83	1.188	1.191	1.195	1.198	1.201	1.204	1.208	1.211	1.214	1.218
.84	1.221	1.225	1.228	1.231	1.235	1.238	1.242	1.245	1.249	1.253
.85	1.256	1.260	1.263	1.267	1.271	1.274	1.278	1.282	1.286	1.290
.86	1.293	1.297	1.301	1.305	1.309	1.313	1.317	1.321	1.325	1.329
.87	1.333	1.337	1.341	1.346	1.350	1.354	1.358	1.363	1.367	1.371
.88	1.376	1.380	1.385	1.389	1.394	1.398	1.403	1.407	1.412	1.417
.89	1.422	1.427	1.432	1.437	1.442	1.447	1.452	1.457	1.462	1.467
.90	1.472	1.478	1.483	1.488	1.494	1.499	1.505	1.510	1.516	1.522
.91	1.528	1.533	1.539	1.545	1.551	1.558	1.564	1.570	1.576	1.583
.92	1.589	1.596	1.602	1.609	1.616	1.623	1.630	1.637	1.644	1.651
.93	1.658	1.666	1.673	1.681	1.689	1.697	1.705	1.713	1.721	1.730
.94	1.738	1.747	1.756	1.764	1.774	1.783	1.792	1.802	1.812	1.822
.95	1.832	1.842	1.853	1.863	1.874	1.886	1.897	1.909	1.921	1.933
.96	1.946	1.959	1.972	1.986	2.000	2.014	2.029	2.044	2.060	2.076
.97	2.092	2.110	2.127	2.146	2.165	2.185	2.205	2.227	2.249	2.273
.98	2.298	2.323	2.351	2.380	2.410	2.442	2.477	2.515	2.555	2.599
.99	2.647	2.700	2.759	2.826	2.903	2.994	3.106	3.250	3.453	3.800

TABLE X CRITICAL VALUES $U_{1-p} = U_{1-\alpha}$ or $U_{1-\alpha/2}$[a]

n_1	p	$n_2 = 2$	3	4	5	6	7	8	9	10	11	12	13	14	15	16	17	18	19	20
2	0.001	0	0	0	0	0	0	0	0	0	0	0	0	0	0	0	0	0	0	0
	0.005	0	0	0	0	0	0	0	0	0	0	0	0	0	0	0	0	0	1	1
	0.01	0	0	0	0	0	0	0	0	0	0	0	1	1	1	1	1	2	2	2
	0.025	0	0	0	0	0	0	1	1	1	1	2	2	2	2	2	3	3	3	3
	0.05	0	0	0	1	1	1	2	2	2	2	3	3	4	4	4	4	5	5	5
	0.10	0	1	1	2	2	2	3	3	4	4	5	5	5	6	6	7	7	8	8
3	0.001	0	0	0	0	0	0	0	0	0	0	0	0	0	0	0	1	1	1	1
	0.005	0	0	0	0	0	0	0	1	1	1	2	2	2	3	3	3	3	4	4
	0.01	0	0	0	0	0	1	1	2	2	2	3	3	3	4	4	5	5	5	6
	0.025	0	0	0	1	2	2	3	3	4	4	5	5	6	6	7	7	8	8	9
	0.05	0	1	1	2	3	3	4	5	5	6	6	7	8	8	9	10	10	11	12
	0.10	1	2	2	3	4	5	6	6	7	8	9	10	11	11	12	13	14	15	16
4	0.001	0	0	0	0	0	0	0	0	1	1	1	2	2	2	3	3	4	4	4
	0.005	0	0	0	0	1	1	2	2	3	3	4	4	5	6	6	7	7	8	9
	0.01	0	0	0	1	2	2	3	4	4	5	6	6	7	8	8	9	10	10	11
	0.025	0	0	1	2	3	4	5	5	6	7	8	9	10	11	12	12	13	14	15
	0.05	0	1	2	3	4	5	6	7	8	9	10	11	12	13	15	16	17	18	19
	0.10	2	2	4	5	6	7	8	10	11	12	13	14	16	17	18	19	21	22	23
5	0.001	0	0	0	0	0	0	1	2	2	3	3	4	4	5	6	6	7	8	8
	0.005	0	0	0	1	2	2	3	4	5	6	7	8	8	9	10	11	12	13	14
	0.01	0	0	1	2	3	4	5	6	7	8	9	10	11	12	13	14	15	16	17
	0.025	0	1	2	3	4	6	7	8	9	10	12	13	14	15	16	18	19	20	21
	0.05	1	2	3	5	6	7	9	10	12	13	14	16	17	19	20	21	23	24	26
	0.10	2	3	5	6	8	9	11	13	14	16	18	19	21	23	24	26	28	29	31
6	0.001	0	0	0	0	0	1	2	3	4	5	5	6	7	8	9	10	11	12	13
	0.005	0	0	1	2	3	4	5	6	7	8	10	11	12	13	14	16	17	18	19
	0.01	0	0	2	3	4	5	7	8	9	10	12	13	14	16	17	19	20	21	23
	0.025	0	2	3	4	6	7	9	11	12	14	15	17	18	20	22	23	25	26	28
	0.05	1	3	4	6	8	9	11	13	15	17	18	20	22	24	26	27	29	31	33
	0.10	2	4	6	8	10	12	14	16	18	20	22	24	26	28	30	32	35	37	39

TABLE X   CRITICAL VALUES $U_{1-p} = U_{1-\alpha}$ or $U_{1-\alpha/2}$[a] (Cont.)

n_1	p	$n_2 = 2$	3	4	5	6	7	8	9	10	11	12	13	14	15	16	17	18	19	20
7	0.001	0	0	0	0	1	2	3	4	6	7	8	9	10	11	12	14	15	16	17
	0.005	0	0	1	2	4	5	7	8	10	11	13	14	16	17	19	20	22	23	25
	0.01	0	1	2	4	5	7	8	10	12	13	15	17	18	20	22	24	25	27	29
	0.025	0	2	4	6	7	9	11	13	15	17	19	21	23	25	27	29	31	33	35
	0.05	1	3	5	7	9	12	14	16	18	20	22	25	27	29	31	34	36	38	40
	0.10	2	5	7	9	12	14	17	19	22	24	27	29	32	34	37	39	42	44	47
8	0.001	0	0	0	1	2	3	5	6	7	9	10	12	13	15	16	18	19	21	22
	0.005	0	0	2	3	5	7	8	10	12	14	16	18	19	21	23	25	27	29	31
	0.01	0	1	3	5	7	8	10	12	14	16	18	21	23	25	27	29	31	33	35
	0.025	1	3	5	7	9	11	14	16	18	20	23	25	27	30	32	35	37	39	42
	0.05	2	4	6	9	11	14	16	19	21	24	27	29	32	34	37	40	42	45	48
	0.10	3	6	8	11	14	17	20	23	25	28	31	34	37	40	43	46	49	52	55
9	0.001	0	0	0	2	3	4	6	8	9	11	13	15	16	18	20	22	24	26	27
	0.005	0	1	2	4	6	8	10	12	14	17	19	21	23	25	28	30	32	34	37
	0.01	0	2	4	6	8	10	12	15	17	19	22	24	27	29	32	34	37	39	41
	0.025	1	3	5	8	11	13	16	18	21	24	27	29	32	35	38	40	43	46	49
	0.05	2	5	7	10	13	16	19	22	25	28	31	34	37	40	43	46	49	52	55
	0.10	3	6	10	13	16	19	23	26	29	32	36	39	42	46	49	53	56	59	63
10	0.001	0	0	1	2	4	6	7	9	11	13	15	18	20	22	24	26	28	30	33
	0.005	0	1	3	5	7	10	12	14	17	19	22	25	27	30	32	35	38	40	43
	0.01	0	2	4	7	9	12	14	17	20	23	25	28	31	34	37	39	42	45	48
	0.025	1	4	6	9	12	15	18	21	24	27	30	34	37	40	43	46	49	53	56
	0.05	2	5	8	12	15	18	21	25	28	32	35	38	42	45	49	52	56	59	63
	0.10	4	7	11	14	18	22	25	29	33	37	40	44	48	52	55	59	63	67	71

11						12						13						14						15					
0.001	0.005	0.01	0.025	0.05	0.10	0.001	0.005	0.01	0.025	0.05	0.10	0.001	0.005	0.01	0.025	0.05	0.10	0.001	0.005	0.01	0.025	0.05	0.10	0.001	0.005	0.01	0.025	0.05	0.10
38	49	54	63	70	79	43	55	61	70	78	87	49	61	68	77	85	95	55	68	74	84	93	103	60	74	81	91	101	111
35	46	51	59	66	74	41	52	57	66	73	82	46	58	64	73	81	90	51	64	70	79	88	98	56	70	76	86	95	105
33	43	48	56	62	70	38	48	54	62	69	78	43	54	60	68	76	85	47	59	66	75	83	92	52	65	71	81	89	99
30	40	45	52	58	66	35	45	50	58	65	73	39	50	56	64	71	80	44	55	61	70	78	86	48	61	67	76	84	93
28	37	42	48	55	62	32	42	47	54	61	68	36	46	52	60	66	75	40	51	57	65	72	81	44	56	62	71	78	87
25	34	38	45	51	58	29	38	43	50	56	64	33	43	48	55	62	69	37	47	52	60	67	75	41	52	57	65	73	81
23	31	35	41	47	53	26	35	39	46	52	59	30	39	44	51	57	64	33	43	48	56	62	70	37	47	52	60	67	75
21	28	32	38	43	49	24	32	36	42	48	54	27	35	40	46	52	59	30	39	44	51	57	64	33	43	48	55	62	69
18	25	29	34	39	45	21	28	32	38	43	50	24	32	36	42	48	54	26	35	39	46	52	59	29	38	43	50	56	64
16	22	26	31	35	41	18	25	29	34	39	45	21	28	32	38	43	49	23	31	35	41	47	53	25	34	38	45	51	58
13	19	23	27	32	37	15	22	25	30	35	40	18	25	28	34	38	44	20	27	31	37	42	48	22	30	34	40	45	52
11	17	19	24	28	32	13	19	22	27	31	36	15	21	24	29	34	39	16	23	27	32	37	42	18	25	29	35	40	46
9	14	16	20	24	28	10	16	18	23	27	31	12	18	21	25	29	34	13	19	23	27	32	37	15	21	25	30	34	40
7	11	13	17	20	24	8	13	15	19	22	27	9	14	17	21	25	29	10	16	18	23	27	32	11	17	20	25	29	34
5	8	10	14	17	20	5	10	12	15	18	22	6	11	13	17	20	24	7	12	14	18	22	26	8	13	16	20	24	28
3	6	8	10	13	16	3	7	9	12	14	18	4	8	10	13	16	19	4	8	11	14	17	21	5	9	12	15	19	23
1	3	5	7	9	12	1	4	6	8	10	13	2	4	6	9	11	14	2	5	7	10	12	16	2	6	8	11	13	17
0	1	2	4	6	8	0	2	4	5	6	9	0	2	3	5	7	10	0	2	3	6	8	11	0	3	4	6	8	11
0	0	0	1	2	4	0	0	0	2	3	5	0	0	1	2	3	5	0	0	1	2	4	5	0	0	1	2	4	6

TABLE X (Cont.)

n_1	p	$n_2 = 2$	3	4	5	6	7	8	9	10	11	12	13	14	15	16	17	18	19	20
16	0.001	0	0	3	6	9	12	16	20	24	28	32	36	40	44	49	53	57	61	66
	0.005	0	3	6	10	14	19	23	28	32	37	42	46	51	56	61	66	71	75	80
	0.01	1	4	8	13	17	22	27	32	37	42	47	52	57	62	67	72	77	83	88
	0.025	2	7	12	16	22	27	32	38	43	48	54	60	65	71	76	82	87	93	99
	0.05	4	9	15	20	26	31	37	43	49	55	61	66	72	78	84	90	96	102	108
	0.10	6	12	18	24	30	37	43	49	55	62	68	75	81	87	94	100	107	113	120
17	0.001	0	1	3	6	10	14	18	22	26	30	35	39	44	48	53	58	62	67	71
	0.005	0	3	7	11	16	20	25	30	35	40	45	50	55	61	66	71	76	82	87
	0.01	1	5	9	14	19	24	29	34	39	45	50	56	61	67	72	78	83	89	94
	0.025	3	7	12	18	23	29	35	40	46	52	58	64	70	76	82	88	94	100	106
	0.05	4	10	16	21	27	34	40	46	52	58	65	71	78	84	90	97	103	110	116
	0.10	7	13	19	26	32	39	46	53	59	66	73	80	86	93	100	107	114	121	128
18	0.001	0	1	4	7	11	15	19	24	28	33	38	43	47	52	57	62	67	72	77
	0.005	0	3	7	12	17	22	27	32	38	43	48	54	59	65	71	76	82	88	93
	0.01	1	5	10	15	20	25	31	37	42	48	54	60	66	71	77	83	89	95	101
	0.025	3	8	13	19	25	31	37	43	49	56	62	68	74	81	87	94	100	107	113
	0.05	5	10	17	23	29	36	42	49	56	62	69	76	83	89	96	103	110	117	124
	0.10	7	14	21	28	35	42	49	56	63	70	78	85	92	99	107	114	121	129	136
19	0.001	0	1	4	8	12	16	21	26	30	35	41	46	51	56	61	67	72	78	83
	0.005	1	4	8	13	18	23	29	34	40	46	52	58	64	70	75	82	88	94	100
	0.01	2	5	10	16	21	27	33	39	45	51	57	64	70	76	83	89	95	102	108
	0.025	3	8	14	20	26	33	39	46	53	59	66	73	79	86	93	100	107	114	120
	0.05	5	11	18	24	31	38	45	52	59	66	73	81	88	95	102	110	117	124	131
	0.10	8	15	22	29	37	44	52	59	67	74	82	90	98	105	113	120	128	136	144
20	0.001	0	1	4	8	12	17	22	27	33	38	43	49	55	60	66	71	77	83	89
	0.005	1	4	9	14	19	25	31	37	43	49	55	61	68	74	80	87	93	100	106
	0.01	2	6	11	17	23	29	35	41	48	54	61	68	74	81	88	94	101	108	115
	0.025	3	9	15	21	28	35	42	49	56	63	70	77	84	91	99	106	113	120	128
	0.05	5	12	19	26	33	40	48	55	63	70	78	85	93	101	108	116	124	131	139
	0.10	8	16	23	31	39	47	55	63	71	79	87	95	103	111	120	128	136	144	152

[a]Adapted from L. R. Verdooren, *Biometrika* (Vol. 50), 1963 by permission of the *Biometrika* trustees.

TABLE XI[a] CRITICAL VALUES OF T IN THE WILCOXON PAIRED DIFFERENCE SIGNED-RANKS TEST

One-Tailed	Two-Tailed	$n = 5$	$n = 6$	$n = 7$	$n = 8$	$n = 9$	$n = 10$
$\alpha = 0.05$	$\alpha = 0.10$	1	2	4	6	8	11
$\alpha = 0.025$	$\alpha = 0.05$		1	2	4	6	8
$\alpha = 0.01$	$\alpha = 0.02$			0	2	3	5
$\alpha = 0.005$	$\alpha = 0.01$				0	2	3

One-Tailed	Two-Tailed	$n = 11$	$n = 12$	$n = 13$	$n = 14$	$n = 15$	$n = 16$
$\alpha = 0.05$	$\alpha = 0.10$	14	17	21	26	30	36
$\alpha = 0.025$	$\alpha = 0.05$	11	14	17	21	25	30
$\alpha = 0.01$	$\alpha = 0.02$	7	10	13	16	20	24
$\alpha = 0.005$	$\alpha = 0.01$	5	7	10	13	16	19

One-Tailed	Two-Tailed	$n = 17$	$n = 18$	$n = 19$	$n = 20$	$n = 21$	$n = 22$
$\alpha = 0.05$	$\alpha = 0.10$	41	47	54	60	68	75
$\alpha = 0.025$	$\alpha = 0.05$	35	40	46	52	59	66
$\alpha = 0.01$	$\alpha = 0.02$	28	33	38	43	49	56
$\alpha = 0.005$	$\alpha = 0.01$	23	28	32	37	43	49

One-Tailed	Two-Tailed	$n = 23$	$n = 24$	$n = 25$	$n = 26$	$n = 27$	$n = 28$
$\alpha = 0.05$	$\alpha = 0.10$	83	92	101	110	120	130
$\alpha = 0.025$	$\alpha = 0.05$	73	81	90	98	107	117
$\alpha = 0.01$	$\alpha = 0.02$	62	69	77	85	93	102
$\alpha = 0.005$	$\alpha = 0.01$	55	61	68	76	84	92

One-Tailed	Two-Tailed	$n = 29$	$n = 30$	$n = 31$	$n = 32$	$n = 33$	$n = 34$
$\alpha = 0.05$	$\alpha = 0.10$	141	152	163	175	188	201
$\alpha = 0.025$	$\alpha = 0.05$	127	137	148	159	171	183
$\alpha = 0.01$	$\alpha = 0.02$	111	120	130	141	151	162
$\alpha = 0.005$	$\alpha = 0.01$	100	109	118	128	138	149

TABLE XI[a] CRITICAL VALUES OF T IN THE WILCOXON PAIRED DIFFERENCE SIGNED-RANKS TEST (Cont.)

One-Tailed	Two-Tailed	$n = 35$	$n = 36$	$n = 37$	$n = 38$	$n = 39$
$\alpha = 0.05$	$\alpha = 0.10$	214	228	242	256	271
$\alpha = 0.025$	$\alpha = 0.05$	195	208	222	235	250
$\alpha = 0.01$	$\alpha = 0.02$	174	186	198	211	224
$\alpha = 0.005$	$\alpha = 0.01$	160	171	183	195	208

One-Tailed	Two-Tailed	$n = 40$	$n = 41$	$n = 42$	$n = 43$	$n = 44$	$n = 45$
$\alpha = 0.05$	$\alpha = 0.10$	287	303	319	336	353	371
$\alpha = 0.025$	$\alpha = 0.05$	264	279	295	311	327	344
$\alpha = 0.01$	$\alpha = 0.02$	238	252	267	281	297	313
$\alpha = 0.005$	$\alpha = 0.01$	221	234	248	262	277	292

One-Tailed	Two-Tailed	$n = 46$	$n = 47$	$n = 48$	$n = 49$	$n = 50$
$\alpha = 0.05$	$\alpha = 0.10$	389	408	427	446	466
$\alpha = 0.025$	$\alpha = 0.05$	361	379	397	415	434
$\alpha = 0.01$	$\alpha = 0.02$	329	345	362	380	398
$\alpha = 0.005$	$\alpha = 0.01$	307	323	339	356	373

[a]From F. Wilcoxon and R. A. Wilcox, "Some Rapid Approximate Statistical Procedures," 1964, 28. Reproduced with the permission of American Cyanamid Company.

TABLE XII   CRITICAL VALUES FOR r′

n	$r'_{0.10}$	$r'_{0.05}$	$r'_{0.025}$	$r'_{0.01}$	$r'_{0.005}$	$r'_{0.001}$
4	.8000	.8000				
5	.7000	.8000	.9000	.9000		
6	.6000	.7714	.8286	.8857	.9429	
7	.5357	.6786	.7450	.8571	.8929	.9643
8	.5000	.6190	.7143	.8095	.8571	.9286
9	.4667	.5833	.6833	.7667	.8167	.9000
10	.4424	.5515	.6364	.7333	.7818	.8667
11	.4182	.5273	.6091	.7000	.7455	.8364
12	.3986	.4965	.5804	.6713	.7273	.8182
13	.3791	.4780	.5549	.6429	.6978	.7912
14	.3626	.4593	.5341	.6220	.6747	.7670
15	.3500	.4429	.5179	.6000	.6536	.7464
16	.3382	.4265	.5000	.5824	.6324	.7265
17	.3260	.4118	.4853	.5637	.6152	.7083
18	.3148	.3994	.4716	.5480	.5975	.6904
19	.3070	.3895	.4579	.5333	.5825	.6737
20	.2977	.3789	.4451	.5203	.5684	.6586
21	.2909	.3688	.4351	.5078	.5545	.6455
22	.2829	.3597	.4241	.4963	.5426	.6318
23	.2767	.3518	.4150	.4852	.5306	.6186
24	.2704	.3435	.4061	.4748	.5200	.6070
25	.2646	.3362	.3977	.4654	.5100	.5962
26	.2588	.3299	.3894	.4564	.5002	.5856
27	.2540	.3236	.3822	.4482	.4915	.5757
28	.2490	.3175	.3749	.4401	.4828	.5660
29	.2443	.3113	.3685	.4320	.4744	.5567
30	.2400	.3059	.3620	.4251	.4665	.5479

[a]Adapted from Glasser and Winter, *Biometrika*, Vol. 48, 1961, by permission of the *Biometrika* trustees.

APPENDIX:
SOLUTIONS
TO EXERCISES

CHAPTER 2

2-1 (a)

Class Lower limit	Upper limit	Frequency	Class mark
1	5	7	3
6	10	5	8
11	15	3	13
16	20	6	18
21	25	3	23
26	30	1	28
31	35	2	33
36	40	2	38
41	45	2	43
46	50	0	48
51	55	3	53
56	60	3	58
61	65	3	63
66	70	2	68
71	75	1	73
76	80	2	78
81	85	3	83
86	90	1	88

2-2 (a)

Class	Frequency
1–5	7
6–15	8
16–20	6
21–25	3
26–45	7
46–65	9
66–90	9

2-3 (a)

Class	Frequency	(b)%
4.25–4.49	3	4.8
4.50–4.74	6	9.5
4.75–4.99	13	20.6
5.00–5.24	8	12.7
5.25–5.49	16	25.4
5.50–5.74	7	11.1
5.75–5.99	6	9.5
6.00–6.24	1	1.6
6.25–6.49	3	4.8

2-4 (a)

Class					
Lower limit	Upper limit	Frequency	Class mark	Lower boundary	Upper boundary
0	9	1	4.5	−0.5	9.5
10	19	2	14.5	9.5	19.5
20	29	2	24.5	19.5	29.5
30	39	4	34.5	29.5	39.5
40	49	4	44.5	39.5	49.5
50	59	11	54.5	49.5	59.5
60	69	32	64.5	59.5	69.5
70	79	23	74.5	69.5	79.5
80	89	7	84.5	79.5	89.5
90	99	1	94.5	89.5	99.5

2-5

Class	Frequency
0–49	13
50–59	11
60–69	32
70–74	13
75–79	10
80–100	8

2-6

Response	Percentage
Florida	36
Bahamas	20
Caribbean cruise	12
Jamaica	12
Barbados	6
Cuba	6
Cayman Islands	4
Puerto Rico	2
Trinidad	2

2-8

Age	Frequency
Less than 1	0
Less than 6	7
Less than 11	12
Less than 16	15
Less than 21	21
Less than 26	24
Less than 31	25
Less than 36	27
Less than 41	29
Less than 46	31
Less than 51	31
Less than 56	34
Less than 61	37
Less than 66	40
Less than 71	42
Less than 76	43
Less than 81	45
Less than 86	48
Less than 91	49

2-9

Price	Frequency
More than 4.24	63
More than 4.49	60
More than 4.74	54
More than 4.99	41
More than 5.24	33
More than 5.49	17
More than 5.74	10
More than 5.99	4
More than 6.24	3
More than 6.49	0

2-11 (b), (c), (e)　　**2-12** No. There is no natural ordering to the classes.　　**2-13** Yes. There is a natural ordering.
2-21 The class intervals are not all equal.　　**2-22** lower　　**2-23** (b)　　**2-24** areas　　**2-25** boundaries
2-26 (a) and (e)　　**2-28** (a) A substantial portion of cases are poor or dangerous. (b) There is a false impression of a very large number of "acceptable" cases. Adjust the histogram so that areas rather than heights are proportional to frequencies.

2-29 (a) and (b)

| | Class | | | | | |
Lower limit	Upper limit	Frequency	Class mark	Lower boundary	Upper boundary
23.0	24.9	3	23.95	22.95	24.95
25.0	26.9	10	25.95	24.95	26.95
27.0	28.9	6	27.95	26.95	28.95
29.0	30.9	6	29.95	28.95	30.95
31.0	32.9	2	31.95	30.95	32.95
33.0	34.9	3	33.95	32.95	34.95
35.0	36.9	1	35.95	34.95	36.95

2-29 (d)　　31 are more than 22.9
28 are more than 24.9
18 are more than 26.9
12 are more than 28.9
6 are more than 30.9
4 are more than 32.9
1 is more than 34.9
0 is more than 36.9

2-30 The 1973 illustration is too large. Its *area* should be 1.73 times that for 1967 (in order to exceed it by 73%). The length and width of the 1973 illustration should each be $\sqrt{1.73}$ (not 1.73) times the length and width for 1967.

CHAPTER 3

3-1 (a) 106.3 (b) 92　　**3-2** $\bar{x} = 56$, $\tilde{x} = 50$, mode = 50　　**3-3** $\bar{x} = 21.3$, $\tilde{x} = 21$　　**3-4** $\bar{x} = 21.52$, mode = 20
3-5 49.79 (a weighted mean)　　**3-6** The first quartile is 57.25. The third quartile is 73.
3-7 "Average" = C; this average is the mode.　　**3-8** 10th = 4.50, 90th = 5.91 or "between 5.90 and 5.95"
3-9 (a) median = "good", mode = "fair" (b) The responses are not numerical.
3-10 (a) $\bar{x} = 8.004$ $\tilde{x} = 5.85$ or "between 5.8 and 5.9" (b) median (c) mean
3-11 (a) 4.17 (b) $s^2 = 29.4545$ $s = 5.43$ (c) 2.4%　　**3-12** range = 2.10, $s = 0.494$, range = $4.25s$
3-13 No, $s = 0.0026$.　　**3-14** For Chebyshev's theorem $k = 2.5$, *at least* 840 must be between 4.80 and 5.20.
3-15 (a) 28 ($k = 1.36$) (b) 50　　**3-16** 0.376 or 37.6% ($s = 1.26066$, $\bar{x} = 3.352$)
3-17 (a) -4.4 (b) 3.35 (c) $s^2 = 16.56$, $s = 4.069$ (d) 11.1　　**3-18** (a) \$33,636 (b) \$63,446 (c) \$74,697
3-19 (a) 19.6 (b) 7.96　　**3-20** (a) 49.15 (b) first quartile = 42.4, third quartile = 56.9
3-21 (a) 205.07 (b) 19.832 (c) 10th percentile = 177.7, 90th percentile = 228.4
3-22 (a) \$5880 (b) \$6869 (c) \$7962 (d) 0.676 or 67.6%　　**3-23** 19,555　　**3-24** 70.875
3-25 coded: $\bar{y} = 0.43$, $s_y = 2.0088$; original: $\bar{x} = 49.15$, $s_x = 10.044$
3-26 (a) 0 (b) coded: $s = 136.45$, original: $s = 0.0013645$
3-27 $y = (x - 1.97) \times 20$ (a) $\bar{y} = -0.1$, $s_y = 2.13$ (b) $\bar{x} = 1.965$, $s_x = 0.106$ (c) 5.4%
3-28 (a) mathematics: $\bar{x} = 67$, $s = 7.83$ geography: $\bar{x} = 70$, $s = 9.39$ (c) Allen's mathematics mark, Robinson's mathematics mark, Gibson's mathematics mark (d) Allen's mathematics mark
3-31 mathematics: $N = 30$, $\mu = 67$, $\sigma^2 = 59.333$, $S^2 = 61.379$; geography: $N = 28$, $\mu = 70$, $\sigma^2 = 85.071$, $S^2 = 88.222$
3-32 $\mu = 17$, $\sigma = 9.84$, $S = 10.3709$; standard scores using σ: -1.22, -0.71, -1.02, -0.20, 1.32, -0.81, -0.51, 1.83, 0.81, 0.51　　**3-36** (sample) range　　**3-37** $\Sigma_{i=1}^6 x_i$　　**3-38** σ^2　　**3-39** 32　　**3-40** (c)　　**3-41** (a)
3-42 class boundaries　　**3-43** 0　　**3-44** $\sqrt{2}$　　**3-45** statistic　　**3-46** $\bar{x} = 28.51$, $\tilde{x} = 28.5$, $s^2 = 11.412$, range = 13.0
3-47 (a) mode = 6, $\tilde{x} = 7$, $\bar{x} = 7.01$ (b) Although the mode (6) differs from the median and mean (both 7 if rounded), it is still appropriate because its frequency (18) is very dominant. (c) 1.92　　**3-48** 1.89
3-49 (a) $y = x - 11.45$; $\bar{y} = -0.2685$, $s_y = 1.716$ (b) $\bar{x} = 11.18$, $s_x = 1.716$, coefficient of variation = 15.3% (c) 11.28 (d) 10th percentile = 8.7, 90th percentile = 13.4 (e) 8.5 becomes -1.56, 13.4 becomes 1.29

3-50 at least 0.81 (i.e., at least 81% of the population is closer than this value to the mean)
3-51 (a) $y = 2(x - 13.2)$; $\bar{y} = -0.24$, $s_y = 2.264$, $\bar{x} = 13.08$, $s_x = 1.132$ (b) 0.57 (c) first quartile $= 12.4$, third quartile $= 13.8$ **3-52** (a) 238.38 (b) 493.72 **3-53** (c) **3-54** (c) **3-55** positive
3-56 $\sum_{i=1}^{n}(x_i - \bar{x})^2$ or simply $\sum(x - \bar{x})^2$ **3-57** (b) **3-58** (b) **3-59** class marks **3-60** (b) **3-61** (b)
3-62 The classes overlap. **3-63** (a) **3-64** The second score may be judged more outstanding because it is converted to a standard score of 2.0, whereas the first is converted to 1.5.
3-65 Although a cumulative distribution polygon (i.e., ogive) could be drawn, it would not be meaningful because the categories are not ordered. Cumulative distributions must be based on ordered categories.
3-66 $-\$1.07$ (i.e., the mean change is a decrease of \$1.07) **3-67** (a) $\bar{x} = 69.1$, $\tilde{x} = 72$ (b) $s = 22.92$, range $= 98$

(c) and (d)

Class Lower	Class Upper	Frequency	Class marks	Lower boundary	Upper boundary
20	29	2	24.5	19.5	29.5
30	39	3	34.5	29.5	39.5
40	49	4	44.5	39.5	49.5
50	59	5	54.5	49.5	59.5
60	69	7	64.5	59.5	69.5
70	79	11	74.5	69.5	79.5
80	89	6	84.5	79.5	89.5
90	99	3	94.5	89.5	99.5
100	109	1	104.5	99.5	109.5
110	119	2	114.5	109.5	119.5
120	129	1	124.5	119.5	129.5

(e)
There are 0 less than 20
There are 2 less than 30
There are 5 less than 40
There are 9 less than 50
There are 14 less than 60
There are 21 less than 70
There are 32 less than 80
There are 38 less than 90
There are 41 less than 100
There are 42 less than 110
There are 44 less than 120
There are 45 less than 130

3-68 (a) and (c)

Distance Lower limit	Distance Upper limit	Frequency	Class mark	Lower boundary	Upper boundary	Class width
0.0	0.9	11	0.45	−0.05	0.95	1.0
1.0	1.9	10	1.45	0.95	1.95	1.0
2.0	2.9	11	2.45	1.95	2.95	1.0
3.0	3.9	11	3.45	2.95	3.95	1.0
4.0	5.9	26	4.95	3.95	5.95	2.0
6.0	7.9	29	6.95	5.95	7.95	2.0
8.0	9.9	12	8.95	7.95	9.95	2.0
10.0	14.9	23	12.45	9.95	14.95	5.0
15.0	19.9	4	17.45	14.95	19.95	5.0
20.0	49.9	9	34.95	19.95	49.95	30.0
50.0	—	4	—	49.95	—	—

(b) There are unequal size classes and there is an open class. (d) 6.36 (e) 10th $= 1.35$, 90th $= 17.45$ (f) The open class does not have a class mark. (g) 8.169 (h) 9.818 (i) 0.27

CHAPTER 4

4-1 (a) There are 25 outcomes—the 25 individual representatives. Two complementary events are "man," "woman." (b) 21/25 (c) 21 to 4   **4-2** 399 to 1   **4-3** 4.65%   **4-4** 5/6   **4-5** 0.04   **4-6** (a) 0.1% (b) 10
4-7 (a) 0.6 or 3/5 (b) 3 to 2   **4-8** 40%   **4-9** 0.8 or 4/5   **4-10** (a) 2/5 (b) 3/5 (c) 7/10 (d) 3/5
4-11 $P(A) + P(B) + P(C) + P(D) + P(E)$ **4-12** 13/60   **4-13** (a) 0.36 (b) 0.16 (c) 0.06 (d) 0.20
4-14 (a) 13/20 (b) 17/20 (c) 4/5   **4-16** The sum of the probabilities exceeds 1.   **4-17** (a) one (b) one (c) 1/3
4-18 17 to 3   **4-19** 1/288   **4-20** 0.546   **4-21** (a) 35/288 (b) 25/864
4-22 (a) 5/8 (b) 5/24 (c) 11/24 (d) 5/11   **4-23** (a) 0.4412 (b) 0.4752 (c) 0.0588
4-24 (a) 9/100 (b) (1) 3/50 (2) 27/200 (3) 9/100 (4) 3/100 (c) 63/200 (d) 81/200
4-25 (a) 0.217 (b) 0.15 (c) 0.20 (d) 0.65 (e) 0.692   **4-26** (a) 0.60 (b) 0.36 (c) 0.93 (d) 0.923   **4-27** 6/11
4-28 (a) 0.80 (b) 0.222 (c) 0.625   **4-29** 222/375 = 0.592
4-30 (b) By symmetry $P(x \leqslant 10) = 1/2$. (c) By symmetry they are equal. (d) rare   **4-31** (b) 2/3
4-32 (b) (1) 0.7 (2) 0.2 (3) 0.95   **4-33** (a) 0.45 (b) 0.40 (c) 0.15   **4-34** (b) (1) 0.225 (2) 0.15 (3) 0.625 (4) 0.7396
4-35 (a) $\mu_{\bar{x}} = 150.00$ $\sigma_{\bar{x}}^2 = 20.229$ (b) 625.63   **4-36** (a) $\mu_{\bar{x}} = 150.00$ $\sigma_{\bar{x}}^2 = 5.63$ (b) 625.63   **4-37** (a) 12.6 (b) 315
4-38 (b) $\mu = 2.75$, $\sigma^2 = 1.4375$, $S^2 = 1.6429$ (c) $\mu_{\bar{x}} = 2.75$, $\sigma_{\bar{x}}^2 = 0.61607$, $\mu_{s^2} = 1.6429$ (d) $\mu_{\bar{x}} = 2.75$, $\sigma_{\bar{x}}^2 = 0.34226$,
$\mu_{s^2} = 1.6429$ **4-40** (a) (1) 0.025 (2) 0.010 (3) 0.005 (b) 25   **4-41** $P(A) + P(B) - P(A \text{ and } B)$ **4-42** (c)
4-43 (a)   **4-44** (c)   **4-45** $P(A) + P(B) + P(C)$ **4-46** mutually exclusive and exhaustive   **4-47** area
4-48 the probability of A given B **4-49** $P(A)P(B|A)$ or $P(B)P(A|B)$ **4-50** (d)   **4-51** (a)
4-52 $P(A) \times P(B) \times P(C)$ **4-53** (a) 4/13 (b) 1/2 (c) 2/13 (d) 17/26 (e) 9/26
4-54 (a) Nothing is wrong. (b) Since $P(A \text{ or } B) = P(A) + P(B) - P(AB)$, $P(A \text{ or } B)$ cannot exceed $P(A) + P(B)$.
(c) If the events are mutually exclusive and exhaustive $P(A) + P(B) + P(C) = 1$, but the given values lead to
$P(A) = 0.30$ $P(B) = 0.40$ and $P(C) = 0.45$, which add to 1.15. (d) A probability cannot exceed 1. (e) Nothing is
wrong if A and B are not mutually exclusive.   **4-55** 1 to 1   **4-56** (a) (1) 8/25 (2) 4/25 (b) (1) 13/33 (2) 2/11
4-57 (a) 0.40 (b) 0.8032 (c) 0.1792 (d) 0.056   **4-58** 13 to 7   **4-59** (a) (1) 0.3654 (2) 0.0037 (3) 0.3144
(b) 0.3779 (c) 0.9623   **4-60** (a) 0.17 (b) 0.36 (c) 0.14 (d) 0.50   **4-61** (b) 4/25 (c) 16/25
4-62 (a) 0.133 (b) 0.526   **4-63** (a) $\mu = 4$, $\sigma^2 = 1.1429$, $S^2 = 1.3$ (b) (2) $\mu_{\bar{x}} = 4$, $\sigma_{\bar{x}}^2 = 0.4762$ (3)
2/7 (c) (2) $\mu_{\bar{x}} = 4$, $\sigma_{\bar{x}}^2 = 0.2540$ (3) 3/35   **4-64** the population mean   **4-65** $P(-0.7 \leqslant x \leqslant 0.3)$ **4-66** (c)
4-67 1   **4-68** the standard deviation   **4-69** (d)   **4-70** (c)   **4-71** 1   **4-72** (c)   **4-73** lower   **4-74** (c)
4-75 (a)   **4-76** *at least* 0.556   **4-77** (b) 4.25 (c) (1) 13/24 (2) 1/6 (3) 1/2 (4) 1/2   **4-78** $28,987
4-79 (a) 34/117 (b) 26/117   **4-80** (a) $\bar{x} = 319.1$, $\tilde{x} = 322.5$ (b) range = 163, $s^2 = 1769.40$

(c) and (d)

Lower limit	Upper limit	Frequency	Class marks	Lower boundary	Upper boundary
240	259	5	249.5	239.5	259.5
260	279	3	269.5	259.5	279.5
280	299	2	289.5	279.5	299.5
300	319	9	309.5	299.5	319.5
320	339	8	329.5	319.5	339.5
340	359	6	349.5	339.5	359.5
360	379	5	369.5	359.5	379.5
380	399	1	389.5	379.5	399.5
400	419	1	409.5	399.5	419.5

(f)   40 values are 240 or more
35 values are 260 or more
32 values are 280 or more
30 values are 300 or more
21 values are 320 or more
13 values are 340 or more
7 values are 360 or more
2 values are 380 or more
1 value is 400 or more

4-81 (a) No (b) odds against B : 3 to 1 odds against C : 3 to 1 (c) odds against A : 5 to 4 odds against B : 2 to 1 odds
against D : 7 to 2   **4-82** (a) (1) 0.56 (2) 0.56 (b) (1) 25 (2) 0.89 (c) (1) 6.25 (2) 0.97
4-84 (a) The true corporation mean (b) 1.986 (c) 0.876

CHAPTER 5

5-1 (a) 20 (b) 2 (c) 6 (d) 6   **5-2** There are 8 schedules.   **5-3** 288   **5-4** 67,600   **5-5** 720   **5-6** 3,307,800
5-7 551,300   **5-8** 2730   **5-9** 1365   **5-11** 1,000,000   **5-12** 600   **5-13** 15   **5-14** 24   **5-15** 30
5-16 4,464,061,875   **5-17** (a) 95,040 (b) 6720   **5-18** (a) 0.00098 (b) 0.0879 (c) 0.2373 (d) 0.1035
5-19 (a) 0.00002 (b) 0.0173 (c) 0.2601 (d) 0.2131   **5-20** 0.2435   **5-21** 0.3394
5-22 (a) 0.3449 (b) 0.3217 (c) 0.2936   **5-23** (a) 2 (b) 0.271 (c) 0.677   **5-24** (a) 0.251 (b) 0.809
5-26 (a) 4/55 (b) 8/55 (c) 6/55 (d) 18/55 (e) 36/55   **5-27** (a) 0.2751 (b) 0.0055 (c) 0.0438
5-28 (a) 0.349 (b) 0.387 (c) 0.930   **5-29** (a) 4×10^{-10} (b) 1.996×10^{-6} (c) 0.0164 (d) 0.2707   **5-30** 0.633
5-31 (a) 1.88×10^{-6} (b) 0.0077 (c) 0.2384   **5-32** (a) 6.45×10^{-5} (b) 0.1937 (c) 0.2935 (d) 0.4402
5-33 (a) $\binom{x-1}{n-1} p^{n-1}(1-p)^{x-n}$ (b) p (c) $\binom{x-1}{n-1} p^{n}(1-p)^{x-n}$
5-34 (a) 0.0081 (b) 0.0397 (c) 0.1941 (d) 0.7297   **5-35** 2.1   **5-36** 7   **5-37** 16%   **5-38** yes
5-39 $\mu = 7, \sigma^2 = 6$ **5-40** 1.25   **5-41** $\mu = 3.3, \sigma^2 = 1.49$ **5-42** 3   **5-43** $\mu = 105, \sigma = 5.47$
5-44 $\mu = 105, \sigma = 5.61$
5-45 (a) (1) $\mu = 0.01, \sigma^2 = 0.00999$ (2) $\mu = 0.1, \sigma^2 = 0.0999$ (3) $\mu = 10, \sigma^2 = 9.99$ (b) $\mu = 10, \sigma^2 = 10$
5-46 $\mu = 6, \sigma^2 = 3.26$ **5-47** $\mu = 15, \sigma^2 = 15$ **5-48** $\mu = 8, \sigma^2 = 7.21$ **5-49** $\mu = 1000, \sigma^2 = 900$
5-50 \$4320   **5-51** (a) 0.3980 (b) 0.4177 (c) 0.9922 (d) 0.7472 (e) 0.0013   **5-52** (a) 0.9732 (b) 0.5584 (c) 0.0721
5-53 (a) 0.0116 (b) 0.7121 (c) 0.0192   **5-54** (a) 0.0244, 2.44%; or with continuity correction 0.0217, 2.17%
(b) 0.0091, 0.91%; or with continuity correction 0.0082, 8.2%   **5-55** (a) 0.3830 (b) 0.9566 (c) 0.9954 (d) 1.000
5-56 0.0040   **5-57** 0.0049   **5-58** 0.0059   **5-59** (a) 0.0212 (b) 0.5429 (c) 0.0039
5-60 (a) 0.5820 (b) 0.7324 (c) 0.1685   **5-61** (a) 0.1314 (b) 0.0029 (c) 0.9676
5-62 (a) 0.9500 (b) 0.2006 (c) 0.0495   **5-63** (a) (1) 0.8664 (2) 0.9544 (3) 0.9974 (b) (1) 0.5556 (2) 0.7500 (3) 0.8889
5-64 (a) 0.9266 (b) 0.9328 (c) 0.9616   **5-65** (a) 0.9372 (b) 0.9412 (c) 0.9676   **5-66** (a) 0.9544 (b) 0.9987 (c) 0.0062
5-67 (a) 0.0456 (b) 0.9370 (c) 0.0548 (d) 0.2119   **5-68** (a) 0.0570 (b) 0.0352   **5-69** (a) 0.0228 (b) 0.4972 (c) 0.1836
5-70 (a) (1) 0.0427 (2) 0.0427 (3) 0.9805 (b) (1) 0.0158 (2) 0.0158 (3) 0.9951
5-71 (a) (1) 0.0475 (2) 0.0475 (3) 0.9768 (b) (1) 0.0207 (2) 0.0207 (3) 0.9929   **5-72** 0.9938   **5-73** 0.9544
5-74 0.9382   **5-75** 0.9977   **5-76** expectation   **5-77** (a)   **5-78** (a)   **5-79** (b)   **5-80** 25.5   **5-81** (b)
5-82 0.7540   **5-83** np **5-84** (c)   **5-85** 0.5000   **5-86** 60   **5-87** (a) 10 (b) 1, 2, 3, 4
5-88 (a) 3/1081 (b) 1/666 (c) 42/1081, 9/222   **5-89** (a) 0.00008 (b) 0.314
5-90 $f(x) \geqslant 0, \Sigma f(x) = 1, \mu = 0, \sigma^2 = 16/15$ **5-91** 0.014   **5-92** (a) 0.10 (b) (1) 0.328 (2) 0.082
5-93 \$0.80   **5-94** \$0.28   **5-95** (a) (1) 0.2088 (2) 65.5 (3) 0.9544 (b) (2) 65.5 (3) 0.9652
5-96 (a) 0.2061 (b) 0.7642 (c) 0.3900   **5-97** (a) 0.930 (b) 0.0025 (c) 0.0401
5-98 (a) 1,947,792 (b) 10,626 (c) 6 (d) 0.2745   **5-99** 0.9361   **5-100** (b)   **5-101** 0 and 1   **5-102** 0   **5-103** (d)
5-104 population or a probability function   **5-105** (a)   **5-106** (c)   **5-107** $\Sigma (x - 100)^2$ **5-108** 1
5-109 $P(1.5 \leqslant x \leqslant 2.7)$ **5-110** 7/27   **5-111** (a) 9999 to 1 (b) 0.677 (or 0.675 using Table II)
5-112 (a) 0.776 (b) 0.0122   **5-113** (a) $f(x) \geqslant 0, \Sigma f(x) = 1$ (b) 34/3
5-114 (a) There are 13 schedules. (b) mean = 3.3, median = 3, mode = 3 (d) 5/13   **5-115** no
5-116 (b) mean = 1.4 mode = 1, variance = 1.33 (c) 0.167   **5-117** (a) 0.03 (b) 0.1255 (c) 0.2271 (d) 0.9983
5-118 0.9505   **5-120** The evidence suggests that the dice may not be balanced. Bet on a number less than 7.

CHAPTER 6

6-1 97   **6-2** 136   **6-3** (a) 423 (b)355   **6-4** 640   **6-5** 16   **6-6** (a) the first (b) 1009   **6-7** 305   **6-8** 1068
6-10 15.18 to 16.48   **6-11** 49.2 to 55.0   **6-12** 0.9859 to 1.0153   **6-13** 0.9787 to 1.0225   **6-14** 81.1 to 110.1
6-15 597.2 to 608.2   **6-16** 35   **6-17** It is between 100.55 and 126.75.   **6-18** Multiply by 4.
6-19 335.2 to 1064.04   **6-20** 0.0158 to 0.0508   **6-21** 6.7 to 9.4   **6-22** 0.340 to 1.046
6-23 6.71 to 10.80   **6-25** 0.450 to 0.705   **6-26** 0.452 to 0.698   **6-27** 0.658 to 0.734   **6-28** 0.295 to 0.705
6-29 \$285 to \$395   **6-30** 19.5% to 33.8%   **6-31** 19.4% to 23.6%   **6-32** 308 to 492
6-34 95% confidence interval for a percentage   **6-35** $1 - \alpha/2$ **6-36** (a)   **6-37** (a)   **6-38** (c)   **6-39** 0.90
6-40 (c)   **6-41** 97   **6-42** (a) 2.99622 to 2.99653 (b) It is assumed that the values are independent and from a
normal distribution with mean equal to the true velocity. (c) Natural variation and errors in measurements produce
different values; hence, several measurements are required.   **6-43** (a) 32.68 to 34.45
(b) (1) 32.56 to 34.57 (2) 1.40 to 2.65   **6-44** (a) 752 (b) 564   **6-45** (a) 688 (b) 527   **6-46** 1.24 to 1.84
6-47 0.13 to 0.32   **6-48** 32.8 to 70.6   **6-49** 601   **6-50** 1960 to 2040   **6-51** (b)
6-52 (c)   **6-53** (c)   **6-54** (a)   **6-55** (a)   **6-56** (c)   **6-57** (c)   **6-58** 0 and 1
6-59 There are no classes for 6–9, 16–19, etc.   **6-60** (c)   **6-61** (a) 0.086 (b) 0.7910 (c) 0.069
6-62 (b) 1/4   **6-63** (a) $\bar{x} = 276.5, \tilde{x} = 282$ (b) range = 126, $s = 29.99$

(c)

Class Lower limit	Upper limit	Frequency	Class mark	Lower boundary	Upper boundary
200	209	1	204.5	199.5	209.5
210	219	2	214.5	209.5	219.5
220	229	1	224.5	219.5	229.5
230	239	2	234.5	229.5	239.5
240	249	4	244.5	239.5	249.5
250	259	3	254.5	249.5	259.5
260	269	5	264.5	259.5	269.5
270	279	6	274.5	269.5	279.5
280	289	7	284.5	279.5	289.5
290	299	6	294.5	289.5	299.5
300	309	9	304.5	299.5	309.5
310	319	3	314.5	309.5	319.5
320	329	2	324.5	319.5	329.5

(e)

Cost	Frequency
200 or more	51
210 or more	50
220 or more	48
230 or more	47
240 or more	45
250 or more	41
260 or more	38
270 or more	33
280 or more	27
290 or more	20
300 or more	14
310 or more	5
320 or more	2
330 or more	0

(f) by Chebyshev's theorem, at least 29; actually 46   **6-64** $\mu = 8.70$, $\sigma^2 = 1.787$ **6-65** 0.5256
6-66 (a) 621 (b) 600   **6-67** 1.89   **6-68** 0.960
6-69 (a) 0.025 (b) 0.95 (c) 0.1587 (d) 0.0124 (e) 0.6600 (f) 0.050 (g) 0.010 (h) 0.050 (i) 0.05 (j) 0.99
6-70 0.23 to 0.35   **6-71** (a) $\mu = 2$, $\sigma^2 = 1.143$ (c) $\mu_{\bar{x}} = 2$, $\sigma_{\bar{x}}^2 = 0.476$ (d) 1/7
6-72 (a) Estimate σ^2 with $s^2 = 0.02669$. (b) 0.113 to 0.283 (c) 12.26 to 12.44 (d) A normal distribution is assumed.
6-73 20   **6-74** 0.677   **6-76** 89

CHAPTER 7

7-1 (a) $H_O: \mu = 100$ $H_A: \mu \neq 100$ (b) $H_O: \mu = 875$ $H_A: \mu > 875$ (c) $H_O: \mu = 180$ $H_A: \mu < 180$
7-2 (a) 0.9372 (b) 0.0628   **7-3** (a) $H_O: \mu = 20.00$, $H_A: \mu \neq 20.00$; reject H_O if $\bar{x} < 19.938$ or $\bar{x} > 20.062$; accept H_O if $19.938 \leqslant \bar{x} \leqslant 20.062$). (b) 0.2843 (c) 0.0329, 0.1151, 0.2843, 0.5239, 0.7580, 0.9034, 0.7580, 0.5239, 0.1151, 0.0329 (e) (1) Accept H_O. (2) Reject H_O.   **7-4** (a) $n \geqslant 16.42$, use $n = 17$. (b) Accept H_O if $19.952 \leqslant \bar{x} \leqslant 20.048$. [7-3 (b)]: 0.0934 (note: < 0.10 since $n > 16.42$); [7-3 (c)]: 0.0015, 0.0162, 0.0934, 0.3121, 0.6293, 0.8723, 0.6293, 0.3121, 0.0162, 0.0015   **7-5** (a) $H_O: \mu = 3.50$, $H_A: \mu > 3.50$ (b) $n \geqslant 35.5$; use $n = 36$. (c) Reject H_O if $\bar{x} > 3.541$. (d) yes   **7-6** (a) 0.0142 (b) (1) 0.1894 (2) 0.1894 (3) 0.0054   **7-7** $H_O: \mu = 80.0$, $H_A: \mu \neq 80.0$; reject H_O if $|z| = |(\bar{x} - 80.0)/(10.0/\sqrt{20})| > 2.576$; for $\bar{x} = 73.1$, $|z| = 3.086$; reject H_O.
7-8 Now reject H_O if $|t| = |(\bar{x} - 80.0)/(s/\sqrt{20})| > 2.861$; for $\bar{x} = 73.1$ and $s = 11.2$, $|t| = 2.755$, accept H_O.
7-9 $H_O: \mu = 75$, $H_A: \mu < 75$, $z = -2.471 < -1.645$; reject H_O. The data do provide proof that the mean percent of methane is less than 75%.   **7-10** $H_O: \mu = 14.9$, $H_A: \mu < 14.9$, $t = -1.248$ is not < -1.729; accept H_O. The data do not prove a decrease. Normality was assumed.

7-11 $|t| = 0.2106 \not> 2.624$; accept H_O. There is no difference.   **7-12** Yes; $t = -2.905$.

7-13 (a) (1) Yes; $t = 1.875 > 1.645 = z_{0.05}$. (2), No; $t = 1.875 < 2.326 = z_{0.01}$.
(b) If noncovered farms show no change, the change in production may be attributed to the program rather than other factors, which should have also affected noncovered farms.

7-14 $H_O: \sigma = 2.5$, $H_A: \sigma > 2.5$; reject H_O if $\chi^2 = 14s^2/2.5^2 > 23.685$. For $s = 3.1$, $\chi^2 = 21.526$; accept H_O. The data do not provide proof that $\sigma > 2.5$. Normality was assumed.

7-15 $\chi^2 = 6.7392 \not< 4.107$. Accept $H_O: \sigma = 0.75$. At the 1% level σ is not proven to be less than 0.75.

7-16 $\chi^2 = 21.2544 < 22.88 = \chi^2_{0.01,\,41}$; Reject H_O. At the 1% level it is proven that $\sigma < 0.75$.   **7-17** yes

7-18 yes   **7-19** (a) $c_1 = 4$, $c_2 = 10$, $\alpha = 0.058$ (b) $c = 10$, $\alpha = 0.019$ (c) $c = 1$, $\alpha = 0.035$ **7-20** no

7-21 yes   **7-22** The claim is not proven because $P(x \geq 10 | n = 15, p = 0.5) = 0.151 \not< 0.05 = \alpha$.

7-23 The claim is proven because $z = 4.000 > 1.645$.   **7-24** There is insufficient evidence because $P(x \geq 12 | n = 18, p = 0.5) = 0.119 \not< 0.025 = \alpha/2$.   **7-25** There is sufficient evidence because $|z| = 2.313 > 1.96$.

7-26 (a) (1) 0.0500 (2) 0.0863 (3) 0.1521   **7-28** Accept H_O.   **7-29** 0.132 to 0.284; reject H_O.

7-30 0.67 to 0.93; accept H_O.   **7-31** (a) 8823 to ∞ (b) Reject H_O.   **7-32** (a) 0 to 0.434 (b) Reject H_O.

7-33 0.002   **7-34** $p < 0.02$ **7-35** (a) 0 to 7.94 (b) $p < 0.01$ **7-36** $p < 0.025$

7-37 $-\infty$ to $\bar{x} + z_\alpha \sigma/\sqrt{n}$; $\bar{x} - z_\alpha \sigma/\sqrt{n}$ to ∞ **7-38** $-\infty$ to $\bar{x} + t_{\alpha,\,n-1}s/\sqrt{n}$; $\bar{x} - t_{\alpha,\,n-1}s/\sqrt{n}$ to ∞

7-39 0 to $(n-1)s^2/\chi^2_{1-\alpha,\,n-1}$; $(n-1)s^2/\chi^2_{\alpha,\,n-1}$ to ∞ **7-40** $\sum_{x=x_0}^{n} \binom{n}{x} p_0^x (1-p_0)^{n-x}$ **7-41** (c)

7-42 (c)   **7-43** $t < -t_{\alpha,\,n-1}$ **7-44** (c)   **7-45** (a)   **7-46** (a)   **7-47** 0.05   **7-48** (b)   **7-49** (b)

7-50 (a) $H_O: \mu = 1.800$, $H_A: \mu \neq 1.800$; reject H_O if $|z| = |(\bar{x} - 1.800)/(0.050/\sqrt{25})| > 1.96$.
(b) Accept H_O if $1.7804 \leq \bar{x} \leq 1.8196$; 0.0012, 0.0207, 0.1492, 0.4840, 0.8300, 0.8300, 0.4840, 0.1492, 0.0207, 0.0012.

7-51 (a) Reject H_O if $\bar{x} < 1.7887$ or if $\bar{x} > 1.8113$. (b) 0, 0, 0, 0.0655, 0.5910, 0.5910, 0.0655, 0, 0, 0

7-52 the data do not provide sufficient evidence.   **7-53** 0 to 0.186   **7-54** 15.953 to 16.093; accept H_O.

7-55 The data do provide sufficient evidence.   **7-56** 0.988 to 1.100; accept the hypothesis.

7-57 $[(z_{\alpha/2} + z_\beta)/k]^2$ **7-58** 13.23 to ∞ **7-59** (a) Reject H_O. (b) $p < 0.025$

7-60 (a) $z = (x \pm 1/2 - np_0)/\sqrt{np_0(1-p_0)(N-n)/(N-1)}$ (b) yes   **7-61** (b)   **7-62** (b)   **7-63** 0.01

7-64 $P(18.5 \leq x \leq 31.5)$ **7-65** An error has been made.   **7-66** (b)   **7-67** (b)   **7-68** (a)

7-69 probability   **7-70** 1 to 5; 1/6   **7-71** (a) 0.48 to 0.71 (b) Accept the hypothesis.

7-72 (a) $\bar{x} = 6.95$, $\tilde{x} = 6$ (b) range $= 23$, $s = 6.514$

(c)

Class Lower limit	Upper limit	Frequency	Class mark	Lower boundary	Upper boundary
0	2	21	1	−0.5	2.5
3	5	7	4	2.5	5.5
6	8	14	7	5.5	8.5
9	11	3	10	8.5	11.5
12	14	2	13	11.5	14.5
15	17	5	16	14.5	17.5
18	20	3	19	17.5	20.5
21	23	3	22	20.5	23.5

(e)   There are 37 values more than 2.
There are 30 values more than 5.
There are 16 values more than 8.
There are 13 values more than 11.
There are 11 values more than 14.
There are 6 values more than 17.
There are 3 values more than 20.
There are 0 values more than 23.

7-73 (a) (1) 0.0516 (2) 0.1401 (b) $z = -2.17 < -1.645$; we do have proof that $\mu < 1000$.

7-74 (a) $H_O: \sigma^2 = 0.0225$, $H_A: \sigma^2 > 0.0225$; $\chi^2 = 30.30 > 26.296 = \chi^2_{0.05,\,16}$. There is evidence at the 5% level that the variance does not meet requirements. (b) 0.0259 to ∞ **7-75** (a) 0.0036 (b) 0.3973

7-76 (a) $H_O: \mu = 300$, $H_A: \mu > 300$; reject H_O if $z = (\bar{x} - 300)/(50/\sqrt{25}) > 1.645$. (b) Type I, 0.05
(c) Type II, 0.1949 (d) 35   **7-77** (a) $p < 0.01$ (b) not significant (c) $p < 0.005$ (d) $p < 0.05$ (e) not significant

7-78 (a) 0.0495 (b) 0.3030 (c) 0.3645 (d) 0.9517   **7-79** 347   **7-80** (a) (1) 20 outcomes—the individual plants (2) ownerships—domestic, foreign, shared or market served—domestic, foreign, both or ownership and market served—domestic, domestic—etc. shared, both (b) (1) 2/5 (2) 2/5 (3) 60% (4) 3 to 1   **7-81** 19

7-82 0.265 (or 0.264 using Table II)

CHAPTER 8

8-1 (a) $|z| = 2.266 > 1.96$; the average assessments differ. (b) -28.3 to -2.1 **8-2** $z = 1.716 > 1.645$; the data do provide proof of the claim. **8-3** It is at least 102 g. **8-4** (a) $m = 25$ d.f., 2.50 to 7.16 (b) Ice thickness differs for the regions on the average. **8-5** (a) 2.47 to 7.19 (b) Again there is a difference. **8-6** $p < 0.005$ **8-7** No; $z = 1.086$. **8-8** Assuming equal variances, the difference "no injection" minus "noradrenalin" is between 17.3 and 41.1 ml/min. **8-9** (a) $t = 3.073$; the claim is substantiated. (b) The difference "current" minus "new" is between 5.13 and 8.07. **8-10** $F = 1.37$; the claim is not proven. **8-11** (a) 0.80 to 2.58 (b) Accept H_O. (c) Accept H_O. **8-12** $F = 1.54$; there is insufficient evidence that the new process is more consistent. **8-13** $F = 2.30$; the results do provide the proof. **8-14** $F = 1.48$; equal variances may be assumed. **8-15** 3.0 **8-16** (a) **8-17** (b) **8-18** (a) **8-19** $n_1 + n_2 - 2$ **8-20** $d =$ "different—one color"; $H_O: \mu_d = 0$, $H_A: \mu_d > 0$; $t = 1.705$; accept H_O. **8-21** (a) If we assume that $\sigma_1 = \sigma_2$, the interval is -15.42 to -0.82. If σ_1 may differ from σ_2, we find $m = 27$ degrees of freedom and the interval is -15.94 to -0.30. (b) Reject $H_O: \mu_1 = \mu_2$. **8-22** $s_A^2 = 103.75 < 152.7 = s_B^2$; we thus cannot prove that $\sigma_A^2 > \sigma_B^2$. Use the cheaper method A. **8-23** $F = 3.18$; demerit points are more varied for drinking drivers. Normality is assumed. **8-24** $m = 13$ d.f.; the mean difference "drinking" minus "no drinking" is between 3.0 and 9.2. **8-25** $t = 1.863$; they do not differ significantly. **8-26** The ratio first to second is between 0.88 and 3.44. **8-27** 0.1963 (interpolated) **8-28** $\bar{d} - t_{\alpha, n-1} s_d / \sqrt{n}$ to ∞; $-\infty$ to $\bar{d} + t_{\alpha, n-1} s_d / \sqrt{n}$ **8-29** $(s_1/s_2) / \sqrt{F_{\alpha; n_1 - 1, n_2 - 1}}$ to ∞ **8-30** 0 **8-31** $d \pm t_{\alpha/2, f} s$ **8-32** (d) **8-33** 0.05 **8-34** 4/45 **8-35** (b) **8-36** (b) **8-37** 10 **8-38** (b) **8-39** $2.50 **8-40** 11 to 4 **8-41** 1629.36 **8-42** (a) $F = 1.49$; variances do not differ. (b) $|t| = 2.759$, $p < 0.02$ **8-43** (a) 188 (b) 158 **8-44** (a) 180 (b) 152 **8-45** (a) (1) 1/8 (2) 1/2 (3) 5/8 (b) (1) 11 to 1 (2) 1 to 3 **8-46** (a) 6.66

(b)

Length of stay	Frequency
1 day or more	367
$2\frac{1}{2}$ days or more	328
$4\frac{1}{2}$ days or more	267
$6\frac{1}{2}$ days or more	195
$8\frac{1}{2}$ days or more	139
$10\frac{1}{2}$ days or more	89
$15\frac{1}{2}$ days or more	26
$20\frac{1}{2}$ days or more	8
$30\frac{1}{2}$ days or more	3

8-47 (a) 3,268,760 (b) 1/3,268,760 (c) 0.3215 **8-48** $z = -1.429$; accept H_O. **8-49** (a) (1) 1/190 (2) 36/190 (3) 153/190 (b) (1) 361/625 (2) 361/500 (3) 361/400 (c) (1) 16/4561 = 0.0035 (2) 3825/4561 = 0.8386 **8-50** There is sufficient evidence; $t = 2.745$. **8-51** (a) 2.0 to 34.6 (b) no (c) The samples were independent samples from normal populations with equal variances. **8-53** (a) $\bar{x} = 50.56$, $s = 23.798$ (b) $\bar{x} = 71.5056$, $s = 0.23798$ (c) 71.407 to 71.604 **8-54** (a) $5.50 (b) $5.00 (c) (1) $500 (2) $5 (3) $50 **8-55** (a) 0.7698 (b) 0.7698 **8-56** 16 **8-57** 1.483 to 1.545 **8-58** (a) Yes; $\chi^2 = 9.028$. (b) 0.0373 to 0.0606 **8-59** (a) 3.5 (b) (1) 0.030 (2) 0.216 (0.214 using Table II) (3) 0.537 (using Table II, 0.533) **8-60** (a) 0.090 (b) (1) 0.702 (2) 0.352 (3) 0.055 **8-61** (a) Reject H_O if $x \geqslant 101$. (b) 0.6950, 0.2119, 0.0119, 0.0000

CHAPTER 9

9-1 $|z| = 3.041$; there is a significant difference. **9-2** 0.02 to 0.18; again, reject H_O. **9-3** Reject H_O. **9-4** (a) 0.07 to 0.25 (b) Accept H_O. **9-5** $z = 1.065$; the claim is not proven. **9-6** $|z| = 2.801$; the drug has an effect. **9-7** The proportion improves by at least 0.141. **9-8** $\chi^2 = 6.236$; there is insufficient evidence to prove that reaction differed. **9-9** $\chi^2 = 17.305$; reject H_O, contingency coefficient = 0.167. **9-10** (a) $\chi^2 = 9.250$; as before, there is a difference. (b) $\chi^2 = 12.843$; as before, reject H_O. **9-11** $\chi^2 = 2.250$; accept H_O. **9-12** $\chi^2 = 4.322$; there is a difference with $p < 0.05$. **9-13** $\chi^2 = 4.947$; there is insufficient evidence.

9-14 (a) (1) μ (2) σ; use $s = 9.94$ (b) Accept H_O that the distribution is normal with mean 100.
9-15 (a) 5.5 (b) Make the last class "12 or more" with frequency 4. $\chi^2 = 22.26$; reject H_O with $p < 0.05$.
9-16 $\chi^2 = 17.4$; the data do not provide proof of a bias.   **9-17** (c)   **9-18** 0.72   **9-19** (b)   **9-20** 12
9-21 they are independent.   **9-22** 4   **9-23** $z = 1.44$; the Toronto proportion is not necessarily greater.
9-24 Expect 17 of each. $\chi^2 = 12.0$; there is evidence that the die is not balanced.
9-25 $\chi^2 = 15.182$; we do not have sufficient evidence that stress affects success.
9-26 There is a difference with $p < 0.05$.   **9-27** $\chi^2 = 3.05$; no significant difference.
9-28 $\chi^2 = 55.48$; there is association.   **9-29** -0.023 to 0.173   **9-30** Accept H_O: normal.   **9-31** (b)
9-32 (b)   **9-33** (b)   **9-34** accepted   **9-35** 1.5   **9-36** (c)   **9-37** (d)   **9-38** three times the old
9-39 (c)   **9-40** (b)   **9-41** (a) 33.1   **9-42** $H_O : \sigma = 2.0$, $H_A : \sigma > 2.0$; $\chi^2 = 17.7375$; accept H_O.
9-43 The claim is substantiated if, in the sample, fewer than 49 subscribers favor the change.
9-44 (a) $\bar{x} = 1.13$ $\tilde{x} = 2$ (b) $s = 6.97$ (c) (1) 0.52 (2) 0.80 (d) 61 to 39   **9-45** Accept H_O.   **9-46** At least 0.03
9-47 (a) Accept H_O if $119.938 \leqslant \bar{x} \leqslant 120.062$. (b) 0.2843 (c) 0.0336, 0.1151, 0.2843, 0.5239, 0.7574, 0.9034, 0.9034,
0.7574, 0.5239, 0.1151, 0.0336   **9-48** (a) 24% (b) (1) 0.35 (2) 0.12 (3) 0.52 (4) 0.48
9-49 $|z| = 1.16$ or $\chi^2 = 1.34$; there is not a significant difference.
9-50 $\chi^2 = 17.73$; there is a significant difference.   **9-51** (a) 0.878 (b) (1) $\mu = 20$, $\sigma^2 = 16$ (2) 0.9957
9-52 $t = 2.418$; reject $H_O : \mu_d = 2$ for $H_A : \mu_d > 2$. The test should be withdrawn. Normality was assumed.
9-53 $2800   **9-54** $z = 1.060$; there is no significant change.
9-55 There is a difference; $t = 6.01$, $p < 0.01$ (two-sided).   **9-57** 271   **9-58** 215
9-59 0.8088 (0.8084 using Table II)   **9-60** 0.77 to 0.83

CHAPTER 10

10-1 $ns_{\bar{x}}^2/\bar{s}^2 = 8.42$; there are differences across regions.

10-2(a)

	I	II	III	IV
$\bar{x}$	225.5	234.5	221.5	219.0
s^2	28.3	35.9	32.3	24.4

(b) $ns_{\bar{x}}^2/\bar{s}^2 = 9.18$; there are differences among preparations.   **10-3** (a) 0.05 (b) 0.01

10-4

		I	II	III	IV
(a)	$\bar{x}$	62.5	64.25	66.625	68.625
(b)	s^2	3.143	2.786	2.839	2.554

(c) $ns_{\bar{x}}^2/\bar{s}^2 = 20.34$ Average numbers of colonies differ for the different solutions.

10-5 $ns_{\bar{x}}^2/\bar{s}^2 = 1.27$; mean age does not differ.   **10-6** independent random samples from normal populations with equal variances   **10-7** $SS_{Tr} = 76.0$, $MS_{Tr} = 25.333$, $SS_E = 226.857$, $MS_E = 9.452$, $MS_{Tr}/MS_E = 2.68$, as before

10-8

Source	d.f.	SS	MS	F
Preparation	3	832.125	277.375	9.18
Error	20	604.5	30.225	
Total	23	1436.625		

10-9

Source	d.f.	SS	MS	F
Group	3	172.75	57.583	20.35
Error	28	79.25	2.830	
Total	31	252.0		

10-10

Source	d.f.	SS	MS	F
Treatment	3	1199.25	399.75	5.75
Error	24	1668.857	69.536	
Total	27	2868.107		

The processes produce different means.

10-11 (a) no effect (b) multiplied by 100^2 (c) no effect

10-12

Source	d.f.	SS	MS	F
Condition	3	831.50	277.17	11.3
Error	20	490.33	24.52	
Total	23	1321.83		

The conditions are not all equal.

10-13

Source	d.f.	SS	MS	F
Procedures	3	133433.250	44477.75	6.57
Error	24	162367.714	6765.32	
Total	27	295800.964		

The procedures are not all equal.

10-14 $F = 2.16$; there are no differences.   **10-15** $F = 14.63$; the programs differ.
10-16 $F = 39.2$; there are differences.   **10-17** $F = 14.12$; there are differences
10-19 Programs I, IV, and V are equivalent and differ from programs II and III, which are equivalent (using Scheffé).   **10-20** (a) 138.714, 173, 184.6, 140.143 (b) "Very light" and "heavy" are equivalent and differ from "moderate" and "light," which are equivalent (using Scheffé).
10-21 Means 2.37 3.41 4.31 the first region differs from the third.
10-22 All differences are significant using Duncan's multiple range.
10-23 The first group differs from the fourth and fifth.

10-24

Source	d.f.	SS	MS	F
Temperature	3	59	19.6	23.6
Duration	1	12.5		
Error	3	2.5	0.83	
Total	7	74		

Significance at 5% but not 1% level.

10-25

Source	d.f.	SS	MS	F
Processes	3	2.149	0.716	2.42
Batch	2	9.215	4.608	15.56
Error	6	1.778	0.296	
Total	11	13.142		

(a) Processes do not differ.
(b) Batches differ.
(c) replication

10-26 (a) $F = 8.22$; brightness has an effect.
    (b) $F = 7.05$; stopping procedures differ.
    (c) $F = 0.66$; no interaction.
10-27 Age and sex have significant effects. There is no interaction.
10-28 No effects are significant at the 1% level. At the 5% level, the drug has an effect and there is interaction.
10-29 means   **10-30** variances   **10-31** (b)   **10-32** 0.05   **10-33** (b)   **10-34** (d)   **10-35** (c)   **10-36** (b)
10-37 means are not all equal.   **10-38** (b)   **10-39** The first differs from the second. The fourth differs from the second and third.

10-40(a)

Source	d.f.	SS	MS	F
Fertilizer	2	284.386	142.193	23.6
Error	18	108.441	6.024	
Total	20	392.827		

(b) The first fertilizer differs from the second and third.

10-41 $F = 1.32$; there are no differences among the insulating materials (two-way analysis).
10-42 $F = 19.514$; there are differences among the treatments (replicated two-way analysis).
10-43 (a) Group means (adjusted) $-9.6, 4.8, -0.6, 1.2, 16.4$; overall 2.44. Group means (original) 3460, 3820, 3685, 3730, 4110; overall 3761 (b) Group variances (adjusted) 70.8, 156.7, 68.8, 150.7, 66.3; $s_{\bar{x}}^2 = 89.048$. Group variances (original) 44,250, 97,937.5, 43,000, 94,187.5, 41,437.5; $s_{\bar{x}}^2 = 55,655$. (c) $F = 4.34$; there are differences. (d) E differs from all but B. (e) The restriction is done in order to attributed differences to cities rather than types of cars.
10-44 For yes–no, $F = 3.05$ is not significant. For groups, $F = 14.18$ is significant. For interaction, $F = 105.9$ is significant.   **10-45** (d)   **10-46** (b)   **10-47** (c)   **10-48** (b)   **10-49** (a)   **10-50** (d)   **10-51** (b)
10-52 a mistake has been made.   **10-53** negligible or zero   **10-54** (b)   **10-55** $\Sigma xf(x)$ **10-56** 2/25
10-57 (a) (the limit is 0)   **10-58** contingency   **10-59** (a) $H_O: p = 0.75$, $H_A: p > 0.75$; reject H_O if 83 or more in the sample of 100 do drink. (b) (1) 0.05, (2) 0.0032 (3) 0.2420   **10-60** 0.47 to 0.67
10-61 $\chi^2 = 18.10$ (pool the last three classes); not Poisson, 2.5.   **10-62** (a) $\chi^2 = 1.27$; the variances do not differ. (b) $|t| = 0.46$; the means do not differ.   **10-63** (a) $F = 5.88$; there are differences. (b) $F < 1$; there are no differences. (c) $F < 1$; there is no interaction. (d) The first and second are equivalent and differ from the third and fourth, which are equivalent.   **10-64** (b) 3040.4 3037 68.872   **10-65** (a) $t = -2.47$; there is proof. (b) 0.0643
10-66 2240.9   **10-67** (a) $\chi^2 = 5.65$; there are no differences. (b) 0.308 to 0.442
10-68 $\chi^2 = 0.92$; there is insufficient evidence of a difference.
10-69 (a) 0.4405 (b) 0.9816 (c) (1) 0.0826 (2) 0.0323   **10-70** (a) 0.8849 (b) 0.0808 (c) 0.7522
10-71 $|z| = 0.913$; there is insufficient evidence of a change.   **10-72** (a) 0.5168 (b) 0.4473
10-73 Gamma-ray determinations are larger since $t = 3.93$ is significant with $p < 0.005$.   **10-74** (a) 62 (b) $\chi^2 = 38.58$, σ is greater than 2.00. (c) 16.26 to 18.92

CHAPTER 11

11-1 (b) slope $\div$ 80/70 = 1.14 (c) $b_0 = -9.68$, $b_1 = 1.12$, $y = -9.68 + 1.12x$ (d) 63.12
11-2 (b) $y = 244.67 + 19.233x$ (c) 360   **11-3** (b) $y = 1.76 - 0.168x$ (c) -0.26; the true relationship is probably not linear but curved.   **11-4** 69.2 ($y = -4.19 + 0.0734x$)   **11-6** (a) (2) It is reasonably linear. (b) (1) 0.934 (2) 0.873 (i.e., 87.3%) (c) (1) $Y = 1.9231 + 0.99775X$, $y = -148.48 + 4.989x$ (2) 150.9   **11-7** 98.8%
11-8 0.891   **11-9** (a) strong nonlinear (b) $r = 0.385$ **11-10** (b) 0.995 (c) (1) $r^2 = 0.99$ (i.e., 99%); the line should be used. (2) 1.50.   **11-11** 0.0046   **11-12** (a) 33.01 to 65.08 (b) $t = 0.136$; accept H_O.
11-13 54.88 to 71.36   **11-14** 344.65 to 375.48   **11-15** $H_O: \beta_1 = 0.05$, $H_A: \beta_1 > 0.05$; $t = 1.53$; accept H_O. There is not sufficient evidence.   **11-16** 143.5 to 158.3   **11-17** (a) 8.23 to 9.40 (b) $H_O: \beta_1 = -0.5$, $H_A: \beta_1 > -0.5$; $t = 0.742$; accept H_O. There is insufficient evidence. (c) 4.55 to 6.03
11-18 (a) $|t| = 0.38$; accept H_O. (b) $|t| = 0.38$; accept H_O again.   **11-19** There is sufficient evidence; $z = 2.137$.
11-20 (a) $r^2 = 0.2209$; not strong (b) (1) $t = 0.922$; not significant (2) $t = 1.506$; not significant (3) $t = 2.259$; not significant (4) $t = 2.818$; significant   **11-21** $t = -3.054$; significant with $p < 0.005$
11-22 -0.198 to -0.859 **11-24** 0.5037 to 0.9734
11-25 (a) (1) not linear (2) $r = -0.7078$, $100r^2\% = 50.1\%$ (b) (1) looks linear (2) $r = 0.9828$, $100r^2\% = 96.6\%$ (3) $y = 18.645 + 1.0689x_{\text{new}}$ (i.e., $y = 18.645 + 1.0689/x$)   **11-26** (a) $y = 36.843 + 5.064x_1 - 1.1015x_2$ (b) $F = 15.12$; reject H_O.   **11-27** (c)   **11-28** (a)   **11-29** (d)   **11-30** (c)   **11-31** 36%
11-32 divided by 2   **11-33** (c)   **11-34** (a)   **11-35** (a)   **11-36** (b)   **11-37** $H_O: |\rho| = \sqrt{2/3}$, $H_A: |\rho| > \sqrt{2/3}$
11-38 (b) $y = 4.692 - 0.261x$; $Y = e^{4.692 - 0.261x}$, which becomes $Y = 109.1e^{-0.261x}$
11-39 $z = 4.68$; there is sufficient evidence.   **11-40** 4858 to 5275   **11-41** 0.500 to 0.980
11-42 (a) $y = 86.2707 + 0.4789x_1 + 14.4329x_2$ (b) 83.96% (c) $F = 28.8$; reject H_O.
11-43 (a) $y = 3.416 + 0.16476x$ where x = year (1 through 8) (b) 0.1448 to 0.1847   **11-44** (b)   **11-45** (b)
11-46 (b)   **11-47** (b)   **11-48** (c) or (d)   **11-49** 0   **11-50** smaller   **11-51** (a)   **11-52** degrees of freedom
11-53 Poisson   **11-54** (a) $\chi^2 = 48$; the variance is not acceptable. (b) 24.39 to 77.41   **11-55** (a) 1/12 (b) 2/27
11-56 $\chi^2 = 215.35$; there is an association.
11-57 $H_O: \mu = 39.9 \times 1.05 = 41.895$, $H_A: \mu > 41.895$; $z = 2.62$; reject H_O. (2) The additive is worthwhile.

11-58 $F = 7.95$; mean stopping distances differ; $p < 0.01$.　**11-59** 16.38 ($y = 50.10 - 0.1686x$)
11-60 $t = 2.583$; there is a significant adverse effect with $p < 0.025$.　**11-61** (a) 0.03 to 0.12 (b) 0.9793
11-62 A differs from all others. D and C are equivalent and differ from E and B, which are equivalent.
11-63 (a) 7.12 (b) 0.70 (c) 0.04108 (d) 2.85% (e) -1.283　**11-64** (a) $F = 1.88$; insufficient evidence
(b) -1.44 to 5.94; accept H_O.　**11-65** (a) (1) $H_O: \rho = 0$; $t = -5.94$; reject H_O. $H_O: \beta = 0$; $t = -5.94$;
reject H_O. ANOVA; $F = 35.28$; reject H_O. (2) -0.924 to -0.581 (b) 0.95 to 1.27
11-66 (a) 601 (b) 547 (c) 0.24 to 0.36　**11-67** (a) 563 (b) 516 (c) 0.26 to 0.34

11-68 (a)

Class	Frequency
70–74	4
75–79	16
80–84	32
85–89	58
90–94	94
95–99	71
100–104	42
105–109	24
110–114	9

(c) mean = 93.1, median = 92.96, standard deviation = 8.36　**11-69** Coating: $F = 169.7$; they differ. Zinc:
$F = 2.73$; no difference. Interaction: $F = 0.16$; no interaction.　**11-70** $z = 2.502$; opinion has changed.
11-71 (a) $y = 0.2758 + 0.3893x_1 + 0.5266x_2$ (b) 0.0203 (c) $F = 42.0$; reject H_O. (d) 0.830　**11-72** (a) 0.0054
(b) 0.2736 (c) 0.2189　**11-73** (a) No; $P(x \leqslant 4|H_O) = 0.38$. (b) Yes; $z = -1.9$, $p = 0.0287$ (c) Yes; $z = -2.758$,
$p = 0.0029$.　**11-74** (b) $y = 388.87 + 97.94 \log_2 \omega$ (c) 682.7　**11-75** (a) 0.8383 (b) 0.6793

CHAPTER 12

12-1 Reject H_0; $p = 0.038$.　**12-2** $\chi^2 = 12$; the medians differ; $p < 0.01$.　**12-3** $z = -2.647$; there is sufficient
evidence.　**12-4** $\chi^2 = 5.219$; no significant difference.　**12-5** There is evidence of a difference.
12-6 Accept H_O.　**12-7** There is evidence, $p = 0.0059$.　**12-8** normal test power = 0.9357,
nonparametric test power = 0.797　**12-9** $U = 102$; no difference　**12-10** no difference
12-11 $\chi^2 = 11.086$; significant; $p < 0.025$.　**12-12** $T = 7$; reinforcement increases assessment $p < 0.005$.
12-13 $\chi^2 = 4.29$; no differences.　**12-14** $U = 80.5$; no difference.
12-15 (a) (1) 10 (2) 1/10 (3) 1/10 (4) for $R_1: \mu = 6$, $\sigma^2 = 3$; for $U_1: \mu = 3$, $\sigma^2 = 3$ (b) The rank sum and U
statistics follow probabilities, etc. as in (a).　**12-16** $r' = 0.65$; significant; $p < 0.025$.
12-17 $r' = -0.76$; there is negative association.　**12-18** $r' = -0.007$; not correlated.
12-19 (a) $n!$ (c) $\mu = 1$ $\sigma^2 = 1/(n - 1)$ (d) r' is 1 minus the expression in (c).　**12-20** 18　**12-21** (a)
12-22 (c)　**12-23** (b)　**12-24** symmetric　**12-25** nonparametric　**12-26** median　**12-27** (c)　**12-28** (b)
12-29 (b)　**12-30** Wilcoxon signed ranks test　**12-31** $|z| = 1.43$; accept H_O.
12-32 $z = 3.175$; the claim is proven.　**12-33** The claim is not substantiated.
12-34 $U = 77.5$; previous produced higher performance.　**12-35** $U = 122$; accept H_O.
12-36 $\chi^2 = 11.667$; reject H_O.　**12-37** $\chi^2 = 14.355$; reject H_O.　**12-39** Accept H_O; no difference.
12-40 $r' = 0.737$; significant; $p < 0.005$　**12-41** No effect　**12-42** (a) Group medians 57.5, 62, 59.5;
overall 60. (b) $\chi^2 = 4.64$; no difference.　**12-43** $U = 50.5$; not significant.
12-44 Gamma-ray is significantly higher; $p < 0.005$.　**12-45** Kruskal–Wallis　**12-46** (c)　**12-47** (a)
12-48 (c)　**12-49** $\bar{y}$　**12-50** $1 - \alpha$　**12-51** (b)　**12-52** variances are equal.　**12-53** linear
12-54 analysis of variance　**12-55** sign or Wilcoxon signed ranks　**12-56** variance　**12-57** (c)
12-58 $\chi^2 = 7.32$; there are differences; $p < 0.05$.　**12-59** $\chi^2 = 5.867$; not significant. **12-60** There is evidence;
$P(x \geqslant 10|H_O$ true$) = 0.047$.　**12-61** $|t| = 2.0$; no difference　**12-62** no difference
12-63 (a) $|t| = 0.62$; not significant. (b) not significant　**12-64** (a) $H_O: \tilde{\mu} = 15{,}000$, $H_A: \tilde{\mu} > 15{,}000$; reject H_O
if the number exceeding \$15,000 is 12 or more. (b) 0.55 (c) no evidence of increase
12-65 The sweetener does not receive a lower rating.　**12-66** (a) It can be proven. (b) $y = 37.073 - 2.944x$
(c) -3.46 to -2.53 (d) 3.94 to 16.32　**12-67** \$1.75　**12-68** (a) $F = 5.01$; there is sufficient evidence.
(b) 1.36 to 16.48　**12-69** 0.930　**12-70** (b) 6, 6, 5, 3 (d) mean = 2.75, median = 2 (e) 2/5 (f) 3/4

12-71 There is not sufficient evidence.   **12-72** (a) $1/2$ (b) $3/5$ **12-73** (a) $F = 8.17$; the vaccines are not all equivalent. (b) $F = 0.38$; there are no differences among the additive levels.
12-74 $r' = 0.418$; accept H_O : no correlation.   **12-75** (a) $H_O : \mu = 100$, $H_A : \mu < 100$; reject H_O if $\bar{x} < 99.342$.
(b) (1) 0.9981 (2) 0.6535 (3) 0.1963 (4) 0.0177 (by interpolation)   **12-76** (a) 0.30 (i.e., 30%)
(b) 0.22 to 0.39 (c) 0.0735   **12-77** (a) $8/47$ (b) $7/36$ (c) $9/36$ **12-78** (a) $\chi^2 = 7.57$;
there is insufficient evidence. (b) 0.00141 to 0.00268   **12-79** (a) -0.07 to 4.87 (b) There is insufficient evidence
of difference.   **12-80** (a) $F = 0.5$; there is no evidence of difference between the groups. (b) $F = 7.875$; there is
evidence of differences among the age groups. (c) $F = 2.375$; there is not sufficient evidence that program and age
interact.   **12-81** 62.936   **12-82** (a) mean = 113.8, median = 113 (b) range = 68, variance = 248.70

(c)

| Class limit | | Frequency | Class mark | Boundary | |
Lower	Upper			Lower	Upper
80	89	4	84.5	79.5	89.5
90	99	4	94.5	89.5	99.5
100	109	11	104.5	99.5	109.5
110	119	15	114.5	109.5	119.5
120	129	8	124.5	119.5	129.5
130	139	6	134.5	129.5	139.5
140	149	3	144.5	139.5	149.5

(e)

Class	Frequency
Less than 80	0
Less than 90	4
Less than 100	8
Less than 110	19
Less than 120	34
Less than 130	42
Less than 140	48
Less than 150	51

(f) 31 values should be within 25. There are actually 43.   **12-83** $t = 1.5$; there is insufficient evidence.
12-84 There is again insufficient evidence.   **12-85** (a) 601 (b) 505   **12-86** (a) 530 (b) 454
12-87 (a) 9.5, -6.0, -2.71, 0.0 (b) 26.996 (c) I and II differ. I and III differ.   **12-88** 72.56 to 74.57
12-89 (a) $\chi^2 = 2.087$; there has been no change. (b) 0.10 to 0.25   **12-90** $t = 2.854$; there is evidence of a
difference.   **12-91** sign test; insufficient evidence of a difference. Signed rank test, $T = 12$; there is evidence.
12-92 (a) 0.80 to 16.48 (b) The depths are not equal on the average.   **12-93** (a) (1) $y = 23.889 - 0.5045x$ (2) 0.465
(b) (1) $y = -0.1667 + 8.5258x - 0.7045x^2$ (2) 0.932

INDEX

"Less than" distribution, 23
Level of significance, 249
Limit, 12
Location:
  measures, 50
Long run for probability, 100

M

Mann-Whitney test, 498–504
Mark, 14
Mean, 46–47
  binomial, 180
  frequency distribution, 68
  hypergeometric, 181
  normal distribution, 185
  Poisson, 182
  population, 86
  probability function, 178
  sample, 46–47
  for sample means, 136, 199
  for sample variances, 141–43
  weighted, 47–48
Mean deviation, 60
Mean squares:
  block, 408
  error, 378, 379, 390, 408, 413
  regression, 475
  treatment, 378, 379, 390, 408, 412
Measure:
  central tendency, 50
  dispersion, 56
  location, 50
  variation, 56
Median, 48–49
  frequency distribution, 68–70
  population, 86
  sample, 48–49
Median test, 494–95
Modal class, 70
Mode, 49
Monotonic relation, 511–13
"More than" distribution, 23
Multiple coefficient of determination, 474
Multiple comparisons, 395–400
Multiple regression, 471–76
Multiplication rule:
  general, 113
  independent events, 111, 112
Mutually exclusive events, 103–4, 106–7

N

Negative skew, 36
Net expectation, 177

Nonlinear regression, 470–71
Nonparametric test, 489, 493–94
Normal distribution, 185–86, 188
  approximation for:
    binomial, 191–93
    hypergeometric, 194
    Poisson, 194
  mean, 185
  for sample means, 200–202
  standard, 186
  variance, 185
Normal equations, 434
Null event, 97
Null hypothesis, 245

O

Observed frequencies:
  contingency table, 340
  goodness of fit, 351–54
Odds, 98–99
Ogive, 25
One-sided confidence interval, 290, 294, 325
One-sided test, 258
One-way analysis of variance, 387–91
Open class, 7
Operating characteristic curve, 253
  one-sided test, 261
  two-sided test, 253
"Or less" distribution, 23
"Or more" distribution, 23
Outcome, 95
  in a continuum, 126–28
  numerical, 125

P

Paired comparisons, 311–13
  nonparametric:
    rank test, 505–6
    sign test, 491–93
Parameter, 86
  probability functions, 170
Parametric test, 493–94
Partition of sample space, 115–16
Percentile, 51–52
  χ^2-distribution, 229–31
  F-distribution, 317–18, 371–73
  frequency distribution, 70
  standard normal, 214–15
  t-distribution, 225–26
Permutations, 155–56
Pictograph, 32
Pie chart, 32

(Continued from inside front cover)

SELECTED EXAMPLES (EXM),
EXERCISES (EXR)
AND PRACTICE PROBLEMS (P)
BY AREA OF APPLICATION

GEOGRAPHY

Barrier islands P 2-1, 4
Erosion rates Exr 7-11
Highway travelers Exr 4-60
Ice thickness Exr 3-19
Pebble orientation Exm 2-29
Population density Exr 11-38
Riverbed material Exr 2-7, 8-26
Snow-water equivalents Exm 5-19
Spring floods Exr 4-18
Storm accidents Exr 5-47
Telephone traffic P 4-18, Exm 5-23
Traveler preference Exr 6-31
Woodland P 10-10
Work distance Exr 3-68

HEALTH, MEDICAL

Alcohol effect Exr 8-23, 4
Bleeding Ulcers P 6-10
Blood flow Exr 8-8, 11-12 Exm 11-8
Body fat and age P 3-8, 10-9 Exr 10-27
Drug response Exr 3-49, 7-56, 11-65

Hazardous materials and genetic structure
Exr 11-73
Heat treatment Exr 8-50
Hospital referrals Exr 3-65
Interferon Exr 9-7
Length of hospital stay Exr 8-46
Nasal congestion P 5-7
Parent care Exr 9-55
Placebo use Exr 5-31, P 8-2
Pupil dilation Exr 12-5
Sperm count Exr 12-3
Tumors Exr 5-29, 10-72
Vitamin C Exr 9-49, 50
X-ray treatment Exm 11-16

PHYSICS, CHEMISTRY, ENGINEERING

Beaker breakage Exm 5-25
Corrosion protection Exr 9-46
Experimental error Exm 4-25
Gauge precision Exr 3-13, P 8-1
Metric adjustment Exm 3-37, 40
Oil viscosity Exr 3-11
Shaft distortion P 2-8
Shell penetration P 10-11
Signal distortion P 5-11